Lecture Notes in Computer Science 2778

Edited by G. Goos, J. Hartmanis, and J. van Leeuwen

W0037148

Springer-Verlag Berlin Heidelberg GmbH

Peter Y. K. Cheung   George A. Constantinides
Jose T. de Sousa (Eds.)

Field-Programmable Logic and Applications

13th International Conference, FPL 2003
Lisbon, Portugal, September 1-3, 2003
Proceedings

 Springer

Series Editors

Gerhard Goos, Karlsruhe University, Germany
Juris Hartmanis, Cornell University, NY, USA
Jan van Leeuwen, Utrecht University, The Netherlands

Volume Editors

Peter Y. K. Cheung
George A. Constantinides
Imperial College of Science, Technology, and Medicine
Dept. of Electrical and Electronic Engineering
Exhibition Road, London SW7 2 BT, UK
E-mail: p.cheung@ic.ac.uk; george.constantinides@ieee.org

Jose T. de Sousa
Technical University of Lisbon
INESC-ID/IST
R. Alves Redol, 9, Apartido 13069, 1000-029, Lisboa, Portugal
E-mail: jts@inesc-id.pt

Cataloging-in-Publication Data applied for

A catalog record for this book is available from the Library of Congress

Bibliographic information published by Die Deutsche Bibliothek
Die Deutsche Bibliothek lists this publication in the Deutsche Nationalbibliografie;
detailed bibliographic data is available in the Internet at <http://dnb.ddb.de>.

CR Subject Classification (1998): B.6-7, C.2, J.6

ISSN 0302-9743
ISBN 978-3-540-40822-2   ISBN 978-3-540-45234-8 (eBook)
DOI 10.1007/978-3-540-45234-8

http://www.springer.de

© Springer-Verlag Berlin Heidelberg 2003
Originally published by Springer-Verlag Berlin Heidelberg New York in 2003.

Typesetting: Camera-ready by author, data conversion by Steingräber Satztechnik GmbH
Printed on acid-free paper   SPIN 10931431   06/3142   5 4 3 2 1 0

Preface

This book contains the papers presented at the 13th International Workshop on Field Programmable Logic and Applications (FPL) held on September 1–3, 2003. The conference was hosted by the Institute for Systems and Computer Engineering-Research and Development of Lisbon (INESC-ID) and the Department of Electrical and Computer Engineering of the IST-Technical University of Lisbon, Portugal.

The FPL series of conferences was founded in 1991 at Oxford University (UK), and has been held annually since: in Oxford (3 times), Vienna, Prague, Darmstadt, London, Tallinn, Glasgow, Villach, Belfast and Montpellier. It brings together academic researchers, industrial experts, users and newcomers in an informal, welcoming atmosphere that encourages productive exchange of ideas and knowledge between delegates.

Exciting advances in field programmable logic show no sign of slowing down. New grounds have been broken in architectures, design techniques, run-time reconfiguration, and applications of field programmable devices in several different areas. Many of these innovations are reported in this volume.

The size of FPL conferences has grown significantly over the years. FPL in 2002 saw 214 papers submitted, representing an increase of 83% when compared to the year before. The interest and support for FPL in the programmable logic community continued this year with 216 papers submitted. The technical program was assembled from 90 selected regular papers and 56 posters, resulting in this volume of proceedings. The program also included three invited plenary keynote presentations from LSI Logic, Xilinx and Cadence, and three industrial tutorials from Altera, Mentor Graphics and Dafca.

Due to the inclusive tradition of the conference, FPL continues to attract submissions from all over the world. The accepted contributions were submitted by researchers from 32 different countries:

USA	42	Belgium	6	Brazil	2	Estonia	1
Spain	33	Netherlands	6	Canada	2	Norway	1
UK	29	Mexico	5	Hungary	2	India	1
Germany	14	Greece	4	Iran	2	Slovakia	1
Japan	13	Poland	4	Korea	2	Slovenia	1
Portugal	12	Switzerland	4	Romania	2		
Italy	9	Australia	3	Singapore	2		
Czech Rep.	8	Ireland	3	Austria	1		
France	7	China	2	Egypt	1		

We would like to thank all the authors for submitting their first versions of the papers and the final versions of the accepted papers. We also gratefully acknowledge the reviewing work done by the Program Committee members and many additional reviewers who contributed their time and expertise towards the compilation of this volume. The members of our Program Committee and all other reviewers are listed on the following pages. We are particularly pleased that of the 1029 reviews sought, 95% were completed.

We would like to thank QuickSilver Technology for their sponsorship of the Michal Servit Award, Celoxica for sponsoring the official FPL website www.fpl.org, Xilinx and Synplicity for their early support of the conference, and Coreworks for help in registration processing. We are indebted to Richard van de Stadt, the author of CyberChair. This excellent free software made our task of managing the submission and reviewing process much easier. We are grateful for the help and advice received from Wayne Luk and Horácio Neto. In addition, we acknowledge the help of the following research students from Imperial College London in checking the integrity of the manuscripts: Christos Bouganis, Wim Melis, Gareth Morris, Andy Royal, Pete Sedcole, Nalin Sidahao, and Theerayod Wiangtong.

We are grateful to Springer-Verlag, particularly Alfred Hofmann and Anna Kramer, for their work in publishing this book.

June 2003

Peter Y.K. Cheung
George A. Constantinides
Jose T. de Sousa

Organization

Organizing Committee

Program Chair	Peter Y.K. Cheung,
	Imperial College London, UK
Program Co-chair	George A. Constantinides,
	Imperial College London, UK
General Chair	Jose T. de Sousa, INESC-ID/IST,
	Technical University of Lisbon, Portugal
Publicity Chair	Reiner Hartenstein,
	University of Kaiserslautern, Germany
Local Chair	Horácio Neto, INESC-ID/IST,
	Technical University of Lisbon, Portugal
Finance Chair	Fernando Gonçalves, INESC-ID/IST,
	Technical University of Lisbon, Portugal
Exhibition Chair	João Cardoso, INESC-ID/UA,
	University of Algarve, Portugal

Program Committee

Nazeeh Aranki	Jet Propulsion Laboratory, USA
Jeff Arnold	Stretch, Inc., USA
Peter Athanas	Virginia Tech, USA
Neil Bergmann	Queensland University of Technology, Australia
Dinesh Bhatia	University of Texas, USA
Eduardo Boemo	University of Madrid, Spain
Gordon Brebner	Xilinx, Inc., USA
Andrew Brown	University of Southampton, UK
Klaus Buchenrieder	Infineon Technologies AG, Germany
Charles Chiang	Synopsys, Inc., USA
Peter Cheung	Imperial College London, UK
George Constantinides	Imperial College London, UK
Andre DeHon	California Institute of Technology, USA
Jose T. de Sousa	Technical University of Lisbon, Portugal
Carl Ebeling	University of Washington, USA
Hossam ElGindy	University of New South Wales, Australia
Manfred Glesner	Darmstadt University of Technology, Germany
Fernando Goncalves	Technical University of Lisbon, Portugal
Steven Guccione	Quicksilver Technology, USA
Reiner Hartenstein	University of Kaiserslautern, Germany
Scott Hauck	University of Washington, USA
Brad Hutchings	Brigham Young University, USA
Tom Kean	Algotronix Consulting, UK
Andreas Koch	University of Braunschweig, Germany
Dominique Lavenier	University of Montpellier II, France
Philip Leong	Chinese University of Hong Kong, China
Wayne Luk	Imperial College London, UK
Patrick Lysaght	Xilinx, Inc., USA
Bill Mangione-Smith	University of California at Los Angeles, USA
Reinhard Männer	University of Mannheim, Germany
Oskar Mencer	Bell Labs, USA
George Milne	University of Western Australia
Toshiyaki Miyazaki	NTT Network Innovation Labs, Japan
Fernando Moraes	PUCRS, Brazil
Horacio Neto	Technical University of Lisbon, Portugal
Sebastien Pillement	ENSSAT, France
Dhiraj Pradhan	University of Bristol, UK
Viktor Prasanna	University of Southern California, USA
Michel Renovell	University of Montpellier II, France
Jonathan Rose	University of Toronto, Canada
Zoran Salcic	University of Auckland, New Zealand
Hartmut Schmeck	University of Karlsruhe, Germany
Rainer Spallek	Dresden University of Technology, Germany

Adrian Stoica	Jet Propulsion Laboratory, USA
Jürgen Teich	University of Paderborn, Germany
Lothar Thiele	ETH Zürich, Switzerland
Liones Torres	University of Montpellier II, France
Stephen Trimberger	Xilinx, Inc., USA
Milan Vasilko	Bournemouth University, UK
Ranga Vemuri	University of Cincinnati, USA
Roger Woods	Queen's University of Belfast, UK

Steering Committee

Jose T. de Sousa	Technical University of Lisbon, Portugal
Manfred Glesner	Darmstadt University of Technology, Germany
John Gray	Independent Consultant, UK
Herbert Grünbacher	Carinthia Technical Institute, Austria
Reiner Hartenstein	University of Kaiserslautern, Germany
Andres Keevallik	Tallinn Technical University, Estonia
Wayne Luk	Imperial College London, UK
Patrick Lysaght	Xilinx, Inc., USA
Michel Renovell	University of Montpellier II, France
Roger Woods	Queen's University of Belfast, UK

Additional Reviewers

Anuradha Agarwal
Ali Ahmadinia
Seong-Yong Ahn
Rui Aguiar
Miguel A. Aguirre
Bashir Al-Hashimi
Ferhat Alim
Jose Alves
Hideharu Amano
Jose Nelson Amaral
David Antos
António José Araújo
Miguel Arias-Estrada
Rubén Arteaga
Armando Astarloa
José Augusto
Shailendra Aulakh
Vicente Baena
Zachary Baker
Jonathan Ballagh
Sergio Bampi

Francisco Barat
Jorge Barreiros
Marcus Bednara
Peter Bellows
Mohammed Benaissa
AbdSamad BenKrid
Khaled BenKrid
Pascal Benoit
Manuel Berenguel
Daniel Berg
Paul Berube
Jean-Luc Beuchat
Rajarshee Bharadwaj
Unai Bidarte
Bob Blake
Brandon Blodget
Jose A. Boluda
Vanderlei Bonato
Andrea Boni
Marcos R. Boschetti
Ignacio Bravo

Ney Calazans
Danna Cao
Francisco Cardells-Tormo
Joao Cardoso
Dylan Carline
Luigi Carro
Nicholas Carter
Gregorio Cappuccino
Joaquín Cerdà
Abhijeet Chakraborty
François Charot
Seonil Choi
Bobda Christophe
Alessandro Cilardo
Christopher Clark
John Cochran
James Cohoon
Stuart Colsell
Katherine Compton
Pasquale Corsonello
Tom Van Court
Octavian Cret
Damian Dalton
Alan Daly
Martin Danek
Klaus Danne
Eric Debes
Martin Delvai
Daniel Denning
Arturo Diaz-Perez
Pedro Diniz
Peiliang Dong
Cillian O'Driscoll
Mark E. Dunham
Alireza Ejlali
Tarek El-Ghazawi
Peeter Ellervee
Wilfried Elmenreich
Rolf Enzler
Ken Erickson
Roberto Esper-Chaín Falcón
Béla Fehér
Michael Ferguson
Marcio Merino Fernandes
Viktor Fischer

Toshihito Fujiwara
Rafael Gadea-Girones
Altaf Abdul Gaffar
Federico Garcia
Alberto Garcia-Ortiz
Ester M. Garzon
Manjunath Gangadhar
Antonio Gentile
Raul Mateos Gil
Rafael Gadea-Girones
Federico Garcia
Jörn Gause
Fahmi Ghozzi
Guy Gogniat
Richard Aderbal Gonçalves
Gokul Govindu
Gail Gray
Jong-Ru Guo
Manish Handa
Frank Hannig
Jim Harkin
Martin Herbordt
Antonin Hermanek
Fabiano Hessel
Teruo Higashino
Roland Höller
Renqiu Huang
Ashraf Hussein
Shuichi Ichikawa
José Luis Imaña
Minoru Inamori
Preston Jackson
Kamakoti
Parivallal Kannan
Irwin Kennedy
Tim Kerins
Jawad Khan
Sami Khawam
Daniel Kirschner
Tomoyoshi Kobori
Fatih Kocan
Dirk Koch
Zbigniew Kokosinski
Andrzej Krasniewski
Rohini Krishnan

Georgi Kuzmanov
Soonhak Kwon
David Lacasa
John Lach
Jesus Lazaro
Barry Lee
Dong-U Lee
Gareth Lee
Jirong Liao
Valentino Liberali
Bossuet Lilian
Fernanda Lima
John Lockwood
Andrea Lodi
Robert Lorencz
Michael G. Lorenz
David Rodriguez Lozano
Shih-Lien Lu
Martin Ma
Usama Malik
Cesar Augusto Marcon
Theodore Marescaux
Eduardo Marques
L.J. McDaid
Paul McHardy
Maire McLoone
Bingfeng Mei
Mahmoud Meribout
Uwe Meyer-Baese
Yosuke Miyajima
Sumit Mohanty
Gareth Morris
Elena Moscu
Francisco Moya-Fernandez
Madhubanti Mukherjee
Tudor Murgan
Ciaron Murphy
Takahiro Murooka
Kouichi Nagami
Ulrich Nageldinger
Jeff Namkung
Ángel Grediaga Olivo
Pilar Martinez Ortigosa
Selene Maya-Rueda
Wim Melis

Allen Michalski
Maria Jose Moure
John Nestor
Jiri Novotny
John Oliver
Eva M. Ortigosa
Fernando Ortiz
Damjan Oseli
Jingzhao Ou
Marcio Oyamada
Chris Papachristou
Fernando Pardo
Stavros Paschalakis
Kolin Paul
Cong Vinh Phan
Juan Manuel Sanchez Perez
Stefania Perri
Mihail Petrov
Thilo Pionteck
Marco Platzner
Jüri Pöldre
Dionisios N. Pnevmatikatos
Kara Poon
Juan Antonio Gomez Pulido
Federico Quaglio
Senthil Rajamani
Javier Ramirez
Juergen Reichardt
Javier Resano
Fernando Rincón
Francisco Rodríguez-Henríquez
Nuno Roma
Eduardo Ros
Gaël Rouvroy
Andrew Royal
Giacinto Paolo Saggese
Marcelino Santos
Gilles Sassatelli
Toshinori Sato
Sergei Sawitzki
Pascal Scalart
Bernd Scheuermann
Jan Schier
Clemens Schlachta
Klaus Schleisiek

Herman Schmit
David Schuehler
Ronald Scrofano
Pete Sedcole
Peter-Michael Seidel
Shay Seng
Sakir Sezer
Naoki Shibata
Tsunemichi Shiozawa
Nalin Sidahao
Reetinder Sidhu
Valery Sklyarov
Iouliia Skliarova
Gerard Smit
Raphael Some
Ioannis Sourdis
Lionel Sousa
Ludovico de Souza
François-Xavier Standaert
Henry Styles
Qing Su
Vijay Sundaresan
Noriyuki Takahashi
Shigeyuki Takano
Kalle Tammemäe
Konstantinos Tatas
Raoul Tawel
John Teifel

Yann Thoma
Tim Todman
Jon Tombs
Cesar Torres-Huitzil
Kuen Tsoi
Marek Tudruj
Richard Turner
Fabrizio Vacca
Sudhir Vaka
Eduardo do Valle Simoes
József Vásárhelyi
Joerg Velten
Felip Vicedo
Tanya Vladimirova
Markus Weinhardt
Theerayod Wiangtong
Juan Manuel Xicotencatl
Andy Yan
Keiichi Yasumoto
Pavel Zemcik
Xiaoyang Zeng
Yumin Zhang
Jihan Zhu
Ling Zhuo
Peter Zipf
Claudiu Zissulescu-Ianculescu
Mark Zwolinski

Table of Contents

Cryptographic Applications 1

Place and Route Tools

Multi-context FPGAs

Cryptographic Applications 2

Low-Power Issues 1

Run-Time Configurations

Cryptographic Applications 3

Compilation Tools

Asynchronous Techniques

Biology-Related Applications

Codesign

Reconfigurable Fabrics

Image Processing Applications

SAT Techniques

Application-Specific Architectures

DSP Applications

Dynamic Reconfiguration

SoC Architectures

Emulation

Cache Design

Cellular Applications

High Level Design Tools 2

Technologies and Trends (Posters)

Applications (Posters)

Tools (Posters)

FPGA Implementations (Posters)

Video and Image Applications (Posters)

Reconfigurable and Low-Power Systems (Posters)

Design Techniques (Posters)

Neural and Biological Applications (Posters)

Codesign and Embedded Systems (Posters)

Reconfigurable Systems and Architectures (Posters)

DSP Applications (Posters)

Dynamic Reconfiguration (Posters)

Arithmetic (Posters)

Design and Implementations 1 (Posters)

Design and Implementations 2 (Posters)

Networks on Chip as Hardware Components of an OS for Reconfigurable Systems*

T. Marescaux[1], J-Y. Mignolet[1], A. Bartic[1], W. Moffat[1],
D. Verkest[1,2,3], S. Vernalde, and R. Lauwereins[1,3]

[1] IMEC vzw, Kapeldreef 75, B-3001 Leuven, Belgium,
marescau@imec.be
[2] also Professor at Vrije Universiteit Brussel
[3] also Professor at Katholieke Universiteit Leuven

Abstract. In complex reconfigurable SoCs, the dynamism of applications requires an efficient management of the platform. To allow run-time allocation of resources, operating systems and reconfigurable SoC platforms should be developed together. The operating system requires hardware support from the platform to abstract the reconfigurable resources and to provide an efficient communication layer. This paper presents our work on interconnection networks which are used as hardware support for the operating system. We show how multiple networks interface to the reconfigurable resources, allow dynamic task relocation and extend OS-control to the platform. An FPGA implementation of these networks supports the concepts we describe.

1 Introduction

Adding reconfigurable hardware resources to an Instruction Set Processor (ISP) provides an interesting trade-off between flexibility and performance in mobile terminals. Because these terminals are dynamic and run multiple applications, design-time task allocation is clearly not an option. Additional dynamism may arise from changing bandwidth availability in networked applications and from intra-application computation variation as in MPEG-4. Tasks must therefore be mapped at run-time on the resources. We need an operating system to handle the tasks and their communications in an efficient and fair way at run-time.

In addition to supporting all the functionality of traditional OSes for ISPs, an Operating System for Reconfigurable Systems (OS4RS) has to be extended to manage the available reconfigurable hardware resources. Hardware support for an OS targeting reconfigurable SoCs is required for two reasons. On the one hand, we have to avoid inefficiencies inherent to software management of critical parts of the system, such as inter-task communication. On the other hand, the ISP needs physical extensions to access, in an unified way, the new functions of all components of a reconfigurable SoC. Interconnection networks are the solution we advocate as hardware support for the operating system.

* Part of this research has been funded by the European Commission through the IST-AMDREL project (IST-2001-34379) and by Xilinx Labs, Xilinx Inc. R&D group.

P.Y.K. Cheung et al. (Eds.): FPL 2003, LNCS 2778, pp. 595–605, 2003.
© Springer-Verlag Berlin Heidelberg 2003

In this paper we use a system composed of an ISP running the software part of the OS4RS, connected to an FPGA containing a set of blocks, called tiles, that can be individually reconfigured to run a hardware task, also called an IP-block. The concepts developed herein are not restricted to FPGAs and are meant to be extended to other reconfigurable SoC architectures as well.

The remainder of this paper is organized as follows. Section 2 introduces related work. Section 3 lists the requirements of operating systems for reconfigurable SoCs and introduces a novel Network on Chip (NoC) architecture able to fulfill these requirements. Section 4 describes our implementation of this NoC architecture to give efficient OS4RS hardware support. Section 5 discusses our implementation results. Finally, conclusions are drawn in Section 6.

2   Related Work

Dally advises in [7] the usage of NoCs [10] in SoCs as a replacement for top-level wiring because they outperform it in terms of structure, performance and modularity. Because we target reconfigurable SoCs we have an extra-reason to use NoCs: they allow dynamic multitasking [2] and provide HW support to an OS4RS.

Simmler addresses in [6] "multitasking" on FPGAs. However, in his system only one task is running on the FPGA at a time. To support "multitasking" he sees the need for task preemption, which is done by readback of the configuration bitstream. The state of the task is extracted by performing the difference of the read bitstream with the original one, which has the disadvantages of being architecture dependent and adding run-time overhead. We address the need for high-level task state extraction and real dynamic heterogeneous multitasking.

In [5], Rijpkema discusses the integration of best-effort and guaranteed-throughput services in a combined router. Such a combined system could be an interesting alternative to our physically separated data and control networks (Sect. 3.3).

The following two papers are tightly related to the work presented here. In [1], Nollet explains the design of the SW part of an OS4RS by extending a Real-Time OS with functions to manage the reconfigurable SoC platform we present in this paper. He introduces a two-level task scheduling in reconfigurable SoCs. The top-level scheduler dispatches tasks to schedulers local to their respective processors (HW tiles or ISP). Local schedulers order in time the tasks assigned to them. Task relocation (Sect. 3.1) is controlled in SW by the top-level scheduler.

Finally, Mignolet presents in [3] the design environment that allows development of applications featuring tasks relocatable on heterogeneous processors. A common HW/SW behavior, required for heterogeneous relocation is obtained by using a unified HW/SW design language such as OCAPI-XL [8]. OCAPI-XL allows automatic generation of HW and SW versions of a task with an equivalent internal state representation. Introduction of switch points is thus possible and allows a high level abstraction of task state information.

3   Multiple NoCs Are Required for OS4RS HW Support

This section first lists the requirements of an OS4RS in terms of hardware support. It then recalls how a single NoC enabled us to partially support an OS4RS and demonstrate dynamic multitasking on FPGAs in [2]. A proposal for complete OS4RS HW support is discussed in the last sub-section.

3.1   OS4RS Requirements in Terms of HW Support

In a heterogeneous reconfigurable platform, traditional tasks of operating systems are getting more complex. The following paragraph enumerates typical functions of the OS and explains why hardware support is required when adding reconfigurable hardware computing elements to an ISP.

Task creation/deletion: This is clearly the role of an operating system. In addition to the traditional steps for task setup in an operating system, we need to partially configure the hardware and to put it in an initial state. OS access to the reconfiguration mechanism of the hardware is therefore required.

Dynamic heterogeneous task relocation: Heterogeneous task relocation is a problem that appears when dealing with the flexible heterogeneous systems, that we target (ISP + reconfigurable hardware). The problem is allowing the operating system to seamlessly migrate a task from hardware to software (or vice-versa) at run-time[1] [1]. This involves the transfer of internal state of the task (contents of internal registers and memories) from HW to SW (or vice-versa).

Inter-task communication: Inter-task communication is traditionally supported by the operating system. A straightforward solution would be to pass all communications (HW to HW as well as HW to SW) through the OS running on the ISP. On a heterogeneous system, this solution clearly lacks efficiency, since the ISP would spend most of its time copying data from one location to another. Hardware support for intra-task data transfers, under control of the OS, is a better solution [2].

Debug ability: Debugging is an important issue when working with hardware/software systems. In addition to normal SW debug, the operating system should provide support to debug hardware tasks. This support, in terms of clock stepping, exception generation and exception handling is local to the HW tile and cannot be implemented inside the ISP running the OS. Specific hardware support is thus required.

Observability: To keep track of the behavior of the hardware tasks, in terms of usage of communication resources and of security, the operating system requires access to various parts of the SoC. It is inefficient for the central ISP to monitor the usage of communication resources and check whether the IPs are not creating security problems by inappropriate usage of the

---

[1] HW to HW relocation may also be required to optimize platform resource allocation and keep communications local within an application.

platform. A hardware block that performs this tracking and provides the OS with communication statistics and signals security exceptions is therefore essential.

3.2 Single NoC Allows Dynamic Multitasking on FPGAs, but Has Limitations

In [2] we demonstrated that using a single NoC enables dynamic multitasking on FPGAs. Separating communication from computation enables task creation/deletion by partial reconfiguration. The NoC solves inter-task communication by implementing a HW message-passing layer. It also partially solves the task relocation issue by allowing dynamic task migration thanks to run-time modification of the Destination Look-up Tables (Sect. 4.1) located in the network interface component (NIC)[2]. These concepts have been implemented in the T-ReCS Gecko demonstrator [3, 9].

As explained in [1] dynamic task relocation requires preemption of the task and the transfer of its state information (contents of its internal registers and memories) to the OS. This state information is then used to initialize the relocated task on a different computation resource (another HW tile or a software thread on the ISP [3]) to smoothly continue the application.

Experimentation on our first setup showed some limitations in the dynamic task migration mechanism. During the task-state transfer, the OS has to ensure that pending messages, stored in the network and its interfaces, are redirected in-order to the computation resource the task has been relocated to.

This process requires synchronization of communication and is not guaranteed to work on our first platform [2]. Indeed, OS Operation And Management (OAM) communication and application data communication are logically distinguished on the NoC by using different tags in the message header. Because application-data can congest the packet-switched NoC, there is no guarantee that OS OAM messages, such as those ensuring the communication synchronization during task relocation, arrive timely.

To support general dynamic task relocation, we have to enhance our system described in [2] to allow the OS to synchronize communications within an application. An interesting approach is to physically separate OS communication from application communications by means of separate NoCs and is discussed in the following section. Additional extensions are required to provide full HW support to the OS4RS as defined in Section 3.1. We need mechanisms to retrieve/restore state information from a task, to control communication load, handle exceptions and provide security and debug support. These extensions are discussed in Sections 4.1 and 4.2.

---

[2] This acronym overloads Network Interface Card because our NIC serves the similar role of abstracting a high-level processor from the low level communication of the network.

3.3   Reconfigurable Hardware Multitasking Requires Three Types of Communication

On the reconfigurable platform we presented in [2], the FPGA executes a task per reconfigurable tile and is under the control of an operating system running on the ISP. The OS can create tasks both in hardware and software [1, 3]. For such as system, as Section 3.2 explains, there are two distinct types of communication: OS OAM data and application data. Furthermore, reconfigurable systems have a third logical communication channel to transmit the configuration bitstreams to the hardware tasks.

Each tile in our reconfigurable SoC has therefore three types of communication: reconfiguration data, OS OAM data and application data. Because application data requires high bandwidth whereas OS OAM data needs low latency, we have decided to first implement each communication type on a separate network to efficiently interface the tiles to the OS running on the ISP: a reconfiguration network, a data network and a control network (Fig. 1). The services implemented on these three networks compose the HW support for the OS4RS. In addition to efficiency, a clean logical separation of the three types of communications in three communication paths, ensures independence of application and OS. The OS does not need to care about the contents of the messages carried on the data network and an application designer does not need to take into account OS OAM interactions.

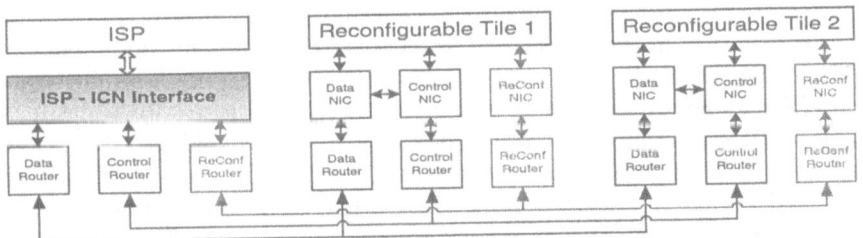

Fig. 1. Three NoCs are required: reconfiguration network, a data network and a control network.

4   Implementation of a Novel NoCs Architecture Providing HW Support to an OS4RS

This section explains how each of the NoCs introduced in section 3.3 plays its role as HW support for an OS4RS and gives the implementation details.

4.1   Application Data Network

By application data, we mean the data transferred from one task to another inside an application. Tasks communicate through message passing. These messages are sent through the Data Network[3] (DN) if the sender and/or the receiver are in a HW tile. A similar message passing mechanism is used for two software tasks residing in the ISP [2]. For performance reasons, application data circulates on the NoC independently of the OS. Nevertheless, the DN must provide hooks for the OS to enable platform management. These hooks, detailed in the next subsections, are implemented in the NIC of the DN and compose a part of the HW support for OS4RS.

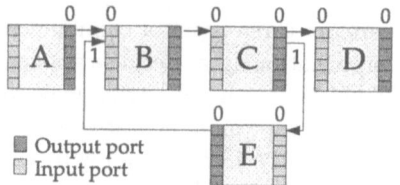

Task	src_out_port	dst_in_port	dst_logic_addr	dst_phys_addr
A	0	0	logic(B)	physical(B)
B	0	0	logic(C)	physical(C)
C	0	0	logic(D)	physical(D)
C	1	0	logic(E)	physical(E)
D	{Ø}	{Ø}	{Ø}	{Ø}
E	0	1	logic(B)	physical(B)

■ Output port
□ Input port

Fig. 2. Application Task Graph showing Input-Output port connections.

Fig. 3. Destination Look-up Tables for every task in the graph.

Data NIC supports dynamic task relocation. Inter-task communication is done on an input/output port basis [2]. Figure 2 shows an example of an application task graph with the input/output port connections between tasks. Each application registers its task graph with the OS upon initialization [1].

For each task in the application, the OS assigns a system-wide unique logic address and places the task on the platform, which determines its physical address (Fig. 3). For every output port of a task the OS defines a triplet (destination input port, destination logic address, destination physical address). For instance, task C (Fig. 2) has two output ports, hence is assigned two triplets, which compose its Destination Look-Up Table (DLT) (Fig. 3). In our system a task may have up to 16 output ports, thus there are 16 entries in a DLT.

The OS can change the DLT at run-time, by sending an OAM message on the Control Network (CN) (Sect. 4.2). Dynamic task relocation in reconfigurable SoCs is enabled by storing a DLT in the data NIC of every tile in the system [2].

Data NIC monitors communication resources. The usage of communication resources on the DN is monitored in the data NIC of every tile. Figures such as number of messages coming in and out of a specific tile are gathered in the NIC in real time and made available to the OS. Another important figure available is the average number of messages that have been blocked due to lack of buffer space in the NIC. These figures allow the OS to keep track of the communication usage on the NoC. Based on these figures and on application priorities, the OS4RS can manage communication resources per tile and thus ensure Quality of Service (QoS) on the platform [1].

---

[3] The data network is very similar to the NoC we described in a previous FPL paper [2].

Data NIC implements communication load control. The maximum amount of messages an IP is allowed to send on the network per unit of time can be controlled by the OS. To this end we have added an injection rate controller in the data NIC. As explained in [2], outgoing messages from an IP are first buffered in the NIC and are then injected in the network as soon as it is free (Best Effort service).

The injection rate controller adds an extra constraint on the time period when the messages may be injected in the NoC. It is composed of a counter and a comparator. The OS allows the NIC to inject messages only during a window of the counter time. The smaller the window, the less messages injected into the NoC per unit of time, freeing resources for other communications. This simple system, introduces a guarantee on average bandwidth[4] usage in the NoC and allows the OS to manage QoS on the platform.

Data NIC adds HW support for OS security. Security is a serious matter for future reconfigurable SoCs. Thanks to reconfiguration, unknown tasks may be scheduled on HW resources and will use the DN to communicate. We must therefore perform sanity checks on the messages circulating on the DN and notify the OS when problems occur. Communication related checks are naturally performed in the NIC. We check whether the message length is smaller than the maximum transfer unit, that messages are delivered in order and especially that IPs do not breach security by sending messages on output ports not configured in the DLT by the OS.

4.2   Control Network

The control network is used by the operating system to control the behavior of the complete system (Fig. 1). It allows data monitoring, debugging, control of the IP block, exception handling, etc. OS OAM messages are short, but must be delivered fast. We therefore need a low bandwidth, low latency CN.

CN uses Message-Based Communication. To limit resource usage and minimize latency we decided to implement the CN as a shared bus, where the OS running on the ISP is the only master and all control network NICs of tiles are slaves. The communication on this bus is message-based and can therefore be replaced by any type of NoC.

The control NIC of every tile is memory-mapped in the ISP. One half of this memory is reserved for ISP to control-NIC communication and the other one for NIC to ISP communication. To send a control OAM message to a tile, the OS first writes the payload data, such as the contents of a DLT (Fig. 3) and finishes by writing a command code on the CN, in this case *UPDATE_DLT*. The control NIC reads the command opcode and processes it. When done, it writes a status opcode in the NIC to NoC memory, to indicate whether the command was successfully processed and posts an interrupt. The OS retrieves this data and clears the interrupt to acknowledge the end of command processing.

---

[4] As long as the data NIC buffers are not permanently saturated.

CN Controls the DN Section 4.1 lists the capabilities of the data NIC in terms of control the OS has over the communication circulating on the DN. The OS commands, to enforce load control or synchronize DN communication, are actually sent over the CN to avoid interference with application data. It is in the control NIC, that statistics and security exceptions from the data NIC are processed and communicated to the OS. It is also through the CN that the OS sends destination look-up tables or injection-rate windows to the data NIC.

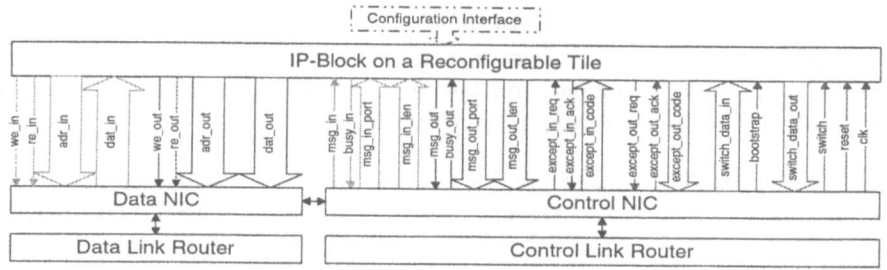

Fig. 4. Reconfigurable Tiles interface to all three NoCs through data and control NICs.

CN implements HW OS support to control IPs Another very important role of the CN is to allow control and monitoring of the IP running on a reconfigurable tile. To clearly understand the need for OS control here, let us consider the life-cycle of a reconfigurable IP block in our SoC platform.

Before instantiating the IP block in a tile by partial reconfiguration, we must isolate the tile from the communication resources, to ensure the IP does not do anything harmful on the DN before being initialized. To this end, the control NIC implements a reset signal and bit masks to disable IP communication (Fig. 4). After reconfiguration, the IP needs to be clocked. However, its maximum clock speed might be less than to that of our DN. Because we do not want to constrain the speed of our platform to the clock speed of the slowest IP (which can always change as new IP-blocks are modified at run-time), the OS can set a clock multiplexer to feed the IP with an appropriate clock rate.

The IP can now perform its computation task. At some stage it might generate an exception, to signal for instance a division by zero. The control NIC implements a mechanism to signal IP exceptions to the OS (Fig. 4). The OS can also send exceptions to an IP, as it can send signals to processes running on the ISP. One usage of these exceptions is to perform IP debugging.

Later on, the OS might decide to relocate the IP to another HW tile or as a process on the ISP [1–3]. The NIC implements a mechanism to signal task switching to the IP and to transmit its internal state information to the OS. The NIC also implements a mechanism to initiate an IP with a certain internal state, for instance when switching from SW to HW.

4.3   Reconfiguration Network

Our reconfigurable SoC targets a Xilinx VIRTEX-2^{TM} PRO as an implementation platform. IPs are instantiated on tiles by partially reconfiguring the chip. In this case, the reconfiguration network is already present on the platform as the native reconfiguration bus of the VII-Pro. The reconfiguration bus is accessed through the internal reconfiguration access port (ICAP) and is based on the technology presented by Blodget in [4]. The main difference resides in the fact that our platform is driving the ICAP through the OS4RS, running on a PowerPC, instead of a dedicated soft core like the MicroBlaze^{TM}.

5   Implementation Results

This section presents results about our enhanced HW support of an OS4RS, in terms of latencies induced by HW OS processing time and in terms of area overhead.

5.1   HW OS Reaction Time

The SW part of our OS4RS [1] is running on an ISP and controls the HW OS extensions located in the data and control NICs, through the control network (Sect. 4). Figure 5 shows the processing in SW and HW, when the OS4RS resets a reconfigurable IP block running on a HW tile. We assume that the control NIC is clocked at $22MHz$ and that the ISP can access the 16-bit wide control network at $50MHz$. The SW part of the OS4RS sends the atomic *RST_IP* command to the control NIC of the IP in $120ns$. A total of $12.8\mu s$ is spent in the control NIC to decode, process and acknowledge the commands issued from the SW part of the OS. Only $320ns$ are spent by the SW OS to send an atomic instruction and request the control NIC to clear the IRQ, acknowledging the command has been processed. The total processing time is under $13.2\mu s$.

In the case of dynamic task relocation from SW to HW (Sect. 4.1), the reconfigurable IP needs to be initialized with the state information extracted from the SW version of the task [3]. Assuming we have 100 16-bits words of

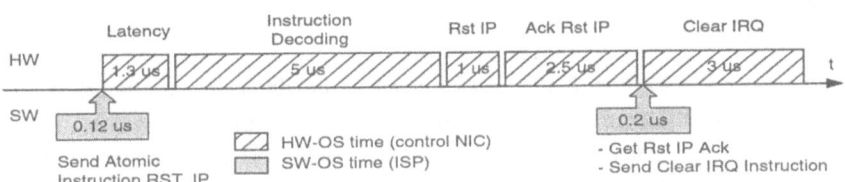

Fig. 5. OS4RS sends a Reset command to an IP. Most of the processing is performed in the control NIC, making it HW support for the OS4RS. Control NIC is clocked at 22MHz and control network is accessed by the ISP at 50MHz.

state information to transfer, the total transaction takes about $440\mu s$ (control NIC transmits a word to the IP in $4.3\mu s$).

In both cases the control NIC abstracts the access to the reconfigurable IP block from the SW part of the OS4RS. Because the NICs offload the ISP from low-level access to the reconfigurable IP blocks, they are considered as the HW part of the OS4RS.

5.2   HW OS Implementation Size

For research purposes, we implement the fixed NoCs together with the reconfigurable IPs on the same FPGA. We report therefore in table 1 the area usage of the NoCs in terms of FPGA logic and consider it as overhead to the reconfigurable IPs they support.

The old single NIC we presented in [2] performed both as control and data interface and is smaller than our new data and control NICs (Tab. 1), but it suffers from the limitations listed in section 3.2. The support of functions required by a full OS4RS such as state transfer, exception handling, HW debugging or communication load control come at the expense of a higher area overhead in the NIC. On our target platform, the Virtex-II Pro 20, this area overhead amounts to 611 slices, or 6.58 percent of the chip per reconfigurable tile instantiated. Nevertheless on a production reconfigurable SoC, the NoCs could be implemented as hard cores, reducing considerably the area overhead on the chip.

Table 1. HW overhead of Data and Control NICs, compared to the single NIC presented in [2].

Element	Virtex-II Slices	$XC2V6000(\%)$	$XC2VP20(\%)$
Control NIC and Router	250	0.74	2.7
Data NIC	361	1.07	3.89
Control+Data NIC	611	1.81	6.58
Old NIC from [2]	260	0.77	2.8

6   Conclusions

This paper presents how NoCs can be used as hardware components of an operating system managing reconfigurable SoCs. To support advanced features, such as dynamic task relocation with state transfer, HW debugging and security, an OS4RS requires specific HW support from the platform. We propose a novel architecture for reconfigurable SoCs composed of three NoCs interfaced to reconfigurable IPs. This approach gives a clean logical separation between the three types of communication: application data, OS control and reconfiguration bitstreams.

NoCs interfaced to reconfigurable IPs provide efficient HW support for an OS4RS. They open the way to future reconfigurable SoC platforms, managed by operating systems that relocate tasks between HW and SW to dynamically optimize resource usage.

References

1. V. Nollet, P. Coene, D. Verkest, S. Vernalde, R. Lauwereins: Designing an Operating System for a Heterogeneous Reconfigurable SoC. Proc. RAW'03 workshop, Nice, April 2003.
2. T. Marescaux, A, Bartic, D. Verkest, S. Vernalde, R. Lauwereins: Interconnection Networks Enable Fine-Grain Dynamic Multi-Tasking on FPGAs. Proc. 12th Int. Conf. on Field-Programmable Logic and Applications, pages 795-805, Montpellier, September 2002.
3. J.-Y. Mignolet, V. Nollet, P. Coene, D. Verkest, S. Vernalde, R. Lauwereins: Infrastructure for Design and Management of Relocatable Tasks in a Heterogeneous Reconfigurable System-on-Chip. Proc. DATE 2003, pages 986-992 , Munich, March 2003.
4. B. Blodget, S. McMillan, P. Lysaght: A Lightweight Approach for Embedded Reconfiguration of FPGAs. Proc. DATE 2003, pages 399-400, Munich, March 2003.
5. Rijpkema and al. E. Rijpkema et al.: Trade Offs in the Design of a Router with both Guaranteed and Best-Effort Services for Networks On Chip. Proc. DATE 2003, pages 350-355, Munich, March 2003.
6. H. Simmler, L. Levinson, R. Männer: Multitasking on FPGA Coprocessors. Proceedings 10th Int'l Conf. Field Programmable Logic and Applications, pages 121-130, Villach, August 2000.
7. W.J. Dally and B. Towles: Route Packets, Not Wires: On-Chip Interconnection Networks, Proc. 38th Design Automation Conference, June 2001.
8. http://www.imec.be/ocapi
9. http://www.imec.be/reconfigurable
10. J. Duato, S. Yalamanchili, L. Ni: Interconnection Networks, An Engineering Approach, September 1997. ISBN 0-8186-7800-3.

A Reconfigurable Platform for Real-Time Embedded Video Image Processing

N.P. Sedcole, P.Y.K. Cheung, G.A. Constantinides, and W. Luk

Imperial College, London SW7 2BT, UK.
pete.sedcole@imperial.ac.uk, p.cheung@imperial.ac.uk,
george.constantinides@ieee.org, w.luk@doc.ic.ac.uk

Abstract. The increasing ubiquity of embedded digital video capture creates demand for high-throughput, low-power, flexible and adaptable integrated image processing systems. An architecture for a system-on-a-chip solution is proposed, based on reconfigurable computing. The inherent system modularity and the communication infrastructure are targeted at enhancing design productivity and reuse. Power consumption is addressed by a combination of efficient streaming data transfer and reuse mechanisms. It is estimated that the proposed system would be capable of performing up to ten complex image manipulations simultaneously and in real-time on video resolutions up to XVGA.

1 Introduction

As advances are made in digital video technology, digital video capture sensors are becoming more prevalent, particularly in embedded systems. Although in scientific and industrial applications it is often acceptable to store the raw captured video data for later post-processing, this is not the case in embedded applications, where the storage or transmission medium may be limited in capacity (such as a remote sensor sending data over a wireless link) or where the data are used in real-time (in an intelligent, decision-making sensor for example). Real-time video processing is computationally demanding, making microprocessor-based processing infeasible. Moreover, microprocessor DSPs are energy inefficient, which can be a problem in power-limited embedded systems. On the other hand, ASIC-based solutions are not only expensive to develop, but inflexible, which limits the range of applicability of any single ASIC device.

Reconfigurable computing offers the potential to achieve high computational performance, at the same time remaining inexpensive to develop and adaptable to a wide range of applications within a domain. For this potential to become of practical use, integrated systems will need to be developed that have better power-performance ratios than currently available FPGAs, most likely by curtailing the general applicability of these devices such that optimisations for the particular application domain can be made. Such systems may be termed 'domain specific integrated circuits'.

This paper outlines a proposed architecture for an integrated reconfigurable system-on-a-chip, targeted at embedded real-time video image processing. The

P.Y.K. Cheung et al. (Eds.): FPL 2003, LNCS 2778, pp. 606–615, 2003.

system is based on the Sonic architecture [1], a PCI-card based system capable of real-time video processing. Our objective is to integrate this system into a single device for embedded video processing applications.

In this paper we identify and discuss the unique issues arising from large scale integration of reconfigurable systems, including:

- How the complexities of designing such systems can be managed. Hardware modules and abstraction levels are proposed to simplify the design process.
- The implications of modularisation on the allocation and management of resources at a physical level.
- Effective connectivity and communication between modules.
- Minimising power consumed in data transmission and data reuse, the two most important factors in low-power design of custom computational platforms [2].

2   Related Work

Reconfigurable custom computing machines, implemented as arrays of FPGAs, have been successfully used to accelerate applications executing on PCs or workstations [3, 4]. Image processing algorithms are particularly suitable for implementation on such machines, due to the parallelisms that may be exploited [1, 5]. As mentioned in section 1, the system described in this paper is based on the Sonic architecture [1], a video image processing system comprising an array of FPGAs mounted on a PCI card.

Research has been conducted into embedded reconfigurable systems, and integrated arrays of processing elements [6–9]. Often these are conceived as accelerators of software-based tasks, and as such are closely coupled with a microprocessor, with the microprocessor forming an integral part of the data-path. As a consequence, these systems are usually adept at exploiting instruction-level parallelism; task-level parallelism is often ignored.

The proposed system has similarities to the DISC [10], which allows relocatable reconfigurable modules to vary in size in one dimension, and also includes the concept of position-independent connections to global signals. The modules in the DISC are very simple however. Our proposed system also incorporates ideas similar to the 'dynamic hardware plugins' proposed by Horta et al. [11] and virtual sockets described by Dyer et al. [12], although in both of these cases communication and interconnect structures are designed ad hoc on an application by application basis. Kalte et al. describe a system-on-a-programmable chip which does include a structured interconnection architecture for connecting dynamically reconfigurable modules [13]. Their system connects multiple masters to multiple slaves using either a multiplexed or crossbar-switch scheme. The interconnection scheme described in our paper allows any module to be a bus master, and is more suitable for streaming data transfers.

3   Managing Design Complexity

As device densities increase, circuit designers are presented with huge amounts of uncommitted logic, and large numbers of heterogeneous resources such as memories, multipliers and embedded hard microprocessor cores. Creating manageable designs for such devices is difficult; attempting to exploit dynamic reconfiguration as well only increases this complexity.

In order to make the design process tractable, we propose a modularised, hierarchical system framework, based on Sonic [1]. Modularisation partitions the system design into conceptually manageable pieces. In addition, high-level algorithmic parallelism can be exploited by operating two or more modules concurrently. To simplify the design, use and reuse of these modules, module interfaces need to be standardised, and abstractions are required in the transfer of data between modules.

3.1   System Architecture

As depicted in Figure 1, modularity is achieved by separating the data path into a variable number of processing elements connected via a global bus. Each processing element (PE) also has unidirectional *chain bus* connections to the adjacent PEs, for fast local data transfer. The left-most and right-most PEs are specialised input and output streaming elements respectively; data stream through the system in a general left-to-right direction.

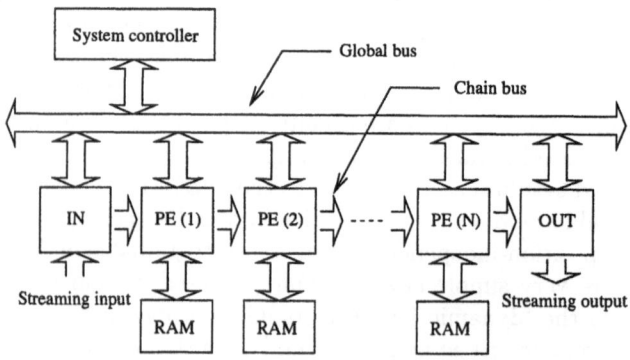

Fig. 1. The proposed system architecture.

Each element in the data path is designed to realise a complex task, such as a filter (2D convolution) or a rotation, implemented physically as a partial configuration instanced within a reconfigurable fabric. The processing applied to the data stream is determined by the type, number and logical sequence of the processing elements. The system designer controls this by programming a microprocessor within the system control module. PE module configurations (stored as bitstreams) are loaded into the fabric and initialised by the control software, which also directs data-flow between PEs. As will be discussed below, the actual transfer of data is handled by a router within each PE. This scheme allows the implementation of a range of resource allocation and reconfiguration scheduling techniques, while realising the advantages of a data-path driven design.

Video processing algorithms can require a large amount of storage memory, depending on the frame size of the video stream. While available on-chip memory is constantly increasing, the size of storage space necessary, coupled with the bit-density (and therefore cost-per-bit) advantage of memory ICs, will ensure that external RAM will remain a necessary feature of video processing systems. The traditional memory bottleneck associated with external memory is avoided in our case by distributing the memory between the processing elements. The connection mechanisms between each PE and memory is discussed further in section 4.

3.2 Processing Elements

The logical composition of a processing element, as shown in Figure 2, comprises an engine, a router and input and output stream buffers. All of these are constructed from resources within the reconfigurable fabric. The PE has connections to the global bus (for communication between any two PEs), to the directly adjacent PEs to the left and right, and (optionally) to two external RAM banks. These interfaces are all standardised, which enables fully developed processing element configurations to be used and reused unchanged within a larger system design.

The core component of each PE is the processing engine; it is in this component that computation is performed on the image pixels. The engine is also the only part of the PE that is defined by the module designer. Serialised (raster-scan) data flow into and out of the engine by way of the buffers, the relevance of

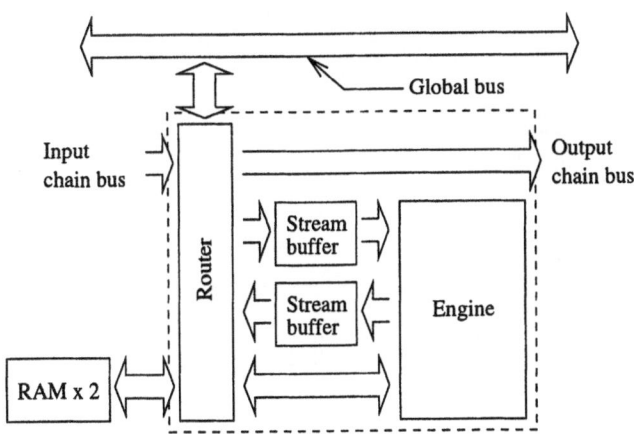

Fig. 2. The structure of a processing element.

which will be discussed in section 5. The key component for data movement abstraction is the router; this is directed by the system controller to transfer data between the buffers and the global bus, the chain bus or external memory. Since the data format is fixed, and all transfer protocols are handled by the router, the module designer is only concerned with the design of the computational logic within the engine.

4 Physical Structure

The mechanisms described in the previous section for controlling design complexity have ramifications for the physical design of the system. As mentioned above, the processing element modules are instanced within a reconfigurable FPGA fabric. The structure of this fabric needs to be able to support variable numbers and combinations of various sized modules. Since modules are fully placed and routed internally at design-time, the completed configurations must be relocatable within the fabric. The provision of mechanisms for connecting PE configurations to the buses and external RAM is required.

The physical system structure is illustrated in Figure 3. The processing elements are implemented as partial configurations within the FPGA fabric. The PEs occupy the full height of the fabric, but may vary in width by discrete steps. The structure and nature of the reconfigurable fabric is based on the Virtex-II Pro FPGA family from Xilinx, Inc. It is heterogeneous, incorporating not only CLBs but RAM block elements and other dedicated hardware elements such as multipliers. However, it exhibits translational symmetry in the horizontal dimension, such that the PEs are relocatable along the length of the fabric. The

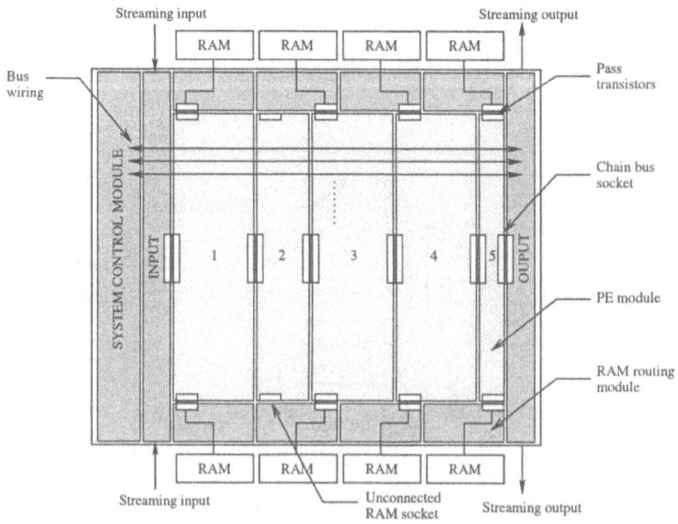

Fig. 3. A diagram representing the physical structure of the proposed system, with the reconfigurable fabric (shaded light grey) configured into five processing elements.

choice of a one-dimensional fabric simplifies module design, resource allocation and connectivity.

The global bus is not constructed from FPGA primitives; it has a dedicated wiring structure, with discrete connection points to the FPGA fabric. The advantage of this strategy is that the electrical characteristics of the global bus wiring can be optimised, leading to high speeds and low power [14, 15]. In addition, the wiring can be more dense than could otherwise be achieved.

Each processing element must have chain bus connections to the neighbouring PEs; this is accomplished through the use of 'virtual sockets', implemented as hard macros. The chain bus signals are routed as 'antenna' wires to specified locations along the left and right edges of the module. When two configurations are loaded adjacently into the array, these wires are aligned, and the signal paths may be completed by configuring the pass transistors (programmable interconnect points) separating the wires. Thus, each module provides 'sockets' into which other modules can plug into. Similar ideas has been proposed previously [11, 12], although the connection point chosen in previous work is a CLB programmed as a buffer.

A similar concept is used in connecting processing modules to the external RAM banks. Since the processing elements are variable-sized and relocatable, it is not possible to have direct-wired connections to the external RAM. The solution to this is to wire the external RAM to 'routing modules' which can then be configured to route the RAM signals to several possible socket points. This allows the registration between the RAM routing module and the processing element to be varied by discrete steps, within limits. If external RAM is not required by a particular processing element, such as PE 2 in Figure 3, the RAM resources may be assigned to an adjacent PE, depending on the relative placements.

5   Data Transfer and Storage

The transfer and storage of data are significant sources of power consumption in custom computations [2], so warrant specific attention. In the preceding system, Sonic, data transfer is systolic; each clock cycle one pixel value is clocked into the engine, and one is clocked out. This limits the pixel-level parallelism possible within the engine, and constrains algorithm design. In particular, data reuse must be explicitly handled within the engine itself, by storing pixel values in local registers. This becomes a significant issue for the engine design when several lines of image data must be stored, which can total tens of kilobytes.

In the proposed architecture the input stream buffer efficiently deals with data reuse. Being constructed from embedded RAM block elements (rather than from CLBs) a high bit density can be achieved. Image data is streamed into the buffer in a serial, FIFO-like manner, filling it with several lines of a frame. The engine may access any valid pixel entry in the buffer; addressing is relative to the pixel at the front of the queue. Buffer space is freed when the engine indicates it has finished with the data at the front of the queue. This system enables

greater design flexibility than a purely systolic data movement scheme while
constraining the data access pattern sufficiently to achieve the full speed and
power benefits of serial streaming data transfer. This is particularly beneficial
when data are sourced from external RAM, where a sequential access pattern can
take advantage of the burst mode transfer capability of standard RAM devices.

The input and output stream buffers are physically constructed from a number of smaller RAM elements for two reasons. Firstly, a wide data-path bit-width
between the buffers and the engine can be achieved by connecting the RAM elements in parallel, enabling several pixels to be processed in parallel within the
PE. The second important benefit is the ability to rearrange the input buffer
RAM elements into two (or more) parallel stream buffers, when the engine requires more than one input data stream, such as in a merge operation. Likewise,
the output buffer may be subdivided into several output streams, if the engine
produces more than one output. We label each stream buffer input or output
from an engine a 'port'.

In addition to allowing efficient data reuse and fine-grained parallelism, the
stream buffers create flexibility in the transfer of data over the global bus. Instead
of a systolic, constant rate data-flow, data can be transfered from an output port
buffer of one PE to the input port buffer of another PE in bursts, which allows
the global bus to be shared between several logical communication channels.
The arbitration between the various logically concurrent channels is handled
by a reconfigurable arbitration unit within the system controller. This enables
a range of arbitration strategies to be employed depending on the application,
with the objective of preventing processing stalls from an input buffer under-run
or output buffer overrun.

6   Performance Analysis

In the previous sections the proposed system was described in qualitative terms.
We will now present a brief quantitative analysis to demonstrate how the system is expected to meet the desired performance requirements. To do this, it is
necessary to make some assumptions about the sizes of various system elements;
these assumptions will be based on the resources available in the Xilinx Virtex
II Pro XC2VP125, and comparison with the latest incarnation of the Sonic architecture: UltraSONIC [16]. The numbers given here are speculative and do not
represent any attempt at optimised sizing. Nevertheless, they are sufficient for
the purpose of this analysis.

Based on the utilised logic cell count of modules in UltraSONIC and the cell
count of the XC2VP125, between four and ten processing elements are possible in
the current technology, depending on the complexity of each PE. Assuming the
same I/O numbers as the Xilinx device (1200) and given the physical connectivity
constraints of the system topology (see Figure 3) it is estimated that external
RAM would be limited to eight banks, implying that not every PE would have
access to external RAM. It is theoretically possible to implement 272 bus lines
spanning the full width of the XC2VP125, however we will assume a bus width

of only 128 bits. The Xilinx device includes 18 columns of Block RAMs, each of which can be arranged as 1024 bits wide by 512 deep memory. Therefore assigning each processing element two 32KB stream buffers is not unreasonable, perhaps configured as 512 wide by 512 deep. A wide buffer facilitates pixel-level parallel computations, although for algorithms that do not exhibit much low-level parallelism a wide buffer would be a disadvantage.

External RAM sizes and system throughput is determined by the video format the system is applied to. Table 1 gives some figures for some representative video formats. In previous work, external RAM was sized such that one video frame would fit one RAM bank [16], which would imply between 330KB to over 3MB per bank in this case.

Using these figures, some calculations on the processing characteristics of the system can be made. With a 512-bit wide buffer, 16 pixels (at 32-bits/pixel) can be operated on in parallel. As Table 2 indicates, several complete lines of image data can be stored in the stream buffers for data reuse. An example engine clock rate is given, assuming each block of 16 pixels is accessed 10 times during the computation, a realistic value for a 2D convolution with a 5x5 kernel. This clock rate is at least an order of magnitude below the state-of-the-art for this technology, which is highly desirable as low clock frequencies correspond with lower operating voltages and very little pipelining, which all translate into lower power consumption. Table 2 also gives the required clock rate of the global bus, for five concurrent channels, all operating at the full throughput rates given in Table 1. Again, these speeds should be easily achievable, especially given that it is a custom structure and not constructed from FPGA primitives.

These calculations demonstrate that the proposed system is expected to be able to perform between four and ten complex image computations simultaneously and in real-time on video data of resolutions up to XVGA.

Table 1. A sample of video formats. Frame sizes are based on 32 bits per pixel, while throughput is calculated at 25 frames per second.

Format	Lines	Columns	Frame size	Throughput
PAL	288	352	330 KB	9.67 MB/s
DVD PAL	576	720	1620 KB	39.55 MB/s
XVGA	768	1024	3072 KB	75.00 MB/s

Table 2. System characteristics. The cycle time is the average time allowed to process 16 pixels in parallel. The transfer capacity of the bus is calculated assuming a 10% overhead for arbitration and control.

Format	Processing element			Global bus	
	Lines/buffer	Cycle time	Clock rate	Transfer capacity	Bus speed
PAL	23	6.3 μs	1.6 MHz	53 MB/s	3.5 MHz
DVD PAL	11	1.5 μs	6.5 MHz	218 MB/s	14.3 MHz
XVGA	8	0.8 μs	12.3 MHz	413 MB/s	27.0 MHz

7   Conclusion and Future Work

A reconfigurable platform suitable for embedded video processing in real-time has been presented. The platform comprises a number of configurable complex processing elements operating within a structured communication framework, all controlled by a central system controller. The design aims to meet the demanding requirements of real-time video processing by exploiting fine-grained (pixel-level) parallelisms within the processing elements, as well as task-level algorithmic parallelisms by operating processing elements concurrently.

The customisation of processing elements enables the system to be adapted to a wide range of applications, and since these elements are dynamically reconfigurable, run-time adaptation is also possible. Moreover, the inherent modularity of the PEs, coupled with the communication infrastructure, facilitates design and reuse.

Finally, power efficiency is addressed by targeting data movement. Dedicated bus wiring can be optimised for low-power transmission, while data movement is minimised by reusing data stored in stream buffers and constraining processing element designs. Parallel processing results in lower average operating frequencies, which can be translated into lower power consumption by reducing the operating voltage.

Our next main objective is to translate the proposal as described in this paper into a prototype, so that we may assess its feasibility. Since the physical structure of the system is based heavily on the Virtex II Pro, it would be logical to implement the prototype in a device from this family. In order to investigate the operational performance of the system, it will be necessary to map one or more actual applications from Sonic to the new platform, which may require the development of tools and processes.

Acknowledgements

The authors would like to thank Simon Haynes, Henry Epsom and John Stone of Sony Broadcast and Professional Europe for their helpful comments. The support from Xilinx Inc., and Patrick Lysaght in particular, is appreciated. N.P. Sedcole gratefully acknowledges the financial assistance from the Commonwealth Scholarship Commission in the United Kingdom.

References

1. Haynes, S.D., Stone, J., Cheung, P.Y.K., Luk, W.: Video image processing with the Sonic architecture. IEEE Computer **33** (2000) 50–57
2. Soudris, D., Zervas, N.D., Argyriou, A., Dasygenis, M., Tatas, K., Goutis, C., Thanailakis, A.: Data-reuse and parallel embedded architectures for low-power, real-time multimedia applications. In: International Workshop - Power and Timing Modeling, Optimization and Simulation. (2000)
3. Arnold, J.M., Buell, D.A., Hoang, D.T., Pryor, D.V., Shirazi, N., Thistle, M.R.: The Splash 2 processor and applications. In: IEEE International Conference on Computer Design: VLSI in Computers and Processors. (1993)

4. Vuillemin, J.E., Bertin, P., Roncin, D., Shand, M., Touati, H.H., Boucard, P.: Programmable active memories: Reconfigurable systems come of age. IEEE Transactions on VLSI Systems **4** (1996) 56–69
5. Athanas, P.M., Abbott, A.L.: Real-time image processing on a custom computing platform. IEEE Computer **28** (1995) 16–24
6. Callahan, T.J., Hauser, J.R., Wawrzynek, J.: The Garp architecture and C compiler. IEEE Computer **33** (2000) 62–69
7. Ebeling, C., Cronquist, D.C., Franklin, P.: RaPiD – Reconfigurable Pipelined Datapath. In: Field–Programmable Logic and Applications. (1996)
8. Goldstein, S.C., Schmit, H., Budiu, M., Cadambi, S., Moe, M., Taylor, R.R.: PipeRench: A reconfigurable architecture and compiler. IEEE Computer **33** (2000) 70–77
9. Waingold, E., Taylor, M., Srikrishna, D., Sarkar, V., Lee, W., Lee, V., Kim, J., Frank, M., Finch, P., Barua, R., Babb, J., Amarasinghe, S., Agarwal, A.: Baring it all to software: RAW machines. IEEE Computer **30** (1997) 86–93
10. Wirthlin, M.J., Hutchings, B.L.: A dynamic instruction set computer. In: IEEE Symposium on FPGAs for Custom Computing Machines. (1995)
11. Horta, E.L., Lockwood, J.W., Taylor, D.E., Parlour, D.: Dynamic hardware plugins in an FPGA with partial run-time reconfiguration. In: Design Automation Conference. (2002)
12. Dyer, M., Plessl, C., Platzner, M.: Partially reconfigurable cores for Xilinx Virtex. In: Field–Programmable Logic and Applications. (2002)
13. Kalte, H., Langen, D., Vonnahme, E., Brinkmann, A., Rückert, U.: Dynamically reconfigurable system-on-programmable-chip. In: Euromicro Workshop on Parallel, Distributed and Network-based Processing. (2002)
14. Benini, L., De Micheli, G.: Networks on chips: A new SoC paradigm. IEEE Computer **35** (2002) 70–78
15. Dally, W.J., Towles, B.: Route packets, not wires: On-chip interconnection networks. In: Design Automation Conference. (2001)
16. Haynes, S.D., Epsom, H.G., Cooper, R.J., McAlpine, P.L.: UltraSONIC: a reconfigurable architecture for video image processing. In: Field–Programmable Logic and Applications. (2002)

Emulation-Based Analysis of Soft Errors
in Deep Sub-micron Circuits*

M. Sonza Reorda and M. Violante

Dip. Automatica e Informatica, Politecnico di Torino, Torino, Italy
{matteo.sonzareorda, massimo.violante}@polito.it

Abstract. The continuous technology scaling makes soft errors a critical issue in deep sub-micron technologies, and techniques for assessing their impact are strongly required that combine efficiency and accuracy. FPGA-based emulation is a promising solution to tackle this problem when large circuits are considered, provided that suitable techniques are available to support time-accurate simulations via emulation. This paper presents a novel technique that embeds time-related information in the topology of the analyzed circuit, allowing evaluating the effects of the soft errors known as single event transients (SETs) in large circuits via FPGA-based emulation. The analysis of complex designs becomes thus possible at a very limited cost in terms of CPU time, as showed by the case study described in the paper.

1 Introduction

The adoption of new manufacturing deep sub-micron technologies is raising concerns about the effects of *single event transients* (SETs) [1] [2] [3], which correspond to erroneous transitions on the output of combinational gates, and several approaches to evaluate them have been proposed, which exploit fault injection [4]. During fault injection execution systems undergo to perturbations mimicking the effects of transient faults. The goal of these experiments is normally to identify portions of the system that are more sensible to faults, to gain statistical evidence of the robustness of hardened systems, or to verify the correctness of the implemented fault tolerance mechanisms.

Simulation-based fault injection [4] may be used to cope with the aforementioned purposes: it indeed allows early evaluating fault effects when only a model of the system is available. Moreover, it is very flexible: any fault model can be supported and faults can be injected in any module of the system. The major drawback of this approach is the high CPU time required for performing circuit simulation.

In the last years, several approaches to simulation-based fault injection have been proposed. Earlier works were based on switch-level simulation tools, such as the one described in [5], which can be adapted to the execution of fault injection experiments targeting the SET fault model. More recently, several authors proposed the use of HDL simulators to perform fault injection campaigns, and several approaches have been presented (e.g., [6]-[9]) for speeding-up the process.

---

* This work was partially supported by the Italian Ministry for University through the Center for Multimedia Radio Communications (CERCOM).

P.Y.K. Cheung et al. (Eds.): FPL 2003, LNCS 2778, pp. 616–626, 2003.

As soon as simulation-based fault injection gained popularity, researches faced the problem of reducing the huge amount of time needed for injecting faults in very complex designs, entailing thousands of gates. Two main approaches were presented. A first one addresses the problem from the simulation efficiency point of view, by minimizing the time spent for simulating each fault [10] [11]. Conversely, the second one focuses on the clever selection of the faults to be injected during simulations, thus reducing simulation time by reducing the total number of faults to be injected [12] [13]. The former approaches that exploit FPGA-based emulation seem to be well suited for analyzing very complex circuits. Indeed, they combine the versatility of simulation, which allows injecting faults in any element of circuits, while providing the performance typical of hardware prototypes.

When soft errors affecting deep sub-micron technologies become of interest, the already available emulation techniques fall short in supporting the analysis of SET effects. As better explained in section 2, SETs are transient modifications of the expected value on gate output, which have duration normally shorter than one clock cycle. As a result, SET effects can be analyzed only through circuit models or tools able to deal with timing information. This can be easily done by resorting to timed simulation tools, but at the cost of very high simulation time. Conversely, FPGA-based hardware emulation can hardly be used unless suitable techniques are available to take into account time-related information.

The main contribution of this paper is in proposing an FPGA-based fault injection environment suitable to assess the effects of SETs via emulation, thus allowing designers analyzing large circuits at low cost in terms of time needed for performing injection experiments. For this purpose we describe a novel technique able to introduce time-related information in the circuit topology. FPGA-based emulation of the time-enriched circuit can thus be exploited to asses SET effects without any loss of accuracy with respect to time-accurate simulations, provided that a suitable fault model is adopted. As a result, we are able to achieve the same accuracy of state-of-the-art approaches based on timed simulations while reducing fault injection execution time up to 5 orders of magnitude.

In order to assess the feasibility of this idea, we developed an emulation-based fault injection environment that exploits a rapid-prototyping board equipped with a Xilinx Virtex 1000E. The experimental results we gathered through this environment, although still preliminary, show the soundness of the proposed approach.

The paper is organized as follows: section 2 reports an overview of the considered fault model, while section 3 describes the circuit expansion algorithm used to embed time-related information in the circuit topology. Section 4 presents the fault injection environment we developed and section 5 reports some experimental results assessing the effectiveness of the proposed approach. Finally, section 6 draws some conclusions.

2   Single Event Transients

Today, the fault model that is normally used during fault injection experiments is the (single/multiple) bit-flip in the circuit storage elements, i.e., registers and embedded memories. However, with the adoption of deep sub-micron technologies, a new fault model is becoming of interest: the *single event transient*.

Single event transients are originated when highly energized particles strike sensible areas within combinational circuits. In deep sub-micron CMOS devices, the most sensible areas are depletion regions at transistor drains [14]. The particle strike produces there several hole-electron pairs that start drifting under the effect of the electric field. As a result, the injected charge tends to change the state of the struck node producing a short voltage pulse. As the depletion region is reformed, the charge-drift process decays, and the expected voltage level at the struck node is restored.

In deep sub-micron circuits the capacitance associated to circuit nodes is very small, therefore non-negligible disturbances can be originated even by small amounts of deposited charge, i.e., when energized particles strike the circuit. As reported in [14], in old 5 Volt CMOS technologies the magnitude of the voltage swing associated to SETs is about 14% greater than the normal voltage swing of the node and thus its impact is quite limited, in terms of both duration and magnitude. Conversely, if the technology is scaled to a 3.3 Volt one, the disturbance becomes 21% larger than a normal swing and thus the transistor that must restore the correct value of the struck node will employ more time to suppress the charge-drift process. As a result, the duration of the voltage pulse the striking particle originates increases with technology scaling. In very deep sub-micron technologies this effect may become a critical issue since the duration of the voltage pulse may become comparable to the gate propagation delay and thus the voltage pulse may spread throughout the circuit, possibly reaching its outputs.

As measurements reported in [14] show, a SET can be conveniently modeled at the gate level as an erroneous transition (either from 0 to 1 or from 1 to 0) on the output of combinational gates.

3 The Circuit Expansion Principle

This section describes the principles at the base of the fault injection technique we developed. Sub-section 3.1 presents the circuit expansion algorithm we exploited, sub-section 3.2 presents the fault list generation algorithm we devised; finally, sub-section 3.3 presents the fault model we defined, which can be fruitfully exploited for assessing the effects of SET in combinational circuits or in the combinational portion of sequential ones.

3.1 Expansion Algorithm

The algorithm we adopted for embedding time-related information in the circuit topology while preserving its original functionality was first introduced in [15] for evaluating power consumption of combinational circuits. Please note that although the original and the expanded circuits are functionally equivalent, i.e., they produces the same output responses to the same input stimuli, their topologies are different.

To illustrate the idea behind the algorithm, let us consider the circuit shown in figure 1, where input and output ports are driven by non-inverting buffers. We assume the variable-delay model, where all the gates in the circuit (including buffers) may have different propagation delays.

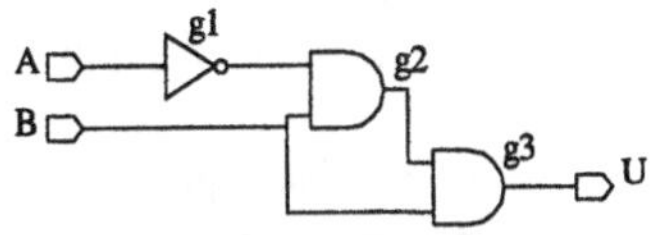

Fig. 1. Circuit example

We assume that the circuit is initially in steady state, i.e., an input vector was applied to inputs A and B and its effects have been already propagated through the circuit to the output U.

The algorithm we exploit modifies the given circuit C by building an expanded circuit C' such that C' is a superset of C. This task is performed according to the following expansion rules:

1. Initially, C' is set equal to C.
2. For each gate $g \in C$ whose output is evaluated at time T from the application of the input vector, a new gate g/T is added to C'. We refer to g/T as a *replica* of g, and to T as the *time frame* of g/T, i.e., g/T belongs to time frame T.
3. The fain-in of $g/T \in C'$ is composed of the replicas of the gates in the fan-in of $g \in C$ which belong to time frames $T' < T$.

Assuming that a new input vector is applied at time T_0, and that each gate i has propagation delay equal to P_i, we compute the time frames of each replica according to the algorithm reported in figure 2.

$TimeFrame[i] = \{ P_i \}$ $\forall i \in$ Circuit Inputs
$TimeFrame[i] = \varnothing$ $\forall i \notin$ Circuit Inputs
foreach($i \in$ Circuit Gates)
  foreach($k \in$ Circuit Gate in the fan-in of i)
   $TimeFrame[i] = TimeFrame[i] \cup (P_i + TimeFrame[k])$

Fig. 2. Algorithm for computing time frames

Any gate i may be associated to a list of time frames ($TimeFrame[i]$), due to the existence of paths of different length connecting gate i with the circuit inputs.

The list of time frames associated to the circuit of figure 1 is reported in table 1; it was computed under the assumption that the propagation delay of each gate (including circuit inputs) is equal to the gate fan-out, while the propagation delay of circuit outputs is assumed to be 1 ns. For instance, in table 1 we have that the output of g3 is first updated at time frame T_0+3 ns in correspondence to the new value of B, and then it is again updated at time frame T_0+4 ns following the update of gate g2.

The obtained expanded circuit embeds time-related information in a structural way: the gates belonging to time frame T are those whose outputs are evaluated T ns after the application of the new input vector.

The application of the expansion rules may lead to replicate the circuit outputs many times. In the considered example we have indeed that the output U is replicated twice. We refer to the replica of the output belonging to the last time frame (i.e., the highest time frame) as the circuit *actual output*. For the considered example U/5 is the actual output of the circuit; conversely, U/4 store the intermediate value the circuit output assumes during the evaluation of the new input vector.

Table 1. Gate update times

Gate	A	B	g1	g2	g3	U
Time frame [ns]	1	2	2	3	3, 4	4, 5

3.2 Fault List Generation Algorithm

The computation of the list of faults to consider during fault injection experiments is a critical issue, since the number of faults in the fault list directly affects the injection time. As a result, the fault list generation algorithm should be able to select the minimum number of faults, i.e., it should select only the SETs (in terms of fault locations and fault injection times) that have chances of affecting the circuit outputs.

The algorithm we developed starts from the expanded version of the circuit whose sensitivity to SETs should be analyzed, and produces the list of faults that should be simulated. In the following we will refer to the gate where the SET originates as the *faulty gate*.

Figure 3 reports an example of SET: the circuit primary inputs are both set to 1, thus the expected output value is 0. Due to a charged particle, the output of g1 is switched to 1 for a period of time long enough for the erroneous transition to propagate through the circuit. As a result, we observe a transition on g3, whose output is set to 1. As soon as the SET effects disappear, the output switches back to the expected value.

SET effects can spread through the fan-out cone of the faulty gate if, and only if, the duration of the erroneous transition is equal to or longer than the faulty gate propagation delay and if the magnitude of the transition is compatible with the device logic levels. In the following, we will concentrate our attention only on those particles that, when hitting the circuit, produce SETs that satisfy the above conditions.

Let T_H be the time when the SET originates, δ be the SET duration, T_s the time when the outputs of the circuit are sampled. Moreover, let Π be the set of propagation delays associated to sensitized paths stemming from the faulty gate to the circuit outputs, i.e., those paths that due to the input vector placed on the circuit inputs allow the SET to spread through the circuit.

Any SET is effect-less, i.e., its effects cannot be sampled on the circuit outputs, if the following condition is met:

$$T_H + \delta + t < T_s \quad \forall\, t \in \Pi \tag{1}$$

If eq. 1 holds, it means that once the SET effects disappear and the expected value is restored on the faulty gate, the correct value has enough time to reach the circuit outputs, and thus the expected output values are restored before they are sampled.

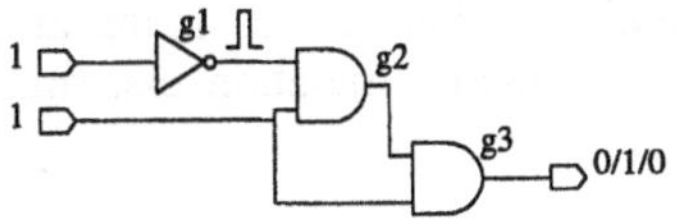

Fig. 3. SET effects

The values T_H and δ are known since they are used to characterize the SET in the fault list. Furthermore, T_s is known a-priori, and it is selected according to the circuit *critical path*, i.e., the maximum propagation delay between the circuit inputs and its outputs. Conversely, Π depends on the circuit topology and its accurate computation normally requires complex and time-consuming algorithms.

In our approach, eq. 1 is exploited in combination with the expanded version of the circuit, thus avoiding the computation of Π for each gate in the circuit. As a result, we can save a significant amount of CPU time that can be devoted to more detailed SET effects analysis.

Thanks to the properties of the expanded circuit, all the gates in the fan-in cone of the circuit *actual outputs* do not satisfy eq. 1. Indeed, the mentioned gates belongs to the circuit *critical path*, and thus any signal skew (like that introduced by SETs) is not acceptable: if allowed to spread through the circuit, the SET will indeed modify the outputs in such a way that an erroneous value will be sampled at time T_s. Moreover, let g/T be a gate in the fan-in cones of the circuit actual outputs. Then, only the effects of those SETs originated on g at time $T_H = T$ may reach circuit outputs.

Each fault in the list of possible SETs can be described as a couple (faulty gate, T_H). The list of faults that should be simulated can thus be easily obtained by traversing the expanded circuit, starting from its *actual outputs*, going toward its inputs and storing all the traversed gates.

3.3   Vector-Bounded Stuck-at

Once the list of possible faults has been computed, fault injection experiments are still required to assess SETs effects. Let (g, T_H) be the SET that we would inject, computed according to the algorithm described in section 3.2. Moreover, let T_v be the time when a new vector is applied to the circuit inputs and T_s be the time when circuit outputs are sampled.

The fault model we propose, called *vector-bounded stuck-at*, consists in approximating any SET with a stuck-at fault on the output of the faulty gate g/T_H in the expanded version of the circuit. During simulation, the vector-bounded stuck-at originates at time T_v and lasts up to T_s. As a result:
- The stuck-at affects the replica g/T_H of the faulty gate g belonging to time frame T_H only during the evaluation of one input vector.
- The stuck-at has no effects on the replicas of the faulty gate belonging to time frames other than T_H. Moreover, it does not affect g/T_H before and after T_V.

Thanks to the properties of the expanded circuit, although the stuck-at on g/T_H is injected at time T_v, it starts influencing the circuit only from T_H. Moreover, by resorting to the time-related information embedded in the expanded circuit, timed-simulation is no longer needed: the vector-bounded stuck-at can indeed be injected as soon as the stimuli are applied to the circuit inputs and fault effects can last for exactly one vector. Zero-delay fault simulation can thus replace timed simulation without any loss of accuracy.

4 The Fault Injection Environment

A typical fault injection environment is usually composed of three modules:

- *Fault List Manager*: it generates the list of faults to be injected, according to a given fault model and the possible indications of designers (e.g., the most critical portions in the system, or a particular time interval).
- *Fault Injection Manager*: it selects a new fault from the fault list, injects it in the system, and then observes the results.
- *Result Analyzer*: it analyzes the data produced by the previous module, categorizes faults according to their effects, and produces statistical information.

For the purpose of this paper we assume that the system under analysis is a VLSI circuit (or part of it). Therefore, we assume that a gate-level description of the system is available that was enriched with time-related information according to section 3.1. We also assume that the input stimuli used during fault injection experiments are already available, and we do not deal with their generation or evaluation. Moreover, the adopted fault model is the vector-bound stuck-at presented in sub-section 3.3.

We implemented the Fault List Manager as a software process that reads both the gate-level system description and the existing input stimuli, and that generates a fault list according to the adopted fault model. Similarly, we implemented the Result Analyzer as a software process.

In our approach the Fault Injection Manager module exploits a FPGA board that emulates an instrumented version of the circuit to support fault injection. The FPGA board is driven by a host computer where the other modules and the user interface run. As a result, the Fault Injection Manager is implemented as a hardware/software system where the software partition runs on the host, and the hardware partition is located on the FPGA board along with the emulated circuit. More in details, the Fault Injection Manager is composed of the following parts:

- *Circuit Instrumenter*: it is a software module running on the host computer that modifies the circuit for supporting fault injection;
- *Fault Injection Master*: it iteratively accesses to the fault list, selects a fault and orchestrates each fault injection experiment by sending to the board the input stimuli. When needed it triggers the fault injection. It finally receives from the board the system faulty behavior. The Fault Injection Master corresponds to a software module running on the host computer.
- *Fault Injection Interface*: it is implemented by the FPGA board circuitry, and it is in charge of interpreting and executing the commands the Fault Injection Master issues.

The architecture of the whole environment is summarized in Figure 4.

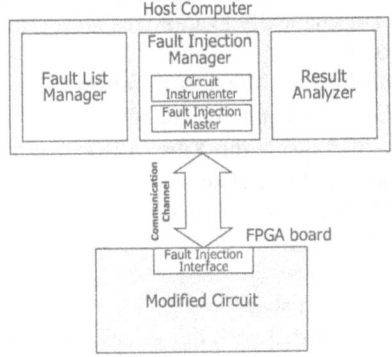

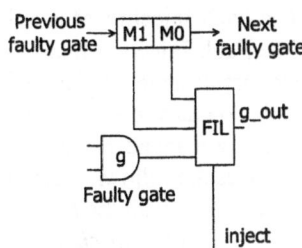

Fig. 4. FI environment architecture

Fig. 5. Instrumented architecture for supporting fault injection

In order to support the injection of the vector-bound stuck-at fault model, we modify the circuit under analysis before mapping it on the FPGA board. For each faulty gate in the circuit the logic depicted in figure 5 is added.

Let suppose the gate *g* in figure 5 be the faulty gate. The following hardware modules are added for allowing fault injection on the output of *g*:

- *inject*: it is an input signal controlled by the Fault Injection Interface. When set to 1 it triggers the injection of the fault. This signal is set to 1 in correspondence of the clock cycle during which the fault should be injected.
- *M1* and *M0*: it is a two-bit shift register, containing the fault the user intends to inject (01 corresponds to the vector-bound stuck-at-0, 10 corresponds to the vector-bound stuck-at-1, while 00 and 11 correspond to the normal gate behavior, i.e., the fault is not injected no matter the value of *inject*). The content of the shift register is loaded with the fault the user intends to inject before starting the application of the input stimuli.
- *FIL*: it is a combinational circuit devised to apply the vector-bound stuck-at to the gate *g*, depending on the value of *inject*, *M1* and *M0*.

In order to simplify the set-up of fault injection experiments, the shift registers corresponding to the considered faulty gates are connect to form a scan chain, which can the Fault Injection Interface can read/modify via three signals (scan-in, scan-out and scan enable). Since the fault that should be injected is defined by simply shifting in the scan chain the proper sequence of bits, the user can easily perform the analysis of single and multiple faults.

5 Experimental Results

We developed the proposed fault injection environment using the ADM-XRC development board [16], which is equipped with a Xilinx Virtex V1000E FPGA. The board hosts a PCI connector and it is inserted in a standard Pentium-class PC, which accesses the board via a memory-mapped protocol.

624   M. Sonza Reorda and M. Violante

In order to insert the proposed hardware fault injection circuitry, we developed a tool that, starting from the original circuit, automatically synthesizes all the modules our architecture requires, according to section 4.

We developed the hardware/software Fault Injector Manager as follows:

- The Fault Injection Master is coded in C++ and it amounts to 500 lines of code.
- The Fault Injection Interface is implemented inside the Xilinx FPGA (along with the circuit under evaluation) and it works as a finite state machine that recognizes and executes the commands the Fault Injection Master issues. We used a memory-mapped protocol to control the operations of the FPGA board, which includes the following commands:

   a. *Reset*: all the registers are set to the start state.
   b. *Load scan chain*: the scan chain is loaded with the values specifying the fault to be injected.
   c. *Apply vector*: the circuit Primary Inputs (PIs) are fed with a given input stimula.
   d. *Fault Inject*: the fault injection takes place.
   e. *Output read*: the values on the circuit Primary Outputs (POs) are read.

The instrumented circuit, together with the Fault Injection Interface, is described in an automatically generated synthesizable VHDL code. We then used typical development tools to obtain the bitstream needed to program the Xilinx device. FPGA Compiler II is used to map the description to the Xilinx device, while Xilinx tools are employed to perform place, route and timing analysis of the implemented design.

We exploited the described fault injection environment to assess the effects of SETs in a circuit implementing a 27-channels interrupt controller. In particular, being interested in assessing the simulation performance the approach attains, we randomly generated several input stimuli of different lengths. We then applied the stimuli to the considered circuit with two different fault injection environments, while considering the same set of faults:

1. Faults were injected in the original circuit comprising 164 gates by exploiting the VHDL-based fault injection environment presented in [17].
2. Faults were injected according to the described emulation-based fault injection environment. For this purpose the considered circuit was first expanded to embed time-related information and then instrumented to support fault injection, obtaining an equivalent but enriched circuit comprising 2,232 gates.

For each fault injection experiment we recorded the CPU time needed for injecting the whole fault list that accounts for 418 faults. VHDL simulations were performed on a Sun Enterprise 250 running at 400 MHz and equipped with 2 GBytes of RAM. The same machine executed the synthesis and placement of the considered circuit in about 20 minutes.

From the results reported in table 2 the reader can observe the effectiveness of the proposed approach, which is able to reduce the time needed for performing fault injection experiments of about 5 orders of magnitude, while providing the same results in terms of fault effect classification of VHDL simulation-based fault injection. When the time for circuit synthesis and placements is also considered for evaluating the overall performance of our approach, we observe that the speed-up with respect to simulation is still very high: about 3 orders of magnitude.

Table 2. Simulation results

Number of input stimuli	VHDL simulation-based fault injection [s]	Emulation-based fault injection [s]
1,000	159,894.72	0.92
2,000	310,688.45	0.92
3,000	459,684.17	0.93
4,000	628,578.90	0.93

6 Conclusions

The paper presented an approach suitable to assess the effects of single event transients via FPGA-based emulation, making thus feasible the analysis of complex deep sub-micron circuits with a very limited cost in terms of time needed for performing fault injection experiments.

The approach we presented is based on a technique able to embed time-related information in the circuit topology, while preserving its original functionality. An FPGA-based fault injection environment is then exploited to perform injection experiments on the enriched circuit.

Experimental results are reported assessing the effectiveness of the proposed approach, which is able to reduce the time needed for performing fault injection by several orders of magnitude.

References

[1] L. Anghel, M. Nicolaidis, "Cost Reduction of a Temporary Faults Detecting Technique", DATE'2000: ACM/IEEE Design, Automation and Test in Europe Conference, 2000, pp. 591-598

[2] C. Constantinescu, "Impact of deep submicron technology on dependability of VLSI circuits", Proc. IEEE Int. Conference on Dependable Systems and Networks, 2002, pp. 205-209

[3] P. Shivakumar, M. Kistler, S. W. Keckler, D. Burger, L. Alvisi, "Modelling the effect of technology trends on the soft error rate of combinational logic", Proc. IEEE Int. Conference on Dependable Systems and Networks, 2002, pp. 389-398

[4] Mei-Chen Hsueh, T.K Tsai, R.K Iyer, "Fault injection techniques and tools", IEEE Computer, Vol. 30, No. 4, 1997, pp. 75-82

[5] P. Dahlgren, P. Liden, "A switch-level algorithm for simulation of transients in combination logic", Proc. Fault Tolerant Computing Symposium, 1995, pp. 207-216

[6] E. Jenn, J. Arlat, M. Rimen, J. Ohlsson, J. Karlsson, "Fault Injection into VHDL Models: the MEFISTO Tool", Proc. Fault Tolerant Computing Symposium, 1994, pp. 66-75

[7] T.A. Delong, B.W. Johnson, J.A. Profeta III, "A Fault Injection Technique for VHDL Behavioral-Level Models", IEEE Design & Test of Computers, Winter 1996, pp. 24-33

[8] D. Gil, R. Martinez, J. V. Busquets, J. C. Baraza, P. J. Gil, "Fault Injection into VHDL Models: Experimental Validation of a Fault Tolerant Microcomputer System", European Conference of Dependable Computing (EDCC-3), 1999, pp. 191-208

[9]   J. Boué, P. Pétillon, Y. Crouzet, "MEFISTO-L: A VHDL-Based Fault Injection Tool for the Experimental Assessment of Fault Tolerance", Proc. Fault-Tolerant Computing Symposium, 1998, pp. 168-173

[10] P. Civera, L. Macchiarulo, M. Rebaudengo, M. Sonza Reorda, M. Violante, "Exploiting Circuit Emulation for Fast Hardness Evaluation", IEEE Transactions on Nuclear Science, Vol. 48, No. 6, December 2001, pp. 2210-2216

[11] L. Antoni, R. Leveugle, B. Fehér, "Using run-time reconfiguration for fault injection in hardware prototypes", Proc. IEEE Int.l Symp. on Defect and Fault Tolerance in VLSI Systems, 2000, pp. 405-413

[12] L. W. Massengill, A. E. Baranski, D. O. Van Nort, J. Meng, B. L. Bhuva, "Analysis of Single-Event Effects in Combinational Logic-Simulation of the AM2901 Bitslice Processor", IEEE Transactions on Nuclear Science, Vol. 47, No. 6, 2000, pp. 2609-2615

[13] L. Berrojo, I. González, F. Corno, M. Sonza Reorda, G. Squillero, L. Entrena, C. Lopez, "New Techniques for Speeding-up Fault-injection Campaigns", Proc. IEEE Design, Automation and Test in Europe, 2002, pp. 847-852

[14] K. J. Hass, J. W. Gambles, "Single event transients in deep submicron CMOS", IEEE 42nd Midwest Symposium on Circuits and Systems, 1999, pp. 122-125

[15] S. Manich, J. Figueras, "Maximizing the weighted switching activity in combinational CMOS circuits under the variable delay model", Proc. IEEE European Design and Test Conference, 1997, pp. 597-602

[16] ADM-XRC PCI Mezzanine card User Guide Version 1.2, http://www.alphadata.co.uk/

[17] B. Parrotta, M. Rebaudengo, M. Sonza Reorda, M. Violante, "New Techniques for Accelerating Fault Injection in VHDL descriptions", IEEE Int.l On-Line Testing Workshop, 2000, pp. 61-66

HW-Driven Emulation
with Automatic Interface Generation

M. Çakır[1], E. Grimpe[2], and W. Nebel[1]

[1] Department of Computer Science
Carl von Ossietzky University Oldenburg, D-26111 Oldenburg, Germany
{cakir,nebel}@informatik.uni-oldenburg.de
[2] OFFIS Research Institute, D-26121 Oldenburg, Germany
grimpe@offis.de

Abstract. This paper presents an approach to automate the emulation of HW/SW-Systems on an FPGA-board attached to a host. The basic steps in design preparation for the emulation are the generation of the interconnection and the description of the synchronization mechanism between the HW and the SW. While some of the related work considers the generation of the interconnection with some manual interventions, the generation of the synchronization mechanism is left to the user as part of the effort to set up the emulation. We present an approach to generate the interconnection and the synchronization mechanism, which allows a HW-driven communication between the SW and the HW.

1 Introduction

During the emulation and prototyping of HW-designs with average complexity, universal FPGA-boards can be used instead of complex logic emulators. These are usually boards, that can be plugged into a host system. The communication between host and board typically uses a standard bus, e.g. via PCI [1]. In contrast to the stand-alone boards, these boards can be used more versatile and not only at the end of the design flow, because the usage of a host allows a flexible emulation of the environment.

Furthermore the combination of host and FPGA makes verification acceleration and validation of HW/SW-systems possible. To accelerate the verification of a HW design a part of the design is emulated on the FPGA and the rest is simulated on the host. For this usage the simulator and a part of the design on the FPGA have to be connected. In order to validate a HW/SW-system the SW communicates with the HW on the FPGA via a standard bus. In this case usually a model of the target processor is used to emulate the behaviour of the SW on the target processor (e.g. ISA models).

This paper concentrates on the interface generation as well as on synchronization mechanism for the emulation. In the next section the problems related with the emulation process are described shortly. In section 3 we give an overview about related work on co-simulation, co-emulation. Section 4 describes our approach to automate

P.Y.K. Cheung et al. (Eds.): FPL 2003, LNCS 2778, pp. 627–637, 2003.
© Springer-Verlag Berlin Heidelberg 2003

the emulation of a HW design with PCI-based FPGA-boards. In section 5 the automatic interface generation is explained in detail. The last section draws a conclusion.

2 Problem Definition

FPGA-based emulation requires complex software flows and manual intervention to prepare the design for the emulation [2]. The preparation process for the emulation on an FPGA-board plugged into a host includes two major steps: i) the generation of the interconnection between the SW and the HW via a physical channel between the host and the FPGA, ii) the generation of the synchronization mechanism (Figure 1).

During the emulation of a HW/SW-system with a PCI-based FPGA-board, the communication between the FPGA and the host occurs via PCI. Although the PCI-core on the FPGA and the PCI-driver on the host are used to abstract the PCI-communication, the user has to design and implement the interconnection between the SW and PCI-driver as well as between the HW and PCI-core manually. This task must be treated differently for different designs, especially if the HW has several communication channels with the SW and if they can request concurrently data transfer.

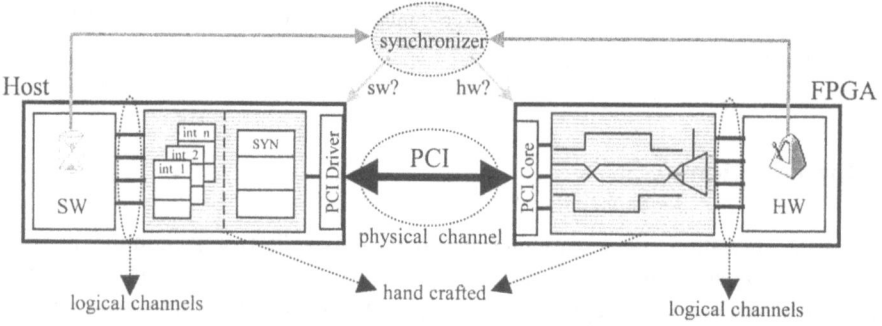

Fig. 1. Emulation on a PCI-board

On the other hand the SW on the host and the HW on the FPGA have to be synchronized with each other. An instance, we call it synchronizer in the following, has to determine the time of the communication between these parts, and this instance has to control the whole system. In general, the whole system would be controlled by the host or the FPGA. In case of the implementation of the synchronizer on the host there are two possibilities: causality based synchronization and timed synchronization. The first solution implicates in principle the sequential execution of the SW and the HW. For the timed synchronization the execution time of the SW on the target processor has to be determined (e.g. by means of instruction set simulator, simulation of the real time kernel). Furthermore the timed synchronization causes a high synchronization overhead. In case of the HW controlled synchronization the synchronizer is implemented as HW on the FPGA and runs in parallel to the HW-part of the design. In this case the synchronizer offers more precise information regarding the timing, since the HW is at least cycle accurate. By means of using this feature and the causality between the HW and the SW it is possible to execute the HW and the SW in parallel, to improve the emulation performance without extra overhead.

In this paper we present a HW-driven approach to reduce the effort of the preparation process for the emulation as well as the synchronization overhead, and to generate (semi-) automatically the interconnection and the synchronization mechanism between the HW on the FPGA and the SW on the host.

3   Related Work

Research on validation/verification of HW-/SW-systems with non-formal methods can be classified in two groups: co-simulation and co-emulation. While in the first group the whole system is simulated at several abstraction levels, the techniques in the second group use an emulation platform either for the whole or for part of the system. [3] is an example to emulate the whole system. However, most of the emulation techniques use a platform with FPGA to emulate the HW and a host to simulate the SW.

In [4] research on co-simulation is classified into three groups: i) co-simulation algorithms for various computation models, ii) automatic generation of co-simulation interface, and iii) co-simulation speed-up. This classification can be seen as valid for the emulation-based validation/verification. While the techniques to generate the interface for the co-simulation can be used for the emulation, the speed-up aspect should be handled differently due to the possibility of the parallelism at the emulation.

In simulation-based approaches HDL-simulators are used to simulate the HW and a host for the SW with processor models at several abstraction levels. To connect two designs with each other an interconnection topology has to be generated and the synchronization of the communication has to be described. In general the related work mostly concentrates on the generation of the interconnection, but the second aspect is usually disregarded and considered as a part of user effort.

[5, 6] describe a similar backplane-based approach. In this approach the connection is established via ports meaning that the user describes the ports which are connected with each other manually. This solution demands the definition of corresponding ports in SW and in HW. The problem in terms of the synchronization is solved partially in timed simulation by global time steps even if it means a certain degree of inflexibility. In case of functional untimed simulation the synchronization has to be described as handshake protocol by the user.

In [7] the connection topology is described with usage of the VCI specification. In a VCI specification the user has to describe the ports of the design, like in the entity of a VHDL-specification, in order to generate the connection topology. In addition to the port declaration a VCI-specification contains a sensitivity clause for every port which is similar to the sensitivity list of a VHDL-process. The sensitivity clause allows the protocol description of an interface (synchronization mechanism). If any

sensitivity clause becomes true, the data transfer between the VHDL-simulator and
the SW is activated. This approach is one of the few approaches that considers the
synchronization of the communication. However, this is suitable only for simple
protocols and the user has to describe the VCI-specification and modify the design
manually.

In emulation-based approaches similar methods are used to connect the HW and
SW with each other. While [8, 9, 10, 11] present simulation-emulation platforms
using an FPGA and an HDL-Simulator, in [12] a co-emulation platform is described.
[13] describes a proposal for a standard C/C++ modeling interface for emulators and
others verification platforms.

The solution of the synchronization problem during the co-emulation in above approaches is either i) a blocking communication mechanism, ii) synchronization with a
global clock or iii) coprocessor-based synchronization with several limitations. While
the approaches with blocking communication allow the execution of only one part of
the system (either SW or HW) at any time [8], the global-clock-based approaches
require a global clock and the SW as well as the HW have to include a time concept
based on this clock [12]. For the co-processor based synchronization the user has to
guarantee that the HW is able to process the last data from the SW before the next
access from the SW to the HW [10, 11]. If this is not the case, sufficient waits have to
be inserted manually in the SW to fulfill this assumption. In spite of this limitation
and manual insertion of the waits into the SW, these approaches support only the
combinational HW-designs, not the sequential designs. All of these approaches provide only the SW-controlled synchronization, which implicates the above-mentioned
limitations and disadvantages.

In summary existing approaches have the following disadvantages: i) only some of
them support the automatic generation of the interconnection, ii) the synchronization
of the communication is mostly disregarded and left the user as part of the effort to
set up the emulation, iii) only a few approaches consider the synchronization of the
communication; however, the generation of the synchronization is in these approaches partial manual and they support only SW-driven synchronization   iv)
FPGA's are only used as fast simulation engines, but not to reduce the overhead of
the synchronization.

4 Our Approach

In this section we shortly describe our approach and the tool ProtoEnvGen which was
implemented to evaluate our approach. A detailed description can be found in [14].

The key objectives of our approach are to reduce the manual intervention, to prepare a design for the emulation and to increase the emulation speed, solving the problems described in section 2. To achieve this our approach i) allows semi-automatic
generation of the interconnection, ii) reduces the overhead for the synchronization by
means of HW-driven synchronization, iii) allows the mixed-level co-emulation without user intervention, iv) hides the implementation details of the emulation platform
from the user.

Our emulation platform consists of a host (PC) and a universal FPGA-board plugged into the host. Our approach is verified on a PCI-based FPGA-board, but it can be adapted easily for any combination of host and FPGA-board. The design, which will be emulated, consists of a SW-part and a HW-part. The SW-part can be in principle the emulation of the environment of a HW-design or the SW-part of a HW/SW-system. However, in this section the SW refers to primarily the emulation of the environment of a HW-design, which will be emulated on the FPGA and called in the following DUT (Design Under Test).

In our approach the whole system is controlled primarily by the HW. However, we allow the controlling of the HW by the SW, too. The control of the system by the HW is based on the fulfillment of the communication requirements from the DUT with the SW. In order to fulfill the communication requirements from the DUT, the communication requests from the DUT are monitored on the FPGA. If any request is detected, the corresponding communication is initiated. If the communication request can not be satisfied, the DUT is stopped and a communication request is sent to the host via an interrupt. The stopping of the DUT means that the clock of the DUT (DUT_clk) stays stable and is not triggered. In case of an interrupt the SW initiates a data transfer to meet the communication request from the HW. The DUT is started again as soon as all requests from the DUT are satisfied.

On our evaluation platform we have two kinds of communication: the DUT-communication and the PCI-communication. The PCI-communication is only part of the emulation and will not be implemented in the real system. A PCI-driver on the host and a PCI-core on the FPGA are used to abstract this communication. The DUT-communication is the communication between the DUT and the SW and is described predominantly in the DUT by the user. Since the whole communication between the DUT and the SW occurs via PCI, the DUT-communication in our emulation platform has to be coupled with the PCI.

We assume that the DUT-communication can have several interfaces (DUT-interfaces) for data transmission to and from the SW on the host. Several DUT-interfaces mean that there are several communication channels between the DUT and the SW, which have to be mapped on the PCI and can request concurrent data transfer. It means that concurrent data transmission requests from the DUT-interfaces will be scheduled and the data transmission through every DUT-interface will be modeled on the PCI, namely on the host between the SW and the PCI-driver, as well as on the FPGA between the DUT and the PCI-core.

The design flow with ProtoEnvGen contains two steps: declaration of the DUT-interfaces and the generation of the emulation environment. While the first step is performed partially user defined, the last step is fully automated. Figure 2 illustrates the functionality of the ProtoEnvGen and its embedding in the emulation setup.

The inputs of the generation process are the VHDL-description of the DUT and the declaration of every DUT-interface. The outputs are a structural VHDL-description of the HW-part (ProtoEnv) and a set of C++-classes for the SW-part to communicate with the HW-part (ProtoEnv-API).

The ProtoEnv, which can be synthesized with a synthesis tool, consists of the DUT, the PCI-core, DUT-interfaces and a controller. In our experiments we have

used a commercial FPGA-board and a soft PCI-core provided by the board manufacturer [15] in order to validate our concept . The controller connects the DUT with the
PCI-core via DUT-interfaces and coordinates the communication between the DUT
and PCI-core, therefore synchronizes the communication between the HW and the
SW. For the HW-driven communication the controller monitors the communication
requests from the DUT-interfaces and manages the communication as well as the
DUT according to these requests. In case of the SW-driven controlling of the DUT,
the controller handles the DUT according to the order of the SW (e.g. stopping, starting of the DUT).

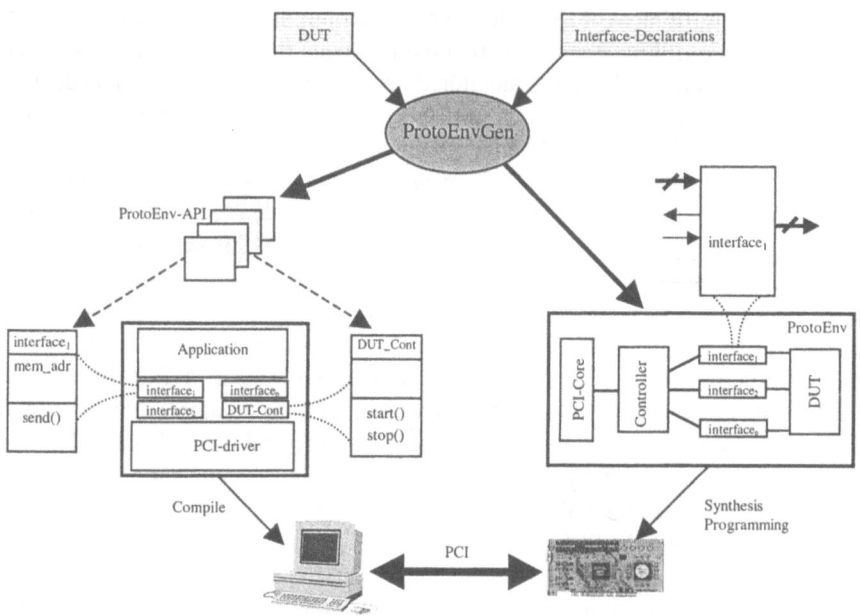

Fig. 2. Designflow with ProtoEnvGen:

The ProtoEnv-API provides access to any DUT-interface, control of the DUT and,
primary, the fulfillment of the requests from the controller in case of the HW-driven
communication. By means of this API the user can send or receive data to/from a
DUT-interface. In order to access to DUT-interfaces for every kind of interface an
interface class with an accessing method (send or receive) is generated and for every
DUT-interface an object of the corresponding class should be instantiated in the SW
(interface$_i$ in Figure 2). If the user wants to control the DUT from the SW, he or she
can reset, start and stop the DUT. For HW-driven communication an interrupt handler
is defined. If an interrupt is set by the controller, which means that a communication
requirement from a DUT-interface has to be satisfied, the interrupt handler determines the reason for the interrupt and calls the corresponding method of the corresponding interface object. The ProtoEnv-API uses a PCI-driver in order to access the
HW. Since the driver development is not the focus of our approach, we use a commercial tool provided by Jungo Ltd. to generate the PCI-driver [16].

5 Interface Generation for the Hardware-Driven Emulation

A DUT-interface is characterized by a port of the DUT and a protocol, and allows the DUT to communicate with the SW. Here, the protocol of a DUT-interface determines when the data transmission occurs, but not how the data transmission is realized physically (e.g. a protocol can characterize that a data transmission via corresponding interface takes place every x cycles, but it does not describe whether the communication is bus-based or point-to-point). A DUT-interface can describe data transmission without any reactivity between the SW and the DUT or a reactive data transmission. Both kinds of data transmission have different requirements regarding the timing of the data flow on the PCI and must be treated differently. To meet the timing constraints required by reactive communication between the DUT and the SW, the timing overhead caused by PCI-communication must be hidden. In contrast to a reactive communication a data transmission without any reactivity is not critical regarding the transfer time on the PCI and can be buffered. To increase the efficiency of the data transfer via PCI every non-reactive DUT-interface has a buffer in the on-board memory. In addition, interfaces are directed – either from HW to SW or vice versa.

In the first version of the ProtoEnvGen we used the Protocol Compiler provided by Synopsys to generate the VHDL-code of the DUT-interfaces with some manual interventions [17], [14]. In the rest of this section we present a method to generate automatically interfaces for the emulation on a PCI-based FPGA-board. In this approach the user intervention is limited only to the declaration of every DUT-interface. During the declaration of a DUT-interface the user should declare the name and whether it is reactive or non-reactive as well as the port, which should be involved in the communication via this interface. The rest of the information for the generation of the interface is extracted from the VHDL-description of the DUT.

The declaration of an interface is followed by the automatic generation of the DUT-interface. It includes the generation of the interconnection and the generation of the synchronization mechanism.

5.1 Generation of the Interconnection

In order to generate the interconnection the following steps are performed after the declaration of all interfaces:

i) Since we use a shared memory communication, the address of the shared-memory for every DUT-interface is determined. While for every non-reactive interface a memory area is assigned, a reactive interface needs only a register.

ii) For every DUT-interface a VHDL-instance is generated. As illustrated in Figure 3, an interface at structural level is an instance with some ports, which are either data ports for the data transfer or control ports for the synchronization, and which allows the connection of the DUT with the controller. The ports *interface_request* and *interface_grant* are intended for synchronization based on hand-shaking. When a data transfer via a DUT-interface is requested, the signal *interface_request* of the corresponding DUT-interface has to be set. The controller sets the signal *interface_grant* as soon as the request of the interface is satisfied. Between the setting of the *interface_request* and the *interface_grant* the DUT is stopped by the controller.

To generate the interconnection in the HW-part the ports of the generated DUT-interface have to be connected to the corresponding ports of either the DUT or the controller.

iii) In the SW-part an interface-object of the corresponding interface-class is instantiated with the address of the shared memory of the DUT-interface as an attribute.

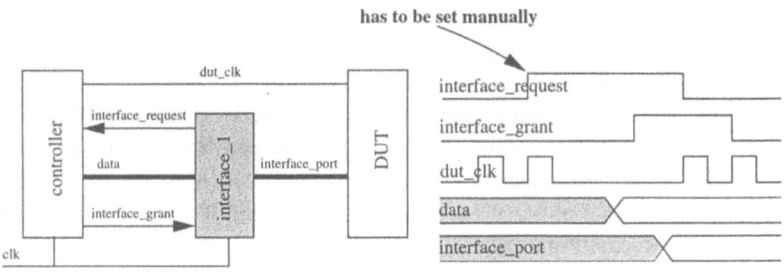

Fig. 3. Communication via a DUT-interface with manual synchronization

The generated interconnection grants access to any DUT-interface from the host by using the accessing methods of the corresponding interface object. However, the synchronization has to occur manually, since there exists no synchronization mechanism yet. This means that either the user has to modify the DUT-interface to set the *interface_request* (as illustrated in Figure 3) or only the SW-driven synchronization is possible.

5.2 Generation of the Synchronization Mechanism

In our approach a DUT-interface is activated when the DUT access to the port which is involved in and connected to the DUT-interface (*interface_port*). This means that every access to the *interface_port* in the DUT has to be detected and for every access the signal *interface_request* of the corresponding DUT-interface has to be set.

For the detection of a port access and the generation of the *interface_request* we search the corresponding statements (e.g. signal assignments) in the VHDL-description of the DUT, and modify them. This process is composed of the following for every DUT-interface: i) A new port, called *interface_enable*, is inserted into the entity of the VHDL-code. ii) On the *interface_enable* an event is generated when the *port_interface* is accessed. Therefore an event statement is inserted before every access statements to the *interface_port*. iii) The port *interface_enable* is connected to the port *interface_request*. Herewith the DUT-interface is activated when it is necessary and waits for the *interface_grant* from the controller.

This approach is problematic if the *interface_port* is a buffered input port of the DUT. For such a port after synthesis a register which is synchronized with the clock of the DUT (*DUT_clk*) is generated. When the *interface_enable* is set, the register of the input port is already loaded. But the register has to be loaded after the generation of the *interface_grant* by the controller, since only after the setting of the *interface_grant* the controller provides the valid data for the actual request. On the other hand the reloading of the register after *interface_grant* causes the changing of the actual state of the DUT. To solve this problem we synchronize the register for the buffered input ports with the master clock of the HW-part instead of the *DUT_clk* and reloaded it as soon as the *interface_grant* is set. The master clock refers to the clock that triggers the PCI-core, the controller and the DUT-interfaces.

Figure 4 illustrates the automatic generation of the synchronization mechanism. Figure 4-a and 4-b show schematically this process on the basis of two block diagrams, while Figure 4-c concretely shows the realization of this concept on the basis of a VHDL-code in a simple example. The original DUT in 4-a has two data ports, and we assume that two interfaces are defined. Both; the ports *in_port* and *out_port* involve in the communication via an interface. The DUT accesses to *in_port* when the condition C_1 is true and to *out_port* when the condition C_2 is satisfied. In modified DUT (4-b) where the modifications are illustrated bold, TFFs are inserted to generate the *interface_enable* events. For example TFF_1 generates an event on the *int_1_en* if C_1 is true, when the *in_port* is accessed. The DFF_1, which is the buffer of the *in_port* and triggered by the *clk* instead of the *dut_clk*, is reloaded when the *int_1_grant* is set. To realize this concept in the example (4-c), the process *proc_new* and the lines marked with asterisk are inserted after the modification of the original DUT.

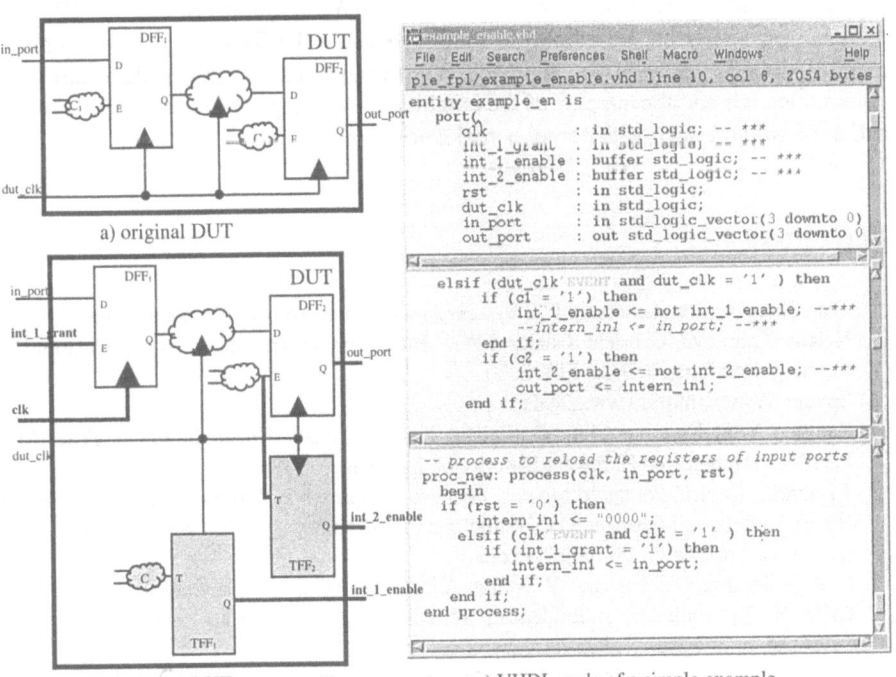

a) original DUT

b) modified DUT

c) VHDL-code of a simple example

Fig. 4. Automatic generation of the synchronization mechanism

The synchronization mechanism presented above implicates the stopping of the DUT after the setting of every *interface_enable*. To avoid this problem the data of every non-reactive DUT-interface can be cached. It means that the controller provides every non-reactive DUT-interface with direction to the DUT valid data before these request new data and sets the signal *interface_grant* of the corresponding interface. Furthermore, this approach supports only the sequential designs, not the combinational designs yet.

6 Conclusion

We have presented an approach to automate the interface generation and the synchronization of the communication between the SW and the HW for the co-emulation. Our tool ProtoEnvGen, which has been implemented and tested with simple examples to evaluate our approach, is also presented shortly. Although in this paper the SW relates to the emulation of the environment of a HW-design, the presented HW-driven emulation approach is also suitable for HW/SW-systems. The HW-driven emulation allows mixed-level emulation regarding time, and speeds up the emulation through the parallel processing of the synchronizer with the DUT. Furthermore, it makes the generation of synchronization mechanism with minimum manual intervention possible.

Our current work is concentrated on the HW-driven emulation of HW/SW-systems, especially on the support of multi-tasking in the SW-part, and the improvement of the efficiency of our approach, e.g. through the caching of data during the communication and the support of the combinational designs, as well as on the verification of our approach with complex real world designs.

References

1. Tom Shanley, Don Anderson: PCI System Architecture, Addision-Wesley, 1999
2. Helena Krupnova, Gabriele Saucier: FPGA-Based Emulation: Industrial and Custom Prototyping Solutions, in Proc. FPL 2000
3. Spyder System, http://www.x2e.de
4. Sungjoo Yoo, Kiyoung Choi: Optimistic Timed HW-SW Cosimulation, in Proc. Asia-Pacific Conference on Hardware Description Language, 1997
5- Alexandre Amory, Fernando Moraes, Leandro Oliveira, Ney Calazans, Fabiano Hessel: A Heterogeneous and Distributed Co-Simulation Environment, in Proc. SBCCI 2002
6. Cosimate, http://www.tni-valiosys.com
7. C. A. Valderrama, F. Nacabal, P. Paulin, A. Jerraya: Automatic Generation for Distributed C-VHDL Cosimulation of Emdedded Systems: an Industrial Experience, in Proc. RSP 1996
8. C. Fritsch, J. Haufe, T. Berndt: Speeding Up Simulation by Emulation – A Case Study, in Proc. DATE 1999
9. N. Canellas, J. M. Moreno: Speeding up HW prototyping by incremental Simulation/Emulation, in Proc. RSP 2000

10. Ramaswamy Ramaswamy, Russel Tessier: The Integration of SystemC and Hardware-Assisted Verification, in Proc. FPL 2002
11. Siavas Bayat Sarmadi, Seyed Ghassem Miremadi, Ghazanfar Asadi, Ali Reza Ejlali: Fast prototyping with Co-operation of Simulation and Emulation, in Proc. FPL 2002
12. W. Bishop, W. Loucks: A Heterogenous Environment for Hardware/Software Cosimulation, in Proc. 30th Annual Simulation Symposium, 1997
13. Functional Requirements Specification: Standard Co-Emulation Modeling Interface (SCE-MI) ver. 1.0 rev. 1.9 2002, http://www.eda.org/itc/scemi19
14. M. Çakır, E. Grimpe: ProtoEnvGen: Rapid ProtoTyping Environment Generator; in Proc. VLSI-SOC 2001
15. PLD Applications, http://www.plda.com
16. Jungo Ltd.: WinDriver User's Guide, http://www.jungo.com
17. Synopsys Inc.: Synopsys Protocol Compiler User Guide, 2000

Implementation of HW$im
– A Real-Time Configurable Cache Simulator

Shih-Lien Lu and Konrad Lai

Microprocessor Research
Intel Labs
Hillsboro, OR 97124
Shih-Lien.l.Lu@intel.com

Abstract. In this paper, we describe a computer cache memory simulation environment based on a custom board with multiple FPGAs and DRAM DIMMs. This simulation environment is used for future memory hierarchy evaluation of either single or multiple processors systems. With this environment, we are able to perform real-time memory hierarchy studies running real applications. The board contains five Xilinx' Virtex^{TM} II-1000 FPGAs and eight SDRAM DIMMs. One of the FPGA is used to interface with a microprocessor system bus. The other four FPGAs work in parallel to simulate different cache configurations. Each of these four FPGAs interfaces with two SDRAM DIMMs that are used to store the simulated cache. This simulation environment is operational and achieves a system frequency of 133MHz.

1 Introduction

Memory hierarchy design is becoming more important as the speed gap between processor and memory continues to grow. In order to achieve the highest performance with the most efficient design simulation is usually employed to assess design decisions. There are several ways to evaluate performance [1]. Backend simulation using address reference traces is one of the methods [2]. Traces can be collected either through hardware or software means. After the traces have been collected, they are stored in some forms. A backend (trace-driven) simulator is employed to evaluate the effectiveness of different cache organizations and memory hierarchy arrangements. This method is repeatable and efficient. However its usefulness is limited by the size of the traces. Another common method used to evaluate memory hierarchy is through complete system simulation. In this approach, every component of a complete system is modeled including the memory hierarchy. Full system simulation is flexible and provides direct performance of benchmarks [3][4]. However, full system simulation is less detail and slower in speed, which may limit the design space explored. It is also much more difficult to setup benchmarks on a simulated machine.

The ideal scenario is to emulate the memory hierarchy with hardware and perform a measurement of the actual system while executing real workloads. Yet the real time

---

^{TM} Virtex is a trademark of Xilinx Corp.

P.Y.K. Cheung et al. (Eds.): FPL 2003, LNCS 2778, pp. 638–647, 2003.

emulation hardware is usually very costly and inflexible. It is difficult to explore all design space interested using emulation. We employed an alternative method called hardware cache simulation (HW$im).

Hardware cache simulator is a real time hardware based simulation environment. It watches the front-side bus and collects traces of an actual system while the system is running real application software. However, instead of just storing the traces and dumping them out after the buffer is full, it processes the traces with hardware in real time and stores only the statistical information. That is, instead of collecting the traces and processing them with a software simulation program later on, the hardware cache simulator simulates the traces immediately. Since traces are processed directly and not stored, we are able to run for a much longer period of time and to observe long-term behavior of benchmarks. It also permits us to explore larger design spaces include large cache effects of multi-processor systems. It removes some artifacts from software simulation having limited trace length.

Hardware cache simulator provides a complementary tool to full system simulation with software. Since it is running at real-time, it allows us to scan different workloads with multiple configurations. When the hardware cache simulator indicates the possibility that a larger cache is significantly beneficial, we can zoom in with the current hardware tracing methodology or full system simulation. Moreover, since we are implementing this simulator with FPGAs we have the flexibility to modify coherency protocol and study migratory sharing behavior. If there are particular behaviors that can benefit from pre-fetching strategies we could also add these pre-fetching algorithms into the simulator to study their impact on performance.

The rest of the paper is organized as follows. Section 2 discusses related work in this area and provides a summary of the HW$im. Section 3 describes the design of HW$im. Section 4 describes our experience in developing the board. Section 5 presents some preliminary results collected using the HW$im. We draw some conclusions in the last section.

2 Related Works and Summary

Much research has been done on using FPGAs for computing [5]. They fall into four general categories – 1) special or general purpose computing machines; 2) re-configurable computing machines; 3) rapid prototyping for verification or evaluation; and 4) emulation and simulation (or accelerators). There are several commercial products that offer boards that can be used for emulation purposes. Some examples are [6][7]. While these products offer flexibility, they are generally used for design verification and software/hardware co-development. Our requirement is a bit different. We need to be at speed with the system we are interfacing with. We are using this system to collect performance information of future systems instead of verifying designs.

A previously published work similar to this work is the IBM MemorIES [8]. The main differences are: (1) Our design uses DRAM instead of SRAM to store simulated cache directory information; (2) We use interleaving (parallel FPGAs) to bridge the bandwidth differences between system bus and SDRAM; (3) Our board is designed as a PCI board which can be installed on a host machine directly. Using DRAM allows larger simulated caches be studied. Interleaving of multiple FPGAs in parallel enable us to perform parallel simulation of different configurations at the same time, thus removing the artifact of any non-deterministic behavior of running applications repeated sequentially. With PCI interface we can download simulation results at a much finer time unit.

Figure 1 illustrates the system setup of HW$im. Two main structures are in the setup. The first is the system-under-test which is the system that runs the application whose behavior we are investigating. The second system is the host system that houses the custom built board with FPGAs. A logic analyzer inter-poser probe is used to capture the signals on the front side bus of the system under test and send them to the HW$im

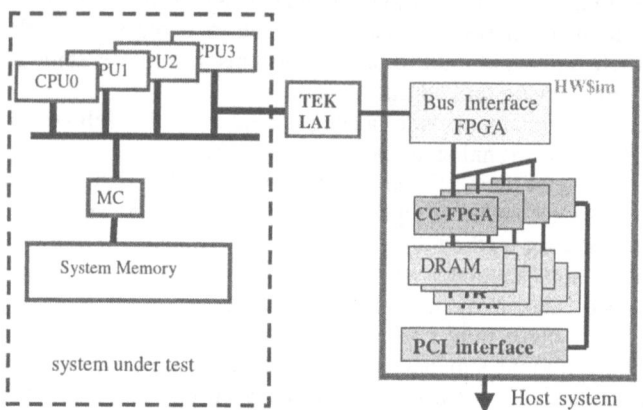

Fig. 1. HW$im Systems Setup

Picture to the right of figure 2 shows the HW$im board plugged into the host system collecting data. The host system is a dual processor machine with Intel Pentium® III Xeon™ Processors. Picture on the left of figure 2 illustrates the system under-test with the Tektronix LAI probe and cables.

Fig. 2. Systems with HW$im

® Pentium is a registered trademark of Intel Corporation or its subsidiaries in the United States and other countries.

™ Xeon is a trademark of Intel Corporation or its subsidiaries in the United States and other countries.

3 Design Description

The key to hardware cache simulatoion is the rapid development of the capabilities of field programmable gate array devices. The amount of gates has grown substantially. They also contain many features that simplify system level designs. To accommodate the need for I/O pins of this design, we adopt the XC2V1000 chips with the FF896 package. This chip provides 432 user I/O pins. We also employ SDRAM DIMMs to store cache directories being simulated. It provides high capacity with a wealth of reference designs available.

3.1 Functional Description of Hardware Cache Simulator

There is total of four main functional blocks in the hardware cache simulator (HW$im) as illustrated on the right hand side of Figure 1. They are 1) Bus Interface (BI); 2) Cache Controller (CC); 3) DRAM (simulated cache directories); and 4) PCI Interface. We will describe the functionality of each block.

3.2 Bus Interface (BI)

This block is responsible for interfacing the front-side bus of the processor and for filtering out transaction requests not emulated such as interrupts and I/O transactions. Transactions are identified and the proper information is queued in the BI according to the bus protocol. Once all of the information for a bus transaction has been collected the transaction is forward to the next block. Since requests on the bus may come in bursts we use an asynchronous queue to decouple the input and output of this block. If the system-under-test is a Pentium® III, the maximum request rate is every three bus clock (BCLK) cycles per request. For Pentium® 4 based systems, the maximum request rate is one request per 2 BCLK.

Since both Pentium® III and Pentium® 4 front-side bus are split transaction buses, the request and response phases can be separated in time to allow overlap transactions. There are distinct phases for front side bus - Arbitration, Request, Error, Snoop, Response and Data. There can be up to 8 total outstanding transactions across

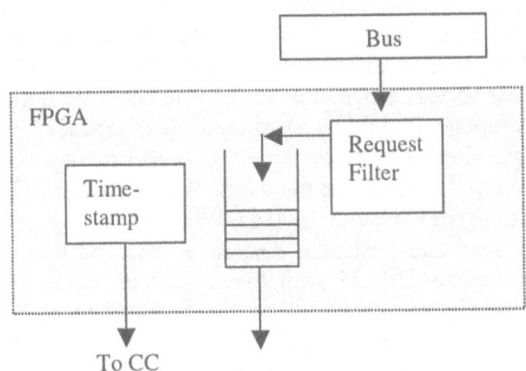

Fig. 3. Bus Interface

all agents at one time. Figure 3 depicts the major sub-blocks within the BI. The Request Filter sub-block in this figure is used to buffer the bus activities. Transaction may be completed out-of-order if they are deferred. This sub-block is also used to filter out I/O requests. Another block in the BI is a FIFO queue. It is used to average out the transaction rate and to buffer the front-side bus frequency from the internal transfer frequency. There is a time stamp unit. It provides the bus cycle field of the request information to the cache controller. With bus cycle time we are able to count the bus utilization rate. The output of the BI is a logical request with the following information:

`BusCycle:ReqType:AgentID:Address`

Transactions are removed from the queue at the request rate which has a maximum frequency of half the speed of the fastest front-side-bus (FSB) frequency. Many new FPGAs contain frequency synthesizers making the design easy and robust.

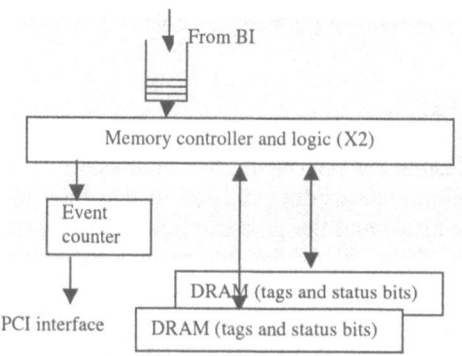

Fig. 4. Cache Controller (X4)

3.3 Cache Controllers and DRAM Interface

The second block is the cache controller. It consists of memory (DIMM) interface logic that performs 1) reading from DRAM; 2) comparing tags, state, etc. in parallel; 3) computing new LRU, state, etc; 4) sending control signals to update statistic counters; and 5) writing modified tags back to DRAM. It also needs to prepare DRAM for the next access. Since most cache references do not alter the state of the cache, the last operation will only be performed when needed.

We estimated the number of cycles used for comparing tags and update states to be 5 cycles (each cycle is 7.5 ns). Since the width of the DIMM is 72 bits we will need to use multiple cycles to read or write the tags. Width of the tags is 8x4x4B. It will take 2 extra cycles to read and write the desired width. The total Read-Modify-Write access with Synchronous DRAM with these 5 cycles delay after read data is 18 cycles. This translates into a total of 135 ns. For P6 bus we may be able to live with 4-way lower bits address interleaving but for Pentium® 4 we must interleave 8-way. We will put two cache controllers on one FPGA ince it is mostly I/O limited. Each DIMM interface will require about 100 pins. Putting 4 controllers on a single FPGA will push the limit of its I/O pins. Instead we put 2 controllers on a FPGA and each controller will interface with a 256MB DIMM. It will be easier to design if we put the shared L3 on a separate memory controller. It will interface to event counters separately to collect statistics for shared L3 configuration. To simulate up to 1GB of L2 with different configuration we need to have 256MB of DRAM for each interleaved way. This setup will simulate from a 512MB direct map up to 4GB 8-way set-associative L2 cache for a quad-processor setup. If we use two 256MB DIMMs for shared L3 we will be able to simulate from 512MB direct map up to 16GB 8-way set-associative L3 cache. Figure 4 depicts the cache controller block diagram.

As mentioned, we collect cache simulation data in real time. Currently the HW$im collects the following information: 1) Total References and total time; 2) Read/Write ratio; 3) Hit/miss rates; 4) Hit/HitM; 5) Deferred/retired; 6) Bus utilization rate; 6) Write backs and Replaced lines; 7) Address range histogram. More counters may be added if desired.

3.4 PCI Interface (PI)

The final block is the PCI (or console) interface. It is used to collect statistics collected during the simulation. Currently the PI is a passive device. Event counters are shadowed and read directly periodically. The period of the read can be adjusted.

4 Design Experiences

4.1 I/O and DCI

Xilinx Virtex™ II parts feature on-chip digitally controlled impedance (DCI) I/O. It enables designers to improve signal integrity without using external resistors for impedance matching on printed circuit boards. We employed DCI as part of the design allowing a local multi-drop bus to operate at speed. The signal level used is LVTTL through out the board since we need to interface with SDRAM as well as a PCI bridge chip. There is a translation from GTL+ to LVTTL at the LAI end. We use a part from Philips (GTL16612) requiring only a 3.3 V supply. The output of the GTL+/LVTTL translator is serial terminated to match the impedance of the cable. The bus connecting four parallel FPGAs is also serial terminated.

4.2 DCM and Clock Domains

It is important for a high performance designs to have efficient clocks. On the HW$im board there are three clocking frequencies. First, there is the front side bus frequency of the system-under-test. All information on the front-side-bus is sampled with this frequency. Since the maximum request rate for Pentium® III is every three bus cycles we divide the bus clock by three. This is the second clock domain. Virtex™ II provides on-chip Digital Clock Manager (DCM) which make this process easy. Due to the length of the cable, signals arriving from the front-side-bus are shifted in phase from the on-chip clock signal generated with the DCM. Again the DCM used provides means to phase shift the clock signal with easy interface. This divided clock is used to transfer data from the BI FPGA to cache controller (CC) FPGAs. On the CC FPGAs there are DCMs used to generate the frequency used to communicate with SDRAM. The flexible clock multiplier as part of the DCM does this. Finally there is the last clock domain used by PCI interface. We use the PCI clock generated by the PCI bus master on the host system to drive the bus between the PCI bridge part and CC FPGAs.

4.3 Timing Constraints

In order to operate at real-time several approaches are employed. First we included many timing constraints in the design User Constraints File (UCF) to ensure that the placed and routed design will meet timing requirements. Some examples are:

```
TIMESPEC "tsinputsspeed" = FROM "PADS" TO "FFS" 7.5 ns;
TIMESPEC "tsoutputspeed" = FROM "FFS" TO "PADS" 7.5 ns;
NET "BCLK" PERIOD = 7.5 ns;
```

Time ignores (TIGs) constraints are also used to remove violations that do not affect the design. When timing still cannot be met, extra pipeline stages are inserted to keep up the frequency. For the BI design we added two extra stages. For the CC FPGAs we added one extra pipeline stages.

4.4 Design Validation

Besides extensive simulation at the functional and timing (after place and route) level, we employed many design-for-test features to validate the correctness of the design. First extra pins were reserved for probing internal nodes. Since FPGA is re-programmable we are able to map internal nodes to these reserved probe pins for detail observation. Second, test-bench codes were programmed into the FPGA for on-line testing. These test-bench codes generate necessary input patterns and check the output for assertions. Many times simple pass or fail signaling is used to give quick feedbacks to the designer. For example we have a internal memory references generator that read from a file and feed the cache controller with known input pattern allowing us to debug the design. Third, a completely re-written set of test codes is used to check the hardware connectivity distinguishing completely hardware bugs from software bugs.

5 Cache Simulator

Just like any trace-driven cache simulator there are several challenges when we are observing behaviors at the front-side-bus level. These challenges are the result of mismatches between the system-under-test and the simulated system. First, a write to an E-state line by a processor cannot be seen on the bus by the simulated cache until a later time. Second, there are two coherence options when a read request hits a modified copy in a remote cache. With one of the option, after an implicit write-back, both lines in the request CPU and the CPU with the modified line go into the shared state. Another option transfers both the data and the ownership from the current owner (with modified line) to the requester. Discrepancy may be created if the host cache and the simulated cache use different options.

5.1 Hidden Write-Shared: Due to Missing Write-Excl

Figure 5 illustrates the sequence of operations that may cause a discrepancy between simulated the cache and the real host system cache. Ci and Cj stand for CPU I and j. Initially both cache lines in Ci and Cj are invalid. A read by CPU Ci causes the line to go into the E state bringing the data into the cache. Another read by CPU Cj cacuses both CPUs to have the data in their respective caches with both lines set to the Share state.

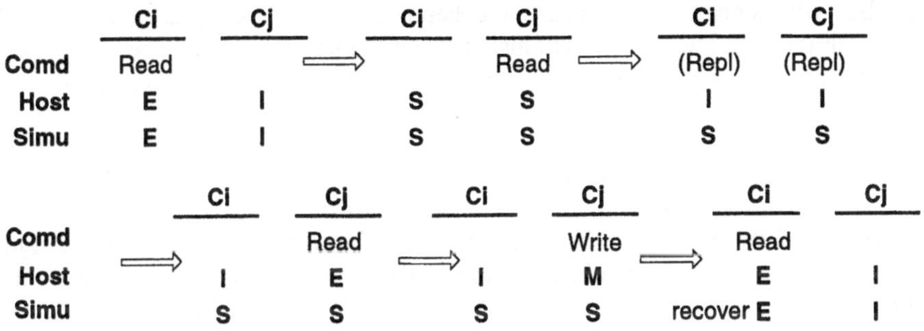

	Ci	Cj		Ci	Cj		Ci	Cj
Comd	Read		⟹		Read	⟹	(Repl)	(Repl)
Host	E	I		S	S		I	I
Simu	E	I		S	S		S	S

		Ci	Cj		Ci	Cj		Ci	Cj
Comd	⟹		Read	⟹		Write	⟹	Read	
Host		I	E		I	M		E	I
Simu		S	S		S	S		recover E	I

Fig. 5. Example of a discrepancy caused by Missed Write to an Estate line

Since the simulated cache is usually larger in size, a line may be replaced in the real CPUs while still resides in the simulated machine. Later a new read follow by a write to the line causes the line to go from I to E to M states. During this sequence the simulated machine maintains the line in Ci and Cj at the S state. Only when a read by Ci later, which causes the real machine to issue a snoop hit on modify, will alert the simulated machine to recover its states. Before this happens all other read by any CPU to this line with the S state will not register the invalidation miss.

5.2 Hidden Read Miss and Write-Shared:
When Host Uses S-S, but Simulated Cache Uses E-I

When the simulated machine uses the E-I coherency scheme while the real system uses the S-S also may cause discrepancy. When a read miss happens, the snoop result indicates that the line requested is at the M-state in another CPU's cache. There are two possible ways to change the state of these two lines. There is the E-I scheme that changed the line of the requested CPU to the E state and invalid the line holding the M-state. There is the S-S scheme that changes both lines to the S-state. When the system under test uses the S-S scheme and the simulated machine uses the E-I scheme there may be problems. This is similar to the case of missing write-excl and can be handled in the same way. The difference is that the E-state is arrived when a read-miss hits a remote modified copy.

5.3 Recovery the Hidden Write-Shared by the Simulated Cache

As mentioned we can recover the hidden write-shared at a later time. Although such recovery can be very accurate, the delay of invalidation may affect the replacement in the simulated cache. In addition, the recovery scheme may add some complexity to the simulated cache. The recovery algorithm is described briefly here. When a read-miss hits multiple S-state lines in the simulated cache with HIT#/HITM# dis-asserted, change the state of the simulated cache to ES (a new state) and keep all other S-state copies. A hidden write-shared is recovered whenever a requested hit an ES-state in any of the simulated cache and a HITM# is asserted. In this case, all the S-state will be invalidated. The new line state in the requester's cache and the change of the ES state line will depend on the underline coherence protocol. A replaced dirty line that hits ES in the simulated cache will also cause a hidden write-shared recovery and an invalidation of all shared copies. This ES copy changes to M-state afterwards.

5.4 HW Cache Simulator Results

We have used the HW$im to capture behaviors of programs. The following figures illustrates some of the capabilities of the HW$im. Figure 6(a) shows the result of running a particular benchmark called SPECJbb [10]. First we can find out the cache miss ratio captured at a particular time unit for a level two cache of size 16MB organized as 8-way set associative cache with line size equals to 32B. We observe a change of behavior at about 60 seconds into the application run. This is due to garbage collection. Once the garbage collection is completed, access to the memory becomes more efficient in that the miss ratio decreased. Another interesting fact is that this application shows that the miss ratio changes dramatically from one time unit to the other. If our software simulation can only capture a small period of real time due to simulation speed then the results may be different depending on which time unit the data was collected.

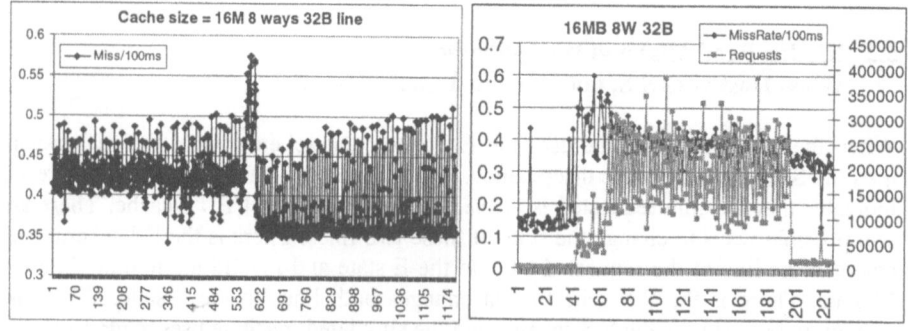

Fig. 6. Miss Ratio per 100 mille-second for - (a) SPECjbb; (b) MpegEnc

Figure 6(b) shows the HW$im result of another application. This application is a multi-threaded media application. Besides the miss ratio we also show the number of requests send to the simulated cache. This program ran from the beginning to the end. Thus, we observe the change of request numbers during the run.

Figure 7 illustrates the effect of different cache size on miss ratio for a large vocabulary continuous speech recognition application software. This graph demonstrate that we are able to simulate very large cache size.

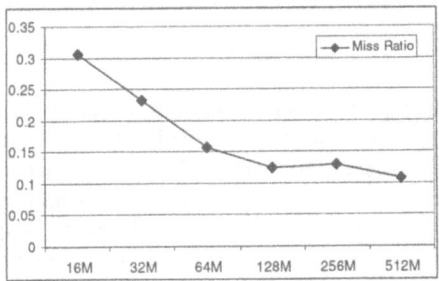

Fig. 7. Overall Miss Ratio of a Speech Recognition Application

6   Conclusion

A custom board with multiple FPGA used for future platform and memory hierarchy study is presented. This board operates at full system front-side-bus speed and can investigate real application behaviors on these possible future designs. This board is able running at full real-time speed due to the use of the state-of-the art FPGAs. It has been used to evaluate cache designs that are difficult to evaluate with current tracing technology.

References

[1] M.A. Holiday, "Techniques for Cache and Memory Simulation Using Address Reference Traces," International Journal in Computer Simulation, 1990.
[2] Richard A. Uhlig and Trevor N. Mudge, "Trace-driven Memory Simulation: A Survey," ACM Computing Surveys, 29 (2), 1997, pp. 128-170.
[3] P. S. Magnusson et. al., "Simics: A Full System Simulation Platform," IEEE Computer, February 2002, pp. 50-58. (http//www.simics.com)
[4] S. A. Herrod, "Using Complete Machine Simulation to Understand Computer System Behavior
     Ph.D. Thesis, Stanford University, February 1998. (http://simos.stanford.edu/)
[5] Steve Guccione, "List of FPGA-based Computing Machines,"
     http://www.io.com/~guccione/HW_list.html
[6] http://www.quickturn.com/
[7] http://www.aptix.com/
[8] Ashwini Nanda et. al., "MemorIES: a programmable, real-time hardware emulation tool for multiprocessor server design," Proceedings of ASPLOS, 2000
[9] http://www.xilinx.com/xlnx/xweb/xil_publications_index.jsp
[10] http://www.spec.org

The Bank Nth Chance Replacement Policy for FPGA-Based CAMs

Paul Berube, Ashley Zinyk, José Nelson Amaral, and Mike MacGregor

Dept. of Computer Science, University of Alberta, Edmonton, Alberta, T6G 2E8, Canada,
{berube,zinyk,macg,amaral}@cs.ualberta.ca

Abstract. In this paper we describe a method to implement a large, high density, fully associative cache in the Xilinx VirtexE FPGA architecture. The cache is based on a content addressable memory (CAM), with an associated memory to store information for each entry, and a replacement policy for victim selection. This implementation method is motivated by the need to improve the speed of routing of IP packets through Internet routers. To test our methodology, we designed a prototype cache with a 32 bit cache tag for the IP address and 4 bits of associated data for the forwarding information. The number of cache entries and the sizes of the data fields are limited by the area available in the FPGA. However, these sizes are specified as high level design parameters, which makes modifying the design for different cache configurations or larger devices trivial.

Key words: field programmable gate array, cache memories, replacement policy, multizone cache, content addressable memories, Internet routing, digital design, memory systems

1   Introduction

The task of an Internet router is to determine which node in the network is the next hop of each incoming packet. Routers use a routing table to determine the next hop based on the destination address of the packet and on the routing policy. Thus a routing table is a collection of destination address/next hop pairs. When adaptive routing policies are employed, routers communicate with each other to update their routing table as the network changes.

Given a routing table configuration, the task of routing a packet consists on consulting the routing table to determine the next hop associated with the destination address. The delay incurred to perform this table lookup affects both the throughput of the network, and the latency to send a packet from a point A to a point B. Several techniques are available to implement the routing table lookup. Storing the table in memory and performing a software lookup may result in a prohibitively high routing delay and unacceptably low throughput. The lookup time can be improved by storing entries associated with frequently seen destinations in a fast hardware cache.

Such caches are standard in Internet routers. The table lookup in a router requires the implementation of the functionality of a content addressable memory (CAM) [2, 11]. While some network processors perform the CAM function using fairly conventional caching technology (NetVortex PowerPlant from Lexra [7], and Intel IXP1200 [6] for

P.Y.K. Cheung et al. (Eds.): FPL 2003, LNCS 2778, pp. 648–660, 2003.

instance), others implement fairly sophisticated CAMs. The IBM PowerNP processor provides an interface for an external implementation of a CAM [12]. The CISCO Toaster 3 network processor implements an on-chip ternary logic CAM [9]. The PMC-Sierra ClassiPI processor implements a very sophisticated CAM that stores not only the next hop associated with each destination address, but also a large set of rules that can affect the routing decisions [8].

Most network processors used in routers store all entries on a single CAM. The destination address in an incoming packet is formed by a prefix followed by wild cards or "don't care" values that match any bit stored in the routing table. This paper is motivated by the idea of separating the entries in the routing table according to the length of the prefix. This separation is expected to be beneficial because incoming packet streams with different prefix length do not interfere with each other. Preliminary studies based on simulation indicate that we should obtain improved hit rates with a multi-zone CAM when compared with a monolithic implementation [3]. The idea of segregating the router table entries by prefix length is analogous to the idea of separating instruction and data streams into separate caches in general purpose processor architectures. In this paper we report the status of our design for a single zone CAM for IP lookup caching.

We developed a prototype for the content addressable memory (CAM) using the Xilinx Virtex-E family of FPGAs. In this paper we describe the implementation of this high-density ternary CAM. Section 2 introduces the overall architecture of our cache. Section 3 details how the Virtex-E FPGA features are used to implement the design. Implementation results are given in section 4. Section 5 describes some related work. Concluding remarks and future work follow in section 6.

2   Cache Architecture

A functional description of our design is presented in Figure 1. The prototype will be tested with a stream of router address lookups collected from network activities in actual routers. The router emulator sends destination addresses to the IP cache in the FPGA. The cache responds with the next-hop information, or indicates that a miss occurred. In the case of a miss, the router emulator sends the missing entry to the cache prototype. The router emulator tracks the number of hits and misses for an entire trace.

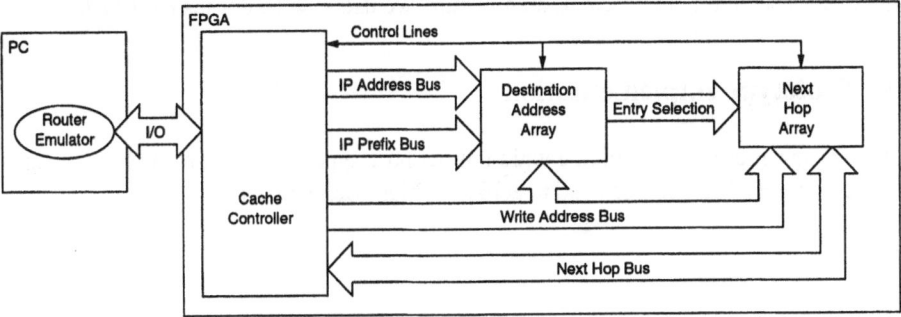

Fig. 1. A functional view of the cache design

2.1 Functional Description

Figure 1 presents a functional description of the IP caching system. The Cache Controller receives requests from the router emulator and returns lookup results. IP addresses are sent to the Destination Address Array (DAA) for matching through the IP Address Bus. The DAA stores IP address prefixes in which several of the least significant bits are set to don't care and thus match any value in the incoming IP address. The DAA is a fully associative ternary CAM. A match in the DAA produces an entry selection to fetch the corresponding next hop information from the Next Hop Array (NHA).[1] The next hop is sent through the Next Hop Bus to the Cache Controller that returns it to the Router Emulator.

If the IP address does not match any of the entries (IP address prefixes) in the DAA, a miss indication is sent to the cache controller, which returns a miss report to the router emulator. An update of the IP cache will be necessary. The emulator responds to the miss report with the length of the prefix that should be stored in the DAA and the next hop information. If there are empty (non-valid) entries in the CAM, one of them will be selected to contain the new address prefix/next hop information. If all the entries in the CAM are valid, the cache controller will select an entry for replacement according to the implemented replacement policy. The prefix of the IP address will be stored in the DAA and the next hop will be stored in the corresponding entry in the NHA.

2.2 Structural Description

Structurally the IP cache is divided into several blocks of entries. In our prototype each block stores 16 address prefix/next hop pairs. For a lookup, the cache controller sends the same IP address to all blocks and expects one of the blocks to reply with the value of the next hop. If a next hop is not found in any of the blocks, an entry will have to be selected for replacement. As shown in Figure 2 each block has a victim selection logic. This victim selection logic has a mechanism to keep track of recent references to CAM entries.

shown in Figure 3 we refer to all the entries that appear at position i on all blocks as the *bank* i. Each cache entry belongs to exactly one block and one bank. Our prototype design is modular at the block level, thus the number of blocks, n, can be varied to produce larger caches when more FPGA "real state" is available. However, to better use the space available in the VirtexE architecture, we use $k = 16$ entries per block.

3 Prototype Implementation

We will now present the implementation of our prototype design, showing how features of the VirtexE FPGA can be used to achieve a high cache density.

---

[1] In the current implementation of the prototype we do not handle the situation in which the address prefix matches multiple entries in the DAA.

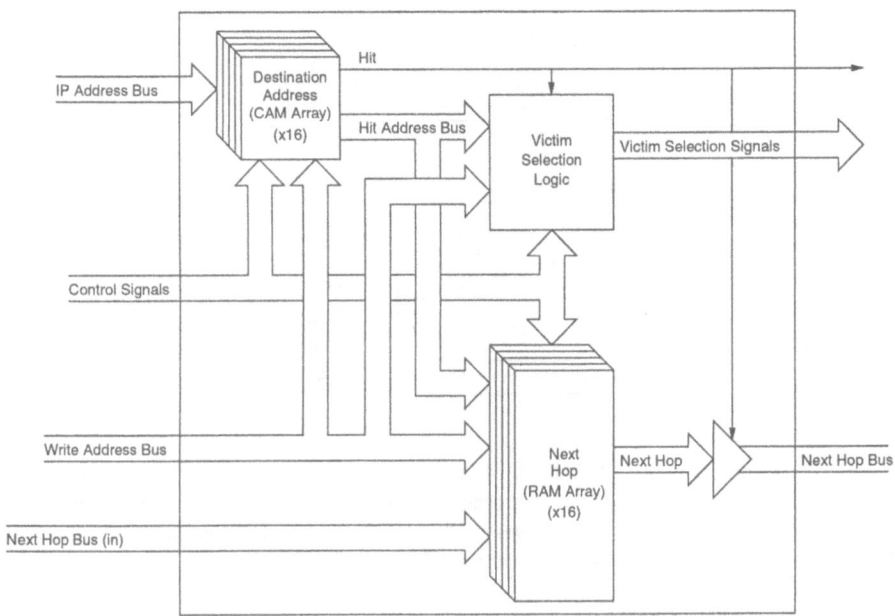

Fig. 2. The structure within a Block. All blocks share common input and output buses.

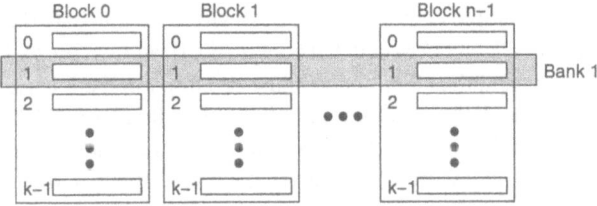

Fig. 3. The relationship between entries, blocks and banks.

3.1   Implementing CAM Cells with Shift Registers

In order to implement a 32-bit CAM cell to store an IP address in the DAA, we use 8 LUTs configured as 16-bit shift registers [1]. Each one of these LUTs will store the value of four of the IP address prefix bits as a one-hot encoding. For instance, if the four most significant bits of the address prefix are 1100, the LUT corresponding to these bits will store the value 0001000000000000. When a lookup occurs we use the incoming IP address to read the value stored in each shift register at the position specified. In this case the value read from the LUT will be 1 only when the incoming IP address has the value 1100 in its 4 most significant bits.

We want to be able to store address prefixes in the CAM cells, not only complete addresses. Let suppose that we want to store an address prefix that has in its 4 most significant bits the value 010X, where X indicate a *don't care* value, *i.e.* it should match either a 0 or a 1 value. In this case, the value stored in the LUT shift register corresponding

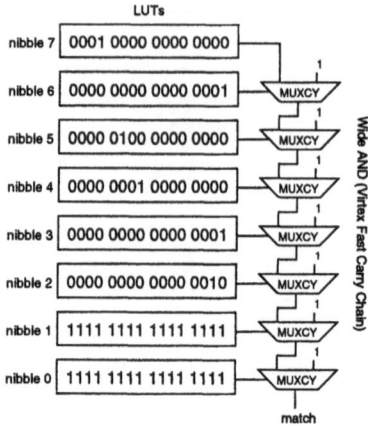

Fig. 4. The IP address prefix 192.168.1.0/24 is stored in the shift register LUTs. The most least significant bits, nibble 0 and nibble 1, correspond to the 8 don't care bits at the end of the prefix. The LUTs that store these don't care values are filled with 1s. A match is computed using a wide AND implemented using the Virtex carry chain MUXs (MUXCY).

to the four most significant bits is 0000000000110000. Now an IP address with either 0101 or 0100 as its 4 most significant bits will read a 1 from that LUT.

A more concrete example is shown in Figure 4 where we illustrate the value stored in each of the 8 LUTs of a 32-bit CAM cell that contain the IP address prefix 192.168.1.0/24. This prefix has 24 bits and thus don't cares must be stored in the 8 least significant bits of the cell. In hexadecimal, the value stored in this cell is 0xC0A801XX. Notice that all the bits stored in the shift registers of LUTs 0 and 1 are 1s. Thus these two LUTs will return a 1 for any incoming address, producing the effect of a match with a don't care.

When a lookup is performed, the value read from each LUT shift register is fed into a wide AND gate implemented using Xilinx's carry chain building block as shown in Figure 4. If the output of the carry chain is a 1, the incoming IP address matches the prefix stored in the CAM cell.

The writing of an IP address prefix requires that the hardware compute the value of each bit to be stored in each LUT shift register. The circuit used to generate the input values for each shift register is shown in Figure 5. Values generated by a 4-bit counter are compared with the value in the IP address nibble corresponding to the LUT. If they match a value 1 is produced to be shifted into the shift register. In order to enable the storage of don't care values, a don't care mask is used to force a match for bits that are specified as don't care. This is done through the selection of either the counter value or the IP address nibble value to be sent to the comparator for each bit. If the IP address is sent, then it is compared with itself, ensuring a match. Since the don't care mask can be arbitrarily generated, the positioning and number of don't care bits is also arbitrary. However, in our application of IP routing, we generate don't care masks which correspond IP prefixes, where only the N least-significant bits contain don't care values.

Notice that a common single counter can be used for all the eight LUT shift registers that store a 32-bit IP address prefix. Moreover a single circuit for the computation of the

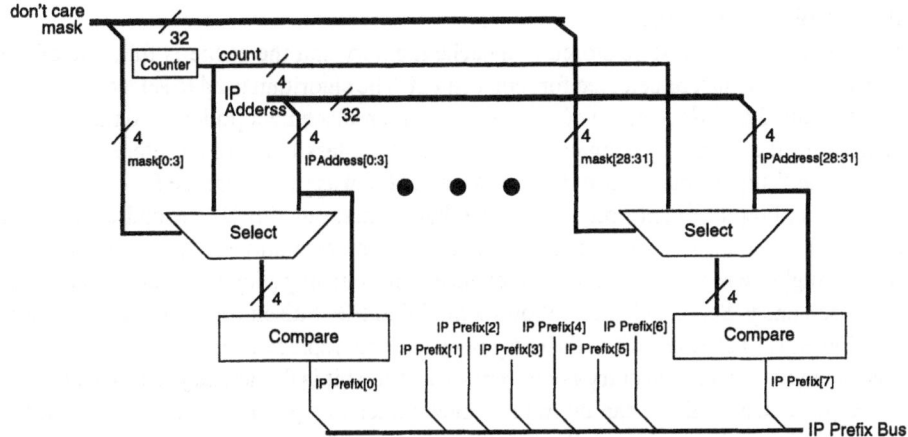

Fig. 5. Computation of the bits to store in the shift register for an IP address prefix nibble.

stored bits is implemented for the entire device. The bits generated by this circuit are written in the IP Prefix bus. The 8 CAM cells for the DAA entry that is performing a replacement will be the only ones selected to write the prefix bits generated.

A great advantage of our design is that we are able to implement a CAM without requiring a comparator for each cell. A simple reading of the 8 shift registers used for the cell will indicate if the IP destination address matched the prefix stored in the DAA. Moreover, each slice in the FPGA contains two flip-flops, thus if we were to use the flip-flops to store ternary values, we would need at least 32 slices for a 32 bit IP prefix. With our technique we can store the same prefix using only 8 LUTs, which requires only 4 slices. Furthermore, this design avoids the two levels of logic required to implement flip-flop based ternary comparisons, resulting in a much faster circuit.

3.2   Implementing the Next Hop Array Using SelectRAMs

The implementation of the NHA is composed of several 16x1 single port SelectRAMs, to achieve the desired data width for 16 entries. In our prototype implementation, we use a data width of 4 bits. When the IP cache is updated, the next hop value is written into the NHA concurrently with the update of the IP address prefix in the DAA. Since the NHA and DAA are co-indexed, they can share address and write-select lines. When a lookup is performed, the output from the NHA is through a high-impedance bus shared by all NHA blocks. Output to this bus is through a controlled buffer, with the control provided by the match signal from the corresponding entry in the DAA. In this design, we do not consider the possibility of multiple hits within a single zone of the cache. However, if multiple hits were possible, additional logic would be required to ensure that only one NHA entry supplied its output to the bus at any given time.

In a flip-flop based design, the NHA would require 4 flip-flops per entry (2 slices), or 32 slices per block. By using SelectRAM to store the next hop data, we require one LUT per bit per block, or 2 slices per block.

3.3   Replacement Policy

When a new entry must be stored in a cache memory and the cache is full, one of the existing entries must be selected for replacement. The algorithm used to select this *victim* entry is called a *replacement policy*. The goal of a replacement policy is to manage the contents of the cache, evicting cached values that are less likely to be referenced again in the near future to make room for a new value that was just referenced.

Standard replacement policies include first in, first out (FIFO), second chance and derivatives thereof. For very small caches a full implementation of least recently used (LRU) might be required. The second chance replacement policy associates a status bit with each entry. Second chance follows a modified FIFO discipline in which the status bit of an entry is checked when the entry is selected for replacement. If the status bit is 1, it is changed to 0 and the entry is not replaced. If the bit is 0, the entry is replaced [14].

Replacement policies can be implemented either as a parallel search, or as a serial search. Second chance is usually described as a serial search that loops through the cache entries until a valid victim is found. While a serial search reduces the amount of hardware components required to implement the cache, performance concerns dictate that a parallel search for a victim is preferable for a hardware IP address cache. However, a hardware implementation of the second chance policy with parallel victim selection is quite expensive. Thus we developed a new replacement policy that we called the *Bank Nth Chance* replacement policy.

The Bank Nth Chance Replacement Policy. The Bank Nth Chance replacement policy is an approximation of LRU and is similar to the second chance algorithm. As shown in Figure 3 the cache entries are organized in blocks and banks. Our replacement policy has a two level selection process. At the first level it uses a simple round-robin mechanism to select the bank that should produce a replacement *victim*. At the second level it uses an n-bit reference field to determine entries in the bank that are *valid victims* (in our prototype we use $n = 3$). A victim is valid if all its reference bits are equal to zero. The reference field of an entry is an n-bit shift register. Whenever the entry is referenced, a 1 is written into the most significant bit of the reference field. The entry is aged by shifting the field to the right by one position and writing a 0 into its most significant bit.

The cache controller maintains a Bank Pointer to select the bank from which the next victim will be chosen. If the bank selected by the bank pointer has no valid victims, all the entries of that bank are aged by shifting a 0 into their reference fields, and the Bank Pointer moves to the next bank. This process of aging the entries in a bank and moving the Bank Pointer will be referred to as an AGEANDADVANCE operation. After this operation is performed, one or more of the entries in the bank might become a valid victim. However, the bank will not be required to perform a victim selection until the next time it is selected by the bank pointer.

When a bank that has a valid victim is selected by the bank pointer, it is required to return the index of one of its entries. This entry will be used for the next replacement. The Bank Pointer is not moved, and thus the next attempt to select a victim will target the same bank. The cache controller selects a victim to be used for a new entry before a cache replacement is necessary. Whenever this pre-selected victim is referenced, the controller starts the selection of a new victim. This pre-selection of the entry reduces the average time required to perform replacements.

Implementation of Bank Nth Chance. The reference fields for the cache entries are stored in the FPGA LUTs configured as 16×1 SelectRAMs. Each block of 16 entries requires three SelectRAMs to store its 16 3-bit reference fields. The Block Reference Field Logic (BRFL), shown in Figure 6, performs victim selection and entry aging for each block of entries. The bank number is used to determine which entry within a block is to be examined, referenced, or aged. Each one of the SelectRAMs stores one bit of the reference field for 16 entries. All the entries of a bank are aged at the same time. Therefore the same bank number is input in all blocks, the Age signal is asserted and the Reference signal is negated causing the reference field to shift and writing a zero in the Field2 SelectRAM. When an entry is referenced, the entry's bank number is inputted in the victim selection logic of the entry's block, the Reference signal is asserted and the Age signal is negated, causing a 1 to be written in the entry's position in the Field2 SelectRAM.

The organization of the cache entries in blocks and banks allows for the simultaneous access of all entries in a bank during the victim selection process. It also allows for the simultaneous aging of all entries in a bank. The victim selection logic is shown in Figure 7. A BRFL contains the Block Reference Field Logic shown in Figure 6. When several blocks contain valid victims for the selected bank, a priority encoder is used to select which one of these blocks will be used for the replacement. The priority encoder also outputs a signal to indicate whether a valid victim was found.

The motivation for the use of the round-robin discipline to visit banks should now be evident. Our design allows for a single bank to be visited at a time. Therefore it makes

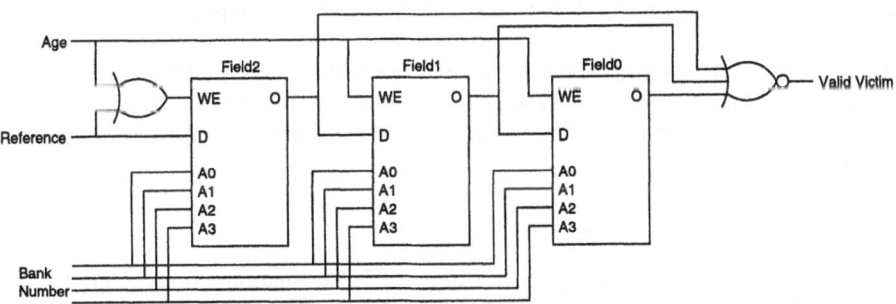

Fig. 6. Block Reference Field Logic (BRFL): Logic for victim selection and entry aging in the reference field of a block.

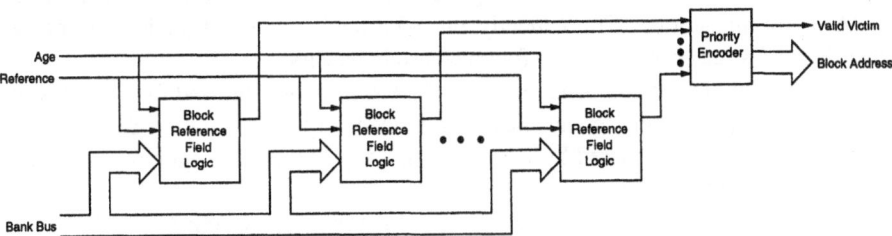

Fig. 7. Victim Selection Logic operating within a bank and across blocks.

sense to not attempt to select a victim from a bank that has recently reported that it had no more valid victims. Using our Bank Nth Chance replacement policy we maximize the inter-visit time for the same bank and thus avoid repeating searches for valid victims in the same set of entries often.

The use of three bits in the reference field builds a better approximation to the LRU policy than the 1-bit second chance algorithm. The round robin selection of banks and the repeated selection of a victim from a bank until the bank has no more victims is an implementation convenience that allows a good trade-off between an expensive fully parallel search for a valid victim and a slow sequential search.

Space Saving of Nth Chance. If flip-flops were used to store the reference fields, then 48 flip-flops would be required for one 16 entry block. Furthermore, the shift register control logic and valid victim check logic, which requires 3 LUTs, would need to be replicated for each entry, using 48 LUTs (24 slices) instead of 3 (2 slices). In total, each block in the flip-flop implementation requires 24 slices, compared to 3 slices when using SelectRAMs, resulting one eight the space requirement.

4   Implementation Results

In this section we present design measurements obtained in the prototype for the Virtex 1000E device. These measurements can be summarized as follows:

Density: When compared with an implementation of a ternary memory based on flip-flops, our LUT/shift register based implementation is 8 times denser.

Cache Size: Our design allows for a 1.5K entry cache to be implemented in a Virtex 1000E FPGA.

Lookups/Second: The largest cache implemented in the Virtex 1000E can perform 16 million lookups per second. Smaller caches have better performance.

Lookup vs. Update Latency: Our cache architecture implements a 2 cycle lookup, and an 18 cycle write.

4.1   Design Density

We attain space saving in our design in the storage of the destination addresses through the use of LUTs configured as shift registers as opposed to flip-flops. With LUTs, each 32-bit entry requires 4 slices compared with 32 slices for a flip-flop based implementation.

Another source of space saving is the implementation of the victim selection logic in the cache controller. Instead of storing the reference bits in flip-flops, we used SelectRAMs for this storage. Consequently, our 3-bit fields requires 1.5 slices per block, rather than 1.5 slices per entry, a 16 fold increase in density.

The simplicity of the Nth bank chance replacement policy allowed for a much simplified logic to update the reference field bits. Since the bits are stored in SelectRAMs, the same control logic is shared by 16 entries. In a flip-flop implementation, each field would require its own logic.

Combining the savings in all these aspects of the design, we achieved a bit storage density that is at least 8 times higher than could be achieved with a straightforward flip-flop based implementation in our Virtex 1000E device.

4.2 Cache Size

In this section we provide an estimate of the maximum cache sizes that could be implemented in several other devices in the Virtex family using our design. To obtain these estimates, we used simple linear regression on our actual implementation results for caches of up to 1024 entries. We found that device utilization could be broken up into 7.6 slices per entry for size-dependent storage and logic, and a fixed cost of approximately 175 slices for interface and control logic. Our estimates are presented in Table 1. We also present an upper bound on achievable cache size for a flip-flop based implementation. This upper bound is calculated based only on space requirements to implement the CAM cells in the DAA, store the contents of the NHA, and implement the reference fields for each entry. Additional logic that would be required for address encoding and decoding, and control, is ignored. Table 2 shows actual device utilization and speeds for various sizes of caches implemented in the Virtex 1000E. Our maximum cache size in the 1000E of 1568 entries closely matches our estimate of 1600 entries.

4.3 Lookup Throughput

Our implementation of a 1568 entry cache in the Virtex 1000E FPGA operates at 33 MHz. A lookup operation requires 2 cycles, so this cache can process 16 Million lookups per

Table 1. Estimates of the maximum cache sizes for each Xilinx VirtexE device. Flip-Flop estimates are based on bit storage space alone. LUT estimates are based on fixed and per-entry device utilization estimates from linear regression analysis of actual implementation results of complete caches.

Device		Maximum Cache Size	
Name	Slices	Flip-Flops	LUTs
50E	768	16	64
100E	1200	32	128
200E	2352	64	272
300E	3072	80	368
400E	4800	128	608
600E	6912	192	880
1000E	12288	336	1600
1600E	15552	432	2016
2000E	19200	528	2512

Table 2. Device usage and clock speed for various cache sizes in the Virtex 1000E.

Cache Entries	Slices Used	Speed (MHz)
64	659	70.2
128	1143	56.1
256	2121	50.5
521	4035	45.2
1024	7934	37.5
1568	21061	33.0

second. Therefore, even in its prototype incarnation, our cache should be able to handle current traffic in Internet edge routers, or the load of an OC-192 interface. When this design is used in a multizone cache design, each zone will be a smaller cache, and so will run at a higher frequency, as shown in Table 2.

4.4 Lookup and Update Latencies

A lookup in our cache requires reading from the Destination Address Array CAM cells. If a match occurs, data is output from the Next Hop Array. Although this lookup datapath is a simple combinatorial circuit, we break the datapath in two stages to allow for a higher clock frequency. This higher clock frequency benefits cache updates and victim selections. As a result, each lookup requires two clock cycles to complete. As our investigation of the design continues, and performance results become available, a single cycle lookup at a lower clock frequency may prove to increase overall performance.

Each write into the DAA requires 18 cycles. One cycle is required for setup, 16 to calculate and shift each of the 16 bits into the LUT shift registers, and a final cycle to complete the operation. The longer delay for updates is justified by the higher storage density, faster lookups due to the lack of comparison logic, and our very inexpensive implementation of ternary logic.

Furthermore, after the initial cycles of the write operation, the cache controller can begin searching for a new victim. In total, three cycles are required to search for a new victim. One cycle is required to age the entries in the current bank, another to advance the bank pointer and a final cycle to check the new bank for a valid victim. Four banks can be searched while the write completes. Initial software simulation indicates that, on average, between 1 and 2 banks will need to be searched to find the next victim. Consequently, victim selection should have a minimal impact on the performance of the cache.

5   Related Work

Our work is based on the prior work of MacGregor and Chvets [3, 13], who introduced the concept of a multizone cache for IP routing. Their work involved developing the statistical foundation for the multizone concept, and supporting simulations. We are building on this foundation by implementing the hardware proof-of-concept.

In the design of their Host Address Range Caches (HARC), Chiueh and Pradha show how the routing table can be modified to ensure that each IP address matches with a single IP prefix [16, 4]. They propose to "cull" the address ranges in the IP address space to ensure that each range is covered by exactly one routing table entry. The use of a similar technique would circumvent the single match restriction of our current design.

Gopalan and Chiueh use programable logic to implement Variable Cache Set Mapping to reduce conflict misses in set-associative IP caches [10]. Their trace-driven simulation studies confirms that IP caches benefit from higher degrees of associativity thus lending support to our decision of designing a fully-associative cache that eliminates all conflict misses.

Ditmar implemented a system for IP packet filtering in firewalls and routers [5]. This system uses a CAM in an FPGA to match IP packets to filtering rules. However, this system has fewer words in the CAM, which changed comparatively slowly over time. Therefore, this CAM could be implemented using software to partially reprogram the device to "hard code" the CAM contents according to a software-based replacement policy, while our cache must be fully implemented in hardware.

Commercial ternary content addressable memories (TCAM) is available from Siber-Core Technologies [15]. Their TCAMs are implement with ASIC technology. The fastest device available can peform 100 million lookup/second. The largest CAM available can store 9 Million Trits.[2]

6 Conclusion

We have nearly completed the implementation of a high density cache in an FPGA. In the Xilinx Virtex 1000E FPGA we can store 1568 32-bit entries, each with 4 bits of associated data and another 3 bits used by the replacement policy. This design runs on a 33MHz clock, and requires only 2 cycles to perform a lookup. Furthermore, we have introduced the new Bank Nth Chance replacement policy, which is well suited to the properties of a LUT-based FPGA. Bank Nth Chance carefully compromises between the speed of a parallel search circuit and the compact size of a sequential search circuit, while maintaining a hit rate similar to that of true LRU.

References

1. Jean-Louis Brelet and Bernie New. *Designing Flexible, Fast CAMs with Virtex Family FPGAs*. Xilinx,
   http://www.xilinx.com/xapp/xapp203.pdf, version 1.1 edition, September 1999.
2. L. Chivin and R. Duckworth. Content-addressable and associative memory. Alternatives to the ubiquitous ram. *IEEE Computer Magazine*, 22(7):51–64, July 1989.
3. Ivan Chvets. Multi-zone caching for ip address lookup. Master's thesis, University of Alberta, Edmonton, AB, 2002. Computing Science.
4. Tzi cker Chiueh and Prashant Pradhan. High performance IP routing table lookup using CPU caching. In *IEEE INFOCOMM (3)*, pages 1421–1428, 1999.
5. Johan M. Ditmar. A dynamically reconfigurable fpga-based content addressable memory for ip characterization. Master's thesis, Royal Institute of Technology, Stockholm, 2000.
6. P. N. Glaskowsky. Intel's new approach to networking: Follow-on to ixp1200 features new cores, new organization. Microprocessor Report - www.MPRonline.com, October 2001.
7. P. N. Glaskowsky. Lexra readies networking chip:new netvortex powerplant augments lexra's ip business. Microprocessor Report - www.MPRonline.com, June 2001.
8. P. N. Glaskowsky. Reinventing the router engine: Pmc-sierra's classipi defies easy classification. Microprocessor Report - www.MPRonline.com, March 2001.
9. P. N. Glaskowsky. Toaster3 pops up at mpf: Cisco details world's first 10 gb/s network processor. Microprocessor Report - www.MPRonline.com, October 2002.
10. Kartik Gopalan and Tzi cker Chiueh. Improving route lookup performance using network. In *Proc. of SC2002 High Performance Networking and Computing*, Baltimore, MD, November 2002.
11. T. Kohonen. *Content-addressable memories*. Springer-Verlag, New York, 2nd edition, 1987.

---

[2] A *trit* is a ternary bit as defined by Herrmann in the late 1980s [17].

12. K. Krewell. Rainier leads powernp family: Ibm's chip handles oc48 today with clear channel to oc192. Microprocessor Report - www.MPRonline.com, January 2001.
13. M. H. MacGregor. Design algorithm for multi-zone ip address caches. In *IEEE Workshop on High Performance Switching and Routing*, Torino, Italy, June 2003.
14. A. Silberschatz and P. Galvin. *Operating System Concepts*. Addison-Wesley, Reading, MA, fifth edition, 1998.
15. SiberCore Technologies. How a tcam co-processor can benefit a network processor. http://www.sibercore.com.
16. Prashant Pradhan Tzi-Cker Chiueh. Cache memory design for network processors. In *Sixth International Symposium on High-Performance Computer Architecture*, pages 409–419, Toulouse, France, January 2000.
17. Joh Patrick Wade. *An Integral Content Addressable Memory System*. PhD thesis, Massachusetts Intitute of Technology, May 1988.

Variable Precision Multipliers
for FPGA-Based Reconfigurable Computing Systems

Pasquale Corsonello[1], Stefania Perri[2], Maria Antonia Iachino[1]
and Giuseppe Cocorullo[2]

[1] Department of Informatics, Mathematics, Electronics and Transportation
University of Reggio Calabria, Loc. Vito de Feo, 88100 Reggio Calabria.
`corsonello@ing.unirc.it`
[2] Department of Electronics, Computer Science and Systems
University of Calabria, Arcavacata di Rende - 87036 - Rende (CS)
`perri@deis.unical.it`
`g.cocorullo@unical.it`
Italy

Abstract. This paper describes a new efficient multiplier for FPGA-based variable precision processors. The circuit here proposed can adapt itself at run-time to different data wordlengths avoiding time and power consuming reconfiguration. This is made possible thanks to the introduction of on purpose designed auxiliary logic, which enables the new circuit to operate in SIMD fashion and allows high parallelism levels to be guaranteed when operations on lower precisions are executed. The proposed circuit has been characterised using VIRTEX XILINX devices, but it can be efficiently used also in others FPGA families.

1 Introduction

The rapid increase in transistor count gives processor hardware the ability to perform operations that previously were only supported in software. The new computing paradigm that promises to satisfy the simultaneous demand for computing performance and flexibility is *Reconfigurable Computing* [1].

Reconfigurable computing utilizes hardware that can be adapted at run-time. The ability to customise the hardware on-demand to match the computation and the data flow of a required specific application has demonstrated significant performance benefits with respect to general-purpose architectures. As an example, thanks to the adaptability, reconfigurable architectures can exploit parallelism available in the matched application. This provides a significant performance advantage compared to conventional microprocessors.

Reconfigurable architectures are mainly based on Field Programmable Gate Arrays (FPGAs). An FPGA is a configurable logic device consisting of a matrix of programmable computational elements and a network of programmable interconnects. However, the development of Systems-on-Chips (SoCs) has been

P.Y.K. Cheung et al. (Eds.): FPL 2003, LNCS 2778, pp. 661–669, 2003.

recently demonstrated in which configurable logic and ASIC components are integrated [2, 3]. Among the existing families of FPGA the SRAM-based ones are the obvious candidates to the realization of evolvable hardware systems. In fact, SRAM-based configurable logic devices are able to adapt their circuit function to the computational demands downloading configuration data (i.e. bit-stream) from an external memory. Changing the bit-stream (i.e. reconfiguring the chip) allows both the functionality of the computational elements and the connections to be rearranged for a different circuit function.

Unfortunately, the flexibility achieved by reconfiguration is not for free. In fact, downloading a new bit-stream onto the chip needs a relatively long time. Moreover, during reconfiguration the logic has to stop computation. To partially solve these problems, FPGA architectures have been developed that support run-time reconfiguration [4, 5, 6]. The latter guarantees two advantages to be obtained: the time required for reconfiguring the chip is reduced; the portion of the chip that is not being reconfigured retains its functionality. Nevertheless, the reconfiguration time still could constitute a bottleneck in the design of an entire reconfigurable system.

Significant improvements can be obtained using an on-chip reconfiguration memory cache [7, 8], in which a few configurations can be stored. In this way, the time required to switch from one configuration to the next one is reduced with respect to the case in which bit-streams are loaded by an external memory.

Multimedia applications are becoming one of the dominating workloads for reconfigurable systems. Many examples of multimedia accelerators based on FPGAs are nowadays available. To achieve real time processing of media signals appropriate architectures supporting parallel operations (SIMD parallelism) on multiple data of 8-,16- and 32-bit are needed [9, 10]. They must be able to run-time adapt data-size and precision to the computational requirements, often at instruction level. This means that, as an example, a 32-bit native precision circuit should be run-time substituted by two 16-bit or by four 8-bit circuits able to perform the same computation, previously executed using the 32-bit precision. In order to do this the hardware should be reconfigured every time the precision changes.

All the above-mentioned solutions do not allow reconfiguring the hardware at instruction level. For these reasons, architectures able to fast run-time adapt themselves to different precisions without loading new bit-streams from memories are highly desired inside FPGA-based designs. The chip could be actually reconfigured just when the running function is not required for a properly long time.

In this paper a new circuit is presented for realizing variable precision multiplication with high sub-word parallelism levels for use within high-density SRAM-based FPGAs. The proposed circuit run-time adapts its structure to different precisions at instruction level avoiding power and time-consuming reconfiguration.

The new multiplier has been characterised using VIRTEX XILINX devices, but it can be efficiently used also in others FPGA families. Moreover, the method demonstrated here can be successfully applied for the realization of others variable precision arithmetic and logic functions.

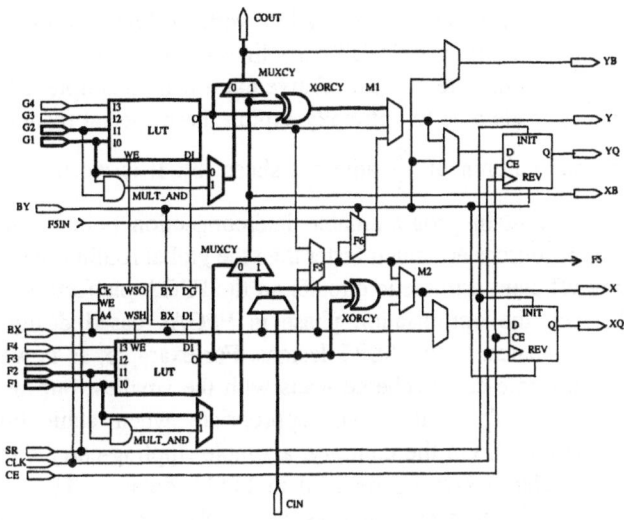

Fig. 1. Schematic diagram of the VIRTEX slice.

2 XILINX Virtex FPGA

VIRTEX, VIRTEX II and VIRTEX II PRO FPGA families developed by XILINX [5] are among the most used architectures in the design of systems for reconfigurable computing. They are SRAM-based FPGAs consisting of configurable logic blocks (CLBs) and programmable interconnects. The former provide the functional elements for constructing the user's logic, whereas the latter provide routing paths to connect the inputs and outputs of the CLBs exploiting programmable interconnection points (i.e. pass-transistor switches).

VIRTEX devices have a certain number of CLBs, ranging from 384 to 6144, each one consisting of two slices (S0 and S1) organized as shown in Fig.1. Each slice contains two look-up tables (LUTs), usable as logic function generators, two flip-flops and additional AND components (MULT_AND), multiplexers (MUXCY) and exclusive OR (XORCY) gates for high-speed arithmetic circuits. MULT_AND gates are used for building fast and small multipliers. MUXCY multiplexers are used to implement 1-bit high-speed carry propagate functions. XORCY gates are special XOR gates usable for generating 1-bit high-speed sum.

In the following, attention will be concentrated on these auxiliary elements to explain how they are used in the realization of high-speed carry chains, which are the basic elements of any arithmetic circuit.

When 1-bit full-adders are implemented, each LUT in the slice receives two operand bits (e.g. a_k and b_k) as input and generates the corresponding propagate signal $p_k = a_k$ XOR b_k. The LUT output drives the selection input of MUXCY and one input of XORCY, which generate the carry-out and the sum bit of the full-adder, respectively. Using this full-adder scheme the length of the carry chain is two bits for slice and the fast carry propagation occurs in a vertical direction through the special routing resources on the CIN and COUT carry lines of the slice.

By means of an appropriate manual relative placement, a simple n-bit ripple-carry adder can be easily made able to optimally exploit these special resources placing its full-adders into one column of $\frac{n}{2}$ adjacent slices [5]. It is worth pointing out that carry lines do not contain programmable interconnection points. Therefore, these dedicated routing resources are much faster than the global routing ones.

For designing efficient arithmetic circuits on the FPGA platform special expertise is required. In fact, the optimisation phase in FPGAs based designs needs very different efforts compared to the ASIC designs. For example, as is well known, for ASIC designs, the ripple-carry scheme leads with the smallest but also the slowest addition circuit [11]. Thus, it is not appropriate when achieving high-speed performance is the target. On the contrary, a simple rippling-carry chain is the best choice when the adder has to be realized in FPGA devices. There, conventional alternative accelerated schemes, such as carry-select and carry-look-ahead, are usually meaningless due to the fact that they require the use of low-speed general routing resources.

3 The Proposed Implementation

Multiplier architectures optimised for ASIC designs are typically inadequate to gain advantages from dedicated carry-propagate logic and fast routing resources available in FPGA devices. As an example, schemes proposed in [12-14] based on the Booth coding result unsuitable for high-performance designs on FPGA platforms. On the contrary, it is well known that the Baugh-Wooley algorithm [15] allows the special resources available into FPGAs to be efficiently exploited. Moreover, no efficient architectures for realising variable precision SIMD multipliers have been yet proposed and optimised for FPGA based systems.

Typically, a NxN variable precision SIMD multiplier is able to execute either one NxN multiplication or two (N/2)x(N/2) or four (N/4)x(N/4) multiplications. The obtained results are usually stored into a 2N-bit output register.

The design of the new variable precision multipliers with N=32 is based on the following considerations. $A_{[31:0]}$ and $B_{[31:0]}$ being two 32-bit operands, the following identity is easily verified:

$$A_{[31:0]} \times B_{[31:0]} = SH16L(SH8L(A_{[31:0]} \times B_{[31:24]}) + S48E(A_{[31:0]} \times B_{[23:16]})) +$$

$$+ S64E(SH8L(A_{[31:0]} \times B_{[15:8]}) + S48E(A_{[31:0]} \times B_{[7:0]})) \tag{1}$$

where SHyL and SzE indicate a left shift by y bits and an extension to z bits, respectively.

It is useful to recall that an n-bit number can be easily extended to (n+h) bits considering that: i) an n-bit unsigned number can be represented as a positive (n+h)-bit signed number having all the h additional bits set to 0; ii) an n-bit signed number can be represented as a signed (n+h)-bit number having all the h additional bits equal to the nth.

The proposed architecture uses four 32x8 multipliers and it can execute one 32x32 or one 32x16 multiplication or two 16x16 or four 8x8 operations.

In order to manipulate both signed and unsigned operands also for the lower precisions, the 32x8 multipliers have to be properly organized. For the target FPGA devices family an extremely efficient kind of automatically core generated macros exists which is useful to this purpose: the controlled by pin x signed multiplier (CBPxS). An nxm CBPxS macro is able to perform signed-signed, unsigned-unsigned or signed-unsigned multiplications extending the operands to (n+1) and to (m+1) bits, respectively. The extension of the n-bit operand is internally executed by the macro itself on the basis of a control signal, which indicates whether that operand is a signed or an unsigned number. On the contrary, the m-bit operand has to be externally extended. This leads with a nx(m+1) multiplier which generates an (n+m)-bit output.

In Fig.2 the block diagram of the new variable precision SIMD multiplier is depicted. It can be seen that its main building modules are:

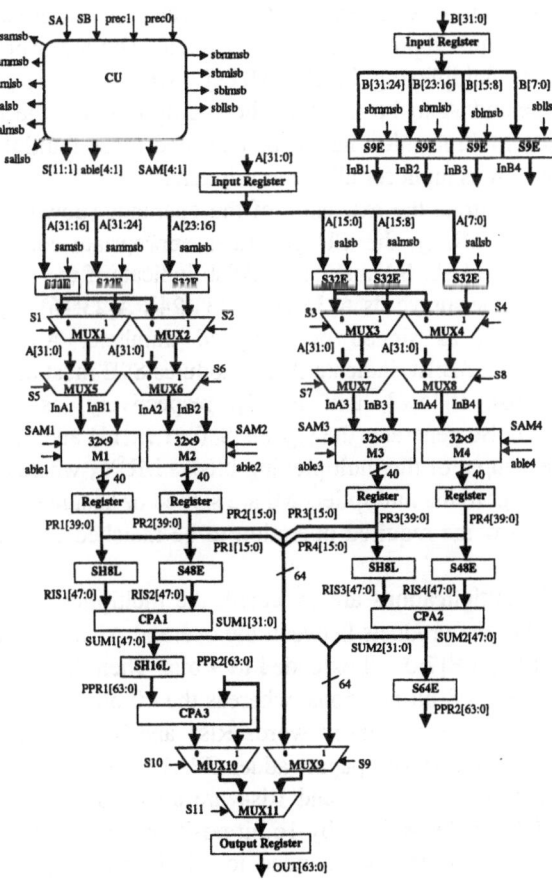

Fig. 2. Architecture of the proposed variable-precision multiplier

- A control unit (CU) needed to properly set the computational flow on the basis of the required precision;
- Four 32×9 CBPxS multipliers (M1, M2, M3 and M4) each generating 40-bit output;
- Two 48-bit and one 64-bit carry propagate adders (CPA1, CPA2 and CPA3).

The CU receives the signals prec1, prec0, SA, and SB as input and generates all the signals needed to orchestrate extensions and data flow. The signal SA indicates if the operand A is a signed (SA=1) or an unsigned (SA=0) number. The same occurs for the operand B on the basis of SB. Conversely, the signals prec1 and prec0 serve to establish the required precision. It is worth underlining that we suppose the above control signals available at runtime. Thus, also the required precision of different operations can be determined at runtime.

From Fig.2, it can be seen that to correctly form the data InAi and InBi input to the multipliers Mi (for i=1,...4), the 16-bit and the 8-bit subwords of operand A are extended to 32 bits, whereas the 8-bit subwords of the operand B are extended to 9 bits. For any required precision the extended subwords of operand B have to be input to the four multipliers. Conversely, for operand A different conditions occur depending on the required precision. When operations on lower precision data have to be executed the multiplexing blocks MUXj, with j=1,...,8, allow the extended subwords of operand A to be input to the multipliers. On the contrary, when the highest precision multiplication has to be executed those multiplexing blocks select the entire operand A as input for all the multipliers.

In order to orchestrate the extension of the operands subwords and the data flow through the blocks MUXj, the CU generates appropriate control signals. SAmsb, SAmmsb, SAmlsb, SAlsb, SAlmsb and SAllsb indicate if the modules for extension S32E have to treat the subwords A[31:16], A[31:24], A[23:16], A[15:0], A[15:8] and A[7:0], respectively, as signed or unsigned numbers. Analogously, SBmmsb, SBmlsb, SBlmsb and SBllsb indicate if the modules S9E must consider the subwords B[31:24], B[23:16], B[15:8], and B[7:0], respectively, as signed or unsigned numbers. The CU also generates the signals S_r (r=1,...,11) and SAM_i. The former are used as the select input of the multiplexing blocks MUX_r, whereas the latter indicate if the data InA_i input to the multipliers Mi are signed or unsigned numbers. The above control signals allow the required operation to be matched and the appropriate data flow to be set.

When 32x32 multiplications are executed, the multipliers M1, M2, M3 and M4 calculate in parallel the products between the operand A and the subwords B[31:24], B[23:16], B[15:8] and B[7:0] all extended to 9 bits. Then, the 40-bit results PR1 and PR3 are left shifted by 8-bit positions, whereas the 40-bit products PR2 and PR4 are extended to 48 bits. The two 48-bit words Ris1 and Ris2 obtained in this way are added by means of the carry-propagate adder CPA1, which generates the 48-bit word SUM1. Contemporaneously, Ris3 and Ris4 are added by CPA2 forming the word SUM2. Then, SUM1 is left shifted by 16 bit positions thus forming the 64-bit partial result PPR1, whereas SUM2 is extended to 64-bit thus forming the partial result

PPR2. To generate the final 64-bit result OUT[63:0], PPR1 and PPR2 are added by CPA3.

It can be noted that the hardware used could support the execution of two 32x16 multiplications. However, a 64-bit output register is used, thus only one 32x16 multiplication result can be accommodated in this register. When this operation is required just two multipliers must operate, whereas the others can be stopped. We have chosen to make M3 and M4 able to go into action and to stop M1 and M2. To control this particular case the signals able1, able2, able3 and able4 are used. When able1 and able2 are low all the 40 bits of PR1 and PR2 (and consequently also the 48 bits of Ris1 and Ris2) are forced to zero. Therefore the whole result Out[63:0] is equal to PPR2, which is directly outputted by means of MUX10 and MUX11, without passing through CPA3.

When operations on 16-bit data are executed, the multiplexing blocks MUX9 and MUX11 allow the less significant 32-bit of the independent results SUM1 and SUM2 to be directly outputted without passing through CPA3. Analogously, when multiplications on 8-bit data are requested, MUX9 and MUX11 allow the less significant 16-bit of the independent results PR1, PR2, PR3 and PR4 to be directly outputted without passing through the addition modules CPA1, CPA2 and CPA3.

It is worth underlining that in order to optimally exploit logic and routing resources available in the target FPGA also for the new multiplier appropriate placement constraints have been used. In this case the RLOC attributes available at the schematic level have been combined with the PROHIBIT directives imposed by means of the .ucf and .pcf files [5]. This allowed the extremely compact layout shown in Fig. 3 to be obtained.

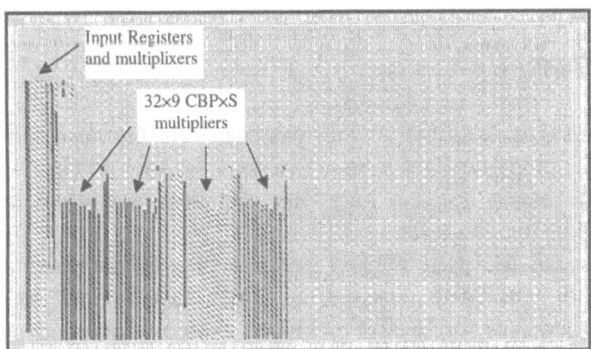

Fig. 3. Layout of the proposed variable-precision multiplier

4   Results

The new variable precision multiplier has been implemented and characterized using a FPGA XILINX VIRTEX XCV400 device. For purposes of comparison, we realized a 32x32 conventional fixed-precision core-generated multiplier. It is worth noting that core-generation tool does not allow the number of registers used in the multiplier pipeline to be arbitrarily chosen. Just minimum and maximum pipelining are allowed. To make an effective comparison minimum pipelining has been chosen. Moreover, two further variable precision schemes have been realized and compared using the design criterion described in the above Section. The first one uses two 32x17 core multipliers and the second one utilizes two 16x17 multipliers. Their characteristics are summarized in Table 1.

Table 1. Performance comparison.

	Delay [ns]	Power [mW/MHz]	Slices	Latency [cycles]	MOPS 8x8	MOPS 16x16	MOPS 32x16	MOPS 32x32
New Mult. 32x9	23	19.5	1108	2	174	87	43.5	43.5
New Mult. 32x17	30.9	15.8	919	2	64.7	64.7	32.4	32.4
New Mult. 16x17	28.7	13.4	729	2	69.7	69.7	34.9	17.5
Reference Mult.	30.7	12.4	680	1	-	-	-	32.6

It can be easily observed that the new multiplier using 32x9 core multipliers assures the fully parallelism on low precision data to be obtained. Therefore, it reaches the maximum attainable computational performance with an acceptable power overhead. It is worth pointing out that it doesn't require latency cycles during precision variations. Thus, the two latency cycles reported in Table 1 only occur when pipeline is filled for the first time.

It is also worth pointing out that the power usage in the new variable precision multiplier varies with the precision settings. For this reason, average power dissipation has been reported in Table 1.

5   Conclusions

SRAM-based FPGAs seem the obvious candidates for the realization of systems for Reconfigurable computing. Reconfiguration provides such FPGAs with an extremely high flexibility, but it needs a long time and does not allow reconfiguring the hardware at the instruction level.

In this paper, we discussed efficient multipliers for FPGA-based variable precision processors. The new circuit operate in SIMD fashion and are able to match computational demands for several precisions avoiding time and power consuming reconfiguration. Thanks to this, they can be very fast run-time adapted to different data-size and precisions at instruction level.

References

[1]   K. Bondalapati, V.K. Prosanna, "Reconfigurable Computing Systems", *Proceedings of IEEE*, Vol.90, n°7, 2002 pp. 1201-1217.
[2]   "Core embeds reconfigurable logic into SoCs",
      http://www.electronicstalk.com/news/lsi/lsi102.html.

[3] S. Davis, C. Reynolds, P. Zuchowski, "IBM licenses embedded FPGA cores from Xilinx for use in SoC ASICs", http://www.xilinx.com/publications/whitepapers/wp164.pdf
[4] XC6200 Field Programmable Gate Arrays, 1996
[5] Virtex Series FPGAs, http://www.xilinx.com
[6] M.J. Wirthlin, B.L. Hutchings, "Improving Functional Density Using Run-Time Circuit Reconfiguration", *IEEE Transactions on VLSI Systems*, Vol. 6, n°2, 1998, pp. 247-256.
[7] Chameleon System, http://www.chameleonsystems.com
[8] E. Cantò, J.M. Moreno, J. Cabestany, I. Lacadena, J.M. Insenser, "A Temporal Bipartitioning Algorithm for Dynamically Reconfigurable FPGAs", *IEEE Transactions of VLSI systems*, Vol. 9, n°1, 2001, pp. 210-218.
[9] A. Peleg, U. Weiser, "MMX Technology", *IEEE Micro*, August 1996, pp. 42-50.
[10] R.B. Lee, "Subword Parallelism with MAX-2", *IEEE Micro*, August 1996, pp. 51-59.
[11] V. Kantabutra, S. Perri, P. Corsonello, "Tradeoffs in Digital Binary Adders", in Layout Optimizations in VLSI Design, Kluwer Academic Publisher, 2001, pp. 261-288.
[12] A. Farooqui, V. G. Oklobdzija, "A Programmable Data-Path for MPEG-4 and Natural Hybrid Video Coding", *Proceedings of 34th Annual Asilomar Conference on signals, Systems and Computers*, Pacific Grove, California, October 29 - November 1, 2000.
[13] M. Margala, J.X. Xu, H. Wang, "Design Verification and DFT for an Embedded Reconfigurable Low-Power Multiplier in System-on-Chip Applications", *Proceedings of 14th IEEE ASIC*, Arlington, September 2001.
[14] S. Kim. M.C. Papaefthymiou, "Reconfigurable Low Energy Multiplier for Multimedia System Design", *Proceedings of 2000 IEEE Computer Society Annual Workshop on VLSI*, April 2000.
[15] K. Hwang, *Computer Arithmetic, Principles, Architecture, and Design*, Wiley, 1979.
[16] J. Fritts, W. Wolf, and B. Liu, "*Understanding multimedia application characteristics for designing programmable media processors*", SPIE Photonics West, Media Processors '99, San Jose, CA, pp. 2-13, Jan. 1999.

A New Arithmetic Unit in $GF(2^m)$ for Reconfigurable Hardware Implementation

Chang Hoon Kim[1], Soonhak Kwon[2], Jong Jin Kim[1], and Chun Pyo Hong[1]

[1] Dept. of Computer and Information Engineering, Daegu University,
Jinryang, Kyungsan, 712-714, Korea
chkim@dsp.taegu.ac.kr
[2] Dept. of Mathematics and Inst. of Basic Science, Sungkyunkwan University,
Suwon, 440-746, Korea
shkwon@math.skku.ac.kr

Abstract. This paper proposes a new arithmetic unit (AU) in $GF(2^m)$ for reconfigurable hardware implementation such as FPGAs, which overcomes the well-known drawback of reduced flexibility that is associated with traditional ASIC solutions. The proposed AU performs both division and multiplication in $GF(2^m)$. These operations are at the heart of elliptic curve cryptosystems (ECC). Analysis shows that the proposed AU has significantly less area complexity and has roughly the same or lower latency compared with some related circuits. In addition, we show that the proposed architecture preserves a high clock rate for large m (up to 571), when it is implemented on Altera's EP2A70F1508C-7 FPGA device. Furthermore, the new architecture provides a high flexibility and scalability with respect to the field size m, since it does not restrict the choice of irreducible polynomials and has the features of regularity, modularity, and uni-directional data flow. Therefore, the proposed architecture is well suited for both division and multiplication unit of ECC implemented on FPGAs.

Keywords: Finite Field Division, Finite Field Multiplication, ECC, VLSI

1 Introduction

Information security has recently become an important subject due to the explosive growth of the Internet, the mobile computing, and the migration of commerce practices to the electronic medium. The deployment of information security procedures requires the implementation of cryptosystems.

Among these cryptosystems, ECC have recently gained a lot of attention in industry and academia. The main reason for the attractiveness of ECC is the fact that there is no sub-exponential algorithm known to solve the discrete logarithm problem on a properly chosen elliptic curve. This means that significantly smaller parameters can be used in ECC than in other competitive systems such as RSA and ElGamal with equivalent levels of security [1]. Some benefits of having smaller key sizes include

P.Y.K. Cheung et al. (Eds.): FPL 2003, LNCS 2778, pp. 670–680, 2003.

faster computations, reductions in processing power, storage space, and bandwidth. Another advantage to be gained by using ECC is that each user may select a different elliptic curve, even though all users use the same underlying finite field. Consequently, all users require the same hardware for performing the field arithmetic, and the elliptic curve can be changed periodically for extra security [1].

For performance as well as for physical security reasons it is often required to realize cryptographic algorithms in hardware. Traditional ASIC solutions, however, have the well-known drawback of reduced flexibility compared to software solutions. Since modern security protocols are increasingly defined to be algorithm independent, a high degree of flexibility with respect to the cryptographic algorithms is desirable. A promising solution which combines high flexibility with the speed and physical security of traditional hardware is the implementation of cryptographic algorithms on reconfigurable devices such as FPGAs [2-3], [10].

One of the most important design rule in FPGAs implementation is the elimination of global signals broadcasting. This is because the global signals in FPGAs are not simple wires but buses, i.e., they are connected by routing resources (switching matrices) having propagation delay [13], [16]. In general, since ECC require large field size (at least 163) to support sufficient security, when the global signals are used, the critical path delay increases significantly [11]. Due to this problem, systolic array based designs, where each basic cell is connected with its neighboring cells through pipelining, are desirable to provide a higher clock rate and maximum throughput performance on fine grained FPGAs [10-12].

In this paper, we propose a new AU, which performs both division and multiplication in GF(2^m) and has systolic architecture, for FPGAs implementation of ECC. The new design is achieved by using substructure sharing between the binary extended GCD algorithm [17] and the most significant bit (MSB)-first multiplication scheme [9]. When input data come in continuously, the proposed architecture produces division results at a rate of one per m clock cycles after an initial delay of $5m$-2 in division mode and multiplication results at a rate of one per m clock cycles after an initial delay of $3m$ in multiplication mode respectively.

Analysis shows that the proposed AU has significantly less area complexity and has roughly the same or lower latency compared with some related systolic arrays for GF(2^m). In addition, we show that the proposed architecture preserves a high clock rate for large m (up to 571), when it is implemented on Altera's EP2A70F1508C-7 FPGA device. Furthermore, the new architecture provides a high flexibility and scalability with respect to the field size m, since it does not restrict the choice of irreducible polynomials and has the features of regularity, modularity, and unidirectional data flow. Therefore, the proposed architecture is well suited for both division and multiplication unit of ECC implemented on fine grained FPGAs.

2 A New Dependence Graph
for Both Division and Multiplication in GF(2^m)

2.1 Dependence Graph for Division in GF(2^m)

Let $A(x)$ and $B(x)$ be two elements in GF(2^m), $G(x)$ be the irreducible polynomial used to generate the field GF(2^m) $\cong$ GF(2)[x]/$G(x)$, and $P(x)$ be the result of the division $A(x)/B(x)$ mod $G(x)$. We can perform the division using the following Algorithm I [17].

[Algorithm I] The Binary Extended GCD for Division in GF(2^m) [17]
Input: $G(x)$, $A(x)$, $B(x)$
Output: V has $P(x) = A(x)/B(x)$ mod $G(x)$
Initialize: $R = B(x)$, $S = G = G(x)$, $U = A(x)$, $V = 0$, *count* = 0, *state* = 0
1. **for** $i = 1$ **to** $2m - 1$ **do**
2.   **if** *state* == 0 **then**
3.   *count* = *count* + 1;
4.   **if** $r_0 == 1$ **then**
5. $(R, S) = (R + S, R)$; $(U, V) = (U + V, U)$;
6.   *state* = 1;
7.   **end if**
8.   **else**
9.   *count* = *count* - 1;
10.   **if** $r_0 == 1$ **then**
11. $(R, S) = (R + S, S)$; $(U, V) = (U + V, V)$;
12.   **end if**
13.   **if** *count* == 0 **then**
14.   *state* = 0;
15.   **end if**
16.   **end if**
17. $R = R/x$;
18. $U = U/x$;
19. **end for**

Based on the Algorithm I, we can derive a new dependence graph (DG) for division in GF(2^m) as shown in Fig. 1.

The DG corresponding to the Algorithm I consists of $(2m-1)$ Type-1 cells and $(2m-1) \times m$ Type-2 cells. In particular, we assumed $m=3$ in the DG of Fig. 1, where the functions of two basic cells are depicted in Fig. 2. Note that, since r_m is always 0 and s_0 is always 1 in all the iterations of the Algorithm I, we do not need to process them. The input polynomials $R(x)$, $U(x)$, and $G(x)$ enter the DG from the top in parallel form. The i-th row of the array realizes the i-th iteration of the Algorithm I and the division result $V(x)$ emerge from the bottom row of the DG in parallel form after $2m-1$ iterations. Before describing the functions of Type-1 and Type-2 cell, we consider the implementation of *count*. From the Algorithm I, since *count* increases to m, we can trace the value of *count* by using m-bits bi-directional shift register (BSR) instead of $\log_2(m+1)$-bits adder (subtractor). The BSR is also used for multiplication, as will be explained in section 2.3.

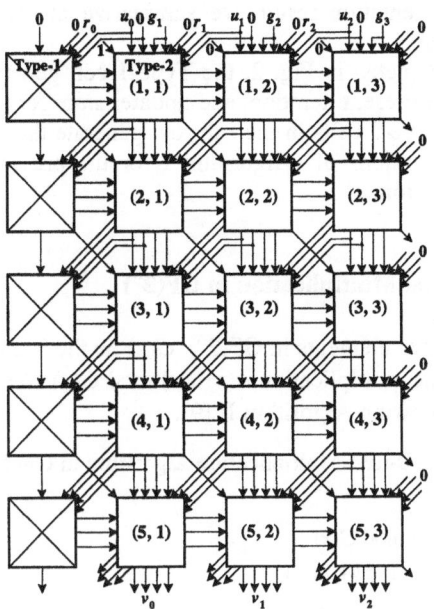

Fig. 1. DG for division in GF(2^3)

In what follows, we add one 2-to-1 multipliexer, and *up* and *down* signals to each Type-2 cell. As shown in Fig. 1, in the first row, only the (1, 1)-th Type-2 cell receive 1, while the others receive 0 for *up*, and all the cells receive 0 for *down*. In the *i*-th iteration, *m*-bits BSR is shifted to the left or to the right according to *state*. When cnt_n in Fig. 2 is 1, it indicates that *count* becomes *n* ($1 \leq n \leq m$). In addition, when *count* reduces to 0, *down* in Fig. 2 becomes 1. As a result, all the cnt_n of Type-2 cells in the same row become 0, and *c-zero* and *state* in Fig. 2 are updated to 1 and 0 respectively. This is the same condition of the first iteration.

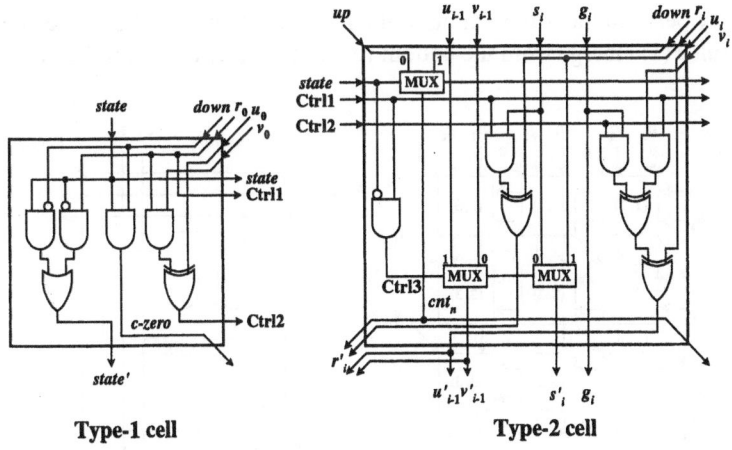

Type-1 cell   **Type-2 cell**

Fig. 2. The circuit of Type-1 and Type-2 cell in Fig. 1.

With *count* implementation result, we summarize the functions of Type-1 and Type-2 cell as follows:

(1) Type-1 cell: As depicted in Fig. 2, the Type-1 cell generates the control signals Ctrl1 and Ctrl2 for the present iteration, and updates *state* for the next iteration.

(2) Type-2 cell: the Type-2 cells in the *i*-th row generate the control signal Ctrl3 and perform the main operations of Algorithm I for the present iteration, and update *count* for the next iteration.

2.2 DG for MSB-First Multiplication in GF(2^m)

Let $A(x)$ and $B(x)$ be two elements in GF(2^m), $G(x)$ be the irreducible polynomial, and $P(x)$ be the result of the multiplication $A(x)B(x)$ mod $G(x)$. We can perform the multiplication using the following Algorithm II [9].

[Algorithm II] The MSB-first Multiplication Algorithm in GF(2^m) [9]

Input: $G(x)$, $A(x)$, $B(x)$

Output: $P(x) = A(x)B(x)$ mod $G(x)$

1. $p_k^{(0)} = 0$, for $0 \leq k \leq m-1$
2. $p_{-1}^{(i)} = 0$, for $1 \leq i \leq m$
3.   for $i=1$ to m do
4.   for $k=m-1$ down to 0 do
5. $p_k^{(i)} = p_{m-1}^{(i-1)} g_k + b_{m-i} a_k + p_{k-1}^{(i-1)}$
6.   end
7.   end
8. $P(x) = p^{(m)}(x)$

Based on the MSB-first multiplication algorithm in GF(2^m), a DG can be derived as shown in the left figure of Fig. 3 [9]. The DG corresponding to the multiplication algorithm consists of $m \times m$ basic cells. In particular, $m=3$ in the DG of Fig. 3, and the right figure of Fig. 3 represents the architecture of basic cell. The cells in the *i*-th row of the array perform the *i*-th iteration of the multiplication algorithm. The coefficients of the result $P(x)$ emerge from the bottom row of the array after m iterations.

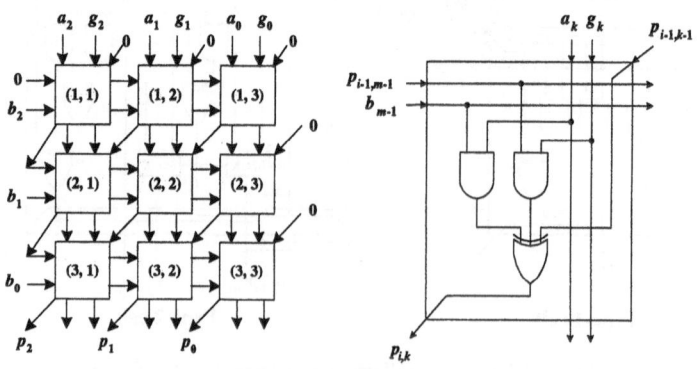

Fig. 3. DG and basic cell for multiplication in GF(2^3) [9].

2.3 A New DG for Both Division and Multiplication in GF(2^m)

By observing the DG in Fig. 1 and the DG in Fig. 3, we can find that the U operation of the division is identical with the P operation of the multiplication except for the input values. In addition, there are two differences between the DG for division and the DG for multiplication. First, in the DG for multiplication, the input polynomial $B(x)$ enters from the left, while, in the DG for division, all the input polynomials enter from the top. Second, the positions of the coefficients of each input polynomial are changed in two DG. In this case, by modifying the circuit of each basic cell, both division and multiplication can be performed using the same hardware.

For the first case, by putting the m-bits BSR in each row of the DG for division, we can make the DG perform both division and multiplication depending on whether $B(x)$ enters from the left or it enters from the top. In other words, after feeding $B(x)$ into m-bits BSR, it is enough to shift it to the left until the multiplication is finished. For the second case, we can use the same hardware by making permutation of the coefficients of each input polynomial. In summary, the DG in Fig. 1 with appropriate modification can perform both division and multiplication. The resulting DG is shown in Fig. 4, and the corresponding Modified Type-1 and Type-2 cell in Fig. 2 are shown in Fig. 5.

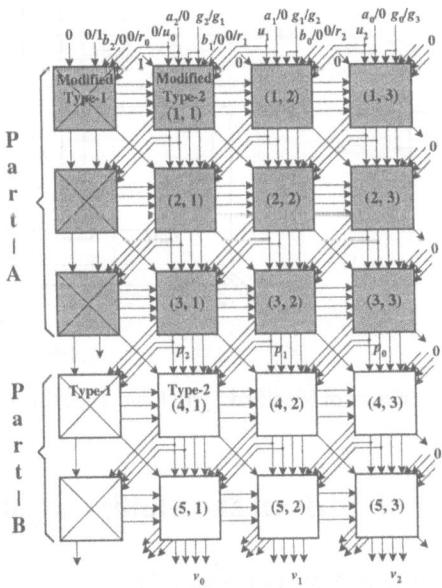

Fig. 4. New DG for both division and multiplication in GF(2^3)

As described in Fig. 4, the DG consists of m Modified Type-1, $m \times m$ Modified Type-2, $m-1$ Type-1, and $(m-1) \times m$ Type-2 cells. The polynomials $A(x)$, $B(x)$, and $G(x)$ enter the DG for multiplication, and the polynomials $R(x)$, $U(x)$, and $G(x)$ enter the DG for division. The multiplication result $P(x)$ emerge from the bottom row of Part-A after m iterations, and the division result $V(x)$ emerge from the bottom row of

Part-B after $2m-1$ iterations. In Fig. 4, Part-A is used for both division and multiplication, and Part-B is only used for division. The modification procedures from Type-1, 2 cell to Modified Type-1, 2 cell are summarized as follows:

(a) Modification of Type-1 cell: We add *mult/div* signal, 2-to-1 AND gate (numbered by 1) and 2-to-1 multiplexer comparing to Type-1 cell. For multiplication mode, *mult/div* is set to 0, and for division mode, *mult/div* is set to 1. As a result, for multiplication, b_{m-i}/Ctrl1 has b_{m-i}, and p_{m-1}/Ctrl2 has p_{m-1} respectively. For division, it has the same function as the Type-1 cell in Fig. 2.

(b) Modification of Type-2 cell: We add *mult/div* signal, 2-to-1 AND gate (numbered by 1), 2-to-1 OR gate (numbered by 2) and 2-to-1 multiplexer comparing to Type-2 cell. In Modified Type-2 cell, since Ctrl3 generates 0 for multiplication mode, a_{m-i}/v_{i-1} is always selected. For multiplication mode, the m-bits BSR is only shifted to the left direction due to the OR gate numbered by 2. In addition, the multiplexer numbered by 4 in Fig. 5 is also added due to the fact that, for multiplication mode, the AND gate number by 3 must receive a_{m-i}/v_{m-1} instead of a_{m-i-1}/v_i. As a result, when *mult/div* is set to 0, the Modified Type-2 cell in Fig. 5 performs as the basic cell in Fig. 3. In addition, when *mult/div* is set to 1, it performs as the Type-2 cell in Fig . 2.

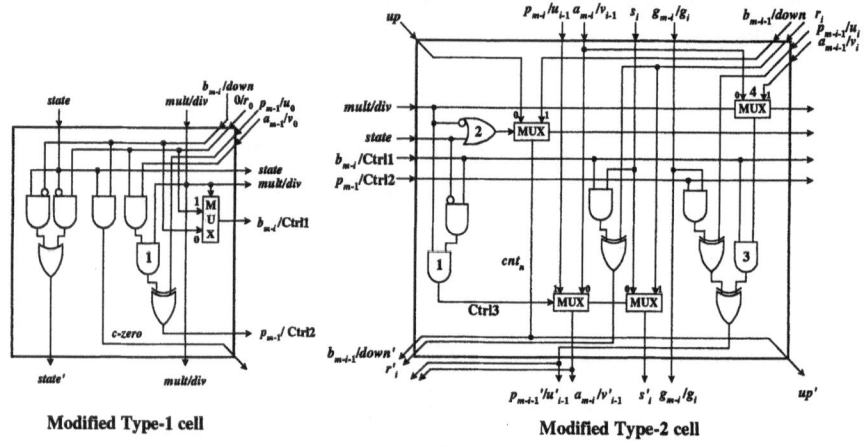

Fig. 5. The circuit of Modified Type-1 and Type-2 cell in Fig. 4

3   A New AU for Both Division and Multiplication in GF(2^m)

By projecting the DG in Fig. 4 along the east direction according to the projection procedure [14], we derive a one-dimensional SFG array as shown in Fig. 6, where the circuit of each processing element (PE-A and PE-B) is described in Fig. 7. In Fig. 6, "•" denotes a 1-bit 1-cycle delay element. The SFG array of Fig. 6 is controlled by a sequence $011...11$ of length m. As described in Fig. 6, two different data set are feed into the array depending on division or multiplication mode.

As described in Fig. 7, the PE-A contains the circuitry of Modified Type-1 and Type-2 cell of Fig. 5. In addition, the PE-B contains the circuitry of Type-1 and

Type-2 cell of Fig. 2. Since two control signals Ctrl1 and Ctrl2, and *state* of the i-th iteration must be broadcast to all the Modified Type-1 and Type-2 cells in the i-th row of the DG, three 2-to-1 multiplexers and three 1-bit latches are added to each PE-A and PE-B. Four 2-input AND gates are also added to each PE-A and PE-B due to the fact that four signals (*i.e.*, b_{m-i-1}/down, r_i, p_{m-i-1}/u_i, and a_{m-i-1}/v_i in Fig. 5) must be fed to each row of the DG from the rightmost cell. When the control signal is in logic 1, these AND gates generate four zeros.

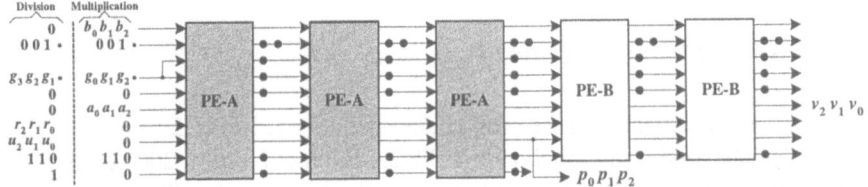

Fig. 6. A one-dimensional SFG array for both division and multiplication in GF(2^3).

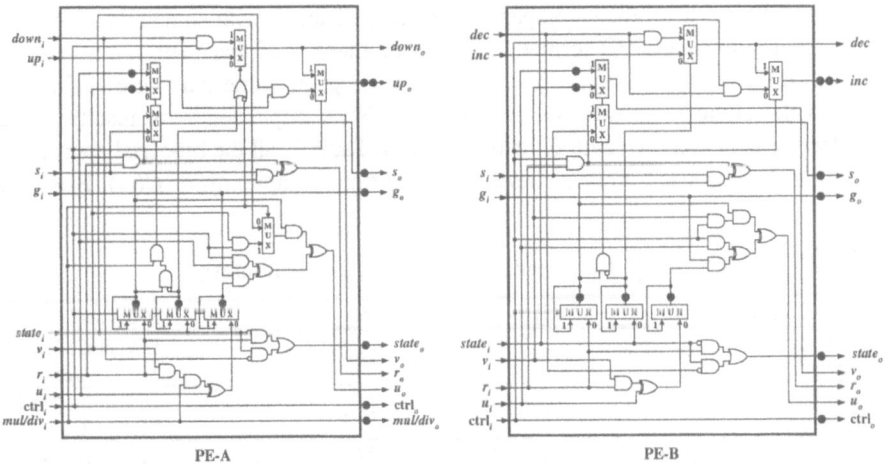

Fig. 7. The circuit of PE-A and PE-B in Fig. 6

The SFG array in Fig. 6 can be easily retimed by using the cut-set systolization techniques [14] to derive a serial-in serial-out systolic array, and the resulting structure is shown in Fig. 8. When the input data come in continuously, this array can produce division results at a rate of one per m clock cycles after an initial delay of $5m-2$ with the least significant coefficient first in division mode and can produce multiplication results at a rate of one per m clock cycles after an initial delay of $3m$ with the most significant coefficient first in multiplication mode.

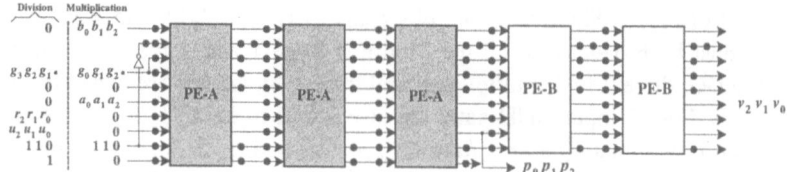

Fig. 8. A new bit-serial systolic array for division and multiplication in GF(2^3)

4   Results and Conclusions

To verify the functionality of the proposed array in Fig 8, it was developed in VHDL and was synthesized using the Synopsis' FPGA-Express (version 2000,11-FE3.5), in which Altera's EP2A70F1508C-7 was used as the target device. The placement and route process and timing analysis of the synthesized designs were accomplished using the Altera's Quartus II (version 2.0).

We summarize theoretical comparison results in Table 1 with some related systolic arrays having the same I/O format. It should be mentioned that there are no circuits for both division and multiplication in GF(2^m) using the same hardware at this moment to the authors' knowledge. In addition, FPGA implementation results of our architecture are given in Table 2. As described in Table 1, all the bit-serial approaches including the proposed array achieve the same time complexity of $O(m)$. However,

Table 1. Comparison with bit-serial systolic arrays

Item \ Circuit	[4]	[5]	[6]	Proposed AU
Throughput (1 /cycles)	$1/(2m-1)$	$1/m$	$1/m$	Division : $1/m$ Mult. : $1/m$
Latency (cycles)	$7m-3$	$5m-1$	$8m-1$	Division : $5m-2$ Mult. : $3m$
Maximum cell delay	$T_{AND2}+T_{XOR2}$ $+T_{MUX2}$	$T_{AND2}+T_{XOR2}$ $+T_{MUX2}$	$2T_{AND2}+2T_{XOR2}$ $+2T_{MUX2}$	$2T_{AND2}+T_{XOR2}$ $+T_{MUX2}$
Basic components and their numbers	AND_2 : $3m^2+3m-2$ XOR_2 : $1.5m^2+1.5m-1$ MUX_2 : $3m^2+m-2$ Latch : $6m^2+8m-4$	AND_2 : $2m-1$ OR_2 : $3m$ XOR_2 : $0.5m^2+1.5m-1$ MUX_2 : $1.5m^2+4.5m-2$ Latch : $2.5m^2+14.5m-6$	Inverter : $2m$ AND_2 : $26m$ XOR_2 : $11m$ MUX_2 : $35m+2$ FILO(m-bit) : 4 Latch : $46m+4m \cdot \log_2(m+1)$ zero-check ($\log_2(m+1)$-bit) : $2m$ adder/subtractor ($\log_2(m+1)$-bit) : $2m$	Inverter : $7m-2$ AND_2 : $26m-12$ OR_2 : $3m-1$ XOR_2 : $8m-4$ MUX_2 : $15m-7$ Latch : $44m-24$
Operation	Division	Division	Division	Division and Multiplication
Area complexity	$O(m^2)$	$O(m^2)$	$O(m \cdot \log_2 m)$	$O(m)$

the proposed systolic array reduces the area complexity from $O(m^2)$ or $O(m \cdot \log_2 m)$ to $O(m)$, and has lower maximum cell delay and latency than the architecture in [6]. Although the circuits in [4-5] have lower maximum cell delay than the proposed array, they can not be applicable to ECC due to their high area complexity of $O(m^2)$. In addition, we do not need additional hardware components described in Table 3 to achieve multiplication, since the proposed array performs both division and multiplication.

Table 2. FPGA Implementation results of the proposed AU

Item \ m	163	233	409	571
# of LE	9920	14209	24926	34950
Clock (MHz)	138.87	127.99	141.8	121.11
Chip Utilization (%)	14.76	21.21	37.09	52.01

LE consists of one 4-to-1 LUT, one flip-flop, fast carry logic, and programmable multiplexers [16]

Table 3. Area-time complexity of the bit-serial systolic array for multiplication in GF(2^m) [9]

Throughput	Latency	Maximum cell delay	Basic components and their numbers
$1/m$	$3m$	$T_{AND2} + 2T_{XOR2}$	$AND_2 : 3m$, $XOR_2 : 2m$, $MUX_2 : 2m$, Latch : $10m$

From Table 2, it is noted that the proposed architecture preserves a high clock rate for large field size m because there are no global signals broadcasting. As a reference, 571 in Table 2 is the largest field size recommended by NIST [15]. Furthermore, since the proposed architecture does not restrict the choice of irreducible polynomials and has the features of regularity, modularity, and unidirectional data flow, it provides a high flexibility and scalability with respect to the field size m. All these advantages of our design lead that, if ECC is implemented on FPGAs to overcome the well-known drawback of ASIC, we can obtain maximum throughput performance with minimum hardware requirement by using the proposed AU.

Acknowledgement

This work was supported by grant No. R05-2003-000-11573-0 from the Basic Research Program of the Korea Science & Engineering Foundation

References

[1] I. F. Blake, G. Seroussi, and N. P. Smart, *Elliptic Curves in Cryptography*, Cambridge University Press, 1999.

[2] G. Orlando and C. Parr, "A High-Performance Reconfigurable Elliptic Curve Processor for GF(2^m)," *CHES 2000*, LNCS 1965, Springer-Verlag, 2000.

[3] M. Bednara, M. Daldrup, J. von zur Gathen, J. Shokrollahi, and J. Teich, "Reconfigurable Implementation of Elliptic Curve Crypto Algorithms," *Proc. of the International Parallel and Distributed Processing Symposium (IPDPS'02)*, pp. 157-164, 2002.

[4] C.-L. Wang and J. -L. Lin, "A Systolic Architecture for Computing Inverses and Divisions in Finite Fields $GF(2^m)$,"• *IEEE Trans. Computers.*, vol. 42, no. 9, pp. 1141- 1146, Sep. 1993.

[5] M.A. Hasan and V.K. Bhargava, "Bit-Level Systolic Divider and Multiplier for Finite Fields $GF(2^m)$," *IEEE Trans. Computers*, vol. 41, no. 8, pp. 972-980, Aug. 1992.

[6] J.-H. Guo and C.-L. Wang, "Systolic Array Implementation of Euclid's Algorithm for Inversion and Division in $GF(2^m)$," *IEEE Trans. Computers.*, vol. 47, no. 10, pp. 1161-1167, Oct. 1998.

[7] J.R. Goodman, "Energy Scalable Reconfigurable Cryptographic Hardware for Portable Applications," PhD thesis, MIT, 2000.

[8] J.-H. Guo and C.-L. Wang,•"Bit-serial Systolic Array Implementation of Euclid's Algorithm for Inversion and Division in $GF(2^m)$", *Proc. 1997 Int. Symp. VLSI Tech., Systems and Applications*, pp. 113-117, 1997.

[9] C. L. Wang and J. L. Lin, "Systolic Array Implementation of Multipliers for Finite Field $GF(2^m)$," *IEEE Trans. Circuits and Syst.*, vol. 38, no. 7, pp. 796-800, July 1991.

[10] T. Blum and C. Paar, "High Radix Montgomery Modular Exponentiation on Reconfigurable Hardware", *IEEE Trans. Computers.*, vol. 50, no. 7, pp. 759-764, July 2001.

[11] S.D. Han, C.H. Kim, and C.P. Hong, "Characteristic Analysis of Modular Multiplier for $GF(2^m)$," *Proc. of IEEK Summer Conference 2002*, vol. 25, no. 1, pp. 277-280, 2002.

[12] R. Tessier and W. Burleson, "Reconfigurable Computing for Digital Signal Processing: A Survey", *J. VLSI Signal Processing*, vol. 28, no. 1, pp. 7–27, May 1998.

[13] K. Compton and S. Hauck, "Reconfigurable Computing: A Survey of Systems and Software", *ACM Computing Surveys*, vol. 34, no. 2, pp. 171-210, June 2002.

[14] S. Y. Kung, *VLSI Array Processors*, Englewood Cliffs, NJ: Prentice Hall, 1988.

[15] NIST, Recommended elliptic curves for federal government use, May 1999. http://csrc.nist.gov/ encryption.

[16] Altera, APEX^{TMII} Programable Logic Device Family Data Sheet, Aug. 2000. http://www.altera.com/ literature/lit-ap2.html.

[17] C.H. Kim and C.P. Hong, "High Speed Division Architecture for $GF(2^m)$", *Electronics Letters*, vol. 38, no. 15, pp. 835-836, July 2002.

A Dynamic Routing Algorithm
for a Bio-inspired Reconfigurable Circuit

Yann Thoma[1,3], Eduardo Sanchez[1],
Juan-Manuel Moreno Arostegui[2], and Gianluca Tempesti[1]

[1] Swiss Federal Institute of Technology at Lausanne (EPFL), Lausanne, Switzerland
[2] Technical University of Catalunya (UPC), Barcelona, Spain
[3] Corresponding author. `yann.thoma@epfl.ch`

Abstract. In this paper we present a new dynamic routing algorithm
specially implemented for a new electronic tissue called POEtic. This reconfigurable circuit is designed to ease the implementation of bio-inspired
systems that bring cellular applications into play. Specifically designed
for implementing cellular applications, such as neural networks, this circuit is composed of two main parts: a two-dimensional array of basic
elements similar to those found in common commercial FPGAs, and a
two-dimensional array of routing units that implement a dynamic routing algorithm which allows the creation of data paths between cells at
runtime.

1   Introduction

Life is amazing in terms of complexity and of adaptability. After the fertilization
of an ovule by a spermatozoid, a simple cell is capable of recursively dividing itself
to create an entire living being. During its lifetime, an organism is also capable of
self-repair in case of external or internal aggressions. Living beings possessing a
neural network can learn tasks which allow them to adapt to their environment.
And finally, at the population level, evolution allows a population to evolve
in order to survive in an ever-changing environment. The aim of the POEtic
project [7][8][10] is to design a new electronic circuit, drawing inspiration from
these three life axes: Phylogenesis (evolution) [6], Ontogenesis (development) [9],
and Epigenesis (learning) [4].

Ontogenetic methods, which are used to develop a self-repair circuit, need
to change the functionality of the circuit at runtime. Epigenetic mechanisms,
using neural networks, could also need to create new neurons, and therefore
new connections between neurons at runtime. As commercially FPGAs usually
don't have any dynamic self-reconfiguration capabilities, a new circuit capable
of self-configuration is essential.

In the next section, we present the general architecture of the POEtic chip,
decomposed into two subsystems. In section 3, we describe the basic elements of
the circuit: the molecules. Section 4 fully explains the dynamic routing algorithm
implemented in order to ease the creation of long distance paths into the chip.

P.Y.K. Cheung et al. (Eds.): FPL 2003, LNCS 2778, pp. 681–690, 2003.
© Springer-Verlag Berlin Heidelberg 2003

2   Structural Architecture

The POEtic circuit is composed of two parts (figure 1): the organic subsystem, which is the functional part of the circuit, and the environmental subsystem. Cells, and thus organisms, are implemented in the organic subsystem. It is composed of a grid of small molecules and of a cellular routing layer. Molecules are the smallest unit of programmable hardware which can be configured by software, while dedicated routing units are responsible for the inter-cellular communication. The main role of the environmental subsystem is to configure the molecules. It is also responsible for the evolution process, and can therefore access and change every molecule's state in order to evaluate the fitness of an organism.

Each cell of an organism is a collection of molecules, which are the basic blocks of our circuit. The size and contents of the cells depend on the application. Therefore, for each application, a developer will have to design cells fitting into the molecules.

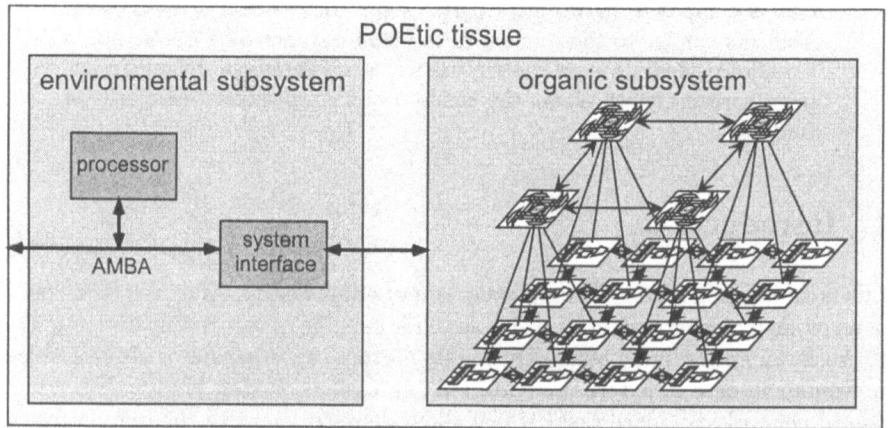

Fig. 1. The POEtic chip, composed of an environmental and an organic subsystems.

2.1   Environmental Subsystem

The environmental subsystem is primairly composed of a micro-controller: a 32-bit RISC-like processor. Its function is to configure the molecules, to run the evolutionary mechanisms, and to manage chip input/output. In order to speed up evolution processes, a random number generator has been added directly in the hardware. An AMBA bus [1] is used to connect the processor to a system interface that takes care of the communication between the processor and the organic subsystem. This bus is also connected to external pins in order to allow multi-chip communication, as well as the use of an external RAM.

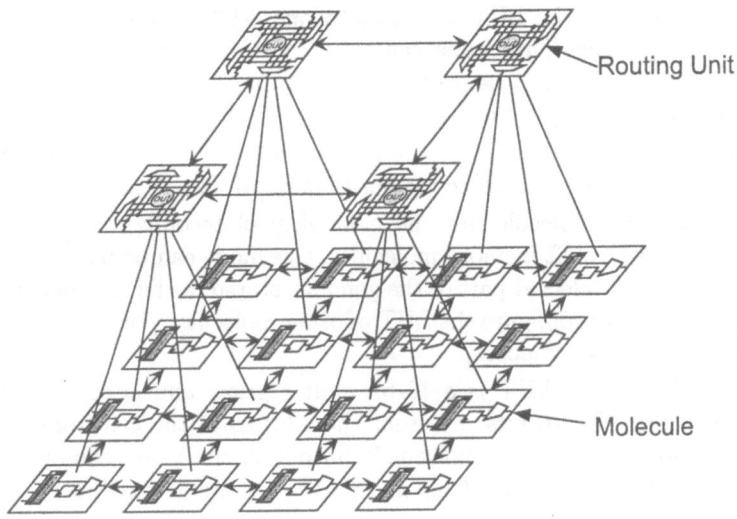

Fig. 2. The organic subsystem is composed of 2 layers: the molecules and the routing units.

2.2   Organic Subsystem

The organic subsystem is made up of 2 layers (cf. figure 2): a two-dimensional array of basic logic elements, called molecules, and a two-dimensional array of routing units. Each molecule has the capability of accessing the routing layer that is used for inter-cellular communication. This second layer implements a dynamic routing algorithm allowing the creation of data paths between cells at runtime.

3   Molecular Structure

As briefly presented above, the first layer of the POEtic tissue is a two-dimensional array of small programmable units, called molecules. Each molecule is connected to its 4 neighbors and to a routing unit (4 molecules for 1 routing unit), and contains a 16-bit look-up table (LUT) and a flip-flop (DFF). This structure, while seemingly very similar to standard FPGAs [2], is however specially designed for POEtic applications: different running modes let the molecule act like a memory, a serial address comparator, a cell input, a cell output, or others. With a total of 8 modes, these molecules allow a developer to build cells that are capable of communicating with each other, of storing a genome, of healing, and so on.

The 8 modes of operation of the molecule are the following:

- *Normal*: the LUT is a simple 16-bit look-up table.
- *Arithmetic*: the LUT is split into two 8-bit look-up tables: one for the molecule output, and one for a carry. A carry-chain physically sends the carry to the south neighbor, allowing rapid arithmetic operations.

- *Communication*: the LUT is split into one 8-bit shift register and one 8-bit look-up table. This mode can be used to implement a packet routing algorithm that will not be presented in this paper.
- *Shift memory*: the LUT is considered as a 16-bit shift register. This mode is very useful to efficiently store the genome in every cell. Shift memories can be chained in order to create memories of depth 32, 48, etc.
- *Configure*: the molecule has the capability of reconfiguring its neighbor. Combined with shift memory molecules, this mode can be used to differentiate the cells. A selected part of the genome, stored in the memory molecules, can be shifted to configure the LUT of other molecules (for instance to assign weights to neural synapses).
- *Input address*: the LUT is a 16-bit shift register and is connected to the routing unit. The 16 bits represent the address of the cell from where the information arrives. The molecule's output is the value coming from the inter-cellular routing layer (this mechanism will be detailed in the next section).
- *Output address*: the LUT is a 16-bit shift register and is connected to the routing unit. The 16 bits represent the address of the cell, and the molecule sends the value of one of its inputs to the routing unit (this mechanism will be detailed in the next section).
- *Trigger*: the LUT is a 16-bit shift register, and is connected to the routing unit. Its task is to supply a trigger every n clock cycles (where n is the number of bits of the addresses), needed by the routing algorithm for synchronization.

To be capable of self-repair and growth, an organism needs to be able to create new cells and to configure them. The configuration system of the molecules can be seen as a shift register of 80 bits split into 5 blocks: the LUT, the selection of the LUT's input, the switch box, the mode of operation, and an extra block for all other configuration bits. Each block contains, as shown in figure 3, together with its configuration, one bit indicating whether the block has to be bypassed in the case of a reconfiguration coming from a neighbor. This bit can only be loaded from the micro-processor, and remains stable during the entire lifetime of the organism.

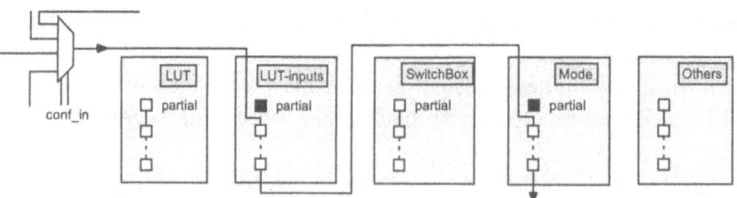

Fig. 3. All configuration bits of a molecule, split up into 5 blocks. The partial configuration bits of blocks 2 and 4 are set, enabling the reconfiguration of the LUT inputs and of the mode of operation by a neighboring molecule.

The special configure mode allows a molecule to partially reconfigure its neighborhood. It sends bits coming from another molecule to the configuration of one of its neighbors. By chaining the configurations of neighboring molecules, it is possible to modify multiple molecules at the same time, allowing, for example, the synaptic weights in a neuron to be changed.

4 Dynamic Routing

As presented above, our circuit is composed of a grid of molecules, in which cells are implemented. In a multi-cellular system, cells need to communicate with each other: a neural network, for example, often shows a very high density of connections between neurons. Commercial FPGAs have trouble dealing with these kinds of applications, because of their poor or nonexistent dynamic routing capacity. Given the purpose of the POEtic tissue, special attention was payed to this problem. Therefore, a second layer was added on top of the molecules, implementing a distributed dynamic routing algorithm. This algorithm uses an optimized version of the dynamic routing presented by Moreno in [5], to which we supplied a distributed control to where there is no global control of the routing process.

4.1 From Software to Hardware

Our dynamic routing algorithm finds the shortest path between two points in the routing layer. In 1959, Dijkstra proposed a software algorithm to find the shortest path between two nodes in a graph in which every branch has a positive length [3]. If we fix all branches to have a weight of 1, we can dramatically simplify the algorithm. It then becomes a breadth-first search algorithm, as follow:

```
1:  paint all vertices white;
2:  paint the source grey and enqueue it;
3:  repeat
4:    dequeue vertex v;
5:    if v is the target, we found a path - exit the algorithm;
6:    paint v black;
7:    for each white neighbor w of v
8:      paint w grey;
9:      set parent w to v;
10:     enqueue w
11: until the queue is empty
12: if we haven't yet exited, we didn't find the target
```

This algorithm acts like a gas expanding in a labyrinth, but in a sequential manner, one node being expanded at a time, with a complexity of $O(V+E)$ where V is the number of vertices and E is the number of edges. Taking advantage of the hardwares' intrinsic parallelism, it is possible, based on the same principle as the breadth-first search algorithm, to expand all grey nodes at the same time.

This dramatically decreases the time needed to find the shortest path between two points, the complexity becoming O(N+M), for a NxM array.

Finding the shortest path is not enough for the POEtic tissue, since we don't have a God telling us which routing unit is the source and which one is the target. In order to have a standalone circuit capable of self-configuration, we need a mechanism to start routings. Molecules, as explained in the previous section, have special modes to access the routing layer. Therefore, input or output molecules have the capability of initiating a dynamic routing.

4.2   Routing Algorithm

The dynamic routing system is designed to automatically connect the cells' inputs and outputs. Each output of a cell has a unique identifier at the organism level. For each of its inputs, the cell stores the identifier of the source from which it needs information. A non-connected input (target) or output (source) can initiate the creation of a path by broadcasting its identifier in the case of an output, or the identifier of its source in the case of an input. The path is then created using a parallel implementation of the breadth-first search algorithm presented above. When all paths have been created, the organism can start operation and execute its task until a new routing is launched, for example after a cell addition or a cellular self-repair.

Our approach has many advantages compared to a static routing process. First of all, a software implementation of a shortest path algorithm, such as Dijkstra's, is very time-consuming for a processor, while our parallel implementation requires a very small number of clock cycles to finalize a path. Secondly, when a new cell is created it can start a routing process without the need of recalculating all paths already created. Thirdly, a cell has the possibility of restarting the routing process of the entire organism if needed (for instance after a self-repair). Finally, our approach is totally distributed without any global control over the routing process, so that the algorithm can work without the need of the central micro-processor.

The routing algorithm is executed in three phases:

Phase 1: Finding a Master

In this phase, every target or source that is not connected to its correspondent partner tries to become master of the routing process. A simple priority mechanism chooses the most bottom-left routing unit to be the master, as shown in figure 4. Note that there is no global control for this priority, every routing unit knows whether or not it is the master. This phase is over in one clock cycle, as the propagation of signals is combinational.

Phase 2: Broadcasting the Address

Once a master has been selected, it sends its address in the case of a source, or the address of the needed source in the case of a target. As shown in section 3, the address is stored in a molecule connected to the routing unit. It is sent

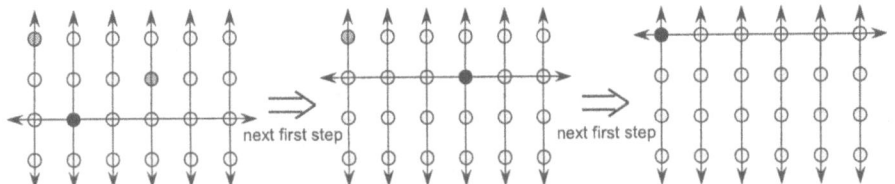

Fig. 4. Three consecutive first steps of the algorithm. The black routing unit will be the master, and therefore perform its routing.

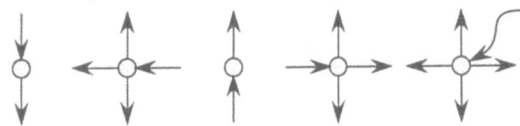

Fig. 5. The propagation direction of the address: north → south ‖ east → south, west, and north ‖ south → north ‖ west → north, east, and south ‖ routing unit → north, east, south, and west.

serially, in n clock cycles, where n is the size of the address. The same path as in the first phase is used to broadcast the address, as shown in figure 5.

Every routing unit, except the one that sends the address, compares the incoming value with its own address (stored in the molecule underneath). At the end of this phase, that is, after n clock cycles, each routing unit knows if it is involved in this path. In practice, there has to be one and only one source, and at least one target.

Phase 3: Building the Shortest Path

The last phase, largely inspired by [5], creates a shortest path between the selected source and the selected targets. An example involving 8 sources and 8 targets is shown in figure 6, for a densely connected network.

A parallel implementation of the breadth-first search algorithm allows the routing units to find the shortest path between a source and many targets. Starting from the source, an expansion process tries to find targets. When one is reached, the path is fixed, and all the routing resources used for the path will not be available for the next successive iterations of the algorithm.

Figure 7 shows the development of the algorithm, building a path between a source placed in column 1, row 2 and a target cell placed in column 3, row 3. After 3 clock cycles of expansion, the target is reached, and the path is fixed, prohibiting the use of the same path for a successive routing.

5   Conclusion

In this paper we presented a new electronic circuit dedicated to the implementation of bio-inspired cellular applications. It is composed of a RISC-like microprocessor and of two planes of functional and routing units. The first one, a

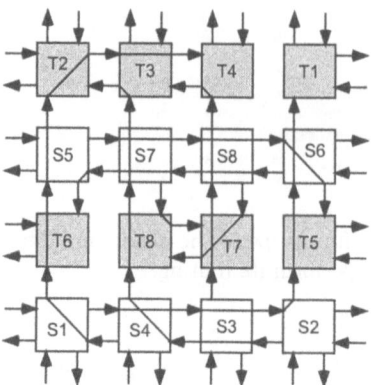

Fig. 6. Test case with a densely connected network.

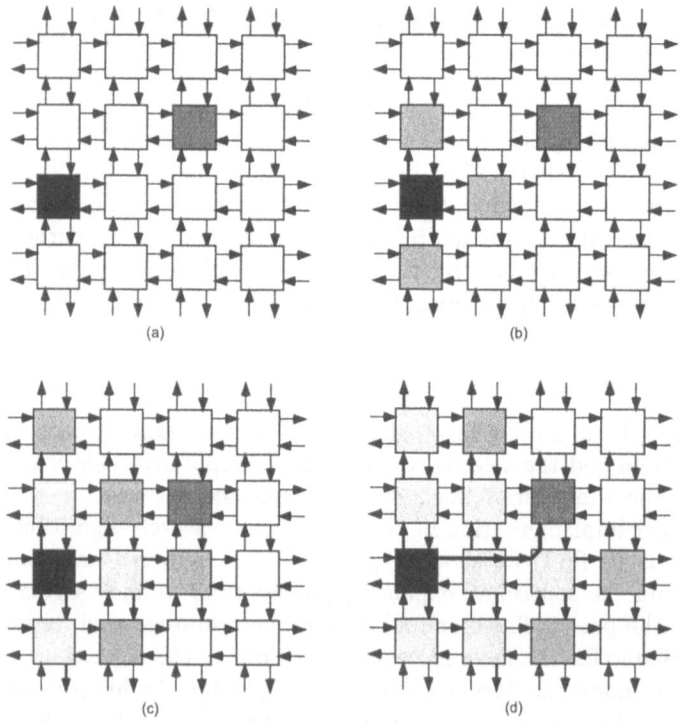

Fig. 7. Step (a) one, (b) two, (c) three and (d) four of the path construction process between the source placed in column 1, row 2 and target cell placed in column 3, row 3.

two-dimensional array of molecules, is similar to standard FPGAs and makes the circuit general enough to implement any digital circuit. However, molecules have self-configuration capabilities that are not present in commercial FPGAs and that are important for the growth of an organism and for self-repair at

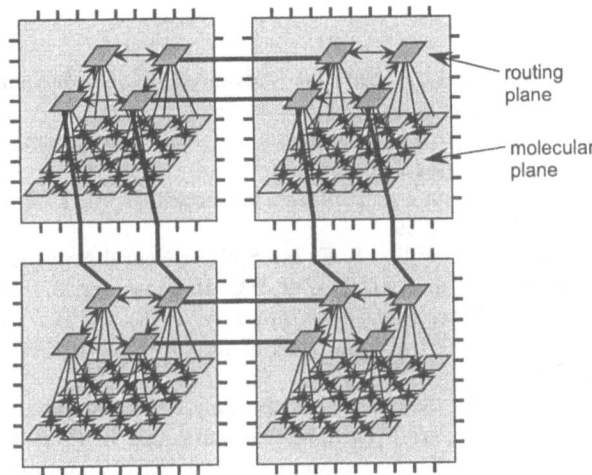

Fig. 8. A multi-chip organism shows the inter-cellular connections.

the cellular level. The second plane is a two-dimensional array of routing units that implement a dynamic routing algorithm. It is used for the inter-cellular communication, letting the tissue dynamically create paths between cells. This algorithm is totally distributed, and hence does not need the control of a micro-processor. Moreover, its scalability allows the creation of cellular networks of any size.

This circuit has been tested with a simulation based on the VHDL files describing the entire system. The next step of the project, which is currently under way and which should be completed by the time the conference will take place, is to develop, from the VHDL files, the VLSI layout and to realize a testchip to validate the design of our circuit.

Due to financial considerations, the first prototype of the POEtic chip will only contain approximately 500'000 equivalent gates. This size will not have enough molecules in one chip for complex designs. It will only be possible to im-plement a very simple organism on such a small number of molecules. Therefore, we included in the design the possibility of implementing a multi-chip organism by seamlessly joining together any number of chips (figure 8).

Acknowledgements

This project is funded by the Future and Emerging Technologies programme (IST-FET) of the European Community, under grant IST-2000-28027 (POETIC). The information provided is the sole responsibility of the authors and does not reflect the Community's opinion. The Community is not responsible for any use that might be made of data appearing in this publication. The Swiss participants to this project are supported under grant 00.0529-1 by the Swiss government.

References

[1]   ARM: Amba Specification, rev 2.0. Advanced RISC Machines Ltd (arm). http://www.arm.com/armtech/amba_spec, 1999.
[2]   Brown, S., Francis, R., Rose, J., Vranesic, Z.: *Field Programmable Gate Arrays.* Kluwer Academic Publishers, 1992.
[3]   Dijkstra, E.W.: A Note on Two Problems in Connexion with Graphs. *Numerische Mathematik*, 1:269–271, 1959.
[4]   Eriksson, J., Torres, O., Villa, A. E. P.: Spiking Neural Networks for Reconfigurable POEtic Tissue. In A.M. Tyrrell, P.C. Haddow, and J. Torresen, editors, *Evolvable Systems: From Biology to Hardware. Proc. 5th Int. Conf. on Evolvable Hardware (ICES '03)*, volume 2606 of *LCNS*, pages 165–173, Berlin, 2003, Springer-Verlag.
[5]   Moreno Arostegui, J. M., Sanchez, E., Cabestany, J.: An In-system Routing Strategy for Evolvable Hardware Programmable Platforms. In *Proc. 3rd NASA/DoD Workshop on Evolvable Hardware*, pages 157–166. IEEE Computer Society Press, 2001.
[6]   Roggen, D., Floreano, D., Mattiussi, C.: A Morphogenetic System as the Phylogenetic Mechanism of the POEtic Tissue. In A.M. Tyrrell, P.C. Haddow, and J. Torresen, editors, *Evolvable Systems: From Biology to Hardware. Proc. 5th Int. Conf. on Evolvable Hardware (ICES '03)*, volume 2606 of *LCNS*, pages 153–164, Berlin, 2003, Springer-Verlag.
[7]   Sanchez, E., Mange, D., Sipper, M., Tomassini, M., Perez-Uribe, A., Stauffer, A.: Phylogeny, Ontogeny, and Epigenesis: Three Sources of Biological Inspiration for Softening Hardware. In T. Higuchi, M. Iwata, and W. Liu, editors, *Evolvable Systems: From Biology to Hardware*, volume 1259 of *LCNS*, pages 33–54, Berlin, 1997. Springer-Verlag.
[8]   Sipper, M., Sanchez, E., Mange, D., Tomassini, M., Perez-Uribe, A.: A Phylogenetic, Ontogenetic, and Epigenetic View of Bio-inspired Hardware Systems. *IEEE Transactions on Evolutionary Computation*, 1:1:83–97, 1997.
[9]   Tempesti, G., Roggen, D., Sanchez, E., Thoma, Y., Canham, R., Tyrrell, A.M.: Ontogenetic Development and Fault Tolerance in the POEtic Tissue. In A.M. Tyrrell, P.C. Haddow, and J. Torresen, editors, *Evolvable Systems: From Biology to Hardware. Proc. 5th Int. Conf. on Evolvable Hardware (ICES '03)*, volume 2606 of *LCNS*, pages 141–152, Berlin, 2003, Springer-Verlag.
[10] Tyrrell, A.M., Sanchez, E., Floreano, D., Tempesti, G., Mange, D., Moreno Arostegui, J.-M., Rosenberg, J., Villa, A.E.P.: Poetic Tissue: An Integrated Architecture for Bio-inspired Hardware. In A.M. Tyrrell, P.C. Haddow, and J. Torresen, editors, *Evolvable Systems: From Biology to Hardware. Proc. 5th Int. Conf. on Evolvable Hardware (ICES '03)*, volume 2606 of *LCNS*, pages 129–140, Berlin, 2003, Springer-Verlag.

A FPL Bioinspired Visual Encoding System to Stimulate Cortical Neurons in Real-Time

Leonel Sousa[1], Pedro Tomás[1], Francisco Pelayo[2],
Antonio Martinez[2], Christian A. Morillas[2], and Samuel Romero[2]

[1] Dept. of Electrical and Computer Engineering, IST/INESC-ID, Portugal
las@inesc-id.pt, pfzt@sips.inesc-id.pt
[2] Dept. of Computer Architecture and Technology, University of Granada, Spain
fpelayo@ugr.es, {amartinez, cmorilas, sromero}@atc.ugr.es

Abstract. This paper proposes a real-time bioinspired visual encoding system for multielectrodes' stimulation of the visual cortex supported on Field Programmable Logic. This system includes the spatio-temporal preprocessing stage and the generation of varying in time spike patterns to stimulate an array of microelectrodes and can be applied to build a portable visual neuroprosthesis. It only requires a small amount of hardware which is achieved by taking advantage of the high operating frequency of the FPGAs to share circuits in time. Experimental results show that with the proposed architecture a real-time visual encoding system can be implemented in FPGAs with modest capacity.

1   Introduction

Nowadays, the design and the development of visual neuroprostheses interfaced with the visual cortex is being tried to provide a limited but useful visual sense to profoundly blind people. The work presented in this paper has been carried out within the EC project "Cortical Visual Neuroprosthesis for the Blind" (CORTIVIS), which is one of the research initiatives for developing a visual neuroprosthesis for the blind [1,2].

A block diagram of the cortical visual neuroprosthesis is presented in fig. 1. It includes a programmable artificial retina, which computes a predefined retina model, to process the input visual stimulus and to produce output patterns to approximate the spatial and temporal spike distributions required for effective cortical stimulation. These output patterns are represented by pulses that are mapped on the primary visual cortex and are coded by using Address Event Representation [3]. The corresponding signals are modulated and sent through a Radio Frequency (RF) link, which also carries power, to the electrode stimulator. This project uses the Utah microelectrode array [4], which consists on an array of 10×10 silicon microelectrodes separated by about 400 μm in each orthogonal direction (arrays of 25×25 microelectrodes are also considered). From experimental measures on biological systems, it can be established a time of 1 ms to "refresh" all the spiking neurons, which means an average time slot of 10 μs dedicated to each microelectrode. The RF link bandwidth allows communication

P.Y.K. Cheung et al. (Eds.): FPL 2003, LNCS 2778, pp. 691–700, 2003.
© Springer-Verlag Berlin Heidelberg 2003

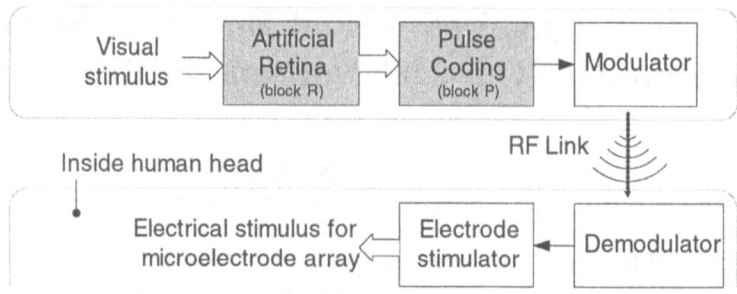

Fig. 1. Cortical visual neuroprosthesis

at a bit-rate of about 1 Mbps, which means an average value of 10 kbps for each microelectrode in a small size array of 10×10 electrodes or about 1.6 kbps for the 25×25 microelectrode array.

This paper addresses the design of digital processors for implementing the shady blocks in fig. 1 in Field Programmable Logic (FPL) technology. The model of the retina adopted is a complete approximation of the spatio-temporal receptive fields' characteristic response of the retina ganglion cells (block R). The neuromorphic pulse coding is based on a leaky integrate-and-fire model of spiking neurons and on the Address Event Representation (AER), that communicates information about the characteristics of spikes and addresses of target microelectrodes without timestamps (block P). The architecture of the system has been designed having in mind the specifications of the problem referred above and the technical characteristics of nowadays Field Programmable Gate Array (FPGA) devices. Experimental results show that a complete artificial model that generates neuromorphic pulse-coded signals can be implemented in real-time even in FPGAs with low-capacity.

This paper is organized as follows. The architecture for modeling the retina and for coding the event lists that will be carried out to the visual cortex is presented in section 2. Section 3 reports the computational architectures designed for implementing the retina model in FPL, discussing their characteristics and suitability to the technology. Section 4 presents experimental results obtained by implementing R and P blocks on a FPGA and section 5 concludes the paper.

2   Model Architecture

The neuron layers of the human retina perform a set of different tasks, which culminate in the spiking of ganglion cells at the output layer [5]. These cells have different transient responses, receptive fields and chromatic sensibilities. The system developed in this paper implements the full model of the retina, which includes all the processing layers, plus the protocol for carrying the spikes to the visual cortex in a serial way through a RF link.

The two main blocks of the system in fig. 1 perform the following tasks: block **R**, the spatiotemporal filtering of the stimulus visual input, with contrast

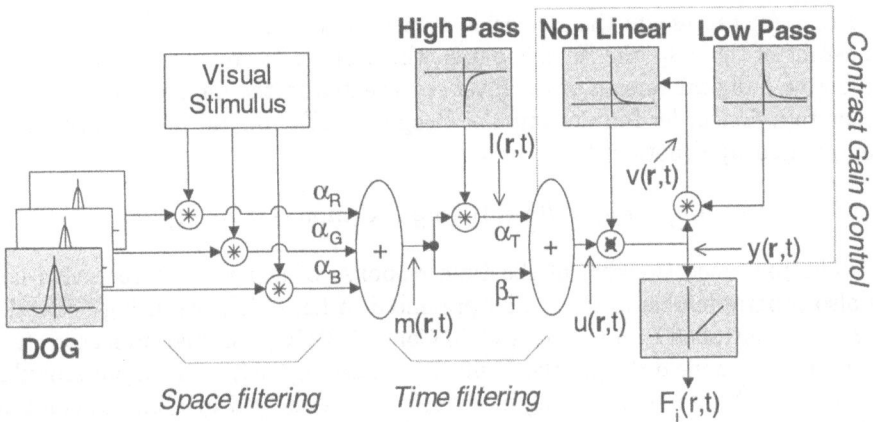

Fig. 2. Retina early layers

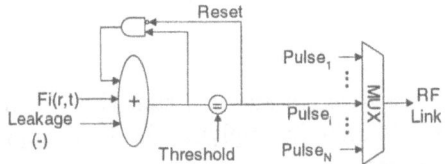

Fig. 3. Neuromorphic pulse coding

gain control and a rectifier circuit for computing the firing rate (fig. 2); block **P**, a neuromorphic pulse coding block which also implements the protocol used to communicate event lists without timestamps (fig. 3).

2.1 Retina Early Layers

The retina computational model used is based on the research published in [6], but it has been extended to the chromatic domain by considering independent filters for the basic colors. Fig. 2 presents the model architecture. It includes a spatial filter for contrast enhancement in the various color components which output is combined with programmable weights to reproduce receptive fields stimulated by specific color channels.

A Difference of Gaussians (DoG) is used to filter the stimulus in space:

$$DOG(\mathbf{r}) = \frac{a_+}{2\pi\sigma^2}e^{-\frac{r^2}{2\sigma^2}} - \frac{a_-}{2\pi\beta^2\sigma^2}e^{-\frac{r^2}{2\beta^2\sigma^2}} \tag{1}$$

where a_+ and a_- represent the relative weights of center and surround, respectively, and σ and $\beta\sigma$ $(\beta > 1)$ are their diameters.

A high-pass temporal filter with the following impulse response is applied to the input signal already filtered in space:

$$h_{HP}(t) = \delta(t) - \alpha H(t)e^{-\alpha t} \tag{2}$$

where $H(t)$ represents the Heaviside step function and α^{-1} is the decay time constant of the response. In this paper, the bilinear approximation was applied to derive a digital version of the filter represented in the Laplace domain [7]. It leads to a first order Infinite Impulse Response (IIR) digital filter which can be represented by equation 3.

$$l[n] = b_{HP} \times l[n-1] + c_{HP} \times (m[n] - m[n-1]) \tag{3}$$

The relevance (weight) of the time response of a particular receptive field is also programmable in the model presented in fig. 2. The resulting activation $u(\mathbf{r}, t)$ is multiplied by a Contrast Gain Control (CGC) modulation factor $g(\mathbf{r}, t)$ and rectified to yield the ganglion cells firing rate response to the input stimulus.

The CGC models the strong modularity effect exerted by stimulus contrast. The CGC non-linear approach is also used in order to model the "motion anticipation" effect observed on experiments with a continuous moving bar [8]. The CGC feedback loop involves a low-pass temporal filter with the following impulse response:

$$h_{LP}(t) = Be^{-\frac{t}{\tau}} \tag{4}$$

where B and τ define the strength and constant time of the CGC. The filter is computed in the digital domain, by applying the same approximation as for the high-pass filter, by equation 5.

$$v[n] = b_{LP} \times v[n-1] + c_{HP} \times (y[n] + y[n-1]) \tag{5}$$

and the output of the filter is transformed into a local modulation factor (g) via the non-linear function:

$$g(t) = \frac{1}{1 + [v(t) \cdot H(v(t))]^4} \tag{6}$$

The output is rectified by using the function expressed in eq. 7:

$$F_i(\mathbf{r}, t) = \psi H(y(\mathbf{r}, t) + \theta)[y(\mathbf{r}, t) + \theta] \tag{7}$$

where ψ and θ define the scale and baseline value of the firing rate $f_i(\mathbf{r}, t)$.

The system to be developed is fully programmable and typical values for all parameters of the model are found in [6]. The model has been completely simulated in MATLAB, by using the *Retiner* environment for testing retina models [9].

2.2   Neuromorphic Pulse Coding

The neuromorphic pulse coding block converts the continuous-varying time representation of the signals produced in the early layers of the retina into a neural pulse representation. In this new representation the signal provides new information only from the moment a new pulse begins. The adopted model is a simplified version of an integrate-and-fire spiking neuron [10]. As represented in fig. 3, the neuron accumulates input values from the respective receptive field (output firing rate determined by retina early layers) until it reaches a given threshold. Then it fires and discharges the accumulated value. A leakage term is included to force the accumulated value to diminish for low or null input values. The amplitude and duration of pulses are then coded by using AER, which is represented in a simplified way in fig. 3 by a multiplexer, to be sent to the addressed microelectrodes via the RF link. An event consists on an asynchronous bus transaction that carries the address of the corresponding microelectrode and is sent at the moment of pulse onset (no timestamp information is communicated). An arbitration circuit is required in the sender side and the receiver has to be listening to the data link with a constant latency.

3   FPL Implementation

This section discusses the implementation in FPL technology of the visual encoding system presented in the previous section. The usage of FPL technology will allow changing the model parameters without the need of projecting a second circuit. This has special importance since different patients have different sight parameters therefore requiring adjustments to the artificial retina. FPL may also allow changing model blocks if new information on how visual stimulus are processed reveals the need to introduce new filters.

For the design of the system, different computational architectures were considered both for the retina early layers and to the neuromorphic coding of the pulses. These architectures lead to implementations with different characteristics, in terms of hardware requirements and speed. The scalability and programmability of the system are also relevant aspects that are taken into account for FPL implementations.

3.1   The Retina Early Layers

To implement the retina early layers there can be multiple approaches which involve different hardware requirements, so the first step would be to analyze the necessary hardware to build the desired processor. Assuming a typical convolutional kernel of 7×7 elements for the DoG spatial filter and that high-pass and low-pass temporal filters are computed by using the difference equations 3 and 5, respectively, then 53 multiplications and 52 additions per image cell are required for just one spatial channel. For a matrix of 100 microelectrodes, the hardware required by a fully parallel architecture makes it not implementable in FPL technology. Therefore, the usage of the hardware has to be multiplexed in time, but the temporal restriction of processing the whole matrix with a frequency up to 100Hz must also be fulfilled. This restriction is however wide enough to consider the processing of cells with a full multiplexing schema inside each main block of the architecture model presented in section 2: DoG spatial filter, High-Pass temporal filter and Contrast Gain Control.

The architectures of these blocks can then be described as shown in figure 4, where each main block includes one or more RAM blocks to save processed

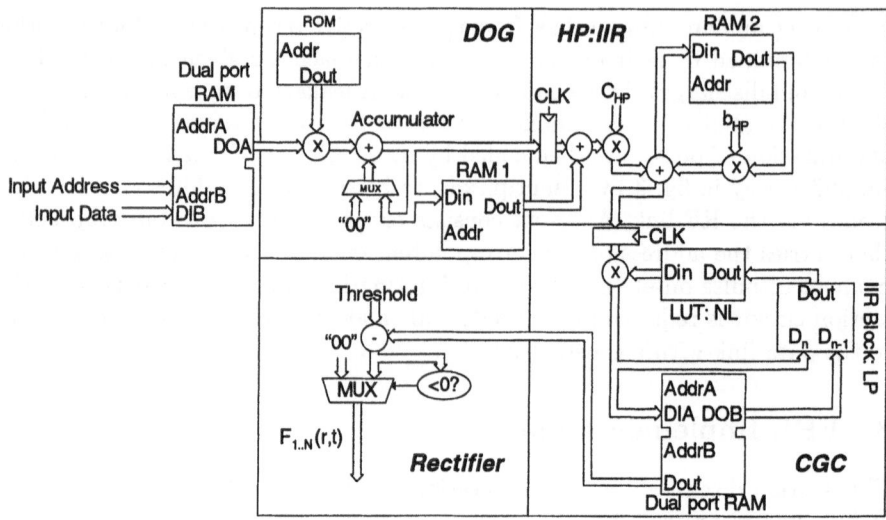

Fig. 4. Detailed diagram for the retina computational architecture.

frames in intermediate points of the system in order to compute the recursive
time filters. At the input, dual-port RAM for the three color channel is used,
while ROM is used to store the three coefficient tables for the spatial DoG
filters. The operation of the processing modules is locally controlled and the
overall operation of the system is performed by integrating the local control
in a global control circuit. The control circuit is synthesized as a Moore state
machine, because there is no need of modifying the output asynchronously with
variations of the inputs. All RAM address signals, which are unconnected in the
figure, are generated by the control circuit.

The DoG filter module calculates the 2D convolution between the input image
and the predefined matrix coefficient stored in ROM. It operates in a sequential
method, where pixels are processed one at a time in a row major order and an
accumulator is used to store the intermediate and final results of the convolution. The local control circuit stores the initial pixel address, row and column,
and then successively addresses all the data, and respective filtered coefficients
stored in the ROM, required for the calculus of the convolution. After multiplying each filter coefficient the resulting value is successively accumulated. When
the calculus is finished for a given pixel, the value is sent both to the internal
RAM1 and to the next high-pass (HP) filter module. This module receives as
operands the space filter results for the actual and the previous images ($m(\mathbf{r}, n)$
and $m(\mathbf{r}, n-1)$) and the previous filter output $l(\mathbf{r}, n-1)$, which is stored in the
RAM2. The filter is computed by first adding both $m[n]$ and $m[n-1]$ inputs,
then multiplies the sum by the coefficient c_{HP} and the product is added with
the previous filter output result $l[n-1]$ multiplied by b_{HP}. The result is both
stored in the RAM2 and communicated to the next processing module that corresponds to the CGC. This module consists on a low pass filter (eq. 5) and a

non-linear function (eq. 6) which is computed by using a lookup table stored in a RAM block. The low-pass filter circuit is not detailed in the figure because it is similar to the high-pass one. The last processing module is a rectifier which cuts the signal whenever its value decreases below a predefined threshold. It is implemented by a simple binary comparator whose output is used to control a multiplexer.

The overall circuit operates in a 3 stage pipeline corresponding to each one of the main processing blocks. Only the first pipeline stage require a variable number of clock cycles depending on the dimension of the filter kernel–49 cycles for a 7×7 kernel. Each of the two other pipeline stages is solved in a single clock cycle.

3.2   Neuromorphic Pulse Coding

Fig. 5 shows two processing modules $i)$ for pulse generation and its representation and $ii)$ to arbitrate the access to the serial bus (the input to the RF modulator in figure 1). This block is connected to the retina early layers through a dual port RAM (CGC block in fig. 4) where one writes data onto one port and the others reads it from the other.

The pulse generation circuit, which converts from firing rate to pulses, can be seen as a simple Digital Voltage Controlled Oscillator (DVCO) working in a two clock cycle stage pipeline. In the first stage the input firing rate is added to the accumulated value. In the second stage a leakage value is subtracted and, if the result is bigger than the threshold, a pulse is fired and the accumulator returns to the zero value (see fig. 5). Note that in this architecture the accumulator circuit is made with a RAM, since it corresponds to a single adder and multiple accumulator registers for the different microelectrodes.

AER was used to represent the pulses while the information is serialized. In a first approach, the architecture consists of an arbitration tree to multiplex the onset of events (spikes) onto a single bus. In this tree, we check if one or more inputs had a pulse to be sent and arbitrate the access to the bus trough

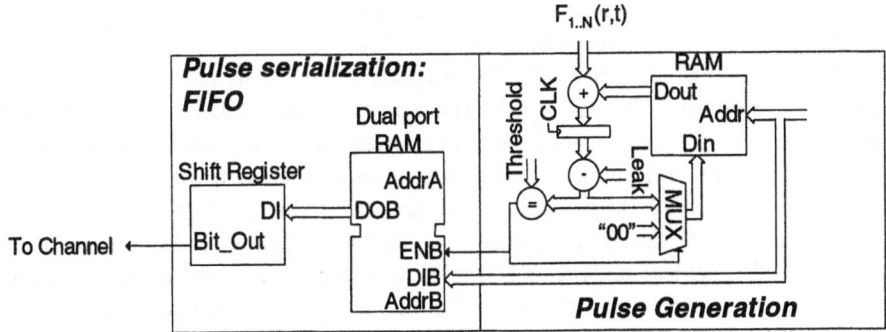

Fig. 5. Block diagram of the circuit for coding the pulses and to implement the address event protocol.

the request/acknowledge signals in the tree [3]. This tree consists of multiple subtrees, where registers can be used for buffering the signals between them. This architecture is not scalable, since it requires a great amount of hardware and is not a solution when arrays with a great number of microelectrodes are used (see section 4). To overcome this problem, and since the circuit for pulse generation is intrinsically sequential, a First In First Out (FIFO) memory is used to register the generated spikes for the microelectrodes. Only one pulse is generated in a clock cycle and information about it is stored in the FIFO memory because requests may find the communication link busy. This implementation has the advantage of not increasing the hardware resources required with the increase in the number of microelectrodes, but its performance is more dependent of the channel characteristics since it increases the maximum wait time to send a pulse.

The FIFO is implemented by using a dual port data RAM, where input is stored in one port and, at the other port, the pulse request is sent to the channel. Also to avoid overwriting when reading data from the FIFO a shift register is used to interface the data output of the RAM with the communication channel.

4   Experimental Results

The visual encoding system was described in VHDL and exhaustively tested on a DIGILAB II board supported on a XILINX SPARTAN XC2S200 FPGA [11] with modest capacity. The synthesis tool used was the one supplied by the FPGA manufacturer, the ISE WebPACK 5. This FPGA has modest capacities, with 56 kbit of block-RAM and 2352 slices, corresponding to a total of 200, 000 system-gates. The functional validation of the circuits was carried out by comparing the results obtained with MATLAB Retiner [9].

For the test and experimental results presented in this section, only one space filter (DoG) is considered and the input signal is represented in 8-bit grayscale. However, internally 12-bit signals are used in order to reduce discretization errors. The considered dimension of the microelectrode array is 100.

Analyzing the results in table 1 it is clearly seen that the retina early layers do not require many FPGA slices but occupies a significant amount of memory. The synthesis of the circuit for an array of 1024 microelectrodes shows a similar percentage of FPGA slice occupation but it occupies all the 14 RAM blocks supplied by the SPARTAN XC2S200. In terms of time restrictions, the solutions works very well since it can process the array of input pixels in a short time, about 0.1ms for 100 pixels, which is much lower than the specified maximum of 10ms. In fact this architecture allows to process movies at frame rate of 100 Hz and a resolution of 10000 pixels without increasing the number of slices used.

The analysis of the AER translation method as however different aspects. If considered a typical matrix of 100 microelectrodes the tree solution (with or without intermediate registers) is a valid one as the resource occupation is low. However, the increase of the number of microelectrodes implies the usage of a great amount of hardware and this solution becomes impracticable for FPL technology. In that case, the FIFO based solution proposed in this paper has the

Table 1. Retina encoding system implemented on a SPARTAN XC2S200 FPGA

Block	Number of microelectrodes	Slice Occupation	Maximum clock frequency	Number of RAM blocks used
Retina early layers	100	18%	47MHz (49 cc*)	6
Pulse Generation	100	2%	51MHz	5
AER Registered Tree	100	9%	99MHz	0
	256	17%	98MHz	0
	512	37%	93MHz	0
AER Unregistered Tree	100	10%	64MHz	0
	256	21%	63MHz	0
	512	45%	57MHz	0
FIFO AER	**	5%	75MHz	1

* 49 clock cycles are required for processing a pixel
** not dependent of the number of micro-electrodes

great advantage of accommodating a great number of electrodes with a small amount of hardware and just one RAM block. With a channel bandwidth of about 1 Mbps, the FIFO based AER circuit does not introduce any temporal restrictions in the firing rate. Even operating at a lower clock frequency than the admitted by the retina early layers circuit, e.g. 40 MHz, the FIFO based circuit is able to attend 100 requests in about 2.5 μs which is much less than the value of 1 ms initially specified and also much less than the time required to send the information through the channel.

In table 1 results are individually presented for the retina early layers and for the neuromorphic pulse coding circuits. The overall system was also synthesized and implemented in a SPARTAN XC2S200 FPGA. A clock frequency greater than 40 MHz is achieved by using about 25% of the slices and 12 of the total of 14 RAM-blocks, for 100 microelectrodes.

5 Conclusions

This paper proposes a computational architecture for implementing a complete model of an artificial retina in FPL technology. The system is devised to stimulate a number of intra-cortical implanted microelectrodes for a visual neuroprosthesis. It performs a bio-inspired processing and encoding of the input visual information. The proposed processing architecture is designed to respect the constraints imposed by the telemetry system to be used, and with the requirement of building a compact and portable hardware solution.

The synthesis results demonstrate how the adopted pipelined and time multiplexed approach makes the whole system fit well on a relatively low complexity FPL circuit, while real-time processing is achieved. The use of FPL is also very convenient to easily customize (re-configure) the visual pre-processing and encoding system for each implanted patient.

Acknowledgements

This work has been supported by the European Commission under the project CORTIVIS ("Cortical Visual Neuroprosthesis for the Blind", QLK6-CT-2001-00279).

References

1. Cortica visual neuro-prosthesis for the blind (cortivis): http://cortivis.umh.es.
2. Ahnelt P., Ammerm ller J., Pelayo F., Bongard M., Palomar D., Piedade M., Ferrandez J., Borg-Graham L., and Fernandez E. Neuroscientific basis for the design and development of a bioinspired visual processing front-end. In *Proc. of IFMBE*, pages 1692–1693, Vienna, 2002.
3. Lazzaro J. and Wawrzynek J. A multi-sender asynchronous extension to the address event protocol. In *Proc. of 16th Conference on Advanced Research in VLSI*, pages 158–169, 1995.
4. Maynard E. The utah intracortical electrode array: a recording structure for potential brain-computer interfaces. *Elec. Clin. Neurophysiol.*, 102:228–239, 1997.
5. Wandell Brian. *Foundations of Vision: Behavior, Neuroscience and Computation.* Sinauer Associates, 1995.
6. Wilke S., Thiel A., Eurich C., Greschner M., Bongard M., Ammermuler J., and Schwegler H. Population coding of motion patterns in the early visual system. *J. Comp Physiol A*, 187:549–558, 2001.
7. Oppenheim A. and Willsky A. *Signal and Systems.* Prentice Hall, 1983.
8. Berry M., Brivanlou I., Jordan T., and Meister M. Anticipation of moving stimuli by the retina. *Nature (Lond)*, 398:334–338, 1999.
9. Pelayo F., Martinez A., Romero S., Morillas Ch., Ros E., and Fernández E. Cortical visual neuroprosthesis for the blind: Retina-like software/hardware preprocessor. In *Proc. of IEEE-EMBS International Conference on Neural Engineering*, Capri, 2003.
10. Gerstner W. and Kistler W. *Spiking Neuron Models.* Cambridge University Press, 2002.
11. XILINX. *Spartan-II 2.5V FPGA Family: Functional Description*, 2001. Product Specification.

Power Analysis of FPGAs:
How Practical Is the Attack ?

François-Xavier Standaert, Loïc van Oldeneel tot Oldenzeel, David Samyde,
and Jean-Jacques Quisquater

UCL Crypto Group
Laboratoire de Microélectronique
Université Catholique de Louvain
Place du Levant, 3, B-1348 Louvain-La-Neuve, Belgium
standaert,vanolden,samyde,quisquater@dice.ucl.ac.be

Abstract. Recent developments in information technologies made the
secure transmission of digital data a critical design point. Large data
flows have to be exchanged securely and involve encryption rates that
sometimes may require hardware implementations. Reprogrammable devices such as Field Programmable Gate Arrays are highly attractive
solutions for hardware implementations of encryption algorithms and
several papers underline their growing performances and flexibility for
any digital processing application. Although cryptosystem designers frequently assume that secret parameters will be manipulated in closed
reliable computing environments, Kocher et al. stressed in 1998 that actual computers and microchips leak information correlated with the data
handled. Side-channel attacks based on time, power and electromagnetic
measurements were successfully applied to the smart card technology,
but we have no knowledge of any attempt to implement them against
FPGAs. This paper examines how monitoring power consumption signals might breach FPGA-security. We propose first experimental results
against FPGA-implementations of cryptographic algorithms in order to
confirm that power analysis has to be considered as a serious threat
for FPGA security. We also highlight certain features of FPGAs that
increase their resistance against side-channel attacks.

1 Introduction

Digital signal processing has traditionally been done using enhanced microprocessors but recent increases in Field Programmable Gate Arrays performance
and size offer a new hardware acceleration opportunity. The last years brought
cryptographic implementations into the field of FPGA designers as several conference and journal publications can witness [10, 11]. These cryptosystem designers frequently assume that secret parameters will be manipulated in closed
reliable computing environments. However, the realities of physical implementations can be extremely difficult to control and may result in the unintended
leakage of side-channel information. This leaked information is often correlated
to the secret keys, thus adversaries monitoring this information may be able to
recover the secret key and breach the security of the cryptosystem.

P.Y.K. Cheung et al. (Eds.): FPL 2003, LNCS 2778, pp. 701–711, 2003.

Side-channel attacks based on time, power and electromagnetic measurements were successfully applied to the smart card technology as witnessed by [1–5]. However, we have no knowledge of any attempt to implement them against FPGAs. Moreover, most major FPGA manufacturers provide no information about the actual security of their devices. This paper presents first experimental results in order to fill that gap. Based on various examples, we discuss the practicability of power analysis attacks against an application-oriented FPGA board but also highlight certain physical features of FPGAs and application boards that make the practical implementation of power analysis significantly harder than in the smart card context.

The paper is structured as follows. Section 2 presents the hardware used to carry out the experiments. Section 3 gives a short description of two cryptographic algorithms: DES and RSA. Section 4 introduces power analysis. We study Simple Power Analysis and Differential Power Analysis in sections 5 and 6. Finally, topics for further researches are in section 7 and conclusions in section 8.

2   Hardware Description

All our experiments were carried out on a VIRTEX-ARM board developed by DICE[1] (Figure 1). This board was developed in 2000 for a multi-purposes use. The board is composed of a control FPGA (Altera® FLEX®10K) and a Xilinx®Virtex®1000 FPGA associated with μ-controllers (Microchip®PIC®, ARM®) and fast access memories. It has multiple compatible PC interfaces (USB, PCI). Practical details about the Virtex1000BG560-4 FPGA that we investigated can be found in [8].

The voltages needed for the board are:

1. 5 volts for the PCI bridge.
2. 2.5 volts for the Virtex core and the ARM μ-processor.
3. 3.3 volts for other devices, including the Virtex I/O blocks.

The usual way to use this board has always been to plug it into a PCI port but to perform a power analysis against a chip, one must have access to its power supply in order to acquire power consumption traces. For this purpose, we insert a small resistance in the supply circuit. As the board has a single ground circuit and only the Virtex chip has to be analyzed (other devices add noise to the measurements) we decided to insert the resistance next to the source supplying the Virtex. We undersupplied certain unnecessary devices and we un-soldered the DC-DC 2.5V convertor (of which internal oscillations generate noise) before carrying out the experiments. Figure 1 illustrates the final test bed where the FPGA is programmed via the JTAG chain[2].

Finally, we used the following hardware to perform our tests :

1. Voltage sources to supply the 2.5 volts path and 3.3 volts path.
2. A waveform generator or a crystal oscillator to generate the clock signal.

---

[1] Microelectronics Laboratory at Université Catholique de Louvain, Belgium.
[2] Boundary-Scan Standard IEEE 1149.1, developed by the Joint Test Action Group.

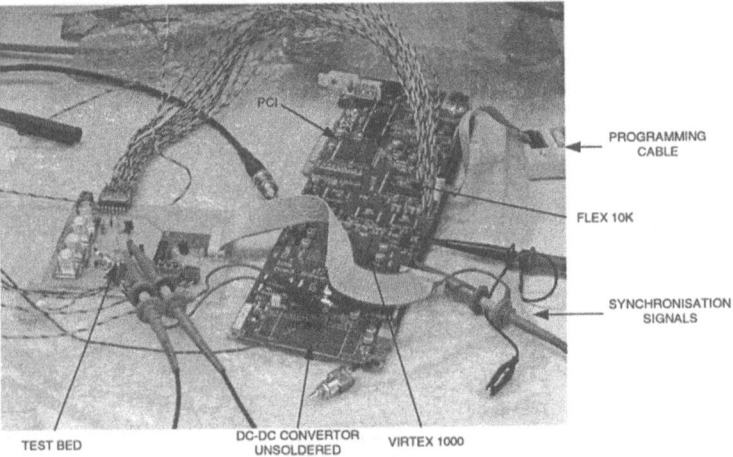

Fig. 1. The FPGA board

3. An oscilloscope to observe the power traces. We used the Tektronix 7140 with a 1 GHz bandwidth.
4. Computer softwares to generate the FPGA programming files and process the data after acquisition.

3 DES and RSA

In 1977, the Data Encryption Standard (DES) algorithm was adopted as a Federal Information Processing Standard for unclassified government communication. Although a new Advanced Encryption Standard was selected in October 2000, DES is still largely in use. DES [6] encrypts 64-bit blocks with a 56-bit key and processes data with permutations, substitutions and XOR operations. It is an iterative block cipher that applies a number of key-dependent transformations called rounds to the plaintext. This structure allows very efficient hardware implementations.

Basically, the plaintext is first permuted by a fixed permutation IP. The result is next split into the 32 left bits and the 32 right bits, respectively L and R that are sent to 16 applications of a round function. The ciphertext is calculated by applying the inverse of the initial permutation IP to the result of the 16-th round.

The secret key is expanded by the key schedule to 16 x 48-bit subkeys K_i and in each round, a 48-bit subkey is XORed to the text. The key expansion consists of known bit permutations and shift operations. As a consequence, finding any subkey bit directly involves that the secret key is corrupted.

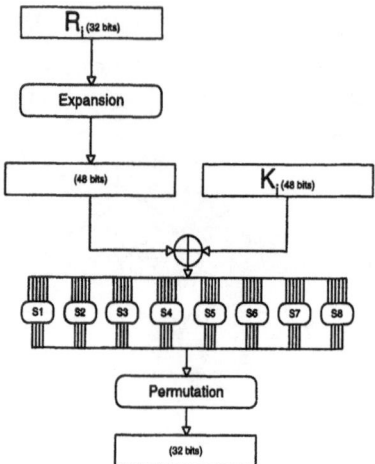

Fig. 2. The function f.

Finally, the round function is easily described by:

$$L_i = R_{i-1} \tag{1}$$
$$R_i = L_{i-1} \oplus f(R_{i-1}, K_i) \tag{2}$$

where f is a nonlinear function detailed in Figure 2: the R_i part is first expanded to 48 bits with the E box, by doubling some R_i bits. Then, it performs a bitwise modulo 2 sum of the expanded R_i part and the 48-bit subkey K_i. The output of the XOR function is sent to eight non-linear S-boxes (S). Each of them has six inputs bits and four outputs. The result is finally permuted in the box P.

The design we used to carry out the experiments is a sequential DES that takes one clock cycle to perform one round.

The RSA cryptosystem [7] is the most widely used public-key cryptosystem worldwide. It may be used to provide both secrecy and digital signatures and its security is based on the intractability of the integer factorization problem. If $n = p \times q$ is a public modulus (p and q are large prime numbers), x the plaintext, and $k = \sum_{i=0}^{l-1} k_i 2^i$ the secret key, the RSA encryption scheme can be viewed as a simple modular exponentiation:

$$y = x^k \ mod \ n \tag{3}$$

The design we used to carry out the experiments is a sequential "square and multiply" algorithm with 14-bit texts and keys. Modular reduction was done with Barrett's reduction rule and one "square and multiply" operation is performed in one clock cycle.

Algorithm 1 Computation of $x^k \; mod \; n$

---

1. $z = 1$;
2. For i $= l - 1$ to 0 loop :
   $z = z^2 \; mod \; n$;
   If $k_i = 1$ then $z = z \times x \; mod \; n$

---

4   Introduction to Power Analysis

Integrated circuits are built out of individual transistors that act as voltage-controlled switches. Current flows across the transistor substrate when charge is applied to (or removed from) the gate. This current then delivers charges to the gates of other transistors, interconnect wires, and other circuit loads. The motion of electric charge consumes power and produces electromagnetic radiations, both of which are externally detectable. Therefore, individual transistors produce externally observable electrical behavior. Because microprocessor logic units exhibit regular transistor switching patterns, it is possible to easily identify macro-characteristics (such as microprocessor activity) by the simple monitoring of power consumption.

In Simple Power Analysis attacks, an attacker directly observes a system's power consumption. The amount of power consumed varies depending on the microprocessor instruction performed. Large features such as DES rounds may be identified, since the operations performed by the microprocessor vary significantly during different parts of these operations.

Differential Power Analysis is a much more powerful attack than SPA, and is much more difficult to prevent. While SPA attacks use primarily visual inspection to identify relevant power fluctuations, DPA attacks use statistical analysis to extract information correlated to the secret key.

Because it was not obvious that power analysis could detect some features of a running design, we performed a simple preliminary tests: we investigated the power consumption of NOT gates applied to bit vectors (all 0s or all 1s) and stored in registers. We clearly observed that the power consumption is correlated to the Hamming weight[3] of these bit vectors (see Figure 3). However, this test gave no indication about a possible dilution of the bit effect when a large design (like DES) is running. Moreover, the power consumption was made clearer because the bit vectors appeared at the outputs of the FPGA, while power analysis usually looks for internal bit switches.

5   Simple Power Analysis of FPGAs

Traditional controllers process the data sequentially and apply a set of instructions to the intermediate states of the computation. They can be viewed as control-oriented designs. As a consequence, an attacker may expect to detect two types of information from their side-channel leakages:

---

[3] Hamming weight: number of one in the bit vector.

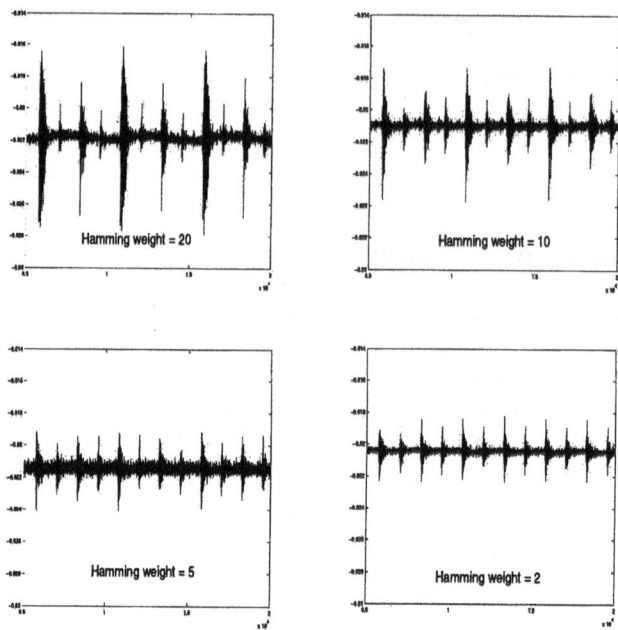

Fig. 3. Hamming weight (5000 traces averaged).

1. The instructions processed.
2. The data processed.

SPA typically tries to take advantages of the sequence of instructions processed. For example, distinguishing the square operation from the multiply operation would allow us to directly recover the key bits in an implementation of RSA. There are numerous examples of programs running on smart cards that allows to distinguish different instructions. However, simple countermeasures usually allows avoiding SPA by masking the instructions.

On the contrary, in most applications, FPGAs are used in order to perform parallel tasks. Cryptographic applications like DES or RSA can be implemented as data-oriented pipeline architectures with several operations running concurrently. Moreover, operations that are spread over several clock edges in smart cards may be reduced to only one clock period in FPGA implementations, which makes distinguishing them unlikely. As a consequence, in these cases, SPA becomes somewhat unpractical and an attacker is limited to information about the data processed as we have in Figure 3.

Exceptions obviously exist. For example applications where enable signals of registers are managed by a control part. Then the activity (or not) of registers may help to distinguish instructions.

We can illustrate these assumptions with an example. We investigated the power consumption of a DES running with weak keys that we can explain as

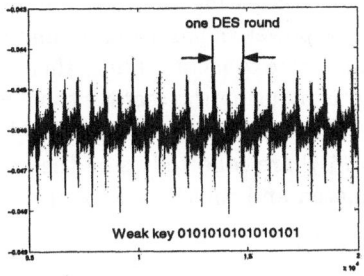 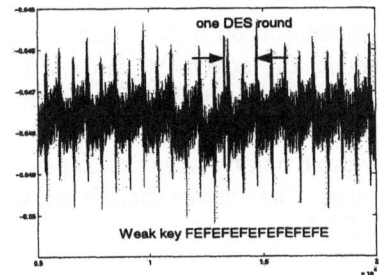

Fig. 4. Weak keys (5000 traces averaged).

follows. Because of the way the initial key is modified to get a subkey for each round of the algorithm, certain initial keys present special properties. Practically, if the subkey used for any round of the algorithm is the same, the inial key is weak. DES has 4 weak keys and we used the following ones (in hexadecimal representation):

Weak Key Value	Actual Subkey
0101010101010101	00000000000000
FEFEFEFEFEFEFEFE	FFFFFFFFFFFFFF

Figure 4 illustrates the power consumption of DES running with weak keys. We observe that:

1. We can clearly identify the rounds of our running DES.
2. The mean power consumed slightly differs between the two cases. One reason could be that the architectures slightly differ because a different key is stored in the VHDL code.
3. The patterns of the power consumed are clearly different. The second is fatter which corresponds to the expected behavior of the device.

This test confirms that the consumed power is strongly correlated with the internal bit switches of FPGAs. It also underlines that SPA-type attacks, where the attacker recovers secret parameters observing the shape of the traces are made difficult by parallel computing (as all components are running concurrently). Moreover, FPGAs offer great opportunities to implement countermeasures against SPA.

6   Differential Power Analysis of FPGAs

As previous section confirmed that the power consumed by FPGAs is correlated with the internal bit switches, Differential Power Analysis is theoretically applicable. This section is devoted to experimental results of DPA implemented against RSA and DES running on a FPGA. We first studied modular implementation.

The basic premise of this attack is that by comparing the power signal of an exponentiation using a known exponent to a power signal using an unknown exponent, the adversary can learn where the two exponents differ, thus learn the secret exponent. The DPA technique begins by using the secret exponent to exponentiate L random values and collect their associated power signals $S_i[j]$ (j is a sample point). Likewise, L power signals $P_i[j]$ are collected using the known exponent. The average signals are then calculated and subtracted to form $D[j]$, the DPA bias signal.

$$D[j] = \frac{1}{L} \sum_{i=1}^{L} S_i[j] - \frac{1}{L} \sum_{i=1}^{L} P_i[j] = \overline{S}[j] - \overline{P}[j] \qquad (4)$$

The portions of the signals $\overline{S}[j]$ and $\overline{P}[j]$ that are dependent on the intermediate data will average out to the same constant as long as the data produced by the RSA computation is equal. We have $D[j] = 0$ if the exponentiation operations are the same and $D[j] \neq 0$ if different.

There are several ways to perform the attack, depending on the assumptions made about the attacker. The simplest one is a "Multiple-Exponent, Single-Data" mode. Then, the attacker guesses the exponent bits (starting from the MSB), decides if the guess was correct by computing $D[j]$ and modifies the exponent bits one by one in order to get $D[j] = 0$ everywhere. Figure 5 shows

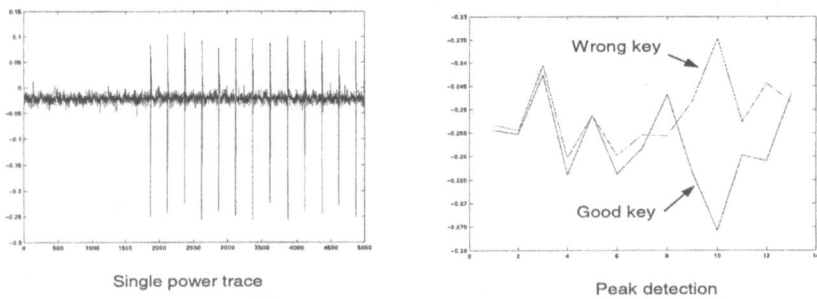

Single power trace   Peak detection

Fig. 5. DPA of RSA (5000 traces averaged).

our practical implementation of the attack. The left picture is a single power consumption trace where we observe the 13 clock edges corresponding to 13 "square and multiply" operations. The right picture shows peaks amplitudes for two keys that are equal until bit number 7. We observe that the consumption traces clearly diverge when exponents differ. Note that the attack depends on how different are the intermediate texts. As a consequence, repeating it with different texts improves its efficiency. Another critical point is that our RSA design was a toy-design with 12-bit vectors. As a consequence the difference between correct and wrong vectors is not large and it was difficult to underline their different power consumption. We increased the power consumption by repeating the RSA computation 20 times on the FPGA. Then we could clearly distinguish the secret exponent.

In the case of DES, the Differential Power Analysis requires a selection function $D(C, b, K_{Sb,16})$ that we define as computing the value of a bit b which is a part of intermediate vector L_{15}. As b results of a partial decryption through the last round of the algorithm, it can be derived from the ciphertext C and the 6 key bits entering in the same s-box as bit b.

To implement the DPA attack, an attacker first observes m encryptions and captures m power traces T_i and their associated ciphertexts C_i. No knowledge of the plaintext is required. With a guessed key K, the function D can be computed for each i and we can obtain two sets of traces: one corresponding with $D_i = 0$ and the other with $D_i = 1$. Each set is then averaged to obtain two average traces A_0 and A_1 and we can compute the difference $\Delta = A_0 - A_1$.

If $K_{Sb,16}$ is correct, the computed value for D will equal the actual value of target bit b with probability 1. As the power consumption is correlated to the data, the plot of Δ will be flat, with spikes in regions where D is correlated to the values being processes. If $K_{Sb,16}$ is incorrect, Δ will be flat everywhere.

The main difference between attacking DES and RSA is that while we have to distinguish the difference between two intermediate vectors in the RSA case, we have to observe the effect of a single bit in the case of DES, what we could not achieve with our low cost equipment. The following practical features of our FPGA board make the implementation of the DPA against DES a challenging task:

1. It should be noted that the application boards usually include several components of which the grounds are connected together. This makes the isolation of the FPGA consumption critical if the power measurements are carried out on the ground pin.
2. The manipulation of the selection bit that is spread over several clock edges in smart cards is reduced to one clock period in our FPGA implementation.
3. FPGAs are running at high work frequencies. Optimal implementations of the DES on the old VIRTEX technology run up to 170 MHz. Recent devices like VIRTEX-2 are much faster. This involve very high sampling rates to catch the consumption details.
4. Contrary to smart cards where the data is managed by 8-bit registers, FPGAs deal with all the bits (64 for DES) at once. This cause a dilution of the desired effect. This is even more critical when the key schedule or other tasks are computed in parallel. As a result, the quantization of power traces may become the bottleneck of the attack, i.e. if the effect of a single bit is out of scale (less than one bit of quantization), the attack becomes unfeasible. Figure 3 illustrates this assessment with a comparison between 20-bit spikes and 2-bit spikes.

7 Further Research

A practical implementation of the DPA against DES is still matter of further research and there are plenty of potential sources for improvements. Nevertheless, there are many others scopes for further research. We propose the following list:

1. Reducing the noise during measurements by isolating the FPGA, using multiple-bit attacks, cooling the devices with nitrogen, ...
2. Applying intrusive attacks to FPGAs: depackaging, layer recovering,...
3. FPGAs usually consists in regular structure. As a consequence, Electro-Magnetic Analysis could be applied in order to focus the acquisition of information leaking to some relevant logic blocks.
4. FPGAs have multiple power sources. Analysis of their distribution inside the logic blocks could help to isolate some components of FPGAs.
5. Studying the security questions raised by the reconfigurability.

8 Conclusions

This work confirmed that power analysis has to be considered as a serious threat for FPGA security. Although certain features of our FPGA board made the practical implementation of power attacks significantly harder than in the smart card context, we have conduced relevant experimental tests. We analyzed the power of a DES running with weak keys and could clearly distinguish both keys. We also implemented a Differential Power Analysis attack against a toy-implementation of RSA. Many solutions would allow to improve our measurements, for example isolating the FPGA from its application board, and a lot of questions concerning the physical security of FPGAs remain open. As a future technological trend seems to be the combination of processors and reconfigurable hardware, there is a field for various research in the coming years.

References

1. P.Kocher, *Timing Attacks on Implementations of Diffie-Hellman, RSA, DSS, and Other Systems*, In the proceedings of CRYPTO 96, Lecture Notes in Computer Science Volume 1109. Springer-Verlag, August 1996.
2. P.Kocher, J.Jaffe, B.Jun, *Differential Power Analysis*, in the proceedings of CRYPTO 99, Lecture Notes in Computer Science 1666, pp 398-412, Springer-Verlag.
3. T.S.Messerges, E.A.Dabbish, R.H.Sloan, *Examining Smart-Card Security under the Threat of Power Analysis Attacks*, IEEE transactions on computers, Vol.51, N5, May 2002.
4. P.Kocher, J.Jaffe, and B.Jun, *Introduction to Differential Power Analysis and Related Attacks*, Cryptography Research 607 Market Street, 5th Floor San Francisco, CA 94102, www.cryptography.com.
5. J.J.Quisquater, D.Samyde, *Electromagnetic Analysis (EMA): Measurements and Countermeasures for Smart Cards*, in Smart Card Programming and Security, Lecture Notes in Computer Science Volume 2140, pp.200-210, Springer-Verlag 2001.

6. National Bureau of Standards. *FIPS PUB 46*, The Data Encryption Standard. U.S. Departement of Commerce, Jan 1977.
7. R.Rivest, A.Shamir, L.Adleman, *A Method for Obtaining Digital Signatures and Public Key Cryptosystems*, Communications of the ACM, 21, pp 120-126, 1978.
8. Xilinx: *Virtex 2.5V Field Programmable Gate Arrays Data Sheet*, http://www.xilinx.com.
9. Altera: *Flex 10K Field Programmable Gate Arrays Data Sheet*, http://www.altera.com.
10. Proceedings of CHES 1999-2002 : Workshop on Cryptographic Hardware and Embedded System, Springer Verlag.
11. Proceedings of FPL 1999-2002 : The Field Programmable Logic Conference, Springer-Verlag.
12. D.Stinson, *Cryptography: Theory and Practice*, CRC Press 2000.

A Power–Scalable
Motion Estimation Architecture
for Energy Constrained Applications

Maurizio Martina, Andrea Molino, Federico Quaglio, and Fabrizio Vacca

Dipartimento di Elettronica – Politecnico di Torino
C.so Duca degli Abruzzi, 24 – 10129 TORINO (ITALY)
{maurizio.martina, andrea.molino, federico.quaglio,
fabrizio.vacca}@polito.it

Abstract. In the research community wireless devices are fostering many design and development activities. The augmented transmission bandwidth supplied by 3G transmission schemes will soon enable an ubiquitous fruition of multimedia content. This paper proposes a reconfigurable, power–scalable architecture for hybrid video coding, suitable for the mobile environment. The complete FPGA design flow shows very interesting performances both in terms of throughput, and power consumption.

1   Introduction

In the last few years, reconfigurable power-aware systems have gained a growing interest in the scientific community. The current global coverage wireless networks are designed specifically to transmit circuit-based voice; however, market analyses foresee that in 2005 50% of wireless traffic, and 80% in 2010 will be non voice, e.g. data transfer and multimedia applications. The major technological challenge will be then to guarantee wireless video access with an acceptable quality, real time response and interactivity. However, behind all these attractive motivations, there are many critical factors that could seriously tackle the design and implementation of multimedia–aware mobile terminals. The design of wireless terminals, in fact, has to deal with limited power budgets and reduced resources availability. Field Programmable Gate Arrays (FPGAs) emerged as the leading technology for their capability of being easily reconfigurable with a quite low effort, allowing, at the same time, to implement even large designs. They became not only powerful prototyping tools for complex designs, but also attracting alternatives to Application Specific Integrated Circuits (ASICs). Thus, FPGAs can be used instead of ASIC where both complex elaborations and certain reconfigurability are the main goals.

Different working profiles could be provided in order to obtain scalable performance (e.g. video quality, frame rate ...) with different levels of power dissipation. Reconfigurable blocks have to be designed to achieve several energy consumption profiles providing, at the same time, the necessary elaboration tasks.

P.Y.K. Cheung et al. (Eds.): FPL 2003, LNCS 2778, pp. 712–721, 2003.

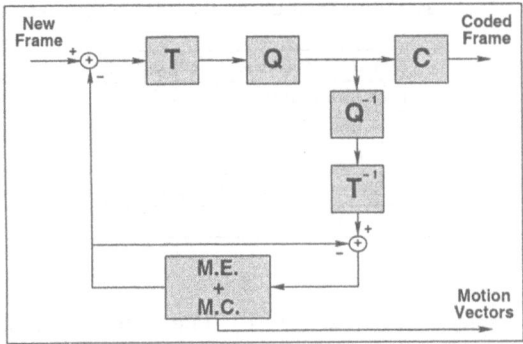

Fig. 1. A block scheme of an hybrid video coder

Good choices for this paradigm can be both DSPs and FPGAs. While the former are able to grant a quite large set of reconfigurable options, they usually suffer from high dynamic power consumption. This figure can seriously tackle their employment in strong power constrained envirorments, expecially when computationally intensive algorithms will be used. On the other hand, FPGAs resemble the proper candidates to obtain a certain degree of reconfiguration and the needed elaboration capabilities with acceptable power consumption, provided that technology improvements can reduce static power consumption.

Power–Scalable solutions must be carefully designed in order to reduce the overall power dissipation. Shut–down policies of the most consuming blocks have to be exploited to reach effective energy savings without compromising the efficiency and the functionality of the entire system. In order to achieve different energy profiles, a Power–Scheduler has to be added to the terminal architecture either as an hardware block or as an Operating System's feature. In this work we investigate the feasibility of a new Motion Estimation IP for hybrid video coding: the main novelty behind the proposed architecture lays in the availability of different power consumption profiles. The paper is organized as follow. In Section 2 hybrid video coder principles and Motion Estimation techniques, for low–power operation, are discussed. In Section 3 the proposed IP for Motion Estimation, based on the previous discussion, is analysed. In Section 4 the IP implementation and its performances are presented; while in Section 5 some conclusions are drawn.

2   Hybrid Video Coders

In figure 1 a simplified description of an actual implementation of an hybrid video coder is sketched. Analysing the coder structure, a *direct path* and a *feedback path* can be identified. The direct path exploits the *spatial correlation* among near pixels in a frame; it is highly probable, in fact, that they can have similar values, thus low–frequencies components are the most relevant in the frame spectrum. To identify them, every new frame produced by the image sensor,

Table 1. Percentage of video coder blocks usage

Building Block	% Usage
Transformation	39.3 %
ME block	47.2 %
Quantiser	5.4 %
Entropy Coder	0.8 %
Other	7.3 %

passes through a *transformation* stage (**T**). The transformed image, then, passes through a *quantiser* (**Q**), that reduces the amount of bits needed for the frame representation. Then, the remaining bits are coded via an *entropy coder* (**C**) to adapt the information to communication medium's capacity. All these processes, typical also for still pictures, produce the so called Intra Frame (I).

The feedback path, on the other hand, tries to detect the *temporal correlation* among frames, i.e. successive frames have, with high probability, little scene variations with respect to the formerly coded ones. The objective is to *predict* the new frame, based on the knowledge of the previous ones, and then to transmit only the differences between the predicted and the real frame. Such this *error frame* passes through the T, Q and C stages obtaining then a so called Inter Frame (P) that requires few bits to be represented. The main hint in this step is that the feedback path has to "simulate" the receiver's steps before trying to predict the next frame. This must be done to avoid differences between the transmitter and the receiver, the prediction stage must work on the same quality frames than the receiver does. The image has to pass through a *de–quantiser* (Q^{-1}) and an *anti–transformation* stage (T^{-1}) to produce a frame similar to the one produced by the image sensor. This reconstructed frame is then stored in a local buffer. After, it is compared with a previously stored image to identify the differences, due to moving objects. This step, called *Motion Estimation* (**ME**), detects the motion occurred between the images in terms of *Motion Vectors* (MVs). A subset of these MVs are used to predict the next frame just applying them to the current frame during the *Motion Compensation* (**MC**) step. This prediction is compared with the actual image to obtain the aforementioned error frame. Understanding the computational effort of each block aids in designing a power efficient coder. Thus, we perform a profiling analysis on a software model of an H.263 video coder[1]. As it can be inferred analysing table 1, Motion Estimation is the most demanding task; therefore a proper hardware implementation can reduce the overall consumption achieving better power figures.

Block–based Matching Algorithms,[2], divide the whole frame in $L \times W$ "mosaic" of *macroblocks* (MBs) each made of $N \times N$ pixels. Then each MB of the current frame is compared with a set of macroblocks of the past stored picture, obtaining the most correlated block, with the one of the current frame. Knowing the coordinates of both these MBs, the MV, is evaluated. A suitable metric is the *Sum of Absolute Differences* (SAD) function that reaches a minimum value

---

[1] The C code was developed at the University of British Columbia, [1].

when the correlation function has its maximum, [2], [5]. The SAD function requires, therefore, $N \times N$ operations to evaluate each macroblock, exhibiting a $O(N^2)$ computational complexity. Reducing this function's evaluation implies fewer power–consumption to perform ME. Depending on the search strategy employed, two separated classes can be identified: *Exhaustive Search Class* and *Heuristic Search Class*.

The former class is basically formed by the well–known Full Search (FS) algorithm, [2], [3]. As it can be inferred from the name, in this search strategy each macroblock in the current frame is compared with all MBs of the past image. Thus the best correlated block is straightforwardly identified but a great energy effort is required. In fact, considering a frame divided into $W \times W$ macroblocks the search complexity is $O(W^2)$; each search step has complexity $O(N^2)$ therefore the FS algorithm can require $O(W^2N^2)$ operations.

To reduced the computational needs of the Full Search other sub–optimal search strategies have been purposed, all based on an assumption on the SAD function properties. Thus a reduced number of steps is needed to identify the best correlated frame. One of these sub–optimal algorithms was the so–called *Three–Step Search* (TSS) algorithm, [2]. In this approach, a constellation of 9 MBs, centred in the past frame, are evaluated identify the block that exhibit the minimum SAD value. This MB becomes the central macroblock of a new constellation where the relative distance among the blocks is halved. The search is stopped when the macroblocks are one–MB–distance one from the others. Considering a frame of $W \times W$ macroblocks where $W = 15$, to perform a complete search requires 25 steps therefore $O(25N^2)$ operation; thus, there is considerable reduction with respect to the Full Search. One of the major drawbacks of TSS is that it is not *centre–biased*, i.e in typical video sequences the main amount of motion is in the central area of the frame, [5]. Higher video quality can be achieved if the search process is performed first in this region. This can be done, with low computational complexity, resorting to the *Four–Step Search* (FSS) algorithm. This algorithm relays on the 9–MBs constellation used by TSS, but, in this approach, the macroblocks are 2–MB–distance long, one from the others, in all the search steps. This strategy is therefore centre–biased, and the algorithm stops when the minimum SAD value block is the one in the centre of the constellation. Then it is performed the so called *Shrinking* step, i.e. the SAD function is evaluated also in the 8 MBs around the central one and 1–MBs–away from it. The shrinking step can be performed anytime during the search process so an analytical evaluation of the computational complexity is not possible. From simulations' results, otherwise, it has been found that FSS has a mean complexity of $O(19N^2)$ operations, [5]. Another algorithm which is an improvement of TSS, is the so–called *Simple and Efficient Search* (SES) algorithm, [4]. Based on the aforementioned assumption on SAD function, it is possible to identify the direction and the verse of the motion vectors without analysing all of the 9 macroblocks of the TSS. The SES algorithm starts to evaluate the SAD function in the central macroblock and in other two blocks that are in two orthogonal directions. The basic idea is to divided the frame into four quadrants; depending on the value assumed by the SAD functions in these three MBs one of these quadrants is selected. This allow to evaluate only other few blocks of the 9–MBs constellation used by TSS. This approach allows a further computational reduction, approximately $O(12N^2)$ operations.

3   The Proposed Architecture

To obtain proper video–quality with reasonable power consumption we designed a Motion Estimation IP based on sub–optimal search strategies. Furthermore, it has to achieve different energy profiles that can be implemented on FPGAs. In this section we present the chosen algorithm and we validate its estimation properties. Heuristic search is needed in order to avoid energy waste due to a full search. Moreover, in video–surveillance applications an high video quality is not required since the main task is to properly detect abnormal situations. This means that video quality, and also the stream frame–rate, can be adapted depending on the environment conditions.

To obtain good video quality with low computational efforts, we have chosen the so–called *Majority Voting Algorithm* (MVA), [6]. This algorithm relays on the aforementioned FSS and TSS ones choosing, depending on a particular merit criterion, which of them has to be used. This is done because near macroblocks tend to exhibit similar motion, so chosen the proper search strategy for a MB is better performed by knowing the motion occurred by its neighbour. This approach can further reduce the overall search without compromising the quality. In fact, FSS, that is centre–based, can properly detect the motion in the centre of the frame, while TSS works well in all the other cases. The choice between TSS and FSS is based on the motion vector values of some of the MBs near the one under elaboration and called "Predictors". These values are compared with a *threshold* that varies depending on the frame format (CIF, QCIF or others).

To achieve further computational reduction, we chose to implement the SES algorithm rather than the original TSS one; in fact SES can be considered like an improvement of the TSS scheme. Moreover, other reductions can be obtained selecting, when it is possible, a sub-sampling strategies called *Pixel Decimation*, [7]. In this approach, the SAD function is evaluated only for $N^2/4$ of the $N \times N$ MB pixels. This means some video quality losses, but great improvements in the power savings. Because of these losses, this strategy can be selected depending on the environmental conditions.

To validate our choices, we wrote a "software model" in C and we compared the estimated motion vectors of our approach with the ones obtained with the Full Search algorithm. In fact, FS has better estimation properties and can be used as *benchmark*. We analysed the differences in term of displacements between the MVs' values obtained by FS and our MVA implementation. The statistical frequencies of these displacements, representing the estimation errors of our model are depicted in figures 2(a) and 2(b). These values have been obtained comparing the estimations of several frames from some well known

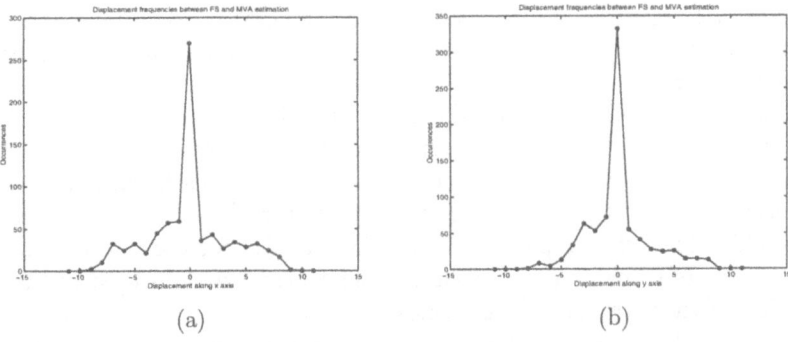

(a)   (b)

Fig. 2. Displacement between FS and MVA MV. (a) Along x-axis, (b) Along y-axis.

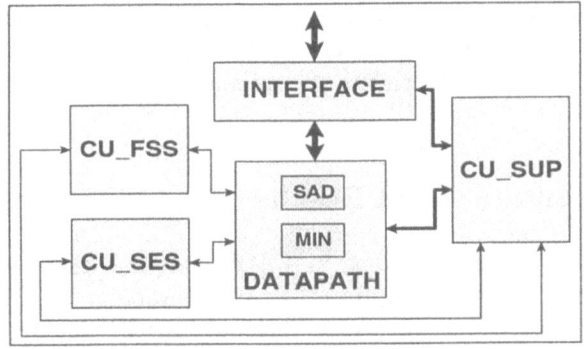

Fig. 3. Block–scheme of the proposed IP

test sequences like *Foreman* and *Coast–guard*. These sequences are all in QCIF format (176 × 144 pixels) with 16 × 16 pixels MBs and without Pixel Decimation.

Analysing these results, we can see that our MVA commits few errors with respect to the Full Search, in fact the statistical distributions, that are similar the well–known Laplace's Distribution, have zero mean and little variance. This has given us good confidence on our algorithm estimation properties; thus it is a good choice for an hardware implementation.

The block–scheme of the proposed IP is depicted in figure 3. As it can be seen this IP is made of several building blocks. The first one is the *Interface* block. As it was observed previously, this architecture is intended to be able to work with different profiles; this means that it has to accept different frame formats in particular CIF (352 × 288 pixels) and QCIF (176 × 144 pixels) sequences. So, an external CPU can select the proper format sending to our IP the frame "dimension", in terms of pixel. Moreover the MBs' format, like 16 × 16 or 8 × 8 pixel, can be chosen during the elaboration time. Also Pixel Decimation can be programmed by the control microprocessor (μP). This one has to provide to our IP the corresponding thresholds, to be used in the MVA algorithm, for the chosen frame format. All these features make our architecture able to be reconfigured "on–line" during the operation time. Moreover the interface block is used to access the external frame buffer in which the current and the past frame are stored.

The *CU_SUP* block, controls the information interchange with the CPU when this one programs the IP. Moreover, it selects the proper MB, of the current frame, on which evaluating the motion estimation. This block is also devoted to select, based on the knowledge of the predictor block MV values and of the threshold values, the proper algorithm, FSS or SES, for the estimation. This is done alternately enabling and disabling the control units devoted to the operation flow of these two search strategies.

In particular, the *CU_SES* block is the control unit that implements the Simple and Efficient Search algorithm. On the other hand, the *CU_FSS* block controls the FSS algorithm's operation flow. In the *Data-path* block, otherwise, all the required analytical elaborations are performed. In particular, the SAD is evaluated for the chosen macroblocks of the search constellations by the *SAD block*. Then, the MB that exhibits the minimum SAD value is chosen by the *MIN* block.

4   Implementation and Results

All of these blocks were described in VHDL code and then synthesised on a Xilinx XCV300E FPGA. To improve the reconfigurability, the code was written by means of `generic` parameters. In this way, our IP can be adapted to the particular operative conditions just varying these parameters and re–synthesising the code. This provides the so–called "off–line" reconfigurability.

The FPGA *speed–grade* for the implementation was set to 8 and 352–pin Ball–Grid Array (BGA) package was chosen. The synthesis step was performed by means of Synplify_Pro tool from Simplicity. The synthesised structure was, then, Placed & Routed on the selected FPGA. This step was accomplished by mean of the WebPack tool provided by Xilinx. From these previous steps, we obtained some preliminary information on the architecture performances, addressed in table 2. The LUT (look–up table) occupation means the percentage of the the total FPGA look–up tables used by our design; while the REG occupation is related with the percentage of non–Input/Output registers used. From table 2 it can be seen that there is a reduction for the maximum frequency achievable by the IP, between the post synthesis and the post Place & Route structures. This reduction is mainly due to some internal *critical paths* within

Table 2. Area & frequency IP performances

	Synthesis	Place & Route
LUT occupation	55 %	55 %
REG occupation	16 %	16 %
Maximum Frequency	38 MHz	23 MHz

the obtained IP. Further studies have to be exploited in order to eliminate them and to achieve better frequency performances.

To be of practical use, our architecture has to achieve very small energy requirements. Thus, we need to have some estimations of the dissipated power in order to have some energy profiles. This task was accomplished using the XPower Analyzer provided by Xilinx. First of all, we evaluated a statistical power consumption value setting the *activity ratios* of all the involved signal to 12.5 %. To obtain a more realistic estimation, we produced a test-bench to validate the overall IP. By means of Modeltech tool from Mentor Graphics, we were able to extract all the actual activity ratios for the IP signal and to store them in a so–called VCD file. This file became an entry for XPower that could perform a more accurate power dissipation estimation. We performed this evaluations doing the motion estimation on frames in QCIF format without applying Pixel Decimation and operating at the maximum frequency. All these results are depicted in table 3. Total P_D represents the overall FPGA power consumption due both to static and dynamic contributions. As it can be seen from table 3, the static power overcame the dynamic one acting as the limiting factor for the use of commercial FPGA in mobile devices. In fact, Static Power is due to internal implementation of the FPGA, therefore is a characteristic of each device family and the designer has no control on it. The dynamic P_D, on the other hand, can be reduced by a proper choice of the architecture that has to be implemented.

Table 3. Power Consumption IP performances

	Statistical estimation	Actual estimation
Total P_D	637 mW	579 mW
Worst Case Static P_D	540 mW	540 mW
Dynamic P_D	97 mW	39 mW

To analyse the power–saving properties of our architecture, we compared its energy consumption with the ones achievable running our C-code "software model" over some *general–purpose* μPs. We also compared the elaboration time needed to perform motion estimation for a QCIF frame for all the considered processors. The results of the comparisons are presented in table 4. As it can be seen, our IP achieves the lowest dissipation and one of the lower elaboration time. StrongArm SA1100 obtains a quite low energy consumption, but needs a long computation time[2]. On the other hand, a reasonable elaboration time is achieved by Mobile Celeron processor, but it has an higher power dissipation, i.e. 23.8 W, and an higher cost that makes its use impractical in an most wireless

---

[2] This results were obtained with the JouleTrack tool, avilable at: http://www-mtl.mit.edu/jouletrack/JouleTrack/index.html

Table 4. Performances Comparisons

Selected processor	Elaboration time	Dissipated energy
Proposed IP	39,6 ms	0.023 J
StrongArm SA1100 206 MHz	1.1 s	0.39 J
Mobile Celeron 1,1 GHz	0,02 s	0.48 J
UltraSparc III	0,11 s	8.8 J
UltraSparc I	0,40 s	11.2 J

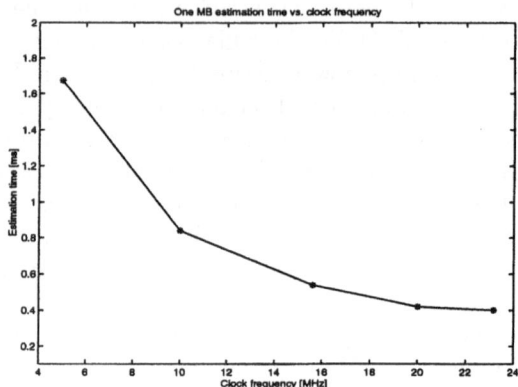

Fig. 4. One MB evaluation time vs. clock frequency

devices. The other processors have been addressed only for the purpose of having some evaluation metrics for the comparisons.

The proposed architecture is able to evaluate a 16 × 16 pixel macroblock (of the current frame) in less than 400 μs operating at maximum frequency; therefore, our IP performs the motion estimation on a 99 MBs' QCIF frames in less than 40 ms. Thus, frame–rates of 25 frames/s are achievable. ME for CIF frames, is exploited with rates of 6 frames/s, but high–rates are not the main target in video–surveillance operations. Furthermore we can applying Pixel–Decimation, at a cost of little losses in video–quality, obtaining higher rates even for CIF sequences[3]. Further power savings can be obtained reducing the operating frequency as was sketched in[8]; this comes at the cost of longer elaboration periods. We exploited this feature, evaluating our IP performances reducing the clock frequency; the obtained results are depicted in figure 4. These values can be used as evaluation metrics for selecting the proper power dissipation/elaboration time trade–off while designing a energy constrained application.

---

[3] The higher resolution of CIF format can compensate the losses due to Pixel Decimation.

5   Conclusions

In this paper we presented a novel Power–Scalable Motion Estimation IP suitable for nomadic use on wireless terminals. The proposed architecture can achieve low dynamic power, good video quality and reasonable frame–rates. Moreover it can operate on different video format and can be reconfigured both off–line and on–line. Further researches have to be accomplished in order to reduce the internal critical path, achieving better maximum frequencies performances. Moreover, new FPGA architectures must be exploited searching low–static–power devices suitable for an actual implementation of our IP on a demonstrator platform.

References

1. M. Gallant, A. Joch, G. Cote and B. Erol, H.263+ Library software codec simulator–Version 0.2, University of British Columbia–Signal Processing and Multimedia Group
2. C. Stiller and J. Konrand, Estimating motion in image sequences – A tutorial on modeling and computation of 2D motion, *IEEE Signal Processing Magazine* **16, 4** (1999) 70–91.
3. V. L. Do and K. Y. Yun, A low–power VLSI architecture for Full–Search block–matching motion estimation, *IEEE Transaction on Circuits and Systems for Video Technology* **8, 4** (1998) 393–398.
4. J. Lu and M. L. Liou, A Simple and Efficient Search algorithm for block–matching motion estimation, *IEEE Transaction on Circuits and Systems for Video Technology* **7, 7** (1997) 429–433.
5. L. Po and W. Ma, A novel Four–Step Search algorithm for fast block motion estimation,
   IEEE Transaction on Circuits and Systems for Video Technology **6, 3** (1996) 313–317.
6. D. S. Turaga and T. Chen, Estimation and mode decision for spatially correlated motion sequences, *IEEE Transaction on Circuits and Systems for Video Technology* **11, 10** (2001) 1098–1107.
7. A. Zaccarin and B. Liu, Fast algorithms for block motion estimation, *Proceedings of the 1992 IEEE International Conference on Acoustic, Speech and Signal Processing* **3** (1992) 449–452.
8. A. Sinha and A. Chandrakasan, Dynamic power management in Wireless Sensor Networks, *IEEE Design & Test of Computer* **18, 2** (2001).

A Novel Approach for Architectural Models Characterization. An Example through the Systolic Ring

P. Benoit[1], G. Sassatelli[1], L. Torres[1], M. Robert[1], G. Cambon[1], and D. Demigny[2]

[1] LIRMM, UMR UM2-CNRS C5506,
161 rue Ada, 34392 Montpellier Cedex 5, France
(33)(0)4-67-41-85-69
name@lirmm.fr
[2] ETIS, UMR-CNRS 8051,
6 avenue du Ponceau, 95014 Cergy Pontoise Cedex, France
(33)(0)1-30-73-66-10
demigny@ensea.fr

Abstract. In this article we present a model of coarse grained reconfigurable architecture, dedicated to accelerate data-flow oriented applications. The proliferation of new academic and industrial architectures implies a large variety of solutions for platform-based designers. Thus, efficient metrics to compare and qualify these architectures are more and more necessary. Several metrics, *Troughput Density*[3][12], *Remanence*[4] and *Operative Density* are then used to perform comparisons on different architectures. Architectures are often customisable and purpose several parameters. Therefore, it is crucial to characterize the architectural model according to these parameters. This paper proposes as a case study the Systolic Ring, and gives a set of metrics as functions of the architecture parameters. The methodology illustrated is generic and proved very efficient to highlight architectural properties such as the scalability.

1 Introduction

A System on Chip (SoC) allows the integration of a whole system on the same silicon die by combining different IP (Intellectual Property) cores. This technology provides significant benefits such as decreasing device's cost and power consumption and is therefore suitable for embedded communication products. Thanks to process geometries dropping, these systems are more and more complex bringing attractive functionalities such as multimedia abilities. However, design techniques must fit silicon technologies evolution in order to cope with the time to market constraints: it is obvious that it is more and more difficult to validate the whole functionality of an entire system. In order to reduce the designing times, SoC providers purpose customisable platform-based designs. Thus, the provider can adapt the platform to different customer needs (size of memory and buses, hardwired accelerators etc.) with a customisable architectural model. This methodology allows to significantly decrease the circuits' time to market.

In this context, customisable cores seem to be very attractive: by changing few parameters of the architectural model, different kind of optimisation could be

P.Y.K. Cheung et al. (Eds.): FPL 2003, LNCS 2778, pp. 722–732, 2003.
© Springer-Verlag Berlin Heidelberg 2003

imagined: for example, the processing power/area trade-off. In a SoC context, this ability allows to cope with different customer needs at the core level; this means that the core architectural model must have been fully characterised according to the parameters of the architecture. From the SoC provider point of view, the most competitive and attractive platform will be the one made of the most evolutionary, customisable and scalable soft cores.

The lack of flexibility of specific hardware has motivated the integration of reconfigurable cores. These architectures provide hardware-like acceleration style while retaining most of the software flexibility : a simple bit-stream defines the functionality. Among the last couple of years lots of new approaches appeared [1]. Real innovations like coarse grain reconfigurable fabrics [2] or dynamical reconfiguration have brought numerous improvements, solving several weaknesses of traditional FPGA architectures. Besides this point, several recurrent issues remain, and the proliferation of architectures lays on an additional problem for platform-based designers: choose the right IP core for a given set of specifications. Some works have already proposed useful tools to directly compare reconfigurable computing architectures, like the Dehon criterion[3][12]. Nevertheless, none addresses the characterisation of architectural models. This characterisation supposes to evaluate metrics as a function of the architecture parameters.

Efficiency and comparison of architecture is always a major problem to address. In this paper, we distinguish two orientations. The first one considers the application implementation efficiency. The second one is based on intrinsic characterization of the architectural model. The goal of this paper is to propose a method to characterize an architectural model: the approach proposed here is performed on a case study, but could be easily extended to any coarse grained reconfigurable architecture.

The next section presents a model of coarse-grained dynamically reconfigurable architecture, the Systolic Ring. This architecture already presented in [4][5], will be used as an example in our characterisation approach.. The third section presents metrics dedicated to compare different architectures. Finally, we propose a novel approach of using intrinsic metrics to characterise an architectural model.

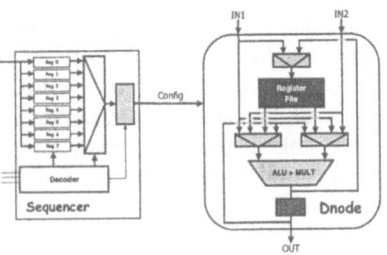

Fig. 1. The Dnode architecture

2 The Systolic Ring Architectural Model

The objective of this section is to briefly describe the Systolic Ring architecture, already presented in [4][5]. This architectural model will be used as an example in the fifth section to present the way to characterize a model of reconfigurable architecture.

The Systolic Ring architectural model features a highly optimised DSP-like coarse grain reconfigurable block, the Dnode (figure 1). The Dnode, a processing element, is the building block of the architectural model. This component is configured by a fixed-size microinstruction code. The configuration comes from a local sequencer. This sequencer can manage the Dnode configuration autonomously (up to eight instructions) or can be managed at a higher level by a global sequencer.

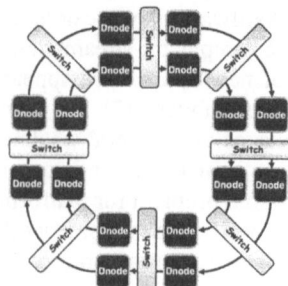 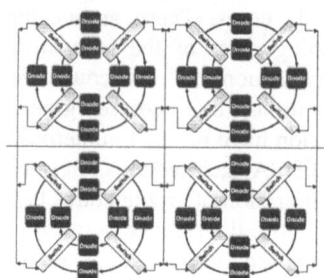

Fig. 2. . The operative layer   **Fig. 3.** Multi - Systolic Ring instances

In the Systolic Ring model, at a higher processing elements are brought together in clusters. A cluster is called "layer". A layer is made of a number "N" of Dnodes; N is the first parameter of the architectural model. In order to interconnect these layers, a configurable switch component is used to establish a full connectivity between two layers and to inject data from the memory. The number "C" of layers used in the model is the second parameter. The global interconnect structure of the processing elements is depicted on figure 2 : this example of the Systolic Ring model is made of 8 layers and 2 Dnodes per layer. The last layer is connected to the first layer with a switch (each switch is connected to the data memory via dedicated FIFO inputs). Thus, the topology is circular, allowing an efficient implementation of pipelined datapaths of different sizes through the successive layers. The Systolic Ring also provides reversed datapaths through a feedback network made of pipelined registers. A reversed datapath is associated to each switch. In addition, a Systolic Ring bus connecting all switches in the architecture and the global sequencer is also available. An other way to increase the processing power in a Systolic Ring architecture consists in using multiple instances of the circular datapaths. Figure 3 depicts this third interconnection level. The Systolic Ring bus is then used to interconnect the number S (S=4 in the example) of Systolic Ring instances. S is the third parameter of the Systolic Ring architectural model.

3 Metrics for Architecture Comparisons

The main goal of metrics consists of evaluating quantitatively and relatively the performances of the architecture to implement an application (application oriented metrics) and of quantifying the structure intrinsic characteristics (intrinsic oriented metrics).

3.1 Application Oriented Approach

The most used metric to compare computing machines to implement an algorithm is the *Throughput Density* [3], also presented in [12]. This metric is computed as follows: $D = 1/ (Area*Time)$. The area is expressed in relative unit, λ^2, where λ is the half size of the minimal width gate in a given technology (for more details, refer to [3]) and the time is expressed in absolute unit, *i.e.* seconds. This metric highlights the architecture which uses the most efficiently the silicon (smallest time and area) for closed technologies and a given application.

3.2 Intrinsic Approaches

3.2.1 Remanence

Remanence has been introduced in [6]. It characterises the architecture dynamism and is computed as follows: $R= (Na.Fe) / (Nc.Fc)$. *Na* is the number of the architecture building block (operators for coarse grain reconfigurable architectures and processors, CLBs for FPGA). *Nc* is the number of configurable building blocks per cycle. *Fe* is the building blocks operating frequency and *Fc* the configuration frequency. The architecture granularity and the technology do not influence the resulting *Remanence* due to the fact that *Remanence* is a ratio.

Remanence represents the number of cycles to (re)configure the whole architecture. For instance, a "very dynamic" architecture has a *Remanence* value equal to one (all the building blocks are configured in only one cycle) and on the opposite, "a highly static" architecture has a *Remanence* value very important.

3.2.2 Scalability and Operative Density

An architectural model is scalable when the number of building blocks increase implies a linear increasing of the architecture area. In this case, the area "A" can be expressed as a linear function of the number of the building blocks "N" as follows: $A(N)=k*N \; \forall N$. Consequently, the derived function of A(N) is a constant : its value is equal to k, the increasing factor. The value of k can be then computed for any value of N and determined as follows: $k=A(N)/N \; \forall \; N$. The inverse of k, 1/k, is an interesting characteristic of an architectural model, because it represents the *Operative Density* "Do". *Do* is computed as follows: $Do(N) = N/A(N) = 1/k$. The *Operative Density* allows to characterise the number of building blocks per area unit.

In order to avoid the influence of the technology, we propose to use a relative unit, λ^2, to express the area of the whole architecture as a function of the number of the building blocks. When the architecture is scalable, *Do* is a constant whatever the value of N. However, *Do* can also be used to illustrate a non-scalable architectural model, to characterise the evolution of the operative density as a function of the architectural model parameters. A complete illustration of this is given in the following section.

3.3 Comparisons

For these comparisons, we have selected three types of architecture dedicated to digital data processing: FGRA or Fine Grained Reconfigurable Architectures (Xilinx

FPGA Virtex E[7] and ARDOISE, a platform based on Atmel FPGA[8]), CGRA or Coarse Grained Reconfigurable Architecture (The Systolic Ring architecture, DART [9], MorphoSys [10]) and a VLIW DSP from Texas Instrument (Very Long Instruction Word Digital Signal Processor), the TMS320 C62 [11].

3.3.1 Throughput Density Comparisons

The results of throughput densities are summarized in table 1. The application considered is a classical algorithm of digital signal processing: the Discrete Cosine Transform (DCT). It is supposed that this algorithm is applied to a 64*64 image, on 8*8 blocks (two-dimensional DCT).

Table 1. Throughput densities comparisons

Name	Type	Techno.	Implement.	N	# cycles	f (MHz)	Time (μs)	Area (Mλ²)	D
Virtex E	FGRA	0.18μm	Signal Flow	N/A	4171	80	52.14	8000	$2.4.10^{-6}$
Systolic Ring	CGRA	0.18μm	Matrix	24	6826	200	34.13	500	$5.8.10^{-6}$
DART	CGRA	0.18μm	N/A	24	9536	130	73.35	300	$4.5.10^{-6}$
MorphoSys	CGRA	0.35μm	Signal Flow	128	1344	100	13.44	5500	$1.3.10^{-6}$
TMS320C62	DSP VLIW	0.15μm	Matrix	8	10240	300	34.13	12300	$2.4.10^{-6}$

Firstly, we can notice that different algorithm implementations are used. In order to compare the most objectively *Throughput Densities* (D), the same implementation would have to be mapped on each architecture. Indeed, it is well known that a signal flow implementation allow better performances. An other point to consider is the CMOS technology. Time is given in absolute units (seconds): thus, only close feature sizes can be objectively compared. Consequently, we can suppose that MorphoSys would reach greater performances with similar silicon technologies as the other architectures presented in the table. Then, we observe that the DSP and the FPGA have a very close D value (around $2.4 .10^{-6}$). The two other examples of CGRA reach a better throughput density due to the fact of the parallel and pipelined topology of the operators allowing a better implementation.

Even if the results proposed here are interesting and useful to qualify these architectures, they are also limited because targeting only one algorithm with several implementations and different feature sizes. In order to compare efficiently the five architectural models, it would be necessary to produce results as a function of the architectural model parameters on several applications and closed feature sizes.

3.3.2 Remanence Comparisons

For the five architectures presented in the table 2, configuration and operating frequencies are equal. The building block granularity can differ from an architecture to an other one. Thus, for the ARDOISE architecture based on an partially dynamically reconfigurable AT40K FPGA [8], the building block grain is the CLB: consequently, 2304 CLBs are available. Seven cycles are needed to configure one CLB, therefore 0.14 CLB is reconfigured in one cycle. Concerning the four others architectures, the grain is coarse: for the Systolic Ring, DART and MorphoSys, the building blocks are made of one multiplier and one ALU. We will consider for each one that the total number of operators is the product of the number of building blocks by 2.

These *Remanence* values show the dynamism of the architecture or the number of cycles needed to configure the whole structure. Hence, the DSP is the most dynamic architecture because it is able to configure the 8 operators in only one cycle. The CGRA architectures can also behave very dynamically by reconfiguring the whole structure in only one cycle thanks to SIMD control modes. They also propose more elevated *Remanence* allowing to configure more slowly the whole architecture but then can fix their configuration reducing consequently the consumed power by the configuration. The highest *Remanence* value is reached by the FPGA showing that if it is needed to reconfigure the architecture, 16457 cycles are necessary.

Table 2. Remanence comparisons

Name	Type	Na	Nc (min)	Nc (max)	F	R (min)	R (max)
ARDOISE	FGRA	2304	0.14	0.14	33	16457	16457
Systolic Ring	CGRA	24	0.5	24	200	1	48
DART	CGRA	24	1	24	130	1	24
MorphoSys	CGRA	128	1	128	100	1	128
TMS320C62	DSP VLIW	8	8	8	300	1	1

3.3.3 Operative Density Comparisons

As stated in the previous parts, multipliers and ALUs are separated to compute the total number of operators(N). Here, in order to compare the most objectively different architectures, we only consider coarse grained operators. By the way, for the ARDOISE architecture, the number of operators (26) is the number of synthesized operators on the whole FPGA architecture (2304 CLBs).

Table 3. Operative densities comparisons

Name	Type	N	Area(Mλ^2)	Do (N)
ARDOISE	FGRA	26	12300	$0.21 . 10^{-2}$
Systolic Ring (S=1, C=6, N=2)	CGRA	24	500	$4.80 . 10^{-2}$
Systolic Ring (S=4, C=8, N=2)	CGRA	128	2700	$4.74 . 10^{-2}$
DART	CGRA	24	300	$8.00 . 10^{-2}$
MorphoSys	CGRA	128	5500	$2.32 . 10^{-2}$
TMS320C62	DSP VLIW	8	12300	$0.06 . 10^{-2}$

The *Operative Densities* results show that CGRA use more efficiently the silicon area due to the fact of the hardwired operators. The area dedicated to control units in the DSP involves for this one the worst operative density value. The ARDOISE architecture operative density, because of the fine grained granularity, is worse than CGRAs. Concerning CGRA, DART has the greatest *Do* value, meaning that the number of operators per Mλ^2 is the most elevated, conferring a greater processing power per Mλ^2. The important interconnection resources available for the Systolic Ring and MorphoSys results in a worse *Do* value.

4 Characterize a Model: An Example through the Systolic Ring

The metrics previously described provide very useful results in order to compare different architectures. However, architectural models such as the Systolic Ring, proposing parameterization abilities, are not fully characterized. Therefore, it seems to be very important to propose a novel way of using previous metrics to reach this objective of characterization. The idea consists in computing the metrics as a function of the architectural parameters. By the way, a fully characterized core could be integrated in platform-based design and parameterized following the customer needs. For a platform-based design provider, this full characterization could be used as a tool to compare different architectural models in order to choose the best one according to his needs.

4.1 Remanence Analysis

The Systolic Ring architectural model is based on four management modes. Thus, we will provide four *Remanence* formalisations. Because the number of Systolic Ring instances is not an influencing factor, it will not appear in the *Remanence* results.

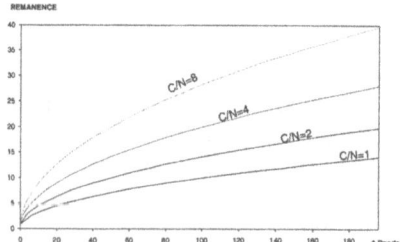 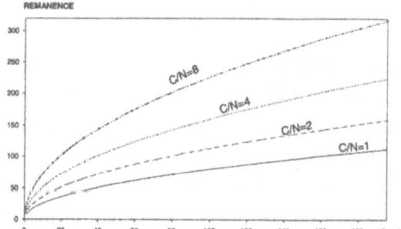

Fig. 4. Global Mode *Remanence*   **Fig. 5.** Maximum of Local Mode *Remanence*

In global mode, one layer of the architecture is configured each cycle, therefore N Dnodes are configured each cycle (Nc=N). The total number of Dnodes being C.N (Na=C.N), we easily deduce that the *Remanence R* is equal to C. In order to represent the evolution of the *Remanence* value as a function of the number Dnodes, we take four examples of customized Systolic Ring. This customization is based on the C/N ratio. This ratio represents the trade-off made on the pipeline degree versus spatial parallelism. To increase the pipeline degree, the C/N ratio must be elevated. At the opposite, to increase the parallelism potential, the C/N ratio must be low. The four ratios chosen are 1, 2, 4 and 8 : we can observe on the figure 4 that the higher the pipeline, the higher the *Remanence*. We also remark that this *Remanence* is increasing quickly for low C.N values (few Dnodes), and more slowly for higher C.N values. We can also highlight that the most elevated value of R is less than 40 for more than 180 Dnodes, meaning that a 180 Dnodes Systolic Ring can be totally reconfigured in less than 40 cycles.

In local mode, local configuration registers of the Dnode sequencer are pre-programmed to execute a maximum of 8 instructions corresponding to 8 Dnodes configuration. One layer being addressed during the configuration phase, the number of Dnodes configured each cycle is N divided by the number of registers preprogrammed. Consequently, supposing that all the Dnodes are preprogrammed with the same number of instructions (from two instructions to eight instructions), we easily deduce that the minimum of the *Remanence* will be 2C and the maximum, 8C.

These results are shown on the figure 5. As a consequence, we can observe that local mode involves higher *Remanence* values meaning that the whole architecture will need more cycles to be reconfigured. However, the maximum reached is only 320 cycles, and is ever lower than a dynamic FPGA *Remanence* (16457 for ARDOISE).

An other means to manage the Systolic Ring is the SIMD mode. In SIMD global mode, all the Dnodes receive the same configuration instructions. This is performed in only one cycle. Therefore, the *Remanence* value is one, meaning that the whole architecture is reconfigured in only one cycle. In SIMD local mode, each Dnode local sequencer register is pre-programmed with the same instruction. Therefore, the *Remanence* value in this mode is equal to the number of pre-programmed registers (from two to eight). With these low *Remanence* values SIMD mode is the most dynamic mode to manage the Systolic Ring.

4.2 Scalability Analysis

This analysis is based on the operative density evaluation. In order to compute this metric, we need to determine the area of the architectural model as a function of the number of processing elements. The total area is approximated by the sum of the 4 constituting elements of our model (PE : Processing Elements ; Config : Configuration memory):

$$A_{TOTAL} = S \cdot [A_{Config}(C,N) + A_{Control}(C,N) + A_{Interconnect}(C,N) + A_{PE}(C,N)] = f(S,C,N)$$

The total area A_{TOTAL} is a parameterised function depending on the three architecture parameters. The number of Dnode in the structure is simply the product of these three parameters: *# Dnodes= S*C*N*. The following results of operative density evaluations are expressed as a function of the total number of Dnodes. The systolic ring architectural model has been fully placed and routed in a C=4, N=2 version. The 0.35μ CMOS AMS technology has been used and the resulting area has allowed us to calibrate this area evaluation.

The figure 6 illustrates the operative density limits for one Systolic Ring instance. Three customized C/N ratios, from 1/8 (low pipeline degree versus spatial parallelism) to 4 (high pipeline degree versus spatial parallelism), were selected. Firstly, we can observe that *Do(N)* decreases quickly for the lowest C.N values and

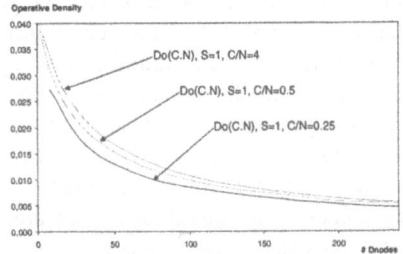

Fig. 6. Operative Density limits for S=1

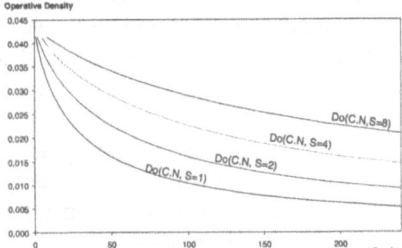

Fig. 7. Scaling Operative Density (C=N)

then decreases more slowly for the highest values. This figure shows that the Systolic Ring is not so scalable. The most influencing factor lays on the interconnection resources. Indeed, the full connectivity allowed between each architecture layer involves a quadratic increasing of the switches' area.

The third interconnection scheme, multi Systolic Ring instantiations (refer to the third paragraph) allows to improve the scalability. The figure 7 illustrates this other level of the Systolic Ring customisability. The operative densities results presented here were performed for C/N ratio equal to one (balanced pipeline degree and parallelism potential), and for one, two, four and eight Systolic Ring instances. The curves represented on the figure show that for a fixed number of Dnodes (proportional to a given processing power), choosing several Systolic Ring instances will significantly improve the scalabilty compared to one Systolic Ring instance. This means that the silicon area needed to integrate the same number of Dnodes is lower when using several Systolic Ring instances. Thus, this scheme allows to balance area versus interconnect resources.

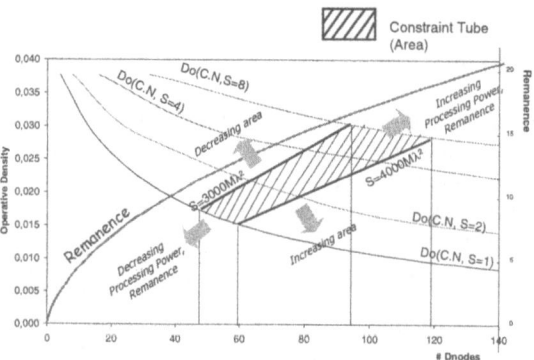

Fig. 8. Characterization and customisation illustration

4.3   Characterization and Customization of the Model

Let suppose that the silicon area available is between $3000M\lambda^2$ and $4000M\lambda^2$. This design space defines the constraint tube. As mentioned in the previous paragraph, the C/N ratio must be selected. This ratio will be fixed following the targeted applications. We choose here for example a C/N ratio equal to 4. Following the area evolution as a function of C and N for one instance of the Systolic Ring, corresponding curves are plotted for several instances of the Systolic Ring. The figure 8 represents the corresponding curves. The constraint tube is then plotted following the area constraints (the two lines were extrapolated from the Systolic Ring area formalisation). In order to show the architecture dynamism, we also plot the *Remanence* curve and add it to the graphic. The figure 8 shows how a Systolic Ring user (a platform-based designer for example) can tune his core with the architectural parameters. Indeed, in the constraint tube, many solutions are possible. For an area around $3000M\lambda^2$, the number of Dnodes can take the values from around 45 to 90 meaning that within the same silicon area, the processing power can be multiplied by a factor two. This characterizes an increased operative density. This is allowed by the way of using eight instances of the Systolic Ring instead of only one. However, this multi-instantiation implies a reduced connectivity between the Dnodes of the architecture and an increased *Remanence*. For an area around $4000M\lambda^2$, the processing power can also be doubled by the same means. Between these two areas, many solutions are possible and the Systolic Ring user can easily compare different possible trade-offs.

5 Conclusion

After having compared different architectures and show the limitations of classical comparing approaches, we have presented a general methodology for the characterization of architectures dedicated to digital signal processing. This methodology is based on evaluating metrics, *Remanence* and *Operative Density*, as functions of the architecture parameters. This methodology helps the designer to choose between several architectural trade-offs, as shown for the Systolic Ring example. This architecture presented in the second section, was used as a case study for both *Remanence* and scalability analysis. These considerations helped to determine architecture feature trade-offs and also contributed to establish the limitations of the architecture considering a set of application-relative constraints (parallelism type, area, processing power). Future works take place in a similar analysis on other crucial factors in a SoC design context such as the power consumption.

References

[1] W. H. Mangione-Smith *et a.* : "Seeking Solutions in Configurable Computing," IEEE Computer, pp. 38-43, December 1997.
[2] R. Hartenstein, H. Grünbacher: "The Roadmap to Reconfigurable Computing" Proc.FPL2000, Aug.27-30, 2000; LNCS,Springer-Verlag
[3] André DeHon : "Comparing Computing Machines", Configurable Computing: Technology and Applications, Proc. SPIE 3526, 2-3 November 1998.
[4] G. Sassatelli *et al* : "Highly Scalable Dynamically Reconfigurable Systolic Ring-Architecture for DSP applications", IEEE Design Automation and Test in Europe (DATE'02) , pp. 553-557, march 2002, Paris, France.
[5] G. Sassatelli : "Architectures reconfigurables dynamiquement pour les systèmes sur puce", Ph.D. thesis, Université Montpellier II, France, April 2002.
[6] D. Demigny *et al* : « La rémanence des architectures reconfigurables, un critère significatif des architectures », proc. of JFAAA, pp. 49-52, december 2002, Monastir, Tunisie.
[7] Xilinx, the Programmable Logic Data Book, 2000.
[8] D. Demigny *et al* : "Architecture à reconfiguration dynamique pour le traitement temps réel des images" Techniques et Science de l'Information Numéro Spécial Architectures Reconfigurables, 18(10) : pp. 1087-1112, december 1999.
[9] R. David *et al* : "DART : A Dynamically Reconfigurable Architecture dealing with Next Generation Telecommunications Constraints", 9th IEEE Reconfigurable Architecture Workshop RAW, April 2002.

[10] H. Singh *et al* : "MorphoSys: An Integrated Re-configurable Architecture"; Proc. of the NATO RTO Symposium on System Concepts and Integration, avril, 1998, Monterey, USA.

[11] "TMS320C62X Image/Video Processing library Programmer's Reference", march 2000, www.ti.com

[12] M. J. Wirthlin and B. L. Hutchings : "Improving Functional Density Using Run-Time Circuit Reconfiguration", IEEE Transactions On Very Large Scale Integration (VLSI) Systems, Vol. 6, pp. 247-256, june 1998.

A Generic Architecture
for Integrated Smart Transducers*

Martin Delvai, Ulrike Eisenmann, and Wilfried Elmenreich

University of Technology, Vienna, Austria,
Institut für Technische Informatik
delvai@vlsivie.tuwien.ac.at, eisenmann@thechilli.net,
wil@vmars.tuwien.ac.at

Abstract. A smart transducer network hosts various nodes with different functionality. Our approach offers the possibility to design different smart transducer nodes as a system-on-a-chip within the same platform. Key elements are a set of code compatible processor cores which can be equipped with several extension modules. Due to the fact that all processor cores are code compatible, programs developed for one node run on all other nodes without any modification. A well-defined interface between processor cores and extension modules ensures that all modules can be used with every processor type. The applicability of the proposed approach is shown by presenting our experiences with the implementation of a smart transducer featuring the processor core and a UART extension module on an FPGA.

1 Introduction

A smart transducer is a sensor or actuator element that is integrated with a processing unit and a communication interface [1]. The processing unit transforms the raw sensor signal to a digital representation, checks and calibrates the signal, and transmits the digital information to its users via a standardized communication interface. In case of an actuator, the smart transducer accepts standardized commands and transforms these into control signals. In many cases, the smart transducer is able to locally verify the control action and provide a feedback at the transducer interface.

A demonstration of a smart transducer network is presented in the DSoS project (*Dependable Systems of Systems* (IST-1999-11585)), where smart transducers are built with commercial embedded 8-bit microcontrollers featuring integrated standard UART communication interfaces.

The miniaturization process of sensors and actuators, however, offers completely new possibilities for system designers. More and more sensor elements are microelectronic mechanical systems (MEMS) that can be integrated on the same silicon die as the associated microcontroller and the communication controller. Such an *integrated smart transducer* promises a number of advantages

---

* This work was supported by the Hochschuljubiläumsstiftung der Stadt Wien via project MOSAIC (H-1147/2002).

P.Y.K. Cheung et al. (Eds.): FPL 2003, LNCS 2778, pp. 733–744, 2003.

over a discrete design: (i) non-linear and electrically weak sensor signals can be generated, conditioned, transformed into digital form, and calibrated without any noise pickup from long external signal transmission lines [2], (ii) power consumption of the smart transducer can be reduced significantly to a level where long-life battery-powered sensors or systems powered by solar cells are possible, (iii) the size of a smart transducer will decrease to a level where the size of the package containing sensor, processing unit, and communication interface is insignificant compared to the size of connectors and casing, (iv) production costs for large lot sizes are considerably lower than for solutions based on a discrete design.

At the same time, a developer faces several problems when designing a smart transducer. Selecting the right microcontroller for a smart transducer is difficult since a smart transducer network can host various different nodes, where the spectrum ranges from simple sensor nodes instrumenting a contact switch up to nodes with control functions or image processing capabilities. Thus, in the majority of cases it will not be possible to select a single processor that economically covers all expected applications [3]. Moreover, if the smart transducer should provide real-time functionalities, commercial processor architectures are often not appropriate, since they are optimized for average throughput. Worst-case analysis for such systems, which is essential for real-time applications, lead to unrealistically pessimistic estimations [4]. Finally, the smart transducers, although different in processing power and interaction capabilities, should support interoperability via a standard communication interface.

It is the objective of this paper to present an architecture for integrated smart transducers that meets the above described requirements on scalability and interoperability. The paper is structured as follows: Section 2 describes the architecture of a smart transducer node in general. Section 3 describes a modular approach to build integrated smart transducers using a microprocessor core with various extension modules. Section 4 provides information about implementations of the processor core hardware, an extension module, that protects processor resources against unauthorised accesses, and a communication extension module. Section 5 gives an overview of related work in the field, while the paper is concluded in Section 6.

2   Smart Transducer Architecture

The generic smart transducer architecture introduces a two-level design approach [5] that reduces the overall complexity of a smart transducer by separating transducer-specific implementation issues from interaction issues between different smart transducers.

Figure 1 depicts the two functionalities as protocol part and local application. The protocol part instruments the network interface. Each network interface of a smart transducer must provide some standard functionalities to transmit data in a temporally deterministic manner in a standard data format, provide means

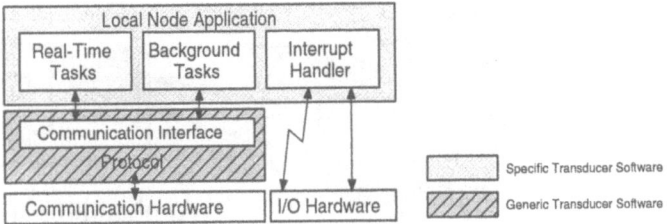

Fig. 1. Generic Smart Transducer Architecture

for fault tolerance, and enable a smooth integration into a transducer network and its application.

For example, the smart transducer interface (STI) [6] standard specifies some services for smart transducers. It comprises a time-triggered transport service within the distributed smart transducer subsystem and a well-defined interface to a CORBA (Common Object Request Broker Architecture) environment. The key feature of the STI is the concept of an Interface File System (IFS) that contains all relevant transducer data. This IFS allows different views of a system, namely a real-time service view, a diagnostic and management view, and a configuration and planning view. The interface concept encompasses a communication model for transparent time-triggered communication. A time-triggered sensor bus will perform a periodical time-triggered communication to copy data from the IFS to the fieldbus and write received data into the IFS. Thus, the IFS is the source and sink for all communication activities. Furthermore, the IFS acts as a temporal firewall [7] that decouples the local transducer application from the communication activities. Data is transmitted in a standard UART format.

The general architecture including the protocol part applies for every smart transducer, thus implying a generic implementation approach. The realization, however, faces the following problems: Due to the differing requirements on code size and processing power for the different local applications, it is not economic to select a single processor architecture. Moreover, each smart transducer type has its own I/O hardware configuration. An architecture should support, on the one hand, the configuration of many types of I/O extensions, and, on the other hand, provide a consistent interface to these modules in order to enable software reuse. Building smart transducers on commercial hardware is possible, but due to the various microcontroller types and I/O interfaces, programming and reuse is complex and error-prone.

3   Implementation Concept

As explained in the previous section a smart transducer network comprises various nodes with different requirements in terms of processing speed, size, interfaces and so on. Due to this fact different standard microcontrollers have to be used within the same network - the programmers have to know instruction

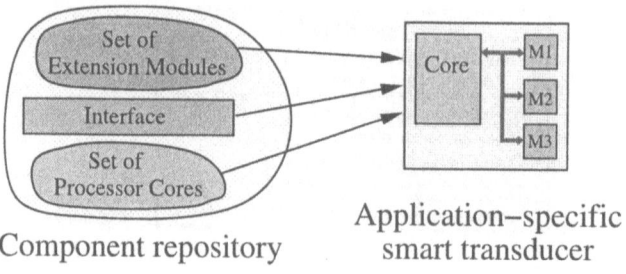

Component repository   Application–specific smart transducer

Fig. 2. Construction of an individual microcontroller

set and peculiarities of each microcontroller type in order to be able to implement, administrate, and maintain such a network. Furthermore, software pieces with the same functionality, e.g. the communication protocol, have to be implemented and tested for each microcontroller type separately. This proves to be especially difficult in the field of real time applications, where not only the correct functionality, but also the temporal behavior has to be considered. Usually, the employment of different hardware components requires a redesign of the software. To overcome these problems we designed a modular construction system, consisting of a set of processor cores and a set of different extension modules. The processor cores are all fully code compatible, but have different features in terms of computational power, required silicon area, memory-size and power consumption. Additionally, every activity within these processors is temporally predictable, which facilitates the design of real-time applications. [8]

All processor cores provide a generic interface to extension modules, which can be used to fit the microcontroller to different requirements. Figure 3 depicts the idea behind that approach.

It difficult to verify the correct functionality of a smart transducer due the fact, that not only the value domain, but also the temporal domain has to be considered. Especially the temporal aspect is critical. To evaluate the functional-

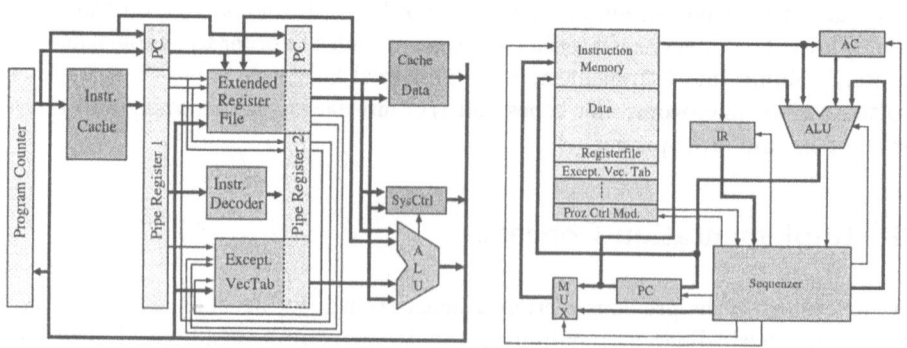

Fig. 3. Architecture of SPEAR and NEEDLE processor cores

ity of such a smart transducer in a network long simulation periods are necessary. Such a simulation would require an unacceptable amount of time. Thus, it is particularly important to get prototypes of smart transducers at early development stages. The use of FPGAs is ideal for this purpose. The smart transducer can be integrated into a real network and tested over longer periods of time. Therefore, we test prototypes of the assembled network nodes on FPGA test boards. Since all employed components of our architecture are described as VHDL models it is easily possible to implement a proved system in a silicon chip.

3.1   Processor Cores

Currently, two processor cores, named SPEAR and NEEDLE [3], have been implemented by our group. While the SPEAR core is designed to support moderate arithmetic performance, NEEDLE pays attention to a compact design. A further processor core LANCE supporting high computational power is planned. All processor cores are absolutely code compatible - this means that they not only use the same instruction set, but also the same exception handling, an identical memory architecture (from a logical point of view) and the same status and configuration registers. Due to this fact the programming code can be interchanged between different processor cores without any modification.

SPEAR Processor Core: SPEAR features a 16 bit processor that executes instructions over a three-stage pipeline. The processor core comprises a set of 32 registers. 26 registers are general purpose registers and 6 registers have a special function: three are coupled with dedicated instructions to efficiently implement stacks and the other three registers are used to save the return address in case of a subroutine call or an exception. Data and instruction memory are both 4 kB in size. It is also possible to add up to 128 kB external instruction memory and 127 kB external data memory. The upper 1 kB of the data memory is reserved for the memory mapped extension modules. In this way an extremely simple and efficient access to these modules is provided. The SPEAR processor supports 32 exceptions, of which 16 are hardware exceptions (= interrupts) and 16 can be activated by software (= trap). The exception vector table contains the pointers to the exception service routines.

NEEDLE Processor Core: NEEDLE is absolutely code compatible with the SPEAR processor, but the implementation is more compact. NEEDLE has no pipeline and executes instructions within several clock cycles. Due the fact that during each clock cycle at most two memory accesses are performed, it was possible to map the entire memory architecture (instruction memory, data memory, register file and exception vector table) into a single 4 kB dual-ported memory block. Moreover, the possibility to bind up to 64 different extension modules to the data memory area still exists. The block diagram of SPEAR and NEEDLE is shown in figure 3.

3.2 Extension Modules

To ensure scalability the processors are easily expandable by different extension modules, such as communication interfaces (sensors, actuators, PS/2, VGA, parallel port, USB, network interfaces, etc.) or application-specific extension modules such as a floating point unit or a protection unit. The extensions are mapped to the top address space of the data memory. For the processors the extension modules are only storage positions that can be accessed with simple load and store instructions. Therefore, from the processor's point of view it makes no difference, whether the extension is a simple sensor, actuator or a complex floating-point unit. Due to the well-defined generic interface [3] modules developed for one processor core can be used by all other cores. It is also possible to instantiate an extension module more than one time. For example, a smart transducer node can be equipped with several UART modules, having different mapping addresses in the data memory.

To illustrate the possibilities offered by the extension modules, two modules, the *protection control unit* and a customized *UART* will be described in the following. The first one extends the functionality of the processor core, the second one shows, how problems existing in commercial UART modules can be solved by a customized UART implementation.

Protection Control Module: The software that runs on a smart transducer node comprises a protocol code and an application part. As explained in section 2 the local application should not have direct access to the network bus. To guarantee that such an illegal access cannot happen, we have to ensure that only the protocol software can access the communication module. Such a protection mechanism is provided by the *protection control* extension module.

The protection control module allows assigning a *protection level* to individual data memory blocks, instruction memory blocks, registers and to extension modules. The module supports four protection levels: zero, low, high, and supervisor protection. As depicted in figure 4 the module comprises four look-up tables, which are are mapped into the data memory of the processor. Via these look-up tables a protection level to each resource can be assigned. The addresses of instruction memory, data memory, register file and extension modules are directly connected to the input of the respective look-up table. The module controls the access by comparing the output of each look-up table (i. e., the assigned protection level) with the current protection level defined in the processor status register (i. e., protection level of the current task) and by evaluating the control signals associated to each address. When a process tries to access a resource with the current protection level being lower than the protection level of the resource, an access-violation exception is generated.

UART Extension Module: In [9], we have examined the applicability of common UARTs for real-time communication and identified the following problems: the arithmetic error, the error due to frequency-drift and the send jitter problem.

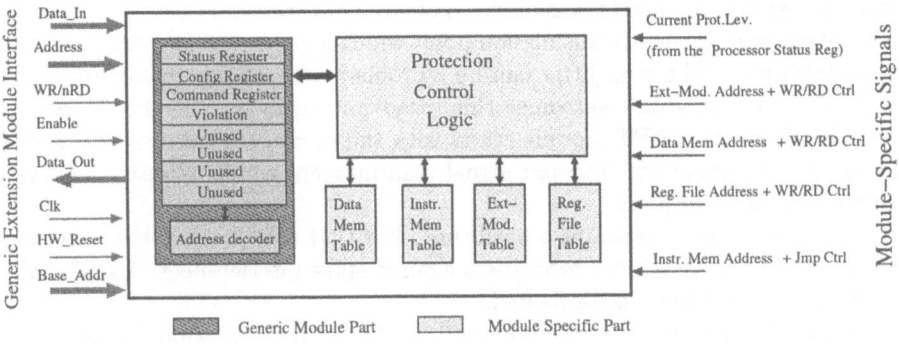

Fig. 4. Protection control module

Therefore, we have developed an alternative VHDL implementation of a UART unit that resolves the intrinsic UART problems and allows the implementation of more efficient protocols, respectively the employment of cheap on-chip oscillators with large drift rates.

The UART module is able to work with an RC-oscillator that provides a frequency of 1 MHz ±50% and a frequency-drift of ±10% per second. As the baud rate is influenced by this drift it has to be continuously adjusted. To deal with the baud rate drift rate problem the UART module offers a synchronization mode. In this mode the UART can be forced to synchronize periodically, like it is necessary in fieldbus protocols such as TTP/A [10] or LIN [11]. For synchronization the UART searches for a regular bit pattern as it is outlined in Figure 5. The synchronization pattern allows the UART to determine the used baud rate.

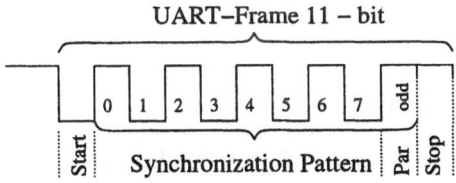

Fig. 5. Synchronization pattern

A substantial divergence from the desired transfer rate is caused by the arithmetic error in baud rate setting, i. e., the accuracy of a selected baud rate depends heavily on the selected clock source. Therefore, it is necessary to select special crystal frequencies (e. g. $1.8432 \cdot 10^6$) to be able to use standard communication rates [9]. Our UART module uses an extrapolation mechanism that minimizes this rounding errors.

As a further problem, many commercial UARTs exhibit intrinsic delays of the sending instant of a transmission because the UART baud rate generator periodically generates potential transmission points. Thus, the start of a transmission is delayed until the next transmission point which is perceived as a send jitter by an outside user. The send jitter can be a problem for real-time communication.

Our UART module overcomes this disadvantage by a different design approach. Thus, the UART module starts with the message transmission immediately after receiving the transmit signal transmission, which almost completely eliminates the send jitter.

The bus interface contains a hardware filter that preprocesses the input signal from the bus to achieve robustness against spike interferences. Additionally, oversampling has been implemented.

Figure 6 illustrates the block diagram of the UART extension module.

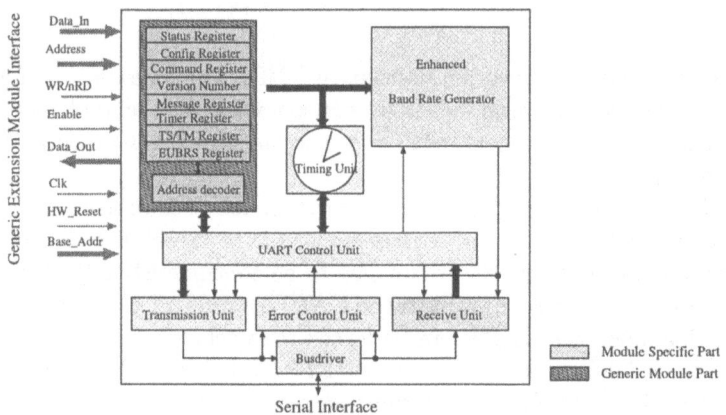

Fig. 6. Block diagram of the UART extension module

3.3   Development Tools

For software development an assembler is implemented. In the next step we will port the GNU C compiler to our processor architecture. Further a *system design tool* and a *debugger* are available to support assembling and testing of smart transducers. The system design tool provides software models of the processor cores and the extension modules. These components can be connected in order to form the desired smart transducer. The entire system could be simulated and verified. It is also possible to define own components in a high level language based on C++. In this way it is possible to evaluate a virtual hardware setup and to decide whether the real hardware should be build or not.

The debugger provides a high degree of flexibility, since it is easily adaptable to different purposes. For example it is possible to set breakpoints not only to a program counter value but also to sensor values or bus signals. Furthermore the debugger allows also the validation of real-time applications. Both, the system design tool and the debugger are controlled via the same graphical user interface. A detailed description of the system design tool and the debugger can be found in [12, 13].

4   Experiences

We have used an Altera APEX 20KE300EQC240-1X FPGA on a Digilab prototyping board for implementing the described architecture. The modular approach has been successfully tested by composing the two processor cores with several extension modules.

The processor core properties are compared in Table 1. While SPEAR computes one instruction per clock cycle at a maximum clock frequency of 40 MHz yielding 40 MIPS (million instructions per second), NEEDLE performs one instruction within either two or three clock cycles leading to an average performance of 10 MIPS at a peak clock frequency of 25 MHz. The size of the processors is given by Logic Elements (these are the smallest hardware units in the APEX FPGA) and the instruction and data memory. Due to the different memory size of the two processor cores, the necessary silicon area for the NEEDLE core will be expected to be one half of that of the SPEAR core. The values for the power consumptions are estimates given by the Synopsys design analyzer tool.

Table 1. Comparison of the SPEAR and the NEEDLE processor

	SPEAR	NEEDLE
Maximum Speed	40 MHz/40 MIPS	25 MHz/10 MIPS
Logic Elements (LE)	1318	1010
Instr-/Data Mem	4/4 KB	2/1,92 KB
Reg File	26 GPR - 6 SPR	26 GPR - 6 SPR
Interrupts/Traps	16/16	16/16
Instructions	80	80
Power consumption	113 mW	62 mW

The size of the extension modules heavily depends on their functionality. For example, the protection control module needs 279 LE, while the UART module acounts for 699 LE.

As communication interface, we have implemented software for a time-triggered UART communication that conforms to the OMG Smart Transducer Interface standard. The implementation is supported by the UART module that, in contrast to most commercial UART units, provides temporal predictability over the send instant of a message and a minimization of the arithmetic error in baud rate setting. The improved UART module greatly reduces the implementation effort for the smart transducer protocol software.

5 Related Work

The core-based system design approach is discussed by Gupta and Zorian in [14]. Their work outlines the advantages of using predesigned, preverified, silicon circuit modules as building blocks of large and complex applications on a single silicon die. ALTERA[1] offers a set of processor cores for embedded applications called Nios® following a similar idea as our proposed approach. The main difference for both approaches lies in its real-time capabilities. WCET analysis for code segments running on the Nios processor version 2.0 or higher is difficult, since the execution time of a single instruction depends on the preceding and the following instruction, the operands used, how recently operands were modified and other additional factors. In contrast our processor cores provide a fully temporal predictable behavior, which make them more suitable for to be used in real-time smart transducer networks.

Related work on sensor integration with VLSI circuitry can be found throughout the literature. Main research focuses on wide bandgap semiconductor materials [15], thin film sensors [16], and MEMS devices [17]. These approaches are related to the work presented in this paper as they outline possibilities for further on-chip I/O modules. The examined literature on smart sensor technologies, however, usually neglects the integration of a transducer with an appropriate communication network interface.

Besides the OMG STI, there are some other communication standards for smart transducers. Many fieldbus protocols, such as Controller Area Network, Local Area Network, Local Interconnect Network, Profibus, Foundation Fieldbus, WorldFip, and Interbus also provide possible solutions for the communication interface. The main differences between these approaches are in real-time features and implementation complexity. We have chosen the OMG STI since it provides hard real-time capabilities while having a low implementation complexity. In general, our architecture supports the easy adaption to a different communication interface by exchanging or adding communication I/O modules.

Another related standard is the IEEE 1451 smart transducer standard, which is not another fieldbus protocol but can be treated in the same way in our case, since IEEE 1451 specifies a 10-wire transducer-independent interface [18], which could be implemented as a communication extension module in our architecture.

6 Conclusion

This paper presented an architecture for the implementation of integrated smart transducers, i.e., a sensor or actuator element, a microcontroller and a network interface on a single silicon die.

The key element of our architecture is a set of fully code compatible microprocessor cores, with a well-specified interface to hardware extension modules that are synthesized on the same semiconductor chip. Due to the modular approach

---

[1] http://www.altera.com/products/devices/nios/nio-index.html

of the proposed architecture, the system is open to various external extension modules such as physical sensors/actuators or network communication modules.

Currently we have implemented two different microprocessor cores and several extension modules. The microprocessor cores are compatible at register and machine language level, so that software can be easily reused. The two extension modules are designed to support a smart transducer implementation. The improved UART module reduces the implementation effort for the smart transducer protocol software, while the protection unit protects resources like communication interfaces from unauthorized access.

References

1. W. Elmenreich and S. Pitzek. Smart transducers – principles, communications, and configuration. In *Proceedings of the 7th IEEE International Conference on Intelligent Engineering Systems (INES'03)*, volume 2, pages 510–515, Assuit – Luxor, Egypt, March 2003.
2. P. Dierauer and B. Woolever. Understanding smart devices. *Industrial Computing*, pages 47–50, 1998.
3. M. Delvai, U. Eisenmann, and W. Huber. Modular construction system for embedded real-time applications. In *Tagungsband of Austrochip 2002*, Vienna, Austria, 2002.
4. P. Puschner and A. Burns. A review of worst-case execution-time analysis. *Journal of Real-Time Systems*, 18(2/3):115–128, May 2000.
5. S. Poledna, H. Angelow, M. Glück, M. Pisecky, I. Smaili, G. Stöger, C. Tanzer, and G. Kroiss. TTP two level design approach: Tool support for composable fault-tolerant real-time systems. *SAE World Congress 2000, Detroit, MI, USA*, March 2000.
6. Object Management Group (OMG). *Smart Transducers Interface Final Adopted Specification*, August 2002. Available at http://www.omg.org as document ptc/2002-10-02.
7. H. Kopetz and R. Nossal. Temporal firewalls in large distributed real-time systems. *Proceedings of the 6th IEEE Workshop on Future Trends of Distributed Computing Systems (FTDCS '97)*, pages 310–315, 1997.
8. M. Delvai, W. Huber, P. Puschner, and A. Steininger. Processor support for temporal predictability - The SPEAR design example. In *Proceedings of the 15th Euromicro Conference on Real-Time Systems*, Porto, Portugal, July 2003.
9. W. Elmenreich and M. Delvai. Time-triggered communication with UARTs. In *Proceedings of the 4th IEEE International Workshop on Factory Communication Systems (WFCS'02)*, Västerås, Sweden, August 2002.
10. H. Kopetz et al. Specification of the TTP/A protocol. Technical report, Technische Universität Wien, Institut für Technische Informatik, Vienna, Austria, March 2000. Available at http://www.ttpforum.org.
11. Audi AG, BMW AG, DaimlerChrysler AG, Motorola Inc. Volcano Communication Technologies AB, Volkswagen AG, and Volvo Car Corporation. LIN specification and LIN press announcement. SAE World Congress Detroit, http://www.lin-subbus.org, 1999.
12. M. Delvai, M. Jankela, and A. Steininger. Towards virtual prototyping of embedded computer systems. In *Proceedings of the 7th World Multiconference on Systemics, Cybernetics and Informatics (SCI'03)*, Orlando, FL, USA, July 2003.

13. M. Delvai, C. El Salloum, and A. Steininger. A generic real-time debugger architecture. In *Proceedings of the 7th World Multiconference on Systemics, Cybernetics and Informatics (SCI'03)*, Orlando, FL, USA, July 2003.

14. R. K. Gupta and Y. Zorian. Introducing core-based system design. *IEEE Design & Test of Computers*, 14(4):15–25, Oct.-Dec. 1997.

15. B. W. Licznerski, K. Nitsch, and H. Teterycz. Polycrystalline wide bandgap materials in sensor technology. In *Abstract Book of the 3rd International Conference on Novel Applications of Wide Bandgap Layers*, pages 32–35, Poland, 2001.

16. S. M. Vaezi-Nejad. Advanced sensors scene in europe. In *Proceedings of the IEEE Instrumentation and Measurement Technology Conference*, volume 2, pages 926–931, Ottawa, Canada, May 1997.

17. G. K. Fedder. Structured design of integrated MEMS. In *Proceedings of the 12th IEEE International Conference on Micro Electro Mechanical Systems*, pages 1–8, January 1999.

18. Institute of Electrical and Electronics Engineers, Inc. *IEEE Std 1451.2-1997, Standard for a Smart Transducer Interface for Sensors and Actuators - Transducer to Micro-processor Communication Protocols and Transducer Electronic Data Sheet (TEDS) Formats*, September 1997.

Customisable Core-Based Architectures for Real-Time Motion Estimation on FPGAs

Nuno Roma, Tiago Dias, and Leonel Sousa

Instituto Superior Técnico / INESC-ID
Dept. of Electrical and Computer Engineering
Rua Alves Redol, 9-1000-029 Lisboa – PORTUGAL
{Nuno.Roma, tdias, las}@inesc-id.p
http://sips.inesc-id.pt

Abstract. This paper proposes new core-based architectures for motion estimation that are customisable for different coding parameters and hardware resources. These new cores are derived from an efficient and fully parameterisable 2-D single array systolic structure for full-search block-matching motion estimation and inherit its configurability properties in what concerns the macroblock dimension, the search area and parallelism level. The proposed architectures require significantly fewer hardware resources, by reducing the spatial and pixel resolutions rather than restricting the set of considered candidate motion vectors. Low-cost and low-power regular architectures suitable for field programmable logic implementation are obtained without compromising the quality of the coded video sequences. Experimental results show that despite the significant complexity level presented by motion estimation processors, it is still possible to implement fast and low-cost versions of the original core-based architecture using general purpose FPGA devices.

1 Introduction

Motion estimation is a fundamental operation in motion-compensated video coding [1], in order to efficiently exploit the temporal redundancy between successive frames. Among the several possible approaches, block-matching is the most used in practice. In this strategy, the current frame is divided into equal sized $N \times N$ pixel blocks that are displaced within a $(N + 2p - 1) \times (N + 2p - 1)$ search window defined in the previous frame. The motion vector is determined by looking for the best matched block of this search window. The Sum of the Absolute Differences (SAD) is the matching criteria that is usually used by most systems, due to its efficiency and simplicity.

Although the Full-Search Block-Matching (FSBM) method exhaustively considers all possible candidate blocks, which guarantees the optimal solution, it requires a lot of computational resources. In fact, FSBM motion estimation can consume up to 80% of the total computational power required by a video encoder. Hence, most of the fast block-matching motion estimation algorithms that have been proposed over the last years restrict the search space by a given search pattern, providing suboptimal solutions (e.g. [2, 3]). However most of these algorithms apply non-regular processing and require complex control schemes, which make their hardware implementation difficult and rather inefficient.

P.Y.K. Cheung et al. (Eds.): FPL 2003, LNCS 2778, pp. 745–754, 2003.
© Springer-Verlag Berlin Heidelberg 2003

Hence, the main objective of this paper is to propose efficient core-based VLSI array architectures based on FSBM to be implemented in Field Programmable Logic (FPL) devices. These architectures are based on a highly efficient core that was recently developed by the authors of this work, that combines both pipelining and parallel processing techniques to design powerful motion estimators based on multiple array architectures and using Application Specific Integrated Circuits (ASIC) [4]. In this paper, this original core architecture is used to derive simpler structures with reduced hardware requirements. The complexity of these architectures is reduced by decreasing the precision of the pixel values or/and the spatial resolutions in the current frame, while maintaining the original resolution in the search space. The pixel precision is configured by defining the number of bits used to represent the input data and by masking or truncating the corresponding Least Significant Bits (LSBs). On the other hand, spacial resolution is adjusted by sub-sampling the blocks of the current frame. By doing so, while the best candidate block in the previous frame is exhaustively searched, the SAD of each candidate block is computed by using only sparse pixels. Considering a typical setup with 16×16 pixels block, by applying $2 : 1$ or $4 : 1$ alternate sub-sampling schemes the number of considered pixels decreases by $1/4$ and $1/16$, respectively.

The efficiency of the proposed structures were evaluated by implementing these customisable core-based architectures in Field Programmable Gate Arrays (FPGA). It is shown that the amount of hardware required by these architectures when subsampling and truncation techniques are applied is considerable reduced, which allows the usage of a common framework for designing a wide range of motion estimation processors with different characteristics. Moreover, experimental results obtained with benchmark video sequences show that the application of those techniques does not introduce a significant degradation in the quality of the coded video sequences. These reduced hardware requirements architectures fit well in actual FPGAs, being a real alternative to those fast motion estimation techniques that require non-regular processing.

This paper is organised as follows. The original FSBM core architecture is presented in section 2. Section 3 proposes reduced hardware architectures to implement motion estimators on FPL devices. Experimental results obtained with FPGAs are presented in section 4 and section 5 concludes the paper.

2   FSBM Core-Based Architectures

Most motion estimation architectures that have been proposed in the last few years for hardware implementation are based on the optimum FSBM algorithm. The main reason for this is not only related to the better performance levels that it generally provides, but it is mainly due to the regularity properties that it also offers. In fact, not only does it conduct to much more efficient hardware structures, but it also provides the usage of significantly simpler control units, which is always a fundamental factor towards a real-time operation based on hardware structures.

Several FSBM structures have been proposed over the last few years (e.g.: [5, 6, 4]). Very recently, it was presented in [4] a new class of parameterisable hardware architectures that is characterised by offering minimum latency, maximum throughput and a full and efficient utilisation of the hardware resources. This last characteristic

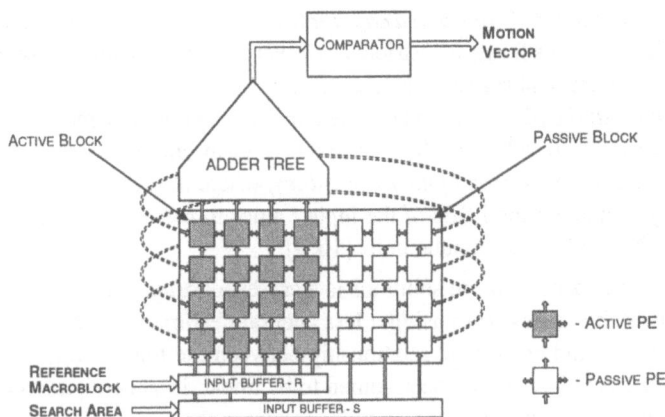

Fig. 1. Processor array proposed in [4] based on an innovative cylindrical structure and adopting the zig-zag processing scheme proposed by Vos [6] ($N = 4$, $p = 2$).

is a fundamental requisite in any FPL system, due to the limited amount of hardware resources. To achieve such performance levels, a peculiar and innovative processing scheme based on a cylindrical hardware structure and on the zig-zag processing sequence proposed by Vos [6] was adopted (see fig. 1). With such a scheme, not only was it possible to minimise the processing time, but it also provided the ability to prevent the usage of some hardware structures (the so called *passive processor elements (PEs)*) that do not carry useful information at some clock cycles.

Moreover, besides this set of performance and implementation characteristics, one other important feature of this class of processors is concerned with its scalable and configurable architecture, making it possible to easily adapt the processor configuration to fulfil the requisites of a given video coder. By adjusting a small set of implementation parameters, processing structures with distinct performance and hardware requirements are obtained, providing the ability to adjust the required hardware resources to the target implementation technology. As an example, while high performance processors requiring more resources are more suited for implementations using technologies such as ASIC or Sea-of-Gates, those low-cost processors that are meant to be implemented in FPL devices with limited hardware resources should use configurations requiring reduced amount of hardware.

3 Reduced Hardware Architectures

Despite the set of configurable properties offered by FSBM architectures and, in particular, by the class of processors proposed in [4], FPL devices often do not provide enough hardware resources to implement such processors. In those cases, the solution that is often adopted is the usage of processing structures based on sub-optimal motion estimation algorithms, that provide faster processing times and require reduced amounts of hardware. Three different categories of sub-optimal motion estimation algorithms have been proposed:

- *Reduction of the set of considered candidate motion vectors*, where the search procedure in the previous frame is restricted to a search pattern within the search window, by using hierarchical search strategies [2, 3];
- *Decimation at the pixel level*, where the considered similarity measure is computed by using only a subset of the $N \times N$ pixels of each reference macroblock [7–9];
- *Reduction of the precision of the pixel values*, where the similarity measure is computed by truncation the LSBs of the input values to reduce the hardware resources required by the used arithmetic units [10, 9].

The main drawback of these solutions is a corresponding increase of the prediction error that inevitable arises from using a less accurate estimation. This tradeoff usually leads to a difficult and non-trivial relationship between the final picture quality and the prediction accuracy that can not be assumed to be linear. In general, a larger prediction error will lead to higher bit rates, which will conduct to the usage of greater quantisation step sizes to compensate this increase, thus affecting the quality of the decoded images.

Up until now only a few VLSI architectures have been proposed to implement fast motion estimation algorithms, by restricting the search positions according to a given search pattern [9]. In general, they imply the usage of non-regular processing structures and require higher control overheads, which difficults their design using efficient systolic structures. Consequently, they have been extensively used in software applications, where such restrictions do not usually apply so strictly.

The set of architectures that are proposed in this paper try to combine the advantages offered by the regular and efficient FSBM structures proposed in [4] with the several strategies to reduce the amount of hardware resources that are offered by sub-optimal motion estimation algorithms. By doing so, it will be possible to implement processors based on this new class of fast motion estimation processors in any FPL device, even in FPGAs with limited resources. To achieve such objective, the original FSBM architecture will be adapted to apply two of the three decimation categories referred above: the decimation at the pixel level and the reduction of the precision of the pixel values.

3.1 Decimation at the Pixel Level

By applying the decimation at the pixel level, the image data of the current frame is subsampled in the orthogonal directions, by considering alternate pixels in each direction. In fact, this scheme corresponds to using a lower resolution version of the reference frame in the search procedure, that is carried out within the previous full-resolution frame. As an example, in a 2 : 1 sub-sampling scheme just one in each pair of consecutive pixels in each direction is considered, giving rise to decimated images with $1/4$ of the area of the original image. In the general case, the SAD similarity measure for a configuration using a 2^S : 1 sub-sampling in each direction is given by (considering N a power of 2):

$$SAD(l, c) = \sum_{i=0}^{\frac{N}{2^S}-1} \sum_{j=0}^{\frac{N}{2^S}-1} \left| x_t(i.2^S, j.2^S) - x_{t-1}(l + i.2^S, c + j.2^S) \right| \qquad (1)$$

The FSBM circuit proposed in [4] can be easily adapted to carry out this type of sub-sampling. In fact, considering that the computation of the SAD similarity measure

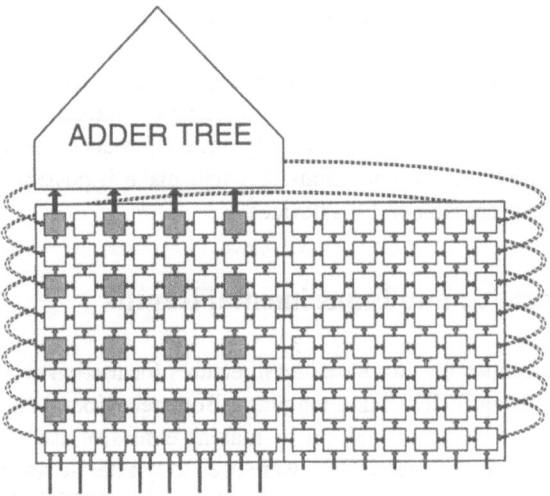

Fig. 2. Modified processing array to carry out a $2 : 1$ decimation function in the computation of the SAD similarity measure using the architecture proposed in [4] ($N = 8, p = 4$).

is performed in the active block of the processor, the decimation can be implemented by replacing the corresponding set of active PEs by passive PEs. By doing so, only the pixels with coordinates $(i.2^S, j.2^S)$ will be considered. In fig. 2 it is presented the block diagram of the modified processing array.

Since the amount of hardware resources required by passive PEs is considerable lower than that required by the active PEs, significant amounts of hardware can be saved. Moreover, by adopting the sub-sampling pattern presented in fig. 2, extra amounts of resources can be saved, since the number of inputs of the adder tree block is also reduced by the same sub-sampling factor (S).

3.2 Reduction of the Precision of the Pixel Values

As it was previously referred, one other strategy to decrease the amount of hardware required by FSBM processors is to reduce the bit resolution of the pixel values considered in the computation of the SAD similarity function by truncating the LSBs. By adopting this strategy alone ($S = 0$) or in conjunction with the previously described sub-sampling method ($0 < S < \log_2 N$), the SAD similarity measure is given by:

$$SAD(l,c) = \sum_{i=0}^{\frac{N}{2^S}-1} \sum_{j=0}^{\frac{N}{2^S}-1} \left| \left\lfloor \frac{x_t(i.2^S, j.2^S)}{2^T} \right\rfloor - \left\lfloor \frac{x_{t-1}(l+i.2^S, c+j.2^S)}{2^T} \right\rfloor \right| \quad (2)$$

$$\equiv \sum_{i=0}^{\frac{N}{2^S}-1} \sum_{j=0}^{\frac{N}{2^S}-1} \left| x_t(i.2^S, j.2^S)_{7:T} - x_{t-1}(l+i.2^S, c+j.2^S)_{7:T} \right| \quad (3)$$

where T is the number of truncated bits and $x_t(i,j)_{7:T}$ are the $(8-T)$ most significant bits of a pixel value of the tth frame.

The adaptation of the original FSBM architecture proposed in [4] to apply this bit truncation scheme is straightforward. In fact, it is only necessary to reduce the operands width of the several arithmetic units implemented in the active PEs, in the adder tree block and in the comparator circuit. Such modification potentially increases the maximum frequency of the pipeline and will significantly reduce the amount of required hardware, thus providing the conditions that will make it possible to implement the motion estimation processors in FPL devices.

4   Implementation and Experimental Results

The proposed customisable core-based architectures for motion estimation were completely described using IEEE-VHDL language. Both behavioural and structural parameterisable descriptions were carried out by making extensive use of "generic" type configuration inputs. These descriptions were used to synthesise several different setups of the proposed core in a general purposed VIRTEX XCV3200E-7 FPGA using the Xilinx Synthesis Tool from ISE 5.2.1. The considered set of configurations assumed each macroblock composed by 16×16 pixels ($N = 16$) and a maximum displacement in each direction of the search area of $p = 16$ pixels. These configurations were thoroughly tested using sub-sampling factors (S) varying between 0 and 2 and a number of truncated bits (T) of 0, 2 and 4. The experimental results of these implementations are presented in tables 1 and 3.

The proposed core-based architectures can provide significant savings of the required hardware resources. From the set of configurations presented in table 1, one can observe that reduction factors of about 75% can be obtained by using a sub-sampling factor $S = 2$ and by truncating the 4 LSBs. However, this relation should not be assumed to be linear. By considering only the pixel level decimation mechanism ($T = 0$), it can be shown that a reduction of about 38% is obtained by using a 2 : 1 sub-sampling factor ($S = 1$), while a 4 : 1 decimation will provide a reduction of about 51%. The same observation can be done by considering only the reduction of the precision of the pixel values ($S = 0$). While using 6 representation bits ($T = 2$) a reduction of the number of CLB slices of about 31% is obtained (a reduction of about 23% of the number of Look-up Tables (LUTs)), if only 4 representation bits are considered ($T = 4$) it provides a reduction of about 51% of the CLB slices (a reduction of about 47% of the number of LUTs). In table 2 it is presented the set of FPGA devices that should be used by each configuration, in order to maximize the efficiency of the hardware resources used by each processor.

In table 3 it is presented the variation of the maximum operating frequency of the considered configurations with the number of truncated bits (T). Contrary to what could be expected, the reduction of the operands width of the several arithmetic units does not significantly influence the processors performance. This fact can be explained if one takes into account the synthesis mechanism that is used by this family of FPGAs to synthesise and map the logic circuits using built in fast carry logic and LUTs.

To assess and evaluate the efficiency of the synthesised processors in an implementation based on FPL devices, they were embedded as motion estimation co-processors in the H.263 video codec provided by Telenor R&D [11], by transferring the estimated mo-

Table 1. Percentage of the CLB slices and LUTs that are required to implement each configuration of the proposed core-based architecture for fast motion estimation in a VIRTEX XCV3200E-7 FPGA ($N = 16; p = 16$).

	T					
	0		2		4	
S	CLB Slices	LUTs	CLB Slices	LUTs	CLB Slices	LUTs
0	90.7%	30.7%	62.6%	23.5%	43.8%	16.3%
1	56.2%	19.0%	38.2%	14.6%	26.6%	10.7%
2	44.7%	14.8%	30.1%	11.4%	21.0%	7.8%

Table 2. Alternative FPGA devices to implement each of the considered configurations ($N = 16; p = 16$).

	T		
S	0	2	4
0	XCV3200	XCV2600	XCV1600
1	XCV2000	XCV1600	XCV1000
2	XCV1600	XCV1000	XCV600

Table 3. Variation of the maximum operating frequency with the number of truncated bits (T) ($N = 16; p = 16$).

T		
0	2	4
76.1 MHz	77.8 MHz	79.8 MHz

tion vectors to the video coder. Peak signal-to-noise ratio (PSNR) and bit-rate measures were used to evaluate the performance of each architecture. These results were also compared with those obtained with a sub-optimal 4-steps logarithmic search algorithm [1], implemented in software.

The first 300 frames of several QCIF benchmark video sequences with different spatial detail and amount of movement were coded in interframe mode, by considering a GOP length of 30 frames and an intermediate quantisation step size of $\Delta = 30$ to keep the quantisation error as constant as possible.

In fig. 3 it is presented the obtained PSNR values for the *carphone* and *mobile* video sequences, characterised by the presence of high amount of movement and high spatial detail, respectively. Several different setups in what concerns the number of truncated bits (T) and the sub-sampling factor (S) for the decimation at the pixel level were considered. For comparison purposes, it was also presented the PSNR values obtained with the 4-steps logarithmic search algorithm. As it can be seen, the PSNR value for the INTER type frames of the *mobile* sequence is almost constant ($25 - 26$ dB). Hence, one can conclude that the degradation introduced by using the reduced hardware architectures is negligible: less than 0.15 dB when the 4 LSBs are truncated and the sub-sampling factor is 2 : 1. Contrasting, the PSNR behaviour for the *carphone* sequence varies significantly along the time. The main reason for this fact is the amount of movement that is also varying. Even so, the reduced hardware architectures proposed in this paper only introduce a slightly degradation in the quality of the coded frames, when compared with the performances of both the reference FSBM architecture ($S = T = 0$) and of the 4-steps logarithmic search algorithm. A maximum reduction of about 0.5 dB is observed

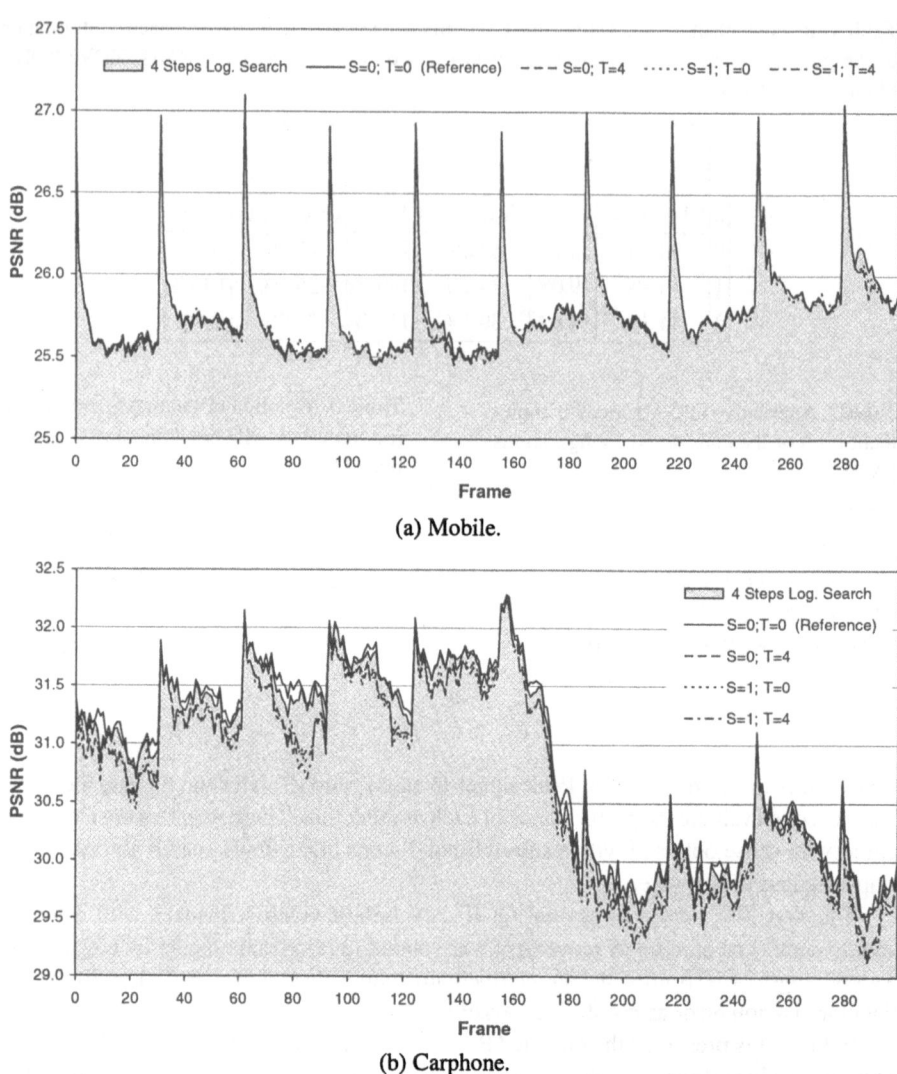

(a) Mobile.

(b) Carphone.

Fig. 3. Comparison between the PSNR measure obtained for three different setups of the proposed reduced hardware architecture, for the original reference configuration and for the 4-steps logarithmic search algorithm.

in a few frames when the PSNR value obtained with the original (reference) architecture is greater than 30 dB.

As it was previously referred, these video quality results were obtained with a constant quantisation step size. The observed small decrease of the PSNR can be explained by the slight increase of the quantisation error, as a consequence of the inherent increase of the prediction differences in the motion compensated block, obtained with these suboptimal matching processors. The obtained bit-rate required to store or transfer each

Table 4. Variation of the output bit rate to encode the considered video sequences by using different setups of the proposed core-base architecture and the 4-steps logarithmic search algorithm.

S	0	2	4
		T	
0	31.9 kbps	−0.8%	+3.4%
1	+1.7%	+3.0%	+3.8%
2	+3.9%	+3.8%	+3.8%

Four-step logarithmic search	+0.3%

(a) Miss America.

S	0	2	4
		T	
0	49.9 kbps	+0.0%	+2.7%
1	+4.5%	+4.1%	+8.0%
2	+10.7%	+8.4%	+8.6%

Four-step logarithmic search	+0.8%

(b) Silent.

S	0	2	4
		T	
0	271.4 kbps	+0.0%	+0.3%
1	+1.1%	+0.3%	+0.5%
2	+6.7%	+0.6%	+0.5%

Four-step logarithmic search	−0.1%

(c) Mobile.

S	0	2	4
		T	
0	79.3 kbps	−0.8%	+4.8%
1	+7.3%	+5.8%	+11.6%
2	+14.5%	+11.6%	+12.0%

Four-step logarithmic search	+1.7%

(d) Carphone.

coded video sequence is presented in table 4 for the two sequences considered above and for two more sequences (*Miss America* and *Silent*). The relative values presented in table 4 reveal the increment of the average bit-rate when the number of truncated bits and the sub-sampling factor increase. This increment reaches its maximum value of 12% for the *carphone* sequence and for a processor setup with only 4 representation bits and a pixel decimation of 2 : 1, due to the presence of a lot of movement. Nevertheless, the obtained values for the other three considered video sequences with less amount of movement are quite smaller and are similar to those obtained with the 4-steps logarithmic search algorithm.

Consequently, one can conclude that the configuration using a 2 : 1 decimation ($S = 1$) and using 4 representation bits ($T = 4$) presents the best compromise between hardware cost (a reduction of about 70%) and video quality. Moreover, this configuration could be implemented by using the lower-cost XCV1000 FPGA, which only has about 40% of the total number of system gates provided by the XCV3200 FPGA.

5   Conclusion

In this paper, new customisable core-based architectures are proposed to implement real-time motion estimation processors on FPGAs. The base core of these architectures is a new 2-D array structure for FSBM motion estimation that leads to an efficient usage of the hardware resources, which is a fundamental requisite in FPL systems. The proposed core-based architectures consist on a wide range of processing structures based on FSBM with different hardware requirements. The reduction of the amount of required hardware is achieved by applying decimation at the pixel and quantisation levels, but still searching all candidate blocks of a given search area.

The proposed core-based architectures were implemented on FPGAs devices from Xilinx and their performance were evaluated by including the motion estimaton processors on a complete video encoding system. Experimental results were obtained by sub-sampling the block of the current frame with $2:1$ and $4:1$ decimation factors and by truncating 2 or 4 LSBs of the representation. From the obtained results it can be concluded that a significant reduction of the hardware resources is achieved with these architectures. Moreover, the quality of the coded video is not compromised and the corresponding bit-rate is not significantly increased. One can also conclude from the results that the configuration using a $2:1$ decimation ($S = 1$) and using 4 representation bits ($T = 4$) presents the best compromise between hardware cost (a reduction of about 70%) and video quality.

Acknowledgements

This work has been supported by the POSI program and the *Portuguese Foundation for Science and for Technology* (FCT) under the research project *Configurable and Optimized Processing Structures for Motion Estimation* (COSME) POSI/CHS/40877/2001.

References

1. Bhaskaran, V., Konstantinides, K.: Image and Video Compression Standards: Algorithms and Architectures. 2nd edn. Kluwer Academic Publishers (1997)
2. Koga, T., Iinuma, K., Hirano, A., Iijima, Y., Ishiguro, T.: Motion-compensated interframe coding for video conferencing. In: Proc. Nat. Telecomm. Conference, New Orleans, LA (1981) G5.3.1–G5.3.5
3. Jain, J.R., Jain, A.K.: Displacement measurement and its application in interframe image coding. IEEE Transactions on Communications **COM-29** (1981) 1799–1808
4. Roma, N., Sousa, L.: Efficient and configurable full search block matching processors. IEEE Transactions on Circuits and Systems for Video Technology **12** (2002) 1160–1167
5. Ooi, Y.: Motion estimation system design. In Parhi, K.K., Nishitani, T., eds.: Digital Signal Processing for Multimedia Systems. Marcel Dekker, Inc (1999) 299–327
6. Vos, L., Stegherr, M.: Parameterizable VLSI architectures for the full-search block-matching algorithm. IEEE Transactions on Circuits and Systems **36** (1989) 1309–1316
7. Liu, B., Zaccarin, A.: New fast algorithms for the estimation of block matching vectors. IEEE Transactions on Circuits and Systems for Video Technology **3** (1993) 148–157
8. Ogura, E., Ikenaga, Y., Iida, Y., Hosoya, Y., Takashima, M., Yamash, K.: A cost effective motion estimation processor LSI using a simple and efficient algorithm. In: Proceedings of International Conference on Consumer Electronics - ICCE. (1995) 248–249
9. Lee, S., Kim, J.M., , Chae, S.I.: New motion estimation algorithm using adaptively-quantized low bit resolution image and its vlsi architecture for MPEG2 video coding. IEEE Transactions on Circuits and Systems for Video Technology **8** (1998) 734–744
10. He, Z.L., Chan, K.K., Tsui, C.Y., Liou, M.L.: Low power motion estimation design using adaptative pixel truncation. In: Proceedings of the 1997 international symposium on Low power electronics and design, Monterey - USA (1997) 167–171
11. Telenor Research Norway: TMN (H.263) encoder/decoder, version 2.0. (1996)

A High Speed Computation System
for 3D FCHC Lattice Gas Model with FPGA

Tomoyoshi Kobori and Tsutomu Maruyama

Institute of Engineering Mechanics and Systems, University of Tsukuba
1-1-1 Ten-ou-dai Tsukuba Ibaraki 305-8573 JAPAN
kobori@darwin.esys.tsukuba.ac.jp

Abstract. In this paper, we describe a new computation method for 3D
FCHC lattice gas model with FPGA. FCHC lattice gas model is a class
of 3D cellular automata and used for simulating fluid dynamics. Many
approaches with FPGAs for cellular automata have been researched to
date. However, practical three dimensional cellular automata such as an
FCHC lattice gas model could not be processed efficiently because they
required large size data for each cell and very complex update rules for
computing cells. We implemented the new method on an FPGA board
with one XC2V6000. The speed gain for FCHC lattice gas model with
$128 \times 128 \times 128$ lattice is about 200 times compared with Athlon processor
1800 MHz.

1 Introduction

Lattice gas model is a class of cellular automata, and used for simulating fluid
dynamics. Because of its simplicity and data independency, lattice gas model has
been widely studied, and many approaches for its high performance computing
with parallel systems and FPGAs have been proposed [4][5].

We have proposed a new computation method for two dimensional (2D)
cellular automata, and showed that 2D lattice gas model can be accelerated on
a small system with limited memory bandwidth[3]. This system for 2D cellular
automata consists of only one desktop computer and one PCI board with one
FPGA, and users are able to compute various kinds of 2D cellular automata
by reconfiguring circuits in the FPGA according to C-like programs written by
the users[2]. We have also proposed a computation method for three dimensional
(3D) cellular automata which is an extension of the 2D computation method and
runs efficiently on the same hardware system[1]. With this extended method, we
showed that we could drastically accelerate the computation of several simple
3D models. This computation method, however, could not efficiently process
practical 3D cellular automata such as a 3D FCHC lattice gas model, because
the number of data input/output width of the FPGA is still limited, and can
not read/write data of many cells at a time when the data width of the cells is
large, which is a common case in practical 3D models.

With the recent improvement of the size and functions of FPGAs (especially
the size of internal RAMs, the capability of dual-port RAMs and more number of

P.Y.K. Cheung et al. (Eds.): FPL 2003, LNCS 2778, pp. 755–765, 2003.

usable I/O pins), it becomes possible to compute practical 3D cellular automata models such as an FCHC lattice gas model with one latest FPGA.

In this paper, we propose an extended computation method for 3D FCHC lattice gas model based on our previous 3D computation method. This computation method is also based on the computation method for 2D lattice gas model we proposed before. In this computation method, status of L cells (decided by the number of I/O pins of the FPGA) are read from the external memory in every clock cycle, and two $x - y$ layers of the lattice and a small part of the next $x - y$ layer are stored in the FPGA. Then, status of L cells in the FPGA are updated for N times (thus, $N \times L$ cells are processed in parallel), and, the new status of the L cells are written back to another external memory. Using this computation method, cells with large data width on large size lattice can be efficiently processed in parallel and pipeline.

In this paper, we, first, introduce the FCHC lattice gas model and the computation method for the model. After that, we describe the implementation of the method and its results.

2 FCHC Lattice Gas Model

2.1 Overview

Many of 2D lattice gas models are based on FHP model with hexagonal cells. Extension to 3D lattice gas models from the FHP models, however, is not straightforward. In 3D lattice gas model, no regular cell with the required symmetries (for 3D lattice gas model) exists. Thus, the *Four dimensional Face centered Hyper-Cubic* (FCHC) cell is used to satisfy the required symmetries.

In FCHC model (Figure1 (a)), to project four dimensional (4D) grid onto 3D grid, the fourth dimension (w axis) is considered as the periodical boundary in three dimensions and takes only 2 values ((1, 0), (-1,1) and (-1, 0)). The Figure 1 (b) shows an example of coordinate system in FCHC lattice model. With the arrangement of cells shown in Figure 1 (b), the other three axes are not influenced by w axis. Each cell in FCHC has 25 particles (on 24 directions to other cells and one position to stay in the cell), and only one particle is allowed to travel in

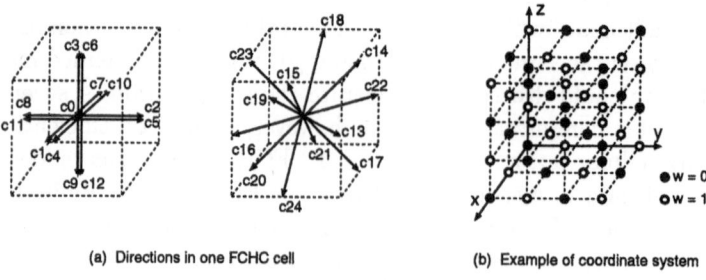

(a) Directions in one FCHC cell   (b) Example of coordinate system

Fig. 1. FCHC Model

each direction. The vectors indicated by each particle on 24 directions are shown as follows:

$C_1 = (1, 0, 0, 1),$ $C_2 = (0, 1, 0, 1),$ $C_3 = (0, 0, 1, 1),$ $C_4 = (1, 0, 0, -1),$
$C_5 = (0, 1, 0, -1),$ $C_6 = (0, 0, 1, -1),$ $C_7 = (-1, 0, 0, -1),$ $C_8 = (0, -1, 0, -1),$
$C_9 = (0, 0, -1, -1),$ $C_{10} = (-1, 0, 0, 1),$ $C_{11} = (0, -1, 0, 1),$ $C_{12} = (0, 0, -1, 1),$
$C_{13} = (1, 1, 0, 0),$ $C_{14} = (0, 1, 1, 0),$ $C_{15} = (1, 0, 1, 0),$ $C_{16} = (1, -1, 0, 0),$
$C_{17} = (0, 1, -1, 0),$ $C_{18} = (-1, 0, 1, 0),$ $C_{19} = (-1, -1, 0, 0),$ $C_{20} = (0, -1, -1, 0),$
$C_{21} = (-1, 0, -1, 0),$ $C_{22} = (-1, 1, 0, 0),$ $C_{23} = (0, -1, 1, 0),$ $C_{24} = (1, 0, -1, 0).$

In FCHC lattice gas model, each cell consists of 25 bits data (for 24 directions to neighbor cells and 1 bit for a staying particle), and 25 bits is necessary (24 bits for particles coming from neighbor cells and 1 bit for a staying particle) to compute new status of each cell. Owing to this large number of parameters to decide new status, the update rule becomes very complex, and even in softwares, tables (called collision table) are used to compute new status of each cell. The size of the table is, however, quite large (the number of entries is $2^{25} bit = 33554432$, and the size of table becomes 512 Mbits). Therefore, the reduction algorithm [7] is used to reduce the size of the table.

2.2   Reduction Algorithm

The reduction strategy is called *momentum normalization*. A basis of the strategy is to specify that the reduced table will contain only the states which have a *normalized momentum*. The coordinates of the momentum q are defined by

$$q_\alpha = \sum_{i=1}^{24} [S_i C_{i\alpha}] \quad (\alpha = 1, ..., 4)$$

where S_i is the boolean variable representing the presence or absence of a particle with velocity C_i as shown in section 2.1. The momentum is said to be normalized if the coordinates satisfy the following

$$q_1 \gg q_2 \gg q_3 \gg q_4 \gg |q_1 - q_2 - q_3| \tag{1}$$

This definition corresponds to a simple sequence of 11 step reduction procedures below. It is not difficult to show that a given input state S is reduced to a state with a normalized momentum by the following sequence. In each step, an *optional isometry* which is applied if the given inequality is satisfied.

1. If $q_1 < 0$, S_1 is applied to q.
2. If $q_2 < 0$, S_2 is applied to q.
3. If $q_3 < 0$, S_3 is applied to q.
4. If $q_4 < 0$, S_4 is applied to q.
5. If $q_1 < q_2$, P_{12} is applied to q.
6. If $q_3 < q_4$, P_{34} is applied to q.
7. If $q_1 < q_3$, P_{13} is applied to q.
8. If $q_2 < q_4$, P_{24} is applied to q.

9. If $q_2 < q_3$, P_{23} is applied to $\boldsymbol{q}$.
10. If $q_1 + q_4 < q_2 + q_3$, apply $\sum_1$.
11. If $q_1 < q_2 + q_3 + q_4$, apply $\sum_2$.

In this sequence,

S_α : symmetry with respect to the plane x_α
$P_{\alpha\beta}$: symmetry with respect to the plane $x_\alpha = x_\beta$
$\sum_1$: symmetry with respect to the plane $x_1 + x_4 = x_2 + x_3$
$\sum_2$: symmetry with respect to the plane $x_1 = x_2 + x_3 + x_4$

If the collision table is invariant under duality, the additional sequence can be executed before the momentum normalization. In the additional sequence, if the number of particles exceeds 12, the input state is replaced by its duality (the input state is bit-inverted). This sequence of procedures brings the number of status down to 316273.

3   Computation Method for FCHC Lattice Gas Model

In this section, we would like to introduce a new computation method for FCHC lattice gas model.

3.1   Basic Idea of the Computation Method

Basic idea of the computation method is based on the computation method for 2D model that we proposed in [3]. The computation method for 2D model consists of two phases: parallel processing of L cells (L is decided by the number of I/O pins and bit-width of one cell), followed by pipeline processing for applying update rule N times continuously (Figure 2). This method can be classified into three strategies by the following relations.

1. I/O width of the FPGA $>=$ lattice width
2. I/O width of the FPGA $<$ lattice width $<=$ FPGA internal memory size
3. I/O width of the FPGA $<$ FPGA internal memory size $<$ lattice width

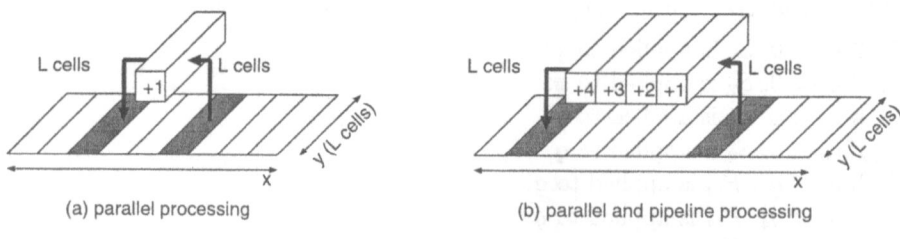

    (a) parallel processing   (b) parallel and pipeline processing

Fig. 2. Basic Strategy of the Computation Method for 2D Cellular Automata

The typical size of lattice in FCHC lattice gas model is $128 \times 128 \times 128$. Thus, the second strategy can be used for a basis of a new computation method for 3D FCHC model when we use the latest FPGAs (with large size internal RAMs). When the size of lattice is larger than FPGA internal memory size, the third strategy can be used though it causes some overhead as described in [3].

3.2   Outline of the Computation Method

Figure 3 shows the computation method for 3D FCHC model. In Figure 3, suppose that an FPGA can read data of L cells at once (along x direction; data on the lattice are arranged to read L cells along x direction in advance) and have three memory planes to store all data of three x-y layers of the lattice (one layer has $x \times y$ cells, typically 128×128 cells).

As shown in Figure 3 (1), first, the FPGA continues to read L cells in each clock cycle. When three layers ($3 \times x \times y$ cells) from the top of the lattice are stored in the three memory planes (Figure 3 (2)), the update rule is applied to L cells (black parts of Figure 3 (2)) in the second plane which stores the second layer. In subsequent clock cycles, new status of L cells are computed in every clock cycle by using the stored cells. During this computation, cells in the fourth layer of the lattice are read in continuously and stored in the first memory plane

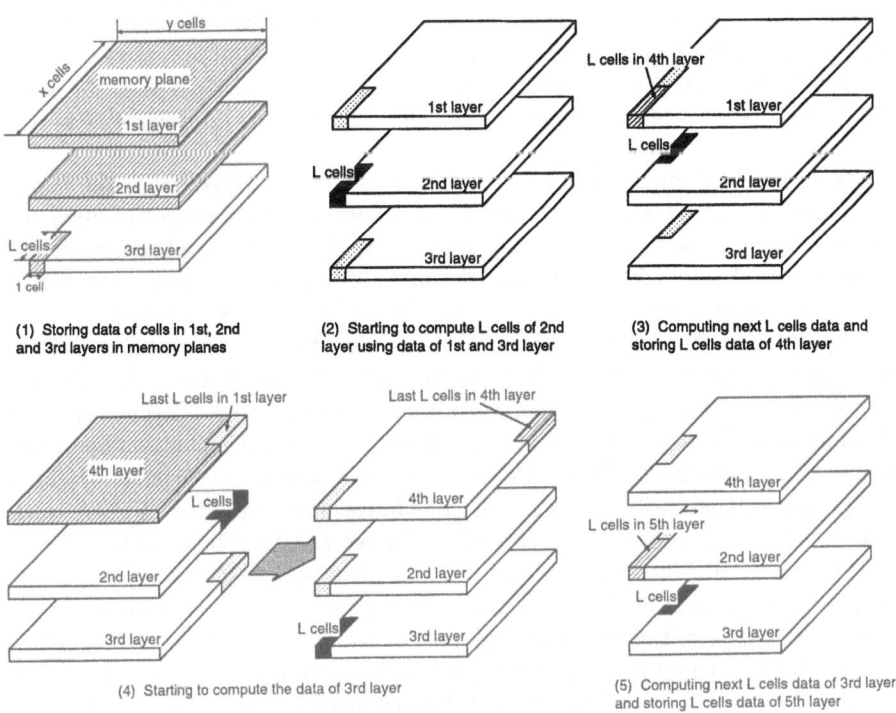

(1) Storing data of cells in 1st, 2nd and 3rd layers in memory planes

(2) Starting to compute L cells of 2nd layer using data of 1st and 3rd layer

(3) Computing next L cells data and storing L cells data of 4th layer

(4) Starting to compute the data of 3rd layer

(5) Computing next L cells data of 3rd layer and storing L cells data of 5th layer

Fig. 3. Outline of the Computation Method for 3D Cellular Automata

because the cells in the first layer are not used again (Figure 3 (3)). In this phase, dual-port access to the memory plane is very important because we can continue the computation which accesses the three memory planes while loading data in the next layer of the lattice into one of the three memory plane (if dual-port access is not supported, we need one more memory plane to process the computation and the data loading in parallel).

When status of all cells in the second layer (in the second memory plane) are updated, L cells in third layer (in the third memory plane) are started to be updated (Figure 3 (4)), and, L cells in the fifth layer of the lattice are beginning to be read into the second memory plane (Figure3 (5)).

In these procedures, status of L cells are updated every clock cycle. The generation of the output by the FPGA is the generation of the inputs +1. By pipelining the circuit and storing data of cells in $N \times 3$ layers, we can process $L \times N$ cells at a time (N is the number for applying the update rule continuously). In this case, the generation of the output by the FPGA becomes the generation of the inputs $+N$, and the effective parallelism is $L \times N$.

3.3   An Improvement to Reduce the Memory Size

Figure 4 shows an improvement of the method to reduce the memory size. In Figure 4, two memory planes to store two x-y layers and a small size memory to store a part of the next layer are used. In the small size memory, two lines along x axis ($x \times 2$ cells), and $L \times 2$ cells in the next line are stored. In our circuit which is described in Section 4, L is 2, and typical sizes of x and y are 128 respectively, therefore, memory size reduction by this improvement is very effective.

As shown in Figure 4 (1), first, an FPGA reads two layers from the top of the lattice in the two memory planes and two lines along x axis in the small size memory by reading L cells in every clock cycle. In practice, when the FPGA finishes to read in the first layer and the first line of the next layer, the FPGA becomes ready to compute new status of cells on boundaries. However, in order to simplify the figure and its explanation, we skip the computation of cells on boundaries. When $2 \times L$ cells of the third line are stored in the small size memory as shown in Figure 4 (2), status of L cells (black part in Figure 4 (2)) are updated (the update of the first $L-1$ cell is skipped again because it includes a boundary computation). The two lines and $2 \times L$ cells are necessary to give correct neighbor cells in the bottom layer to the L cells which are updated. Then, FPGA reads next L cells on the third line, and status of next L cells (black part in Figure 4 (3)) are updated. The first L cells in the first plane are not necessary any more, thus, the first L cells in the small size memory (gray part in Figure 4 (3)) are moved to the part of the first memory plane in which the first L cells in the first layer is stored. In Figure 4 (4), FPGA reads the next L cells on the third line, and status of next L cells (black part in Figure 4 (4)) are updated, and the next L cells in the small size memory (gray part in Figure 4 (4)) are moved to the first memory plane again.

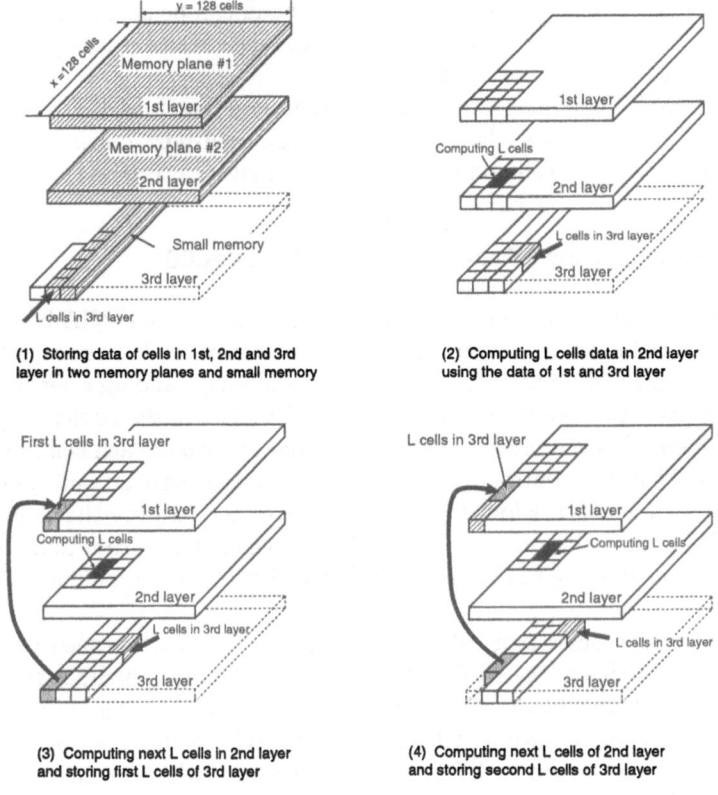

(1) Storing data of cells in 1st, 2nd and 3rd
layer in two memory planes and small memory

(2) Computing L cells data in 2nd layer
using the data of 1st and 3rd layer

(3) Computing next L cells in 2nd layer
and storing first L cells of 3rd layer

(4) Computing next L cells of 2nd layer
and storing second L cells of 3rd layer

Fig. 4. An Improved Method for 3D Cellular Automata

By repeating this cycle of reading L cells into small size memory, updating L cells on the first/second memory plane, and moving L cells from the small size of memory to the second/first memory plane, we can update status of all cells of the whole lattice.

4   Details of the Implementation of the Computation Method

4.1   ADM-XRC-II PCI Mezzanine Board

Before describing the detail of the implementation of the method, we would like to introduce features of the FPGA and the PCI board on which we implemented the method.

The PCI board (ADM-XRC-II by Alpha Data) has one Virtex-II XC2V6000, six 4MB SSRAM banks with 32 bit width (24MB and 196 bit width in total), and flexible front panel I/O. In our implementation, the front panel I/O is used for an interface with two additional SSRAM banks (8MB each). Accordingly, we can use eight SSRAM banks. Each memory bank can be accessed independently from the FPGA and the host computer. DMA data transfer is supported for the memory access by the host computer.

Virtex XC2V6000 has two kinds of memories; distributed RAMs consists of LUTs and block RAMs (dual-port). With the dual-port block RAMs, we can store enough data of cells required by the computation method.

4.2   The Overview of the Circuit on XC2V6000

The total I/O width of the FPGA is 32×8 bits because the FPGA board support eight memory banks. In our implementation, two memory banks are used to store input to the FPGA (data of the lattice of generation g), and another two memory banks are used to store the output of the FPGA (data of the lattice of generation $g + N$, where N is the pipeline depth). The data width of each cell in the lattice is 25 bits. Therefore, data of two cells are read in and written out in every clock cycle (L is 2). The other four memory banks are used to store the collision table. The width of the collision table is also 25 bits, which means that four cells on the lattice can be processed in parallel. Thus, the depth of the pipeline (N) becomes 2 (4/2).

In order to process four cells in parallel and pipeline, we need to store four layers of the lattice in the FPGA (two layers for the first pipeline stage and two layers for the next pipeline stage). Memory planes to store these four layers are implemented using dual-port Block RAMs of XC2V6000. The maximum layer size which can be stored in XC2V6000 is 128×128, which is the typical lattice size in the FCHC lattice gas model. In our computation method, we can take any size for z axis. Thus, we can process any type of $128 \times 128 \times k$ size lattice by exchanging x, y and z axis. When the size of other two axes is larger than 128, we can extend our computation method as described in Section 3.1. The two small size memories (one for the first pipeline stage and another for the second pipeline stage) are implemented using shift registers called *Shift Register Look up table* of XC2V6000.

4.3   Details of the Circuit

Figure 5 shows the block diagram of the circuit. In the circuit, there are four (2×2) data computing units to compute next status of cells and two data control units for storing data of cells. With one set of one data control unit and two data computing units, two cells of the same generation are processed in parallel. By duplicating the set, one more generation of the two cells is computed continuously on the FPGA (pipeline processing). Thus, four cells are processed at a time.

Figure 6 (a) shows the structure of the data control unit. Each data control unit consists of block RAMs to store data of two layers ($128 \times 128 \times 2$ cells), and shift registers (*Shift Register Look up table*) to store two read-in lines (2×128 cells) and 2×2 cells. 50 block RAMs and 450 LUTs are used in one data control unit.

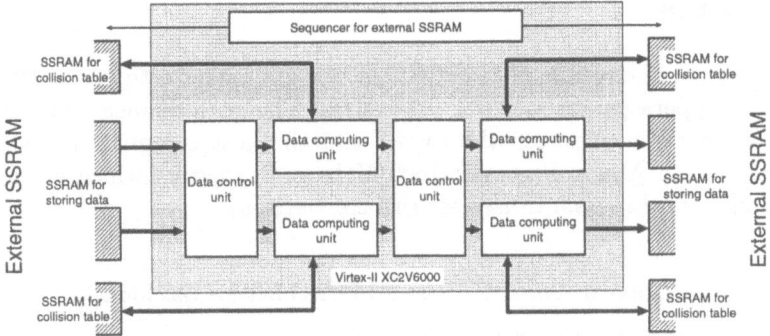

Fig. 5. The Block Diagram of the Circuit

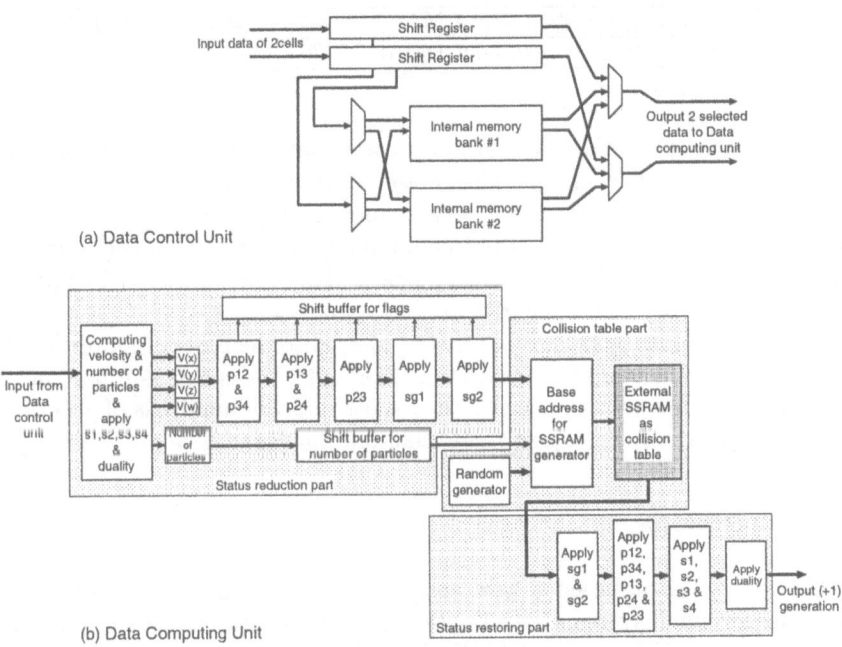

Fig. 6. The Block Diagrams of Sub-units

Figure 6 (b) shows the structure of the data computing unit. It consists of three parts; status reduction part, collision table part and status restoring part. In the status reduction part, input to the unit (25 bits) is reduced to 19 bits, and the collision table in the external SSRAM is accessed using the reduced status as its address. Then, each bit in the status which is read out from the collision table is exchanged by re-executing reduction steps in reverse order to restore correct new status in the status restoring part. In status reduction and restoring parts, some rules are applied in parallel. This data computing unit is completely pipelined, and can process new status of one cell in each clock cycle.

5   Results

Table 1 shows the results of 3D FCHC lattice gas model with 128 × 128 × 128 cells. The operation frequency is 52.7 MHz. The performance gain excluding data transfer time between the FPGA and the host computer is 2345 times compared with Athlon Processor 1800 MHz (a C program which is translated from a Fortran Program is used for this comparison).

Table 1. Speed Gain for 3D FCHC Lattice Gas Model

	Computation time (for 2 generation)	Speed gain
Software	$2 \times 23.8 \times 10^3$ (msec)	1
FCHC lattice gas system with DMA	242.3 (msec)	197
FCHC lattice gas system without DMA	20.3 (msec)	2345

This drastic speed gain comes from parallel and pipelined processing (4 cells are processed in parallel and pipeline, data in 24 neighbors are accessed and processed in parallel, the sequence of state reduction steps are executed in parallel and pipeline), and in 3D cellular automata, many data which spread in large address space are accessed for computing new status of one cell, which suppresses the effectiveness of cache memories in micro-processors.

When all outputs by the FPGA are transferred to the host computer (namely, status of all cells in every two generations are sent), the speed gain becomes 197 times. However, we need to transfer the results in every hundreds of generations in general. Thus, we can expect more than one thousand of speed gain.

6   Conclusions

In this paper, we proposed a new computation method of 3D FCHC lattice gas model for small systems with limited memory bandwidth. We implemented the method on a FPGA board (ADM-XRC-II with one Virtex-II XC2V6000), and the speed gain for a 3D FCHC lattice gas model with 128 × 128 × 128 lattice is about 200 times compared with Athlon Processor 1800 MHz. With this method, we could achieve a high performance computing of FCHC lattice gas model which is equivalent to large parallel/distributed systems.

The circuit which we implemented consists of data control unit and data computing unit, which are completely separated. Therefore, the method can be applied for other 3D cellular automata which requires large data width for each cell and very complex update rules. We are going to develop hardware libraries for those 3D cellular automata. With the libraries, user can process those types of 3D cellular automata efficiently and easily.

References

1. Tomoyoshi Kobori, Tsutomu Maruyama. "High Speed Computation of Three Dimensional Cellular Automata with FPGA" FPL2002, PP.1126-1130, 2002.
2. Tomoyoshi Kobori, Tsutomu Maruyama and Tsutomu Hoshino. "A Cellular Automata System with FPGA" Proc. FCCM '01, IEEE Computer Soc 2001.
3. Tomoyoshi Kobori, Tsutomu Maruyama and Tsutomu Hoshino. "High Speed Computation of Lattice Gas Automata with FPGA" FPL2000, PP.801-804, 2000.
4. Norman Margolus. "An FPGA architecture for DRAM-based systolic computations" Proc. FCCM '97, pp. 2-11, IEEE Computer Soc 1997.
5. C. Adler, B. M. Boghosian, E. G. Flekkoy, N. Margolus and D. H. Rothman, "Simulation Three-Dimensional Hydrodynamics on a Cellular-Automata Machine", Journal of Statistical Physics, 1995
6. U. Frisch, D. d'Humières, B. Hasslacher, P. Lallemand, and Y. Pomeau. "Lattice gas hydrodynamics in two and three dimensions" Complex Systems, 1 PP.649-707 1987.
7. M.Hènon, "Implementation of the FCHC Lattice Gas Model on the Connection Machine", J. Stat. Phys. Vol.68,353-377 1992

Implementation of ReCSiP:
A ReConfigurable Cell SImulation Platform

Yasunori Osana, Tomonori Fukushima, and Hideharu Amano

Dept. of Computer Science, Keio University
3-14-1 Hiyoshi, Kouhoku-ku, Yokohama #223-8522, JAPAN

Abstract. A reconfigurable accelerator for cell simulators called "ReCSiP" is proposed. It consists of both reconfigurable hardware and software platform. For high performance simulation, numerical solution of kinetic formulas, which require a large amount of computation, are processed on the reconfigurable hardware. It also provides programming interface for developing cell simulators. In this paper, Michaelis-Menten solver is designed and implemented on ReCSiP. The result of preliminary evaluation shows that ReCSiP is 8 times faster than Intel PentiumIII 1.13GHz when simple metabolic simulations are executed.

1 Introduction

Simulating cellular process is a big challenge in both of biology and computer science. Whole-cell simulators which are currently under development[1][2] are much more complicated than traditional computational sciences like fluid dynamics or molecular dynamics because it has various kinds of chemical reaction in a system[3].

When a cellular system is simulated by computer software, the system is modeled as a network of chemical reactions. Each reaction is described in the form of differential equation, and optimal numerical solution is selected and applied on each equation. Since a lot of equations must be solved at every time step, simulations take long computation time.

However, traditional parallel systems (e.g. PC/WS clusters, shared memory machine and so on) are not suitable for performance enhancement because of the frequent synchronization as the result of tight dependencies between chemical reactions in the next (or previous) time-steps.

In this paper, we propose a reconfigurable accelerator "ReCSiP", and Michaelis-Menten solvers are implemented on it. Chemical reactions are independent in itself, but across time-steps, they have dependencies to each other. This causes communication bottleneck on traditional parallel systems, but this bottleneck can be avoided by parallel processing of differential equations in a single reconfigurable platform.

ReCSiP consists of hardware platform called ReCSiP board and software executed in the host computer. Parallel processing of differential equations of cell simulation is executed on the board. Software layer provides the device driver and programming interface for easy use of the ReCSiP board.

P.Y.K. Cheung et al. (Eds.): FPL 2003, LNCS 2778, pp. 766–775, 2003.

2   Overview of ReCSiP

The purpose of ReCSiP is to accelerate metabolic simulation in private, desktop computing environment. Biologists can use high performance systems such as supercomputers or clusters in grid environment, but such resources are globally shared, and frequently busy and congested. Desktop super computing environment for personal use should be investigated as an alternative choice for biologists. ReCSiP is designed to achieve high throughput computation by the co-operation of the host CPU and reconfigurable platform consisting of FPGA.

As shown in Fig.1, ReCSiP consists of the hardware layer called ReCSiP board, and the software layer with the device driver and API. By using API codes, users can access ReCSiP board directly from their programs. Cell simulators and other applications can be easily accelerated with ReCSiP.

2.1   Structure of ReCSiP Board

ReCSiP board has an FPGA, a Xilinx Virtex-II (XC2V6000-4) as a core reconfigurable resource, and a QuickLogic QuickPCI (QL5064) for 64bit/66MHz PCI interface. 4 sets of 32bit×1Mwords (4Mbytes) synchronous SRAM and a

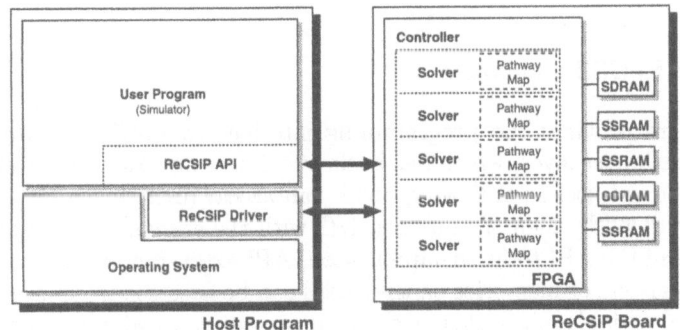

Fig. 1. Overview of ReCSiP

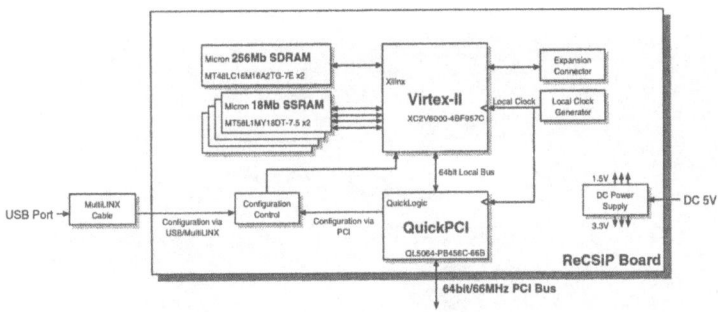

Fig. 2. Structure of ReCSiP Board

Table 1. Components on ReCSiP board

	Vendor	Series	Part No.
Core FPGA	Xilinx	Virtex-II	(XC2V6000-4BF957C)
PCI Interface	QuickLogic	QuickPCI	(QL5064-PB456C-66B)
Memory (SRAM)	Micron	SyncBurst SRAM	(MT58L1MY18DT-7.5)
Memory (DRAM)	Micron	Synchronous DRAM	(MT48LC16M16A2TG-7E)

set of 32bit× 16Mwords (64Mbytes) SDRAM are connected to Virtex-II. The
memory modules can be accessed from the FPGA simultaneously, and remove
the bottleneck of memory access from the FPGA. The specification of the board
is shown in Table 1.

PCI interface in QuickPCI manages both PIO access and DMA transfer at 64bit/66MHz. The local bus between QuickPCI and Virtex-II is also
64bit/66MHz, and the maximum frequency of SDRAM/SSRAM is 133MHz.
The clock frequency of memory modules are controlled by the clock manager on
Virtex-II.

For users who design the logic on FPGA, ReCSiP's standard modules are
available including local bus interface, SDRAM interface, SSRAM interface and
so on.

2.2   The ReCSiP Software

The software provides the programming interface to ReCSiP board. ReCSiP
driver is the lowest layer, which has basic interface using `ioctl()`. ReCSiP API
is an easy-to-use user level library, but it can access ReCSiP board without the
driver's overhead by memory mapped I/O with the system call, `mmap()`.

To extend the API to fit the application, API extension mechanism is implemented. Registers or memories on the board can be accessed from user programs
like as usual variables since this mechanism can automatically generate low level
interface code from address map file using `mmap()` or other system calls. By this
mechanism, users can build up their simulator on ReCSiP easily.

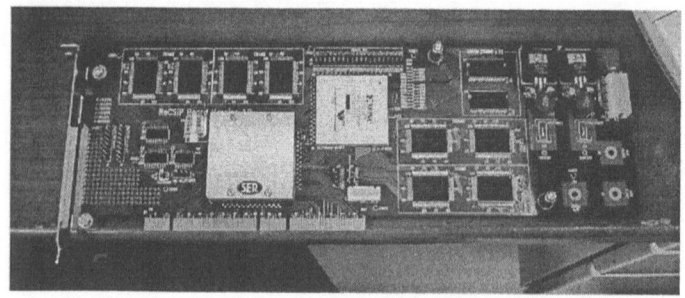

Fig. 3. The ReCSiP Board

2.3   Michaelis-Menten Solver on ReCSiP

As the first example of cell simulation with ReCSiP, a solver of a basic enzyme reaction kinetic model was designed and evaluated.

2.4   Michaelis-Menten Reaction Model

There are various kinds of chemical reactions in a cellular system, and approximate models are used in kinetic simulations. Michaelis-Menten model is the most widely used model of substrate-enzyme reaction is represented as (1).

In this scheme, $[E]$ is concentration of the enzyme and $[S]$ is concentration of the substrate. The enzyme-substrate complex $[ES]$ is formed by collision of enzyme and substrate. Then $[ES]$ quickly changes to $[E]$ and the product, $[P]$. This reaction is catalyzed.

$$[E] + [S] \underset{k_2}{\overset{k_1}{\rightleftharpoons}} [ES] \xrightarrow{k_3} [E] + [P] \tag{1}$$

Formation/degradation velocity of each substance in scheme (1) can be approximated as (2), (3), (4) and (5).

$$\frac{d[S]}{dt} = -k_1[S][E] + k_2[ES] \tag{2}$$

$$\frac{d[P]}{dt} = k_3[ES] \tag{3}$$

$$\frac{d[E]}{dt} = -k_1[S][E] + (k_2 + k_3)[ES] \tag{4}$$

$$\frac{d[ES]}{dt} = k_1[S][E] - (k_2 + k_3)[ES] \tag{5}$$

Therefore, it is possible to calculate concentration of each substance in time series when all initial concentrations and ks (kinetic constants) in the target system are given.

2.5   The First Prototype

The solver of Michaelis-Menten model using the 1st-order Euler's method was designed and implemented. The first prototype was designed to find problems in the design and implementation of metabolic simulation accelerator on the FPGA. It is only the specialized processing unit for Michaelis-Menten model, so it has no interface logic to PCI or any other systems.

Implementation. The first prototype of Michaelis-Menten solver has following components as shown in Fig.4:

- single precision, floating point adder (addsub)
- single precision, floating point multiplier (mult)
- exponent shifter for numerical integration (shift)

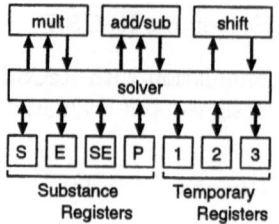

Fig. 4. Structure of the First Prototype

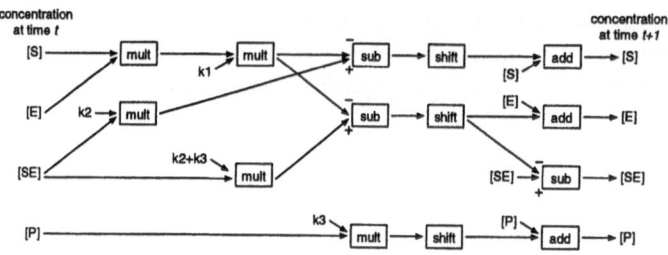

Fig. 5. Steps to Solve the Michaelis-Menten Model

– substance registers which hold concentration of each substances
– temporary registers which hold intermediate data

With these components, the kinetic equations ((2)~(5)) with the integration step can be solved as Fig.5. In the first step of this method, each derivative is calculated. Then, the exponent part of derivative is shifted to get d/dt^1. At last, the difference is added to the value at time t, and the integration completes.

The method shown in Fig.5 was implemented with the modules addsub, mult and shift. The pipeline completes the process of 4 equations in 32 clocks, and it can start a new process every 10 clocks. The schedule of calculation is shown in 6.

Evaluation. The prototype was designed with Verilog-HDL. Synthesis, place and route were done with Synopsys FPGA Compiler-II and Xilinx ISE 4.1. The maximum operating frequency was calculated from the result of place and route, then the throughput was calculated. This evaluation was done with the place and route for multiple units on an FPGA. However, the evaluated system doesn't have any host interface. So it's just for preliminary evaluation.

The result is shown in Table 2. By comparison to throughput of microprocessors in Table 3, 2 units of this prototype have the same throughput as 1.13GHz version of PentiumIII.

---

[1] This method is very fast and easy to implement numerical integration because it does not need a divider. In our second prototype, d/dt is calculated by multiplying dt to k_n before the simulation starts.

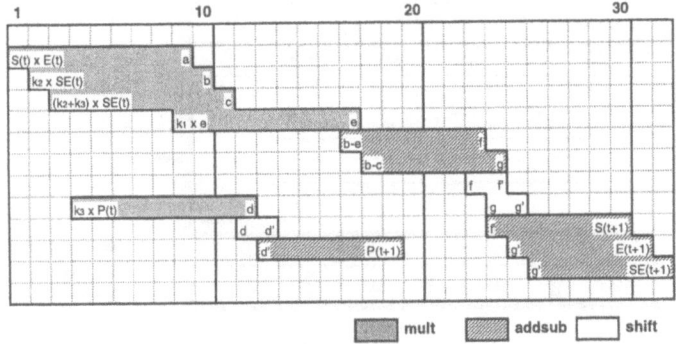

Fig. 6. Pipeline Schedule of the First Prototype

Table 2. The Size, Frequency and Throughput of the First Prototype

Units	Slices (Used %)		Frequency	Throughput
1	2,211	(6%)	62MHz	6.2
2	4,422	(13%)	64MHz	12.8
3	6,644	(19%)	58MHz	17.4
4	8,845	(25%)	56MHz	22.4
8	17,849	(52%)	54MHz	43.2

Table 3. Throughput of Microprocessors

Processor	Frequency (MHz)	OS/Compiler	Throughput (Mop/sec)
UltraSPARC II	300	Solaris8 / gcc-2.95.2	5.21
Pentium III	800	FreeBSD-4.2 / gcc-2.95.2	6.25
Pentium III	1133	FreeBSD-4.2 / gcc-2.95.2	10.15

The Communication Bottleneck. This prototype receives concentration of 4 substances from the host CPU, then sends them back at each time step. The size of this transaction is $4 \times 2 \times 32(\text{bit}) = 256(\text{bit})$. Here, the maximum bandwidth of 64bit/66MHz PCI bus is $64 \times 66 = 4.2(\text{Gbps})$, so $4.2(\text{Gbps}) \div 256(\text{bit/timestep}) = 16.5(\text{timestep/sec})$ is the maximum processing ability. However, as shown in Table 3, desktop CPU will soon reach to 16Mops/sec. Some breakthrough is needed to solve this communication bottleneck, and exploit the power of the FPGA.

2.6 The Second Prototype

Concept. To avoid the communication bottleneck, the second prototype (Fig.7) was designed and implemented. This prototype has a mechanism to program how the solver receives, processes, stores and sends back data. The program for the solver is stored in "Code RAM", while the data is stored in "Data RAM". Data

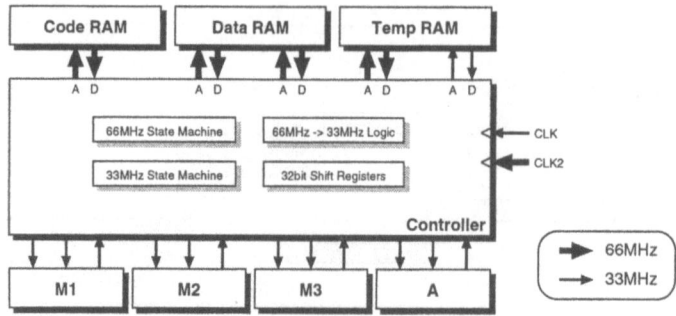

Fig. 7. The Structure of Second Prototype

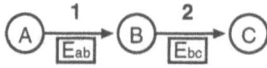

Fig. 8. Sample of Reaction Pathway

in the Data RAM can be used in the succeeding time step, and can be partially modified from the host. Host program does not need to get all data in every time step, since all data in simulation is stored in the Data RAM. By accessing to only essential data, transactions over PCI bus can be reduced drastically.

For example, to process the reaction pathway in Fig.8, data transfer between the solver and host is as follows:

- in the first prototype:
    - to process reaction 1, receive A, B, Eab and A.Eab at time t
    - as the result of reaction 1, send A, B, Eab and A.Eab at time $t + \Delta t$
    - to process reaction 2, receive B, C, Ebc and B.Ebc at time t
    - as the result of reaction 2, send B, C, Ebc and B.Ebc at time $t + \Delta t$
- in the second prototype:
    - receive the program
    - to process reaction 1 and 2, receive A, B, C, Eab, Ebc, A.Eab and B.Eab at time t *if necessary*
    - as the result of the reactions, A/dt, B/dt, C/dt, Eab/dt, Ebc/dt, $A.Eab/dt$ and $B.Eab/dt$ *if necessary*

In this way, transactions over PCI bus can be reduced by using the memory on the accelerator board.

Implementation. As shown in Fig.7, the second prototype of the solver has 3 multipliers (M1, M2 and M3) and an adder (A). The solver itself is fully pipelined, and controlled with a simple finite-state machine as shown in Fig.9. The address for Code RAM is simply incremented by every clock cycle, then Data RAM is accessed as Code RAM points to.

Table 4. Data RAM for the Sample Pathway

Address	Data	Address	Data	Address	Data	Address	Data
00h	A	04h	k_{11}	08h	B	0ch	k_{21}
01h	Eab	05h	k_{12}	09h	Ebc	0dh	k_{22}
02h	A.Eab	06h	k_{13}	0ah	B.Ebc	0eh	k_{23}
03h	B	07h	$k_{12}+k_{13}$	0bh	C	0fh	$k_{22}+k_{23}$

Table 5. Format of Code RAM

Address	Pointer	
$4n\times00$h	S	E
$4n\times01$h	k_2	ES
$4n\times02$h	k_2+k_3	k_1
$4n\times03$h	k_3	—

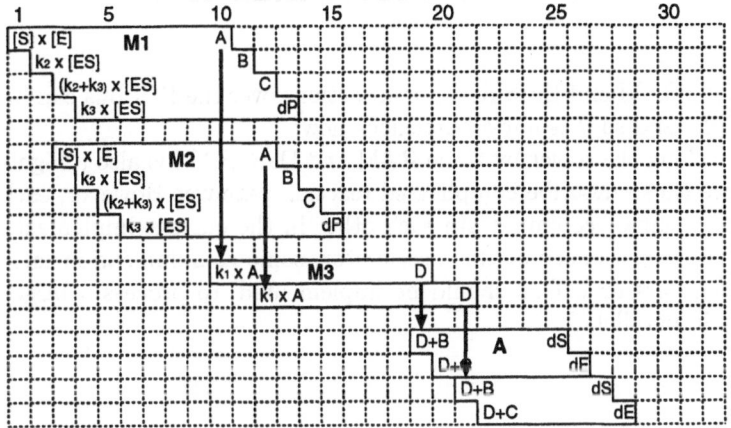

Fig. 9. Pipeline Schedule of the Second Prototype

The solver is fully pipelined, and it can start process each 2 clocks. Most of the logic operates at 33MHz, Data RAM and Code RAM run at 66MHz to ensure sufficient bandwidth. Code RAM is the array of pointer, which points to Data RAM in the format shown in Table 5. The width of Code RAM is 32 bit, and two 16 bit pointers are stored in each address. Some special values are reserved as control commands.

The second prototype is designed so as to be implemented on ReCSiP board. It provides a simple and easy software interface by mapping Code RAM, Data RAM and some control registers on the memory address space of the host processor, via PCI bus. Simulation will start by storing the code and data on each RAM, then kick the control register.

Evaluation. The size, maximum operating frequency and throughput are shown in Table 7. It is about 8 times faster than 1.13GHz version of Intel PentiumIII.

Table 6. Contents of Code RAM for the Sample Pathway

Address	Pointer		Address	Pointer	
00h	A	Eab	04h	B	Ebc
01h	k_{12}	A.Eab	05h	k_{22}	B.Ebc
02h	$k_{12} + k_{13}$	k_{11}	06h	$k_{22} + k_{23}$	k_{21}
03h	k_{13}	—	07h	k_{23}	—
			08h	END	—

Table 7. The Size, Frequency and Throughput of the Second Prototype

Units	Slices (Used %)		Frequency	Throughput
1	3,713	(10%)	48MHz	24
2	7,449	(22%)	44MHz	44
3	11,163	(33%)	43MHz	64
4	14,897	(44%)	42MHz	84

With Code RAM and Data RAM, transactions over the PCI bus can be reduced as possible, and so it is not a bottleneck now.

Controlling the solver by Code RAM and Data RAM is also a highly flexible way, since it can describe complicated reaction pathway. However, each unit on the FPGA cannot communicate each other in the current implementation. To exploit the performance of this accelerator, some communication facilities like shared registers or FIFOs should be implemented, to process a large reaction pathway in parallel.

3   Conclusion

A host-accelerator co-operation environment for bioinformatics, "ReCSiP" is proposed. As the preliminary evaluation, a solver for a basic metabolic simulation using Michaelis-Menten model is implemented. Its performance is 8 times as that of 1.13GHz PentiumIII by the parallel execution in the solver.

Current Michaelis-Menten solver needs further improvement to run various type of simulations with more realistic models. For example, some more stiff algorithms of numerical solution of differential equations should be implemented.

Now, implementation of other applications including microscopic image processing or DNA sequence matching is going on.

Now we're planning to design ReCSiP board version 2. The following new facilities are added on the current version:

- 18Mbit QDR-SRAMs instead of current 18Mbit synchronous SRAM,
- a DDR-SDRAM SO-DIMM slot for running big simulations, and
- a direct board to board communication interface.

References

1. J. Schaff et al. The virtual cell. In *Proceedings of Pacific Symposium on Biocomputing*, volume 4, pages 228–239, Jan. 1999.
2. Masaru Tomita et al. E-cell: software environment for whole-cell simulation. *Bioinformatics*, 15(1):72–84, Jan. 1999.
3. Kouichi Takahashi et al. Computational challenges in cell simulation: A software engineering approach. *IEEE Intelligent Systems*, pages 64–71, Oct. 2002.

On the Implementation
of a Margolus Neighborhood Cellular Automata on FPGA

Joaquín Cerdá[1], Rafael Gadea[1], Vicente Herrero[1], and Angel Sebastià[1]

[1] Group of Digital Systems Design, Dept. Of Electronic Engineering,
Universidad Politécnica de Valencia,
46022 Valencia, Spain
{joacerbo, rgadea, viherbos, asebasti}@eln.upv.es
http://dsd.upv.es

Abstract. Margolus neighborhood is the easiest form of designing Cellular Automata Rules with features such as invertibility or particle conserving. In this paper we propose two different implementations of systems based on this neighborhood: The first one corresponds to a classical RAM-based implementation, while the second, based on concurrent cells, is useful for smaller systems in which time is a critical parameter. This implementation has the feature that the evolution of all the cells in the design is performed in the same clock cycle.

1 Introduction

Cellular Automata (CA) model massively parallel computation and physical fenomena [1]. They consist of a lattice of discrete identical sites called *cells*, each one taking a value from a finite set, usually a binary set. The value of the cells evolve in discrete time steps according to deterministic rules that specify the value of each site in terms of the values of the neighboring sites. This is a parallel, synchronous, local and uniform process [1, 2, 5, 10, 14].

CA are used as computing and modeling tools in biological organization, self-reproduction, image processing and chemical reactions. Also, CA have proved themselves to be useful tools for modeling physical systems such as gases, fluid dynamics, excitable media, magnetic domains and diffusion limited aggregation [13, 3, 4, 5, 8, 12]. CA have been also applied in VLSI design in areas such as generation of pseudo-random sequences and their use in built-in self test (BIST), error-correcting codes, private-key cryptosystem, design of associative memory and testing the finite state machine [9, 10, 11, 15, 16].

1.1 Invertible Cellular Automata and Margolus Neighborhood

A CA is invertible when its global function is a bijection, i.e., if every configuration – which, by definition, has exactly one successor – also has exactly one predecessor [6].

P.Y.K. Cheung et al. (Eds.): FPL 2003, LNCS 2778, pp. 776–785, 2003.

Invertibility is such a desiderable property, because in the context of dynamical systems coincides with microscopic reversibility. One of the most common ways of constructing invertible cellular automata is by using a partitioning schema [6].

Partitioning Cellular Automata (PCA) are based on a different kind of local map that takes as input the contents of a region and produces as output the new state of the whole region (rather than of a single cell). This way, the state space is completely divided into non-overlapping regions. In order to exchange information between regions, partitions must change at the next step. The partitioning format is specially good for many applications because it makes very easy to construct invertible rules.

The most important partitioning scheme is the Margolus Neighborhood, introduced in [3]. In this neighborhood each partition is 2x2 cells as shown in figure 1. We alternate between even partitions (solid lines) and odd partitions (dotted line) in order to couple them all. Periodic boundary conditions are assumed.

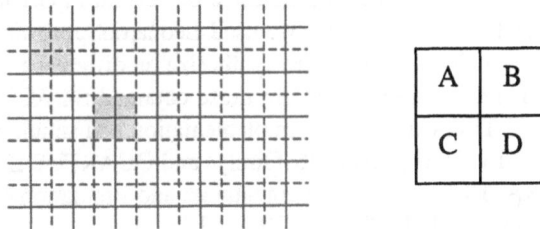

Fig. 1. Margolus Neighborhood (*left*): even (*solid lines*) and odd (*dotted lines*) partitions of a two-dimensional array into 2x2 blocks. The block on the right introduces a notation to refer to the cells into the partition

Several rules based on Margolus neighborhood have been introduced in different areas. Among them we can mention BBMCA introduced by Margolus itself and capable of universal computation [3], rules TM and HPP for modelling gases [8] and the rule DIAG_then_DOWN for simulating particles that fall down and pile up [7].

2   Sequential Implementation of Margolus neighborhood

A classical architecture for implementing Margolus neighborhood at VLSI is presented in figure 2. In this approach, the lattice of cells is stored in a RAM memory disposed as a 2D array of single bits. The control path has to generate the proper signals to read the four positions that form a block and present them to the process unit. The process unit applies the transformation codified by the local map and after it is finished, the control generates the signals to store the result back in the memory.

It is necessary for the control unit to have an input from a parity generator that ensures the correct alternancy between even and odd evolutions, so the directions in the memory to be accesed are different in each case.

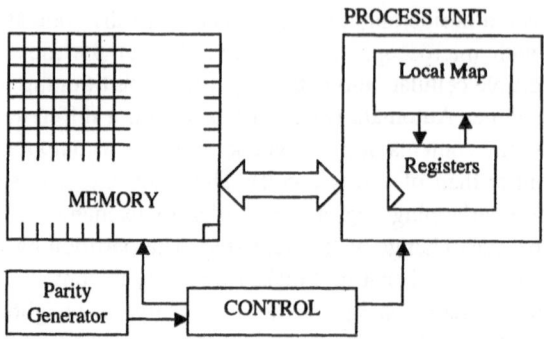

Fig. 2. Sequential implementation of Margolus neighborhood. The value of the cells is stored in a RAM array

The proposed architecture was implemented on several FPGA families to compare performances. For the logic synthesis we used LeonardoSpectrum Version 2002b.21 from Exemplar Logic. This allows us to synthesize on different families from several manufactures. The results shown here are those obtained for ALTERA and XILINX devices. For the ALTERA devices the implementation and simulation was performed with MAX+PLUSII v. 10.0 (FLEX10K family) and QUARTUS II v. 2.2 (APEX20K, APEXII and STRATIX families). Implementation and simulation on XILINX devices was performed using FOUNDATION SERIES 3.1i.

2.1 Sequential Control Implementation

As a first stage, we implemented only the control unit on the FPGA. This is thought to operate on a external RAM in which information about cell states is stored. Table 1 summarizes the results on Maximum frequency and Logic Cells (LCs) needed for ALTERA devices. Implementations are done for a 8x8 matrix of cells.

Similar results can be given for XILINX Devices. XILINX Slice is more complex than ALTERA LC. In fact, it contains 2 LUTs, so used resources seem to be lower in this case, but it is due to the granularity.

Table 1. Results of Sequential Control Implementation in ALTERA devices

Family	FLEX10K	APEX20K	APEXII	STRATIX
Device	EPF10K20 RC240-3	EP20K200 RC240-1	EP2A25 F672C7	EP1S25 F780C6
LCs	72	75	75	61
Maximum Frequency	82,64MHz	87,67MHz	268,24MHz	308,36MHz

Table 2. Results of Sequential Control Implementation in XILINX devices

Family	4000	VIRTEX	VIRTEXII
Device	4020XLPQ160-09	V200BG256-6	XC2V80FG256-4
Slices	25	38	38
Maximum Frequency	82,905MHz	158,806MHz	173,461MHz

2.2 Using Embedded Memory

In most cases we can use the embedded memory included in the devices for storing information about the matrix of cells. This limits the maximum size of the system: The maximum size of the matrix of cells depends on the total amount of RAM embedded on the chip. Table 3 gives the results of implementation of a 8x8 matrix on the listed ALTERA devices. Also we have included the maximum size of the array allowed for a device and for a family[1].

The same results for XILINX devices are given in Table 4. It is important to note that some families, such as 4000, lacks of embedded memory and emulate it using distributed memory (LUTs).

Table 3. Results of Sequential Implementation using embedded memory in ALTERA devices

Family	FLEX10K	APEX20K	APEXII	STRATIX
Device	EPF10K20 RC240-3	EP20K200 RC240-1	EP2A25 F672C7	EP1S25 F780C6
LCs	71	75	76	61
Maximum Frequency	30,58MHz	54,24MHz	113,91MHz	187,21MHz
Maximum Size (Device)	64x64	256x256	512x512	1024x1024
Maximum Size (Family)	128x128	512x512	1024x1024	2048x2048

Table 4. Results of Sequential Implementation using embedded memory in XILINX devices

Family	4000	VIRTEX	VIRTEXII
Device	4020XLPQ160-09	V200BG256-6	XC2V80FG256-4
Slices	31	38	39
Maximum Frequency	55,460MHz	92,200MHz	120,337MHz
Maximum size (Device)	128x128	128x128	256x256
Maximum size (Family)	512x512	512x512	1024x1024

---

[1] Actually the total dimension, if we could use all the memory available, should be a little more, but the design has been thought to have size of $2^n \times 2^n$.

The main drawback of the design is the fact that the time required to perform an evolution increases linearly with the total size of the array. The control unit needs 4 clock cycles to read a 2x2 block from memory and 4 cycles to store them back. Also, the process unit needs one cycle to perform the evolution of the partition, so 9 clock cycles are necessary to evolve each 2x2 block from memory.

2.3 Dual-Port Memories

Evolution rates can be increased by relying on special features of the most powerful families. Concretely, we can get advance of the embedded dual-port RAM that allows the user to perform simultaneous read-and-write operations. This can reduce the number of cycles needed to perform a block evolution from 9 to 5, almost doubling the number of evolutions per second that the system can achieve. But this feature is not supported by all the families under study. Results on the suitable families are given in Table 5 for the 8x8 array.

Even if we rely on most powerful families and take advantage of dual-port embedded memories, sequential implementation leads us to a system that presents poor timing-characteristics as size increases. This effect is due to the fact that memory must be explored and updated sequentially. So, the evolutions per second that the system can achieve decreases with the total size. Figure 3 shows this evolution for the best devices studied.

Table 5. Results of Sequential Implementation using dual-port embedded memory in ALTERA and XILINX devices

Family	APEXII	STRATIX	VIRTEX	VIRTEXII
Device	EP2A25 F672C7	EP1S25 F780C6	V200BG256-6	XC2V80FG256-4
LCs / Slices	91	68	40	40
Maximum Frequency	111,02MHz	187,86MHz	85,690MHz	143,021MHz

3   Concurrent Implementation

In some applications, evolution rates like those presented in the previous paragraph are non acceptable. The main advance of a Celular Automata Structure is precisely its high paralelisation degree, and the implementation previously proposed converts it into a serial scheme to perform the matrix actualisation. If we want to obtain a real concurrent Cellular Automata, new strategies need to be explored.

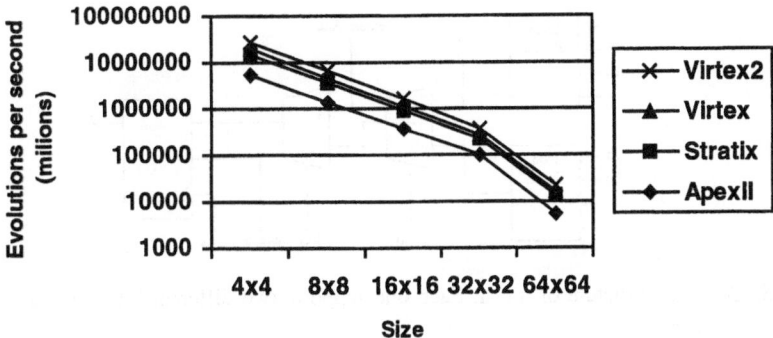

Fig. 3. Effects of the size in number of evolutions per secon in sequential implementation

3.1 Proposed Architecture

If we carefully study the connection schema of a Margolus Neigborhood Cellular Automata, easily we can distinguish between four classes of cells, depending on its position into the global matrix. In figure 4 these four classes are shown. Class 1 Cells are updated in odd cycles as Type D cells (referred to the notation introduced in figure 1) from a dotted line block, and are updated in even cycles as Type A Cells from a solid line block. The rest of classes and their connectivities are easily inferred from the figure.

This connection scheme leads us to reduce all classes of Cells to a common structure that is depicted on figure 5. For each cell in the matrix we need to define two different functions: the even one and the odd one. A multiplexer selects between results on even or odd branches depending on the present cycle. Also, to easily supply initial data to the circuit, we have included a second multiplexer for data synchronous load. Finally, an Enable terminal has been added to hold and start the evolution.

If we fix the proposed architecture for all the classes of cells in the matrix, the only difference between classes is the way the neighbor cells are connected to the inputs of the even/odd functions.

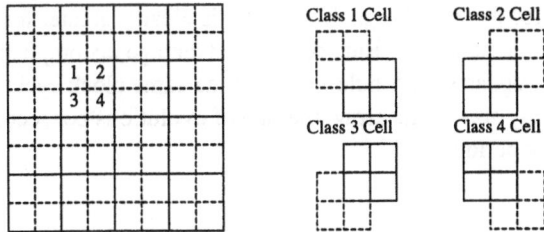

Fig. 4. Classes of cells in Margolus neighborhood. Each class of cells behaves as one type of cell of those introduced in figure 1 different for each even/odd evolution

782   J. Cerdá et al.

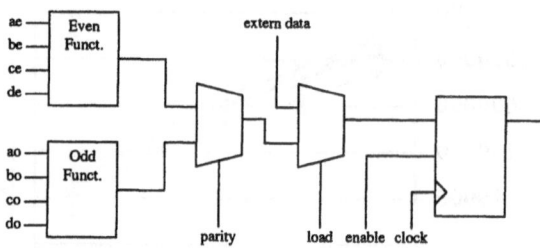

Fig. 5. Common structure of a cell. Each one supports two different functions: the even one and the odd one

3.2   Implementation and Results

This concurrent architecture was implemented on several FPGA families to compare performances. For the logic synthesis we used LeonardoSpectrum Version 2002b.21 from Exemplar Logic on different families from several manufactures. The results shown here are those obtained for ALTERA and XILINX devices. For the ALTERA devices the implementation and simulation was performed with MAX+PLUSII v. 10.0 (FLEX10K family) and QUARTUS II v. 2.2 (APEX20K, APEXII and STRATIX families). Implementation and simulation on XILINX devices was performed using FOUNDATION SERIES 3.1i.

ALTERA Logic Cell contains a four-input look-up-table that is specially indicated for implementing one of the even/odd functions. So, in every device from ALTERA, the complete Cell showed in figure 5 needs 4 LCs to be implemented: two for the two functions and the rest for the multiplexers and register.

Table 6 summarizes the results of the implementations for a 8x8 matrix in different ALTERA devices. It is important to remark that even though in the synthesis step every device needs 4 LCs per Cell, implementation changes that number, due to the mapping process. In family STRATIX LC is a little more complex, so mapping can be optimized and the number of LCs per Cell is reduced to 3, allowing us to include more Cells in the device. Table 6 also lists the maximum size of the matrix, considering the number of LCs in the device and in the family[2]. Finally, the maximum frequency is listed for the 8x8 design. This value is clearly greater than that obtained in the sequential implementation. But the main advantage accomplished is the way the circuit evolves: it updates the whole matrix in the same clock cycle, independently of the size. So, for any size of the array, the rate of evolutions per second is constant and equal to one clock period.

---

[2] In this case, design is not limited to sizes of 2^nx2^n. With this implementation the only limitation, characteristical of Margolus neighborhood, is that size must be the square of an even number.

Table 6. Results of Concurrent Implementation in ALTERA devices

Family	FLEX10K	APEX20K	APEXII	STRATIX
Device	EPF10K20 RC240-3	EP20K200 RC240-1	EP2A25 F672C7	EP1S25 F780C6
LCs (Cell)	4	4	4	4
LCs (Full Design)	257	257	257	193
Maximum Frequency	81,96MHz	152,05MHz	152,07MHz	303,12MHz
Maximum size (device)	16x16	40x40	68x68	92x92
Maximum size (family)	54x54	100x100	114x114	194x194

Table 7. Results of Concurrent Implementation in XILINX devices

Family	4000	VIRTEX	VIRTEXII
Device	4020XLPQ160-09	V200BG256-6	XC2V80FG256-4
Slices (Cell)	2	2	2
Slices (Full Design)	96	97	97
Maximum Frequency	40,596MHz	103,082MHz	165,618MHz
Maximum size (device)	22x22	39x39	18x18
Maximum size (family)	74x74	90x90	176x176

Similar results can be obtained for XILINX devices. As we mentioned before, XILINX Slice contains 2 LUTs, so one single Cell requires only 2 Slices to be implemented. However, after the implementation the mapping can be optimized, reducing the number to approximately 1,5 Slices per Cell. Table 7 summarizes these results for XILINX devices. Again, given maximum frequency corresponds to the 8x8 design.

Seeing these results one questions how the total size affects maximum frequency. In practice, we obtain little variation as the total size increases, due to the locality connection of cells in the overall design. This effect can be seen in Figure 6, in which only the most powerful families were used.

3.3   Comparison between Implementations

The two main differences between both strategies are evident from the given results: sequential implementation permits large matrix sizes, thus having the drawback of that the time per evolution increases linearly with the size of the matrix. On the other hand, concurrent implementation is more adequated when time is a critical parameter, even though the sizes obtained are small. This could be indicated in some VLSI applications such as random number generation or BIST.

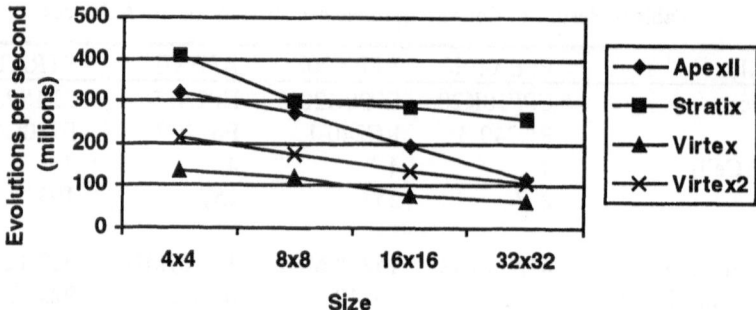

Fig. 6. Effects of the size in number of evolutions per second in the concurrent implementation

4   Conclusions

We have studied Margolus neighborhood Cellular Automata. We propose two implementations for Margolus neighborhood Cellular Automata, the first one RAM based that allows large sizes but offers poor timing characteristics as processing times increases linearly with memory size. This processing time can be reduced to the half using dual-port memories to speed-up data access. The other implementation, that we call concurrent, gives support to much small systems in which time is a critical factor, performing a complete evolution in only one clock cycle.

References

1. Wolfram, S.: Cellular Automata. Los Alamos Science, 9 (1983) 2–21
2. Wolfram, S.: Statistical mechanics of cellular automata. Reviews of Modern Physics, 55 (1983) 601–644
3. Margolus, N.: Physics-Like models of computation. Physica 10D (1984) 81-95
4. Toffoli, T.: Cellular Automata as an alternative to (rather than an approximation of) Differential Equations in Modeling Physics. Physica 10D (1984) 117-127
5. Toffoli, T.: Occam, Turing, von Neumann, Jaynes: How much can you get for how little? (A conceptual introduction to cellular automata). The Interjournal (October 1994)
6. Toffoli, T., Margolus, N.: Invertible cellular automata: a review. Physica D, Nonlinear Phenomena, 45 (1990) 1–3
7. Gruau, F. C., Tromp, J. T.: Cellular Gravity. Parallel Processing Letters, Vol. 10, No. 4 (2000) 383–393
8. Smith, M. A.: Cellular Automata Methods in Mathematical Physics. Ph.D. Thesis. MIT Department of Physics (May 1994).
9. Wolfram, S.: Cryptography with Cellular Automata. Advances in Cryptology: Crypto '85 Proceedings, Lecture Notes in Computer Science, 218 (Springer-Verlag, 1986) 429–432
10. Sarkar, P.: A brief history of cellular automata. ACM Computing Surveys, Vol. 32, Issue 1 (2000) 80–107

11. Shackleford, B., Tanaka, M., Carter, R.J., Snider, G.: FPGA Implementation of Neighbor-hood-of-Four Cellular Automata Random Number Generators. Proceedings of FPGA 2002 (2002) 106–112

12. Vichniac, G. Y.: Simulating Physics with Cellular Automata. Physica 10D (1984) 96-116

13. Popovici, A., Popovici, D.: Cellular Automata in Image Processing. Proceedings of MTNS 2002 (2002)

14. Wolfram, S.: A New Kind of Science. Wolfram media (2002)

15. Corno, F., Rebaudengo, M., Reorda, M.S., Squillero, G., and Violante, M.: Low power BIST via hybrid cellular automata. 18th IEEE VLSI Test Symposium, 2000.

16. Corno, F., Reorda, M.S., Squillero, G.: Exploiting the selfish gene algorithm for evolving hardware cellular automata. Proceedings of Congress of Evolutionary Computation (CEC2000).

Fast Modular Division for Application in ECC on Reconfigurable Logic

Alan Daly[1], William Marnane[1], Tim Kerins[1], and Emanuel Popovici[2]

[1] Dept. of Electrical & Electronic Engineering,
University College Cork, Ireland
{aland,liam,timk}@rennes.ucc.ie
[2] Dept. of Microelectronic Engineering,
University College Cork, Ireland
e.popovici@ucc.ie

Abstract. Elliptic Curve Public Key Cryptosystems are becoming increasingly popular for use in mobile devices and applications where bandwidth and chip area are limited. They provide much higher levels of security per key length than established public key systems such as RSA. The underlying operation of elliptic curve point multiplication requires modular multiplication, division/inversion and addition/subtraction. Division is by far the most costly operation in terms of speed. This paper proposes a new divider architecture and implementation on FPGA for use in an ECC processor.

1 Introduction

Elliptic Curve Cryptosystems (ECC) were independently proposed in the mid-eighties by Victor Miller [1] and Neil Koblitz [2] as an alternative to existing public key systems such as RSA and DSA. No sub-exponential algorithm is known to solve the discrete logarithm problem on a suitably chosen elliptic curve, meaning that smaller parameters can be used in ECC with equivalent security to other public key systems. It is estimated that an elliptic curve group with 160-bit length has security equivalent to RSA with a bit length of 1024-bit [3].

Two types of finite field are popular for use in elliptic curve public key cryptography: $GF(p)$ with p prime, and $GF(2^n)$ with n a positive integer. Many implementations focus on using the field $GF(2^n)$ due to the underlying arithmetic which is well suited to binary numbers [4]. Addition is simply bitwise XOR in $GF(2^n)$, whereas carry adders must be used when operating in $GF(p)$. However, most $GF(2^n)$ processors are limited to operation on specified curves and key sizes. An FPGA implementation of a $GF(2^n)$ processor which can operate on different key sizes without reconfiguration has previously been presented in [5]. ECC Standards define different elliptic curves and key sizes which ECC implementations must be capable of utilising [6][7]. With a $GF(p)$ processor, any curve or key length up to the maximum size, p, can be used without reconfiguration.

Few implementations of ECC processors over $GF(p)$ have been implemented on hardware to date due to the more complicated arithmetic required[8]. The

P.Y.K. Cheung et al. (Eds.): FPL 2003, LNCS 2778, pp. 786–795, 2003.

modular division operation has been implemented in the past by modular inversion followed by modular multiplication. No implementations to date have implemented a dedicated modular division component in an ECC application.

This paper proposes a modular divider for use in an ECC processor targeted to FPGA which avoids carry chain overflow routing. The design proposed has a bit length of 256-bits and thus would provide security well in excess of RSA-1024 when used in an ECC processor.

2   Elliptic Curve Cryptography over $GF(p)$

An elliptic curve over the finite field $GF(p)$ is defined as the set of points (x, y), which satisfy the elliptic curve equation

$$y^2 = x^3 + ax + b$$

where x, y, a and b are elements of the field, and $4a^3 + 27b^2 \neq 0$.

To encrypt data, it is represented as a point on the chosen curve over the finite field. The fundamental encryption operation is point scalar multiplication, i.e. a point P_1 is added to itself k times.

$$
\begin{aligned}
Q &= kP \\
&= \underbrace{P + P + \ldots + P}_{k \ times}
\end{aligned}
$$

In order to compute kP, a double and add method is used and k is represented in binary form and scanned right to left from LSB to MSB, performing a double at each step and an addition if k_i is 1. Therefore the multiplier will require $(m - 1)$ point doublings and an average of $(\frac{m-1}{2})$ point additions, where m is the bitlength of the field prime, p.

The operations of elliptic curve point addition and point doubling are best explained graphically as shown in Fig.1 and Fig.2. To add two distinct points, P_1 and P_2, a chord is drawn between them. This chord will intersect the curve at exactly one other point, and the reflection of that point through the x-axis is defined to be the point $P_3 = P_1 + P_2$.

In the case of adding point P_1 to itself (doubling P_1), the tangent to the curve at P_1 is drawn and found to intersect the curve again at exactly one other point. The reflection of this point through the x-axis is defined to be the point $2P_1$. (Note: The *point at infinity*, $\mathcal{O}$ is taken to exist infinitely far on the y-axis, and is the identity of the elliptic curve group.)

Point Multiplication operations on elliptic curves over $GF(p)$ are performed by modulo addition, subtraction, multiplication and inversion. Different coordinate systems can be used to represent elliptic curve points, and it is possible to reduce the number of inversions by representing the points in *projective coordinates*. However, this results in a large increase in the number of Multiplications required per point operation (13 for addition, 6 for doubling) [3][8].

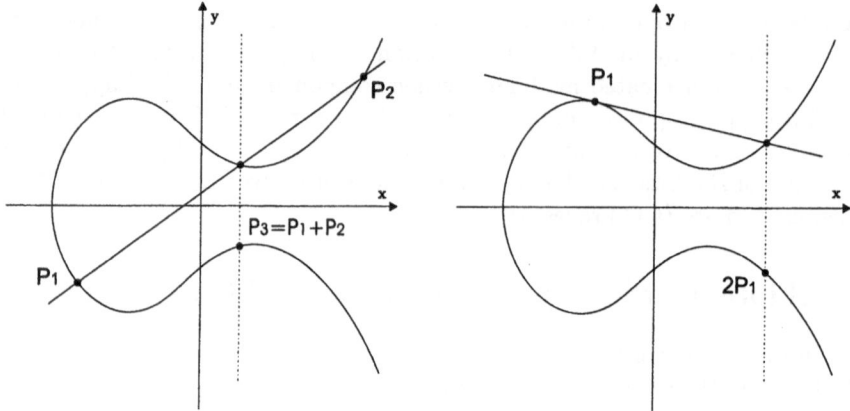

Fig. 1. Point Addition   **Fig. 2.** Point Doubling

The point addition/doubling formulae in *affine coordinates* are given below. Let $P_1=(x_1,y_1)$ and $P_2=(x_2,y_2)$, then $P_3=(x_3,y_3)=P_1+P_2$ is given by:

$$x_3 = \lambda^2 - x_1 - x_2$$
$$y_3 = \lambda(x_1 - x_3) - y_1$$
$$\lambda = \begin{cases} \dfrac{y_2 - y_1}{x_2 - x_1} & \text{if } P_1 \neq P_2 \\[2ex] \dfrac{3x_1^2 + a}{2y_1} & \text{if } P_1 = P_2 \end{cases}$$

The performance of an ECC processor depends on the efficiency of the finite field computations required to perform the point addition and doubling operations. Point addition requires 2 multiplications, 1 division and 6 addition/subtraction operations. Point doubling requires 3 multiplications, 1 division and 7 additions/subtractions. Division is the critical operation and the speed of the overall processor will depend on a high-speed divider.

3   Modular Inversion

One method to perform the division operation is to perform an inversion followed by a multiplication.

The multiplicative inverse of an integer a (mod M) exists if and only if a and M are relatively prime. There are two methods to calculate this inverse. One is based on Fermat's Theorem which states that $a^{M-1}(\text{mod } M) = 1$ and therefore $a^{M-2}(\text{mod } M) = a^{-1}$ (mod M). Using this fact, the inverse may be calculated by modular exponentiation. However this is an expensive operation requiring on average $1.5 \log_2 m$ multiplications. Another method to calculate the inverse is by knowing that the greatest common divisor of any two integers may be expressed as a linear combination of the two.

Table 1. Area and speed results for the Inverter designs.

Design	Area (Slices)	% of XCV2000e	Equivalent Gates	Max Freq. (MHz)
64-bit	735	3.8	13, 855	45
128-bit	1, 259	6.6	24, 628	38
256-bit	2, 453	12.8	48, 226	28

Since a and M are relatively prime, the following expression may be solved for r and s:

$$ar + Ms = 1 \ (\text{mod } M)$$

This linear equation implies that:

$$ar \equiv 1 \ (\text{mod } M)$$

Therefore, r is the inverse of a (mod M). The values of r and s may be derived by reversing the steps of the Euclidean algorithm. This procedure is known as the Extended Euclidean algorithm.

Kaliski proposed a variation of the extended Euclidean algorithm to compute the Montgomery Inverse, $a^{-1}2^m$ (mod M) of an integer a [9]. The output of this algorithm is in the Montgomery domain, which is useful when performing further Montgomery multiplications [10]. The Montgomery multiplier is an efficient algorithm for computing modular multiplication, however inputs must first be mapped into the Montgomery domain, and then re-mapped before output. Therefore, the Montgomery multiplier is useful only when repeated multiplications are necessary, such as in the RSA cryptosystem where modular exponentiation is the underlying operation. Efficient Montgomery multipliers for implementation on FPGA have previosly been presented in [11]. The performance of ECC point addition and doubling could be improved by performing all operations in the Montgomery domain.

Montgomery inverters have previously been presented in [12][13][14][15][16]. Inverter speed and area results based on designs presented in [14] are given in table 1. The architecture requires a maximum of $(3m + 2)$ clock cycles to perform an m-bit Montgomery inversion, or $(4m + 2)$ cycles to perform a regular integer inversion.

4 Modular Division

A new binary add-and-shift algorithm to perform modular division was presented recently by S.C. Shantz in [17]. This provides an alternative to division by inversion followed by multiplication. The algorithm is given here in algorithm 1. It computes $U = \frac{y}{x}$ (mod M) in a maximum of $2(m - 1)$ clock cycles where m is the bitlength of the modulus, M.

Algorithm 1 : Modular Division

Input : $y, x \in [1, M - 1]$ and M
Output : U, where $y = U * x \pmod{M}$

    $A := x,\ B := M,\ U := y,\ V := 0$
    while $(A \neq B)$ do
      if $(A$ even$)$ then
          $A := A/2$;
          if $(U$ even$)$ then $U := U/2$ else $U := (U + M)/2$;
      else if $(B$ even$)$ then
          $B := B/2$;
          if $(V$ even$)$ then $V := V/2$ else $V := (V + M)/2$;
      else if $(A > B)$ then
          $A := (A - B)/2$;
          $U := (U - V)$;
          if $(U < 0)$ then $U := (U + M)$;
          if $(U$ even$)$ then $U := U/2$ else $U := (U + M)/2$;
      else
          $B := (B - A)/2$;
          $V := (V - U)$;
          if $(V < 0)$ then $V := (V + M)$;
          if $(V$ even$)$ then $V := V/2$ else $V := (V + M)/2$;
    end while
    return U;

---

5   Basic Division Architecture

As can be seen from algorithm 1, the divider makes decisions based on the parity and magnitude comparisons of m-bit registers. To determine the parity of a number, only the LSB need be examined (0 implies *even*, 1 implies *odd*). However, true magnitude comparisons can only be achieved through full m-bit subtractions, and thus introduce delay before decisions can be made.

The first design presented here uses m-bit carry propagation adders to perform the additions/subtractions. At each iteration of the while loop in algorithm 1, U_{new} is assigned one of the following values, depending on the parity and relative magnitude of A and B: U, $\frac{U}{2}$, $\frac{(U+M)}{2}$, $\frac{(U-V)}{2}$, $\frac{(U-V+M)}{2}$ or $\frac{(U-V+2M)}{2}$.

The basis for the proposed designs is to calculate all 6 possible values concurrently, and use multiplexors to select the final value of U_{new}. This eliminates the time required to perform a full magnitude comparison of A and B before calculation of U and V can even commence. The same is true for the determination of V_{new}. These architectures are illustrated in Fig.3 and Fig.4.

The architecture for the determination of A_{new} and B_{new} is simpler as illustrated in Fig.5. The U, V, A and B registers are also controlled by the parity and relative magnitudes of A and B, and are not clocked on every cycle.

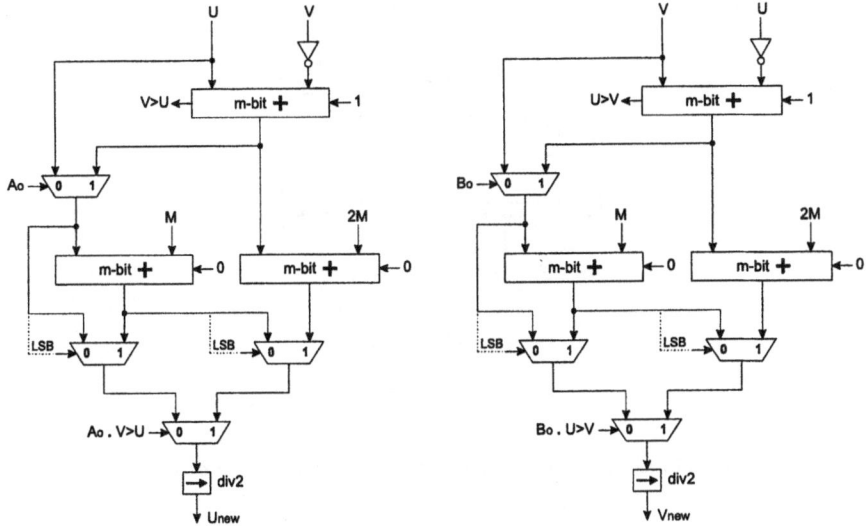

Fig. 3. Determination of U_{new} **Fig. 4.** Determination of V_{new}

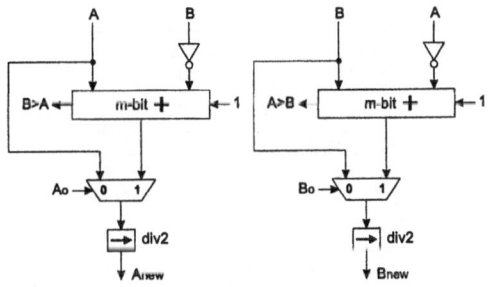

Fig. 5. Determination of A_{new} and B_{new}

6   Proposed (Carry-Select) Division Architecture

The clock speed is dependant on the bitlength of the divider since the carry chain of the adders contributes significantly to the overall critical path of the design. When the carry chain length exceeds the column height of the FPGA, the carry must be routed from the top of the column to the bottom of the next. This causes a significant decrease in the overall clock speed. The carry-select design proposed here halves this adder carry chain at the expense of extra adders and control, but in doing so improves the performance of the divider.

This architecture is similar to a carry-select inverter design proposed in [14]. The values of $(U-V)$ and $(A-B)$ are determined by splitting the m-bit registers U, V, A, and B into $(\frac{m}{2})$-bit registers U_L, U_H, V_L, V_H, A_L, A_H, B_L and B_H. The values of $(U-V)_L$ and $(U-V)_H$ are then determined concurrently to produce $(U-V)$ as illustrated in Fig.6 and Fig.7.

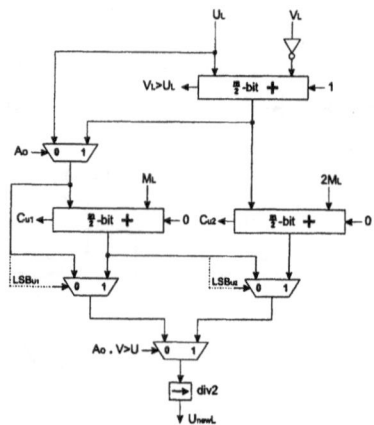

Fig. 6. Carry-Select Architecture: Determination of U_{newL}

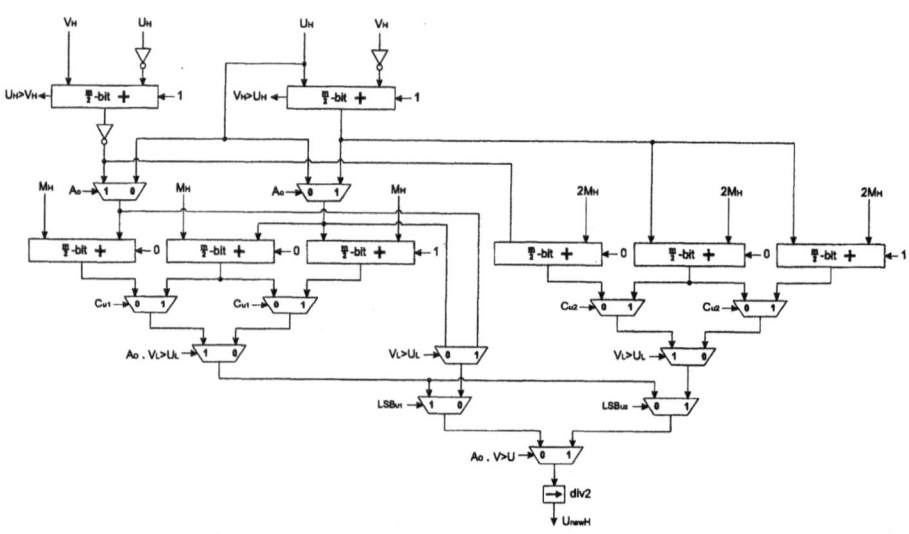

Fig. 7. Carry-Select Architecture: Determination of U_{newH}

When calculating $(U - V)$, if $(V_L > U_L)$, then one extra bit must be "borrowed" from U_H. (i.e. The value of $(U - V)_H$ will actually be $(U_H - V_H - 1)$) It is observed that $(U_H - V_H - 1)$ is actually equal to $(\overline{V_H - U_H})$, and therefore only a bitwise inverter is needed to produce this value as seen in Fig.7. However, it is also possible that a carry from the addition of M_L or $2M_L$ will affect the final value of U_{newH}. Therefore carries of -1, 0 and 1 must be accounted for in the determination of U_{newH}. To allow for this, an extra $(\frac{m}{2})$-bit adder with a carry-in of 1 is required to calculate $(U_H - V_H + M_H + 1)$.

The determination of V_{new} is similar to that of U_{new}. The values of A_{new} and B_{new} are determined as illustrated in Fig.8.

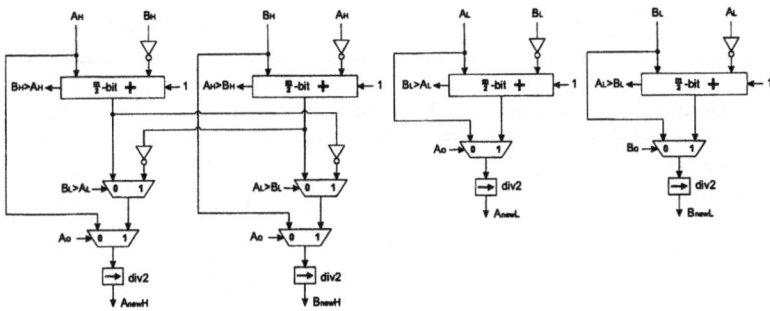

Fig. 8. Carry-Select Architecture: Determination of A_{new} and B_{new}

Table 2 compares the proposed carry-select divider and the basic divider in terms of adder and multiplexor area. The adders and multiplexors in the carry-select divider are half the size of those in the basic design, so overall there is a 50% increase in the size of adders required. The multiplexors are actually instantiated as 4-input Look-Up Tables (LUT's) on the FPGA, and so the 80% increase is not reflected in the actual mapped area results.

7 Results

Speed and area comparisons of 64, 128 and 256-bit dividers for both designs are given in Table 3. The percentage increase in speed and area for each design are given in Table 4. VHDL synthesis and place and route were performed on Xilinx ISE. The results are post place and route with a top level architecture to load the data in 32-bit words. The target FPGA device for this research is the Xilinx Virtex XCV2000e-6bg560 which has 80 CLB's per column. Therefore the maximum unbroken carry chain length is 160 bits.

Little improvement in performance is achieved for the 64 and 128-bit dividers due to increased control logic and multiplexing. However, once the carry chain exceeds the FPGA column height, the basic design suffers a considerable deterioration in clock speed. The proposed carry-select design performs over 50% faster for the 256-bit divider, which is a realistic bit-length for secure ECC communications. A 50% increase in area is also observed, thereby keeping the (time × area) product similar for both designs.

Table 2. Number of adders/multiplexors required for both designs.

Divider	Adders	2:1 Multiplexors
Basic	8 x n-bit	10 x n-bit
Carry-Select	24 x $\frac{n}{2}$-bit	36 x $\frac{n}{2}$-bit
Area Increase	50 %	80 %

Table 3. Area and speed results for the two designs.

Design	Area (Slices)	% of XCV2000e	Equivalent Gates	Max Freq. (MHz)
64-bit	1,212	6.3	20,858	45
32-bit x 2	1,472	7.7	26,566	45
128-bit	2,215	11.5	39,622	31
64-bit x 2	3,217	16.8	56,236	38
256-bit	3,872	20.2	72,610	17
128-bit x 2	5,849	30.5	104,918	27

Table 4. Percentage increase in area/speed of carry-select design over basic design.

Divider Bitlength	Increase in Area (%)	Increase in Speed (%)
64-bit	21.5 %	0 %
128-bit	45.2 %	22.5 %
256-bit	51 %	58.8 %

Comparing these results to the inversion architecture results presented in Section 3, it is observed that both inverter and divider have almost identical maximum operating clock frequencies.

The divider requires half the number of clock cycles to perform the operation, however it requires significantly more area. It is estimated that using this new architecture, division can be performed twice as fast as by the alternative invert and multiply architecture.

8   Conclusions

Modular division is an important operation in elliptic curve cryptography. In this paper, two new FPGA architectures, based on a recently published division algorithm [17] have been presented and implemented. The basic design computes all possible outcomes from each iteration and uses multiplexors to select the correct answer. This avoids the necessity to await the outcome of a full m-bit magnitude comparison before computation can begin. The second, carry-select divider design splits the critical carry chain into two, and again performs all calculations before the magnitude comparison has been completed. For a 256-bit divider, an improvement in speed of over 50% was achieved with the proposed carry-select divider at a similar area cost over the basic design. The operation speed of the proposed divider is almost identical to that of an inverter, and needs only half the number of clock cycles . Since an additional modular multiplication is required when using inversion in ECC point multiplication, this division architecture is better suited for implementation in an ECC processor.

Acknowledgement

This work is funded by a research innovation project from Enterprise Ireland.

References

1. V. S. Miller. "Use of Elliptic Curves in Cryptography". *Advances in Cryptography Crypto'85*, (218):417–426, 1985.
2. N. Koblitz. "Elliptic Curve Cryptosystems". *Math Comp*, 48:203–209, 1987.
3. I. Blake, G. Seroussi, and N. Smart. *"Elliptic Curves in Cryptography"*. London Mathematical Society Lecture Note Series 265. Cambridge University Press, 2000.
4. M. Ernst, M. Jung, F. Madlener, S. Huss, and R. Blümel. "A Reconfigurable System on Chip Implementation for Elliptic Curve Cryptography over GF(2^n)". *Cryptographic Hardware and Embedded Systems - CHES 2002*, (LNCS 2523):381–398, Aug 2002.
5. T. Kerins, E. Popovici, W. Marnane, and P. Fitzpatrick. "Fully Parameterizable Elliptic Curve Cryptography Processor over GF(2^m)". *12th Intl Conference on Field-Programmable Logic and Applications FPL2002*, pages 750–759, Sept 2002.
6. IEEE Standards Department. IEEE 1363/D13 Standard Specifications for Public Key Cryptography, 2000.
7. ANSI X9.62. Public Key Cryptography for the Financial Services Industry. The Elliptic Curve Digital Signature Algorithm (ECDSA), 1999.
8. G. Orlando and C. Paar. "A Scalable GF(p) Elliptic Curve Processor Architecture for Programmable Hardware". *Cryptographic Hardware and Embedded Systems - CHES 2001*, (LNCS 2162):348–363, May 2001.
9. B. S. Kaliski Jr. "The Montgomery Inverse and it's applications". *IEEE Trans. on Computers*, 44(8):1064–1065, Aug 1995.
10. P. L. Montgomery. "Modular Multiplication without Trial Division". *Math Computation*, 44:519–521, 1985.
11. A. Daly and W. Marnane. "Efficient Architectures for Implementing Montgomery Modular Multiplication and RSA Modular Exponentiation on Reconfigurable Logic". *10th Intl Symposium on FPGA (FPGA 2002)*, pages 40–49, Feb 2002.
12. A. Gutub, A. F. Tenca, and C. K. Koc. "Scalable VLSI Architecture for GF(p) Montgomery Modular Inverse Computation". *IEEE Computer Society Annual Symposium on VLSI*, pages 53–58, April 2002.
13. A. Gutub, A. F. Tenca, E. Savas, and C. K. Koc. "Scalable and unified hardware to compute Montgomery inverse in GF(p) and GF(2^n)". *Cryptographic Hardware and Embedded Systems - CHES 2002*, (LNCS 2523):484–499, Aug 2002.
14. A. Daly, W. Marnane, and E. Popovici. "Fast Modular Inversion in the Montgomery Domain on Reconfigurable Logic". *Irish Signals and Systems Conference ISSC2003*, To Appear : July 2003.
15. E. Savas and C. K. Koc. "The Montgomery Modular Inverse - Revisited". *IEEE Trans. on Computers*, 49(7):763–766, July 2000.
16. T. Kobayashi and H. Morita. "Fast Modular Inversion Algorithm to Match any Operation Unit". *IEICE Trans. Fundamentals*, E82-A(5):733–740, May 1999.
17. S. C. Shantz. "From Euclid's GCD to Montgomery Multiplication to the Great Divide". Technical Report TR-2001-95, Sun Microsystems Laboratories, 2001.

Non-uniform Segmentation
for Hardware Function Evaluation

Dong-U Lee[1], Wayne Luk[1], John Villasenor[2], and Peter Y.K. Cheung[3]

[1] Department of Computing, Imperial College, London, UK
{dong.lee,wl}@ic.ac.uk
[2] Electrical Engineering Department, University of California, Los Angeles, USA
villa@icsl.ucla.edu
[3] Department of EEE, Imperial College, London, UK
p.cheung@ic.ac.uk

Abstract. This paper presents a method for evaluating functions in hardware based on polynomial approximation with non-uniform segments. The novel use of non-uniform segments enables us to approximate non-linear regions of a function particularly well. The appropriate segment address for a given function can be rapidly calculated in run time by a simple combinational circuit. Scaling factors are used to deal with large polynomial coefficients and to trade precision with range. Our function evaluator is based on first-order polynomials, and is suitable for applications requiring high performance with small area, at the expense of accuracy. The proposed method is illustrated using two functions, $\sqrt{-\ln(x)}$ and $\cos(2\pi x)$, which have been used in Gaussian noise generation.

1 Introduction

The evaluation of functions is often the performance bottleneck of many compute-bound applications. Examples of these functions include elementary functions such as $\ln(x)$ or $\sqrt{x}$, and compound functions such as $\sqrt{-\ln(x)}$ or $\tan^2(x) + 1$. Computing these functions quickly and accurately is a major goal in computer arithmetic; software implementations are often too slow for numerically intensive or real-time applications. The performance of such applications depends on the design of a hardware function evaluator. Advanced FPGAs enable the development of low-cost and high-speed function evaluation units, customizable to particular applications. The principal contribution of this paper is a fast and efficient hardware function evaluator using polynomial approximations. The key novelties of our work include:

- a method for polynomial approximations with non-uniform segments;
- hardware architecture and implementation of the proposed method;
- evaluation of this method with a logarithmic function and a cosine function.

The rest of this paper is organized as follows. Section 2 covers background material and previous work. Section 3 explains our segmentation technique. Section 4 describes the hardware architecture. Section 5 presents a method for determining the placement of segment boundaries. Section 6 discusses evaluation and results, and Section 7 offers conclusion and future work.

P.Y.K. Cheung et al. (Eds.): FPL 2003, LNCS 2778, pp. 796–807, 2003.
© Springer-Verlag Berlin Heidelberg 2003

2 Background

Polynomial approximation [13], [14] involves approximating a continuous function f with one or more polynomials p of degree n on a closed interval $[a, b]$. The aim is to minimize a distance $\|p - f\|$. There are two kinds of approximations: least squares approximations that minimize the average error, and least maximum approximations that minimize the worst-case error [15]. In both cases, the aim is to minimize a distance $\|p - f\|$. For least squares approximations, that distance is:

$$\|p - f\|_2 = \sqrt{\int_a^b w(x)(f(x) - p(x))^2 dx},\qquad(1)$$

where w is a continuous weight function for selecting parts of $[a, b]$ where we want the approximation to be more accurate. For least maximum (minimax) approximations, the distance is:

$$\|p - f\|_\infty = \max_{a \leq x \leq b} |f(x) - p(x)|.\qquad(2)$$

Our work is based on minimax polynomial approximations, which involve minimizing the worst-case error. Since we are interested in fixed-point number representation in our work, we will be concerned with the worst-case absolute errors. A recent study of minimax polynomial approximation on FPGAs can be found [21].

Much of the work on function evaluation is generally concerned with producing highly accurate approximation with complex designs. Instead, we will focus on applications that require very high speed and small area but not high accuracy. Examples of such applications include Gaussian noise generation [3] and belief propagation in LDPC decoding [20]. We will focus in this paper on first-order polynomials of the form $p(x) = c_1 \times x + c_0$, where c_1 is the gradient and c_0 is the y-intercept, which can be computed by two table lookups, a multiplication and an addition.

Previous work on polynomial approximations involves equally sized segments [4], [5], [6], [7], [8], [9], [10], [11], [12]. Approximations using such uniform segments are suitable for functions with linear regions, but they can be inefficient for non-linear functions. It is desirable to choose the boundaries of the segments to cater for the non-linearities of the function. Highly non-linear regions may need smaller segments than linear regions. This approach minimizes the amount of storage required to approximate the function, leading to more compact and efficient designs.

3 Function Evaluation Based on Non-uniform Segmentation

The interval of approximation $[a, b]$ is divided into a set of sub-intervals, called segments. The best-fit straight line, in a minimax sense, to each segment is found. A lookup table is used to store the coefficients for each line segment, and the functions can then be evaluated using a multiplier and an adder to calculate the linear approximation [1].

Using well-known methods that compute elementary functions such as CORDIC [2], the evaluation of compound functions is a multi-stage process. Consider the evaluation of the function $\sqrt{-\ln(x)}$ over the interval $(0, 1]$. Using CORDIC, the computation of this function is a two-stage process: the logarithm of x followed by the square root. With our approach, we look at the entire function over the given domain, and therefore we do not need to have two stages.

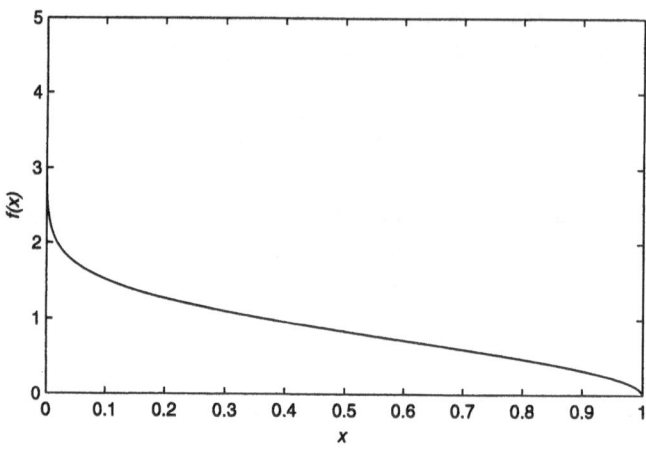

Fig. 1. $\sqrt{-\ln(x)}$ over $(0, 1]$.

As shown in Figure 1, the greatest non-linearities of the function $\sqrt{-\ln(x)}$ occur in the regions close to zero and one. If uniform segments are used, a large number of small segments would be required to get accurate approximations in the non-linear regions. However, in the middle part of the curve where it is relatively linear, accurate approximation can be obtained using relatively few segments. It would be efficient to use small segments for the non-linear regions, and large segments for linear regions. Arbitrary-sized segments would enable us to have the least error for a given number of segments; however, the hardware to calculate the segment address for a given input can be complex. Our objective is to provide near arbitrary-sized segments with a simple circuit to find the segment address for a given input.

We have developed a novel method which can construct piecewise linear approximation such that: (a) the segment lengths used in a given region depends on the local linearity, with more segments deployed for regions of higher non-linearity; and (b) the boundaries between segments are chosen such that the task of identifying which segment to use for a given input can be rapidly performed. The proposed method consists of five steps.

1. Determine optimal placement of segment boundaries (see Section 5) – this would include dividing into regions such that in each region the function either monotonically increases or decreases.
2. For a non-linear region, if the non-linearity is monotonically increasing, then increase segment size by a factor of two or more at each step; if the non-linearity is

monotonically decreasing, then reduce segment size by a factor of two or more at each step.

3. The segment addresses can be obtained by computing the prefixes [16] with a simple combinational or pipelined circuit.

4. If necessary, divide the function into several intervals, then apply step 1–3 (see the function $\cos(2\pi, x)$ in Section 6).

5. If necessary, repeat the above steps with higher-order terms.

As an example to illustrate our approach, consider approximating $\sqrt{-\ln(x)}$ with an 8-bit input (Figure 1). Using the traditional approach, the most-significant bits of x are used to index the uniform segments. For instance if the most-significant four bits are used, 16 uniform segments are used to approximate the function. Using our approach, it is possible to use small segments for non-linear regions (regions near 0 and 1), and large segments for linear regions (regions around 0.5). The idea is to use segments that grow by a factor of two from 0 to 0.5, and segments that shrink by a factor of two from 0.5 to 1 in the x-axis of Figure 1. We use segment boundaries at locations 2^{n-8} and $1 - 2^{-n}$ where $0 \leq n < 8$. Up to 14 segments can be formed this way. A circuit based on prefix computation can be used for calculating segment addresses (Figure 2) for a given input x. It checks the number of leading zeros and ones to work out the segment address. A cascade of OR gates is used for segments that grow by factors of two, and a cascade of AND gates is used for segments that shrink by factors of two; these circuits can be pipelined and a circuit with shorter critical path but requiring more area can be used [16]. Note that the choice of segments does not have to be factors of two, it could be more. The appropriate taps are taken from the cascades depending on the choice of the segments and are added to work out the segment address. In Figure 2, the maximum available taps are taken, giving 14 segment addresses. Some taps would not be taken if the segments grow or shrink by more than a factor of two. It can be seen that the critical path of this circuit is the path from x_6 or x_7 to the output of the adder. By introducing pipeline registers between the gates, higher throughput can be easily achieved.

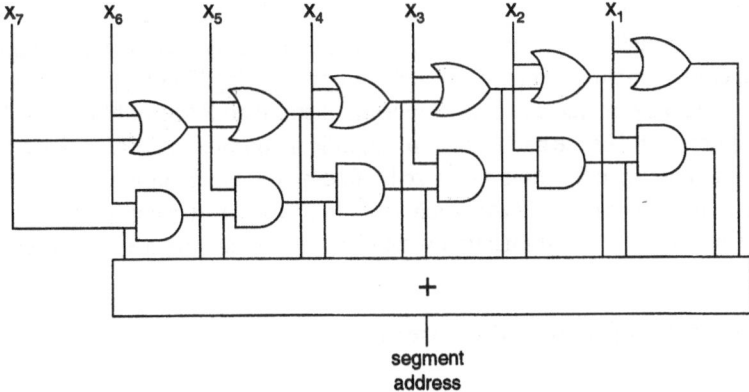

Fig. 2. Circuit to calculate the segment address for a given input x. The adder counts the number of ones in the output of the two prefix circuits.

When approximating $\sqrt{-\ln(x)}$ with 32-bit inputs based on polynomials of the form $p(x) = c_1 \times x + c_0$, the gradient of the steepest part of the curve is in the order of 10^8, thus large multipliers would be required. To overcome this problem, we use scaling factors of multiples of two to reduce the magnitude of the gradient, essentially trading precision for range. This is appropriate since the larger the gradient, the less important precision becomes. The use of scaling factors provides the user the ability to control the precision for both c_1 and c_0, resulting in variation of the size of the multiplier and adder. Hence for each segment four coefficients are stored: c_1 and its scaling factor, c_0 and its scaling factor.

It is also possible to divide the input interval into uniform or non-uniform intervals, and have uniform or non-uniform segments inside each interval. In this case, the most-significant bits are used to address the intervals, and the least-significant bits are used to address the segments inside each interval. It can be seen that one can have any number of nested combinations of uniform and non-uniform segments. This hybrid combination of nested uniform and non-uniform segments provides a flexible way to choose the segment boundaries. Currently, this segmentation step is done by hand, which is slow and far from optimal. A possible approach to automate this step is discussed in Section 5.

4   Hardware Architecture

The architecture of our function evaluator shown in Figure 3 is based on polynomials of the form $p(x) = c_1 \times x + c_0$. The most-significant bits are used to select the interval, and the least-significant bits are passed through the segment address calculator which calculates the segment address within the interval. The design shown is developed for the common cases, and has been used in the examples of this paper. For other cases, one could divide the input bits into more than two parts and apply the segment address calculation depending on whether the parts use uniform or non-uniform segments.

The ROM outputs the four coefficients for the chosen interval and segment. c_1 is multiplied by the input x and c_s_1 is used to scale the output. The scaling circuit involves shifters, which increase or decrease the value by powers of two. This scaled multiplication value is added to the scaled c_0 coefficient to produce the final result.

For high throughput applications, the segment address calculator, the multiplier and the adder can be pipelined. For typical applications targeting FPGAs, the ROM would be small and could be implemented on-chip using distributed RAM or block RAM. Often the multiplier would be the part taking up a significant portion of the area. Therefore it is important to minimize the multiplier size by finding out the minimum bit width for the coefficient c_1. Also recent FPGAs, such as Xilinx Virtex-II devices, provide dedicated hardware resources for multiplication which can benefit the proposed architecture.

5   Placement of Segment Boundaries

Let f be a continuous function on $[a, b]$, and let an integer $m \geq 2$ specify the number of contiguous intervals into which $[a, b]$ has been partitioned: $a = u_0 \leq u_1 \leq ... \leq u_m = b$. Let n_i and $d_i(i = 1, ..., m)$ be non-negative integers and let P_i denote the set

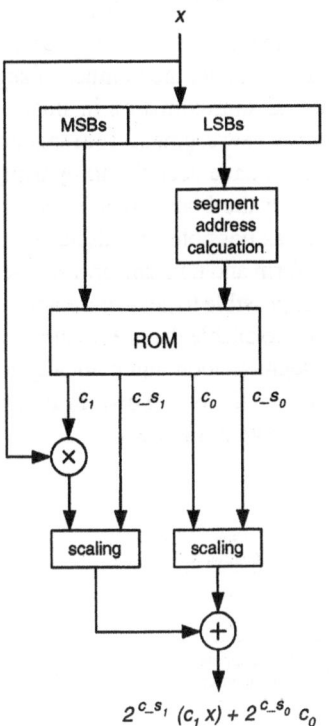

$$2^{c\_s_1}(c_1x) + 2^{c\_s_0}c_0$$

Fig. 3. Our function evaluator architecture.

of rational functions p_i whose numerators and denominators are polynomials of degrees less or equal to n_i and d_i, respectively. For $i = 1, ..., m$, define

$$h_i(u_{i-1}, u_i) = \min_{p_i \in P_i} \max_{u_{i-1} \leq x \leq u_i} |f(x) - p_i(x)|. \tag{3}$$

Let $\mu = \mu(u) = \max_{1 \leq i \leq m} h_i(u_{i-1}, u_i)$. Lawson states in his paper [18] that the segmented rational minimax approximation problem is that of minimizing μ over all partitions u of $[a, b]$. It can be shown that if the error norm is a non-decreasing function of the length of the interval of approximation, that the function to be approximated is continuous and that the goal is to minimize the maximum error norm on each interval, then a balanced error solution is optimal; the term "balanced error" means that the error norms on each interval are equal.

Pavlidis and Maika present an iterative scheme for segmentation in their paper [19] which results in a suboptimal balanced error solution. The scheme is based on an iteration of the form

$$u_m^{k+1} = u_m^k + c(e_{m+1}^k - e_m^k), \quad m = 1, ..., n-1. \tag{4}$$

Here u_m^k is the value of the m-th point and the k-th iteration, e_m^k is the error on $(u_{m-1}^k, u_m^k]$ and c is an appropriate small positive number. It can be shown that for sufficiently small

c the scheme converges to a solution [19]. In this algorithm, the number of segments is fixed and this determines the maximum error. However in many cases, it may be more useful to fix the accuracy desired and let the number of segments vary. Starting from a or b one could apply polynomial approximation in small increments, until the desired accuracy is reached. Then start a new segment from that point.

Once the segment boundaries have been found by using one of the two approaches above, the next step is to match the boundaries based on our addressing scheme as close to the suboptimum ones as possible. As discussed in Section 3, our addressing scheme is based on nested uniform and non-uniform segments. By carefully using these combinations of segments, it is possible to get a close approximation to the suboptimum segment boundaries. Our aim is to enable the user to input constraints such as maximum error norm and to apply the segmentation automatically to produce lookup tables and the corresponding circuits such as the one shown in Figure 3. A possible approach of such an automated method is shown in Figure 4.

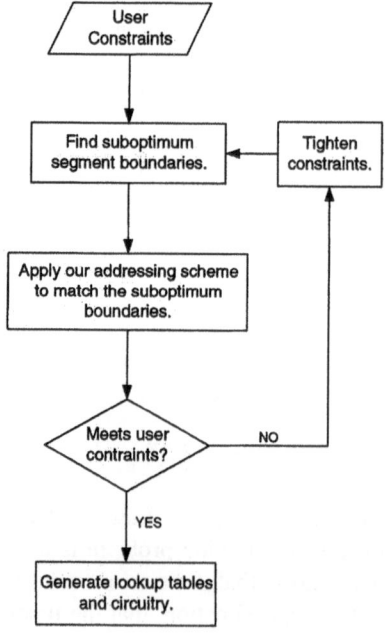

Fig. 4. Steps for automating segmentation.

6   Evaluation and Results

Our function evaluator has been successfully implemented for the Gaussian noise generator presented in [3]. Three functions are approximated: $\sqrt{-\ln(x)}$, $\cos(2\pi x)$ and

$\sin(2\pi x)$ over $[0, 1]$. 32-bit inputs are used for $\sqrt{-\ln(x)}$ and 16-bit inputs are used for $\cos(2\pi x)$ and $\sin(2\pi x)$.

We first consider the function $\sqrt{-\ln(x)}$. As stated earlier, the greatest non-linearities of this function occur in the regions close to zero and one. To be consistent with the change in linearity, we use line segment locations to boundaries at locations 2^{n-32} for $0 < x \le 0.5$, and $1 - 2^{-n}$ for $0.5 < x \le 1$, where $0 \le n < 32$. A total of 59 segments are used to approximate this function as shown in Figure 5. Since $\sqrt{-\ln(x)}$ approaches infinity for x values close to zero, the smallest x value is $1/2^{32}$, resulting in a maximum output value of around 4.7.

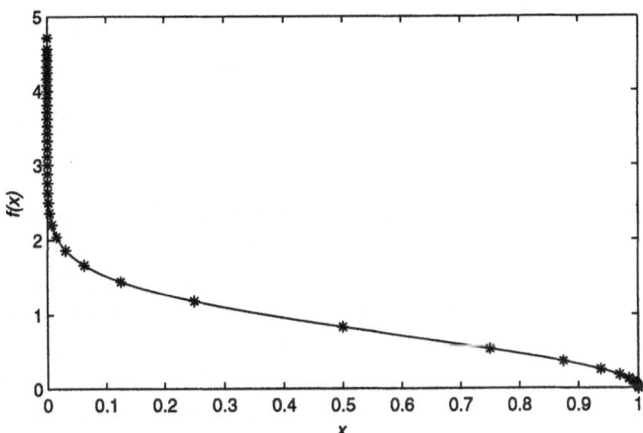

Fig. 5. The segments used to approximate $\sqrt{\ln(x)}$ with 32-bit inputs. The asterisks indicate the segment boundaries of the linear approximations.

The maximum absolute error of this approximation is 0.020. However this is the case only if we have infinite precision for the coefficients, which is not realistic. Multipliers take significant amount of resources on FPGAs, therefore the coefficients for the gradient should be as small as possible. Tests are carried out to find the optimum number of bits for the gradient coefficients that provides the least absolute error. Figure 6 shows how the maximum absolute error varies with the number of bits used for the gradient of $\sqrt{-\ln(x)}$. The figure indicates that six bits are sufficient to give a maximum absolute error of 0.031. The approximation should differ from the true value by less than one unit in the last place (ulp) [17]; the least significant bit of the fraction of a number in its standard representation is defined to be the last place. With this error, it is sufficient to give an output accuracy of eight bits (three bits for integer and five for fraction). If uniform segments are used, small segment size would be needed in order to cope with the highly non-linear parts of the curve. In fact, one would require around 617 million segments to get the same maximum absolute error with uniform segments. This is a good example to demonstrate the effectiveness of our non-uniform approach. It is clear that our approach works well especially for functions with exponential behavior.

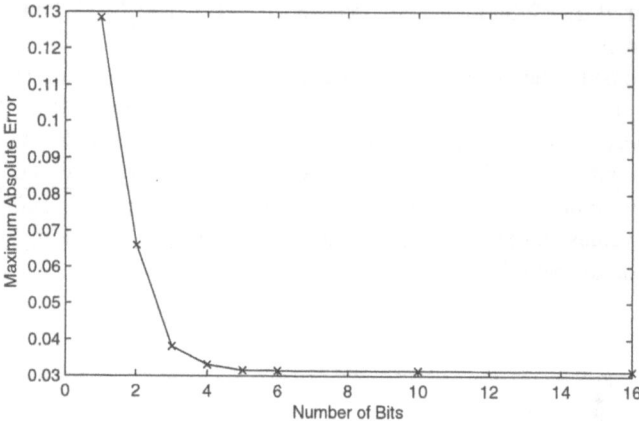

Fig. 6. Variation of function approximation error with number of bits for the gradient of $\sqrt{-\ln(x)}$.

To evaluate the functions $\cos(2\pi x)$ and $\sin(2\pi x)$, due to the symmetry of the sine and cosine functions, only the input range $[0, 1/4]$ for $\cos(2\pi x)$ needs to be approximated [15]. The specific axis-partitioning technique for $\sqrt{-\ln(x)}$ is unsuitable for $\cos(2\pi x)$, since the non-linearities of the two functions are different. If the same technique is used, there would be many unnecessary segments near the beginning and end of the curve, and not enough segments in the middle regions. As before we consider both the local linearity of the curve, and the computational concerns with respect to choosing specific segment boundary locations, leading to the approximations shown in Figure 7. The curve is divided into four uniform intervals and within each interval, non-uniform segmentation is applied. Note that for each interval, not all taps are taken from the segment address calculator. We use a total of 21 segments to approximate this function.

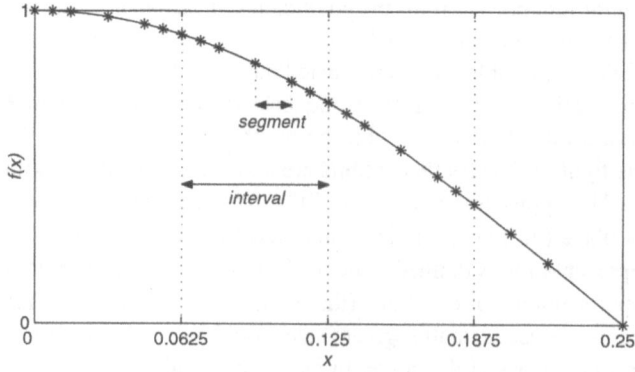

Fig. 7. Approximation for $\cos(2\pi x)$ over $[0, 1/4]$. The asterisks indicate the segment boundaries of the linear approximations.

With finite precision on the coefficients, the maximum absolute error of this approximation is 0.0035, which is sufficient to give an output accuracy of eight bits (all eight bits for fraction). Using uniform segments, the same error can be obtained with a slightly larger number of segments; this is because the curve does not have high non-linearities.

Table 1 shows a comparison of the number of segments for the two functions for non-uniform and uniform segmentation in order to achieve the same worst-case error. Note that for uniform segmentation, the number of segments needs to be a power of two. This is because the most-significant n bits are used for addressing. For instance, the actual number of uniform segments needed for the $\sqrt{-\ln(x)}$ function is 617 million, but 1 billion segments are used which is the next power of two (2^{30}). We do not have this kind of a restriction with our non-uniform addressing scheme. The table also shows the number of bits used for each coefficient in the look-up tables. The lookup tables for the three functions $\sqrt{-\ln(x)}$, $\cos(2\pi x)$ and $\sin(2\pi x)$ have a total size of just 3504 bits. With such small lookup table size, all the coefficients can be stored on-chip for fast access.

Table 1. Second column shows the comparison of the number of segments for non-uniform and uniform segmentation. Third column shows number of bits used for the coefficients to approximate the $\sqrt{-\ln(x)}$ and $\cos(2\pi x)$ functions.

function	non-uniform	uniform	c_1	c_s_1	c_0	c_s_0
$\sqrt{-\ln(x)}$	59	1 billion	6	5	32	5
$\cos(2\pi x)$	21	32	8	4	16	4

The function evaluators for the three functions are written using the Handel-C hardware compiler from Celoxica [25], and are mapped and tested on a Xilinx Virtex-II XC2V4000-6 device [24]. The design occupies 1864 slices, four block multipliers and two block RAMs, and takes up around 7% of the device. A fully pipelined version of our design operates at 133 MHz with a latency of 14 clock cycles, and the function evaluators are capable of 133 million operations per second; the completion time for each input is given by 14 / 133 million = 105 ns. The design has also been implemented on a low cost Xilinx Spartan-IIE XC2S300E-7, which occupies 70% of the chip and is capable of 62 million operations per second. Our hardware implementations have been compared with software implementations (Table 2). The Virtex-based FPGA implementation is 158 times faster than the Athlon-based PC in terms of throughput, and 11 times faster in terms of completion time.

Well-known function evaluation methods, such as SBTM [5], [6], deal with the approximation of elementary functions over a fixed input range where the function is linear. Range reduction techniques such as those presented in [22] and [23] are used to bring the input within the linear range. However, range reduction is not possible for most compound functions. Our approach caters for both non-linear and linear regions, which makes it suitable for both elementary and compound functions. Currently, our approach tends to produce small lookup table sizes with low accuracy; we hope to improve accuracy by further work on automatic segmentation.

Table 2. Performance comparison: computation of $\sqrt{-\ln(x)}$, $\cos(2\pi x)$ and $\sin(2\pi x)$. All PCs are equipped with 512MB DDR RAM. The XC2V4000-6 FPGA belongs to the Xilinx Virtex-II family, while the XC2S300E-7 belongs to the Xilinx Spartan-IIE family. The software implementations are written in C generating single precision floating point numbers, and are compiled with the GCC 3.3 compiler [26].

platform	clock speed (MHz)	latency (clock cycles)	area (slices)	throughput (operations / second)	completion time (ns)
XC2V4000-6 FPGA	133	14	1864	133 million	105
XC2S300E-7 FPGA	62	14	2129	62 million	226
AMD Athlon PC	1400	-	-	0.84 million	1187
Intel Pentium 4 PC	2400	-	-	0.79 million	1261

7 Conclusion

This paper presents a novel method for evaluating functions using polynomial approximations by employing non-uniform segments. The non-uniform segments deal with the non-linearities of functions which occur frequently. A simple cascade of AND and OR gates can be used to rapidly calculate the segment address for a given input. Scaling factors are used to deal with large polynomial coefficients, trading precision with range. Two functions developed for the generation of Gaussian noise are used as examples to illustrate and to evaluate our approach. Results show the advantages of using non-uniform segments over uniform ones. Current and future work includes automating the selection of boundaries, and exploring the use of higher order polynomials for more accurate approximations. This would enable us to apply our approach to a wide range of functions and to obtain detailed comparison with other methods. We will also look at how our function evaluator can be used to speed up addition and subtraction functions in logarithmic number systems [12], which are highly non-linear functions.

Acknowledgment

The authors thank Jun Jiang and Shay Ping Seng for their assistance. The support of Celoxica Limited, Xilinx Inc., the U.K. Engineering and Physical Sciences Research Council (Grant number GR/N 66599, GR/R 55931 and GR/R 31409), and the U.S. Office of Naval Research is gratefully acknowledged.

References

1. O. Mencer, N. Boullis, W. Luk and H. Styles, "Parameterized function evaluation for FPGAs", *Field-Programmable Logic and Applications*, LNCS 2147, pp. 544–554, 2001.
2. J.E. Volder, "The CORDIC trigonometric computing technique", *IEEE Trans. on Elec. Comput.*, vol. EC-8, no. 3, pp. 330–334, 1959.
3. D. Lee, W. Luk, J. Villasenor and P.Y.K. Cheung, "A hardware Gaussian noise generator for channel code evaluation", *Proc. IEEE Symp. on Field-Prog. Cust. Comput. Mach.*, 2003.

4. D. Das Sarma and D.W. Matula, "Faithful bipartite rom reciprocal tables", *Proc. 12th IEEE Symp. on Comput. Arith.*, pp. 17–28, 1995.
5. M.J. Schulte and J.E. Stine, "Symmetric bipartite tables for accurate function approximation", *Proc. 13th IEEE Symp. on Comput. Arith.*, vol. 48, no. 9, pp. 175–183, 1997.
6. M.J. Schulte and J.E. Stine, "Approximating elementary functions with symmetric bipartite tables", *IEEE Trans. Comput.*, vol. 48, no. 9, pp. 842-847, 1999.
7. J.A Pineiro, J.D. Bruguera and J.M. Muller, "Faithful powering computation using table look-up and a fused accumulation tree", *Proc. 15th IEEE Symp. on Comput. Arith.*, 2001.
8. J. Cao, B.W.Y. We and J. Cheng, " High-performance architectures for elementary function generation", *Proc. 15th IEEE Symp. on Comput. Arith.*, 2001.
9. V.K. Jain, S.A. Wadecar and L. Lin, "A universal nonlinear component and its application to WSI", *IEEE Trans. Components, Hybrids and Manufacturing Tech.*, vol. 16, no. 7, pp. 656–664, 1993.
10. H. Hassler and N. Takagi, "Function evaluation by table look-up and addition", *Proc. of the IEEE 12th Symp. on Comp. Arith.*, pp. 10–16, 1995.
11. J. Detrey and F. de Dinechin, "Multipartite tables in JBits for the evaluation of functions on FPGAs", *Proc. IEEE Int. Parallel and Distributed Processing Symp.*, 2002.
12. D.M. Lewis, "Interleaved memory function interpolators with application to an accurate LNS arithmetic unit", *IEEE Trans. Comput.*, vol. 43, no. 8, pp. 974–982, 1994.
13. J.F. Hart et al., *Computer Approximations*, Wiley, 1968.
14. J.R. Rice, *The Approximation of Functions*, vol. 1,2, Addison-Wesley, 1964, 1969.
15. J.M. Muller, *Elementary Functions: Algorithms and Implementation*, Birkhauser Verlag AG, 1997.
16. R.E. Ladner and M.J. Fischer, "Parallel prefix computation", *JACM*, vol. 27, no. 4, pp. 831–838, 1980.
17. I. Koren, *Computer Arithmetic Algorithms*, Prentice Hall, 1993.
18. C.L. Lawson, "Characteristic properties of the segmented rational minimax approximation problem", *Numer. Math.*, vol. 6, pp. 293–301, 1964.
19. T. Pavlidis and A P Maika, "Uniform piecewise polynomial approximation with variable joints", *Journal of Approximation Theory*, vol. 12, pp. 61–69, 1974.
20. C. Jones, E. Vallés, C. Wang, M. Smith, R. Wesel and J. Villasenor, "High throughput Monte Carlo simulation for error floor testing in capacity achieving channel codes", *Proc. IEEE Symp. on Field-Prog. Cust. Comput. Mach.*, 2003.
21. N. Sidahao, G.A. Constantinides and P.Y.K. Cheung, "Architectures for function evaluation on FPGAs", *Proc. IEEE Int. Symp. on Circ. and Syst.*, 2003.
22. J.S. Walther, "A unified algorithm for elementary functions", *Proc. Spring Joint Comput. Conf.*, 1971.
23. N.W. Cody and W. Waite, *Software Manual for the Elementary Functions*, Prentice-Hall, 1980.
24. Xilinx Inc., *Virtex-II User Guide v1.5*, 2002.
25. Celoxica Limited, *Handel-C Language Reference Manual*, ver. 3.1, document no. RM-1003-3.0, 2002.
26. GNU Project, *GCC 3.3 Manual*, http://gcc.gnu.org, 2003.

A Dual-Path Logarithmic Number System Addition/Subtraction Scheme for FPGA

Barry Lee and Neil Burgess

Cardiff school of Engineering, Cardiff University, Queen's Buildings,
The Parade, Cardiff. CF24 3TF U.K.
{Leebr2,BurgessN}@cf.ac.uk

Abstract. A new architecture for calculating the addition/subtraction function required in a logarithmic number system (LNS) is presented. A substantial logic saving over previous works is illustrated along with similarities with the dual-path floating-point addition method. The new architecture constrains the lookups to be of fractional width and uses shifting to achieve this. Instead of calculating the function $\log_2(1\pm2^{M\text{-}K})$ in two lookups the function arithmetic is performed (i.e. the two functions $2^{M\text{-}K}$ and $\log_2(\ )$, plus a correction function) as this allows logic sharing that maps well to FPGA. Better-than-floating-point (BTFP) accuracy is used to enable a future comparison with floating-point.

1 Introduction

The logarithmic number system (LNS) is a number system based on fixed-point arithmetic and has a high dynamic range comparable to floating-point. The LNS has the benefits of simplified multiplication, division and powering (including: square root, cube, square and reciprocal square root to name a few special cases) but at the cost of a complex addition/subtraction function. This paper describes the design of an FPGA core to evaluate the addition/subtraction function as a basis to efficiently port the LNS benefits to FPGA. The function is calculated with an error that is equivalent to floating-point to enable a fair comparison of the two systems (not covered in this work).

The paper is organised as follows. Section 2 introduces the LNS and gives some background details. Section 3 outlines the function partition method adopted. Section 4 describes the reasoning behind the BTFP accuracy. Section 5 gives details of the FPGA implementation including the function approximation methods and macro structures. Section 6 presents the results and finally section 7 concludes.

2 LNS Background

A base-2 LNS number is represented by the couple $<s_a, e_a>$ where s_a is a 1-bit sign and e_a is an n-bit fixed-point number. Typically e_a is of two's complement form with integer (I) and fractional (F) sections as shown in figure 1. The real value R_v of an LNS number is given as,

P.Y.K. Cheung et al. (Eds.): FPL 2003, LNCS 2778, pp. 808–817, 2003.
© Springer-Verlag Berlin Heidelberg 2003

$$R_v = (-1)^{s_a} \times 2^{e_a}. \tag{1}$$

This can be considered as a floating-point value that has an exponent with fractional precision and a significand of 1. A bias can be added to e_a to simplify comparisons similarly to the bias added to a floating-point exponent.

$$e_a = $$

Fig. 1. The integer and fractional sections of the magnitude e_a of an LNS number

Consider two logarithmic values K and M, equations (2) and (3) respectively. The equivalent LNS operators to the real operators of multiplication, division, powering and addition are shown in (4), (5), (6) and (7) respectively.

$$K = \log_2(X). \tag{2}$$

$$M = \log_2(Y). \tag{3}$$

$$\log_2(X \times Y) = K + M. \tag{4}$$

$$\log_2(X / Y) = K - M. \tag{5}$$

$$\log_2(X^P) = X \times P. \tag{6}$$

$$\log_2(X \pm Y) = K + \log_2(1 \pm 2^{M-K}), \text{ where, } K \geq M. \tag{7}$$

Multiplication and division are fixed-point addition and subtraction operations with only simple overflow detection overheads and are exact operations. Powering is fixed-point multiplication, which can be considered as shifting for square root and squaring. The addition/subtraction function, a graph of which is shown in figure 2, is somewhat more complicated to perform. From figure 2 it can be seen that the addition function is well behaved and has a small range of (0,1]. Approximating the addition function is straight forward and a piecewise polynomial approach has been used in previous works [1] and [2]. The subtraction function is not so simple to approximate, which is primarily due to the singularity where the function value tends to -∞ as the argument tends to 0. The region where the function tends to minus infinity is the lookup region required to calculate the difference of two numbers that are very similar and where catastrophic cancellation occurs (for floating-point). The difference of two very similar numbers produces a very small result and in the LNS this is represented by a very large (in magnitude) negative value.

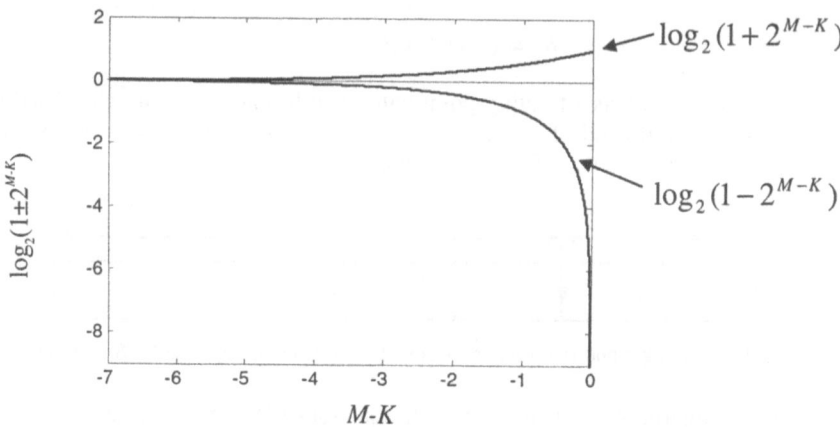

Fig. 2. A graph of the functions to be computed to enable LNS addition and subtraction

3   Function Partitioning

To arrive at the method implemented in this work the addition/subtraction function was first split into three sections. The three sections A, B and C are shown in figure 3.

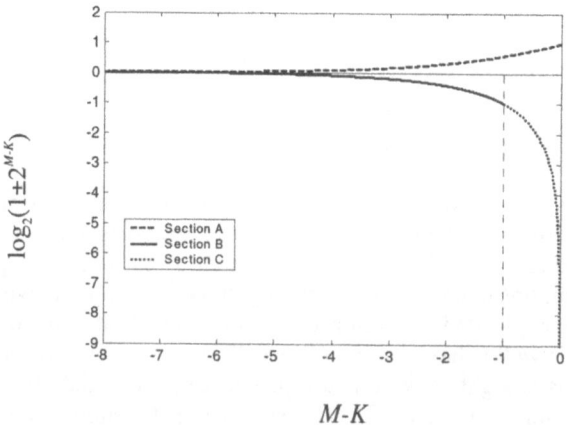

Fig. 3. The three section split of the addition/subtraction functions

To reduce the lookup sizes needed for the function approximations we calculate section A of the addition function and section B of the subtraction function by calculating the power function 2^{M-K} and the logarithm function $\log_2()$. This idea was first suggested in [3] although it is applied in a different manner. For section A the addition function is calculated on the range $[T,0]$, where T is the smallest value of $M-K$ that does not cause the addition function to evaluate to a number smaller than the smallest number in the number system. This is analogous to the addition of a very small number and a very large number where the small number is shifted out of range of the large number and effectively an addition of zero produces the same result after rounding. The diagram of the required hardware to calculate the function in section A is shown in figure 4(a).

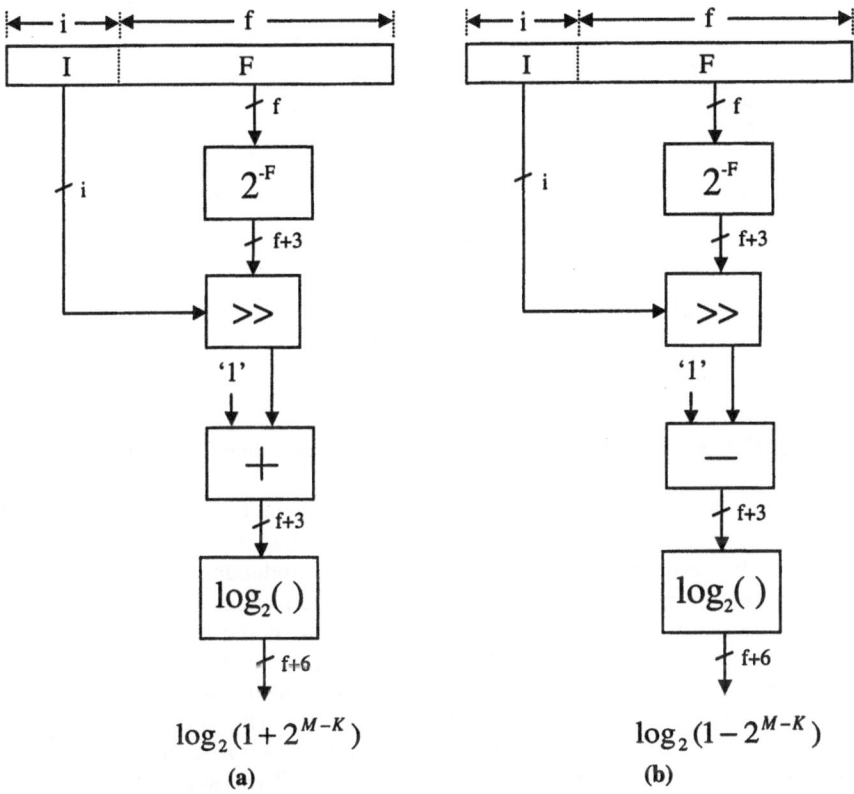

$$\log_2(1+2^{M-K})$$

(a)

$$\log_2(1-2^{M-K})$$

(b)

Fig. 4. (a) The required hardware to calculate the addition function on the range [T,0]. **(b)** The required hardware to calculate the subtraction function on the range [T,1]

For section B the subtraction function is calculated on the range [T,1] where T has the same definition as for the addition function above. The hardware required for section B is illustrated in figure 4(b). For section C, the subtraction function on the range (1,0), the hardware model of figure 4(b) cannot be used. This is due to the loss of accuracy, which can only be corrected by using very high precision approximations. To solve the accuracy dilemma we use an identity (9) introduced in [4].

$$\text{Let, } R = M - K. \tag{8}$$

$$\log_2(1-2^R) = \log_2\left(\frac{1-2^R}{-R}\right) + \log_2(-R). \tag{9}$$

The identity consists of two parts. The first part, which we will call the 'correction function', is a correction to the $\log_2(-R)$ function. A plot of the subtraction function, the $\log_2(-R)$ function and the correction function is shown in figure 5.

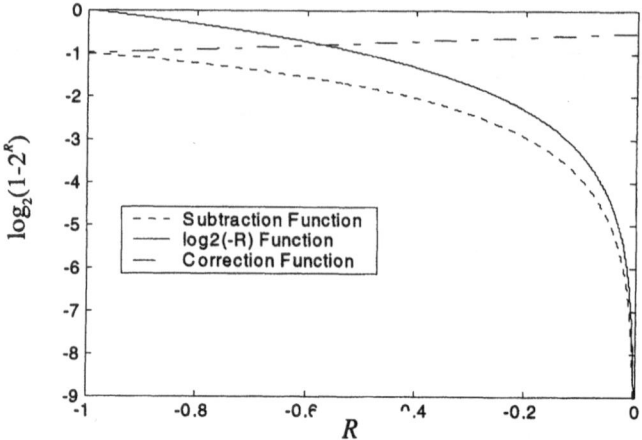

Fig. 5. A plot of the components that make up the subtraction function identity

From figure 5 we can see that the correction function plots a smooth regular curve of restricted range, which lends itself well to polynomial approximation. The $\log_2(-R)$ function can be easily evaluated using a shifting range reduction technique to reduce the approximation range to a single 'binade' and then adding the binary encoded shifting quantity to the approximation. The diagram of the required hardware to calculate the subtraction function for the range (1,0) is shown in figure 6.

To forge a complete hardware solution, similarities in the hardware diagrams are combined and hardware sharing is used where possible. The same hardware can be used to produce sections A and B with a minor modification to allow the addition component to perform a subtraction as necessary. The range of the function approximations is the same for each section, i.e. 2^{-F} is evaluated on the range [0,1) and $\log_2( )$ is evaluated on the range [1,2). Sections A and B need to calculate $\log_2( )$ over the range [1,2) as does section C so the $\log_2( )$ approximation can be shared between all sections. This will not have a great impact on the delay as the function need only be calculated once for any approximation input. The correction function and the 2^{-F} function can share the same arithmetic hardware (multipliers and adders) to calculate their approximations because again, only one of the functions is required for any approximation input. The diagram of the hardware model to calculate the whole LNS addition/subtraction function is shown in figure 7.

4   Better-than-FP Accuracy

This section on accuracy is adapted from the work by Lewis [1] and gives a bound on the permissible approximation error to produce BTFP results, a term coined by

Arnold [2]. Floating-point has a relative error $2^{-f-2} < \varepsilon_{FP} \le 2^{-f-1}$, so the worst-case error is 2^{-f-1}. The LNS has an absolute accuracy 2^{-f-1} corresponding to a relative accuracy of $\varepsilon_{LNS} = 2^{2^{-f-1}} -1 \approx 2^{-f-1.528}$ and giving a ½ ulp accuracy improvement over floating-point. The worst-case relative error is achievable for a floating-point number system and should be for IEEE compliant systems [5]. The worst-case relative error is only achievable for LNS if exact arithmetic is possible.

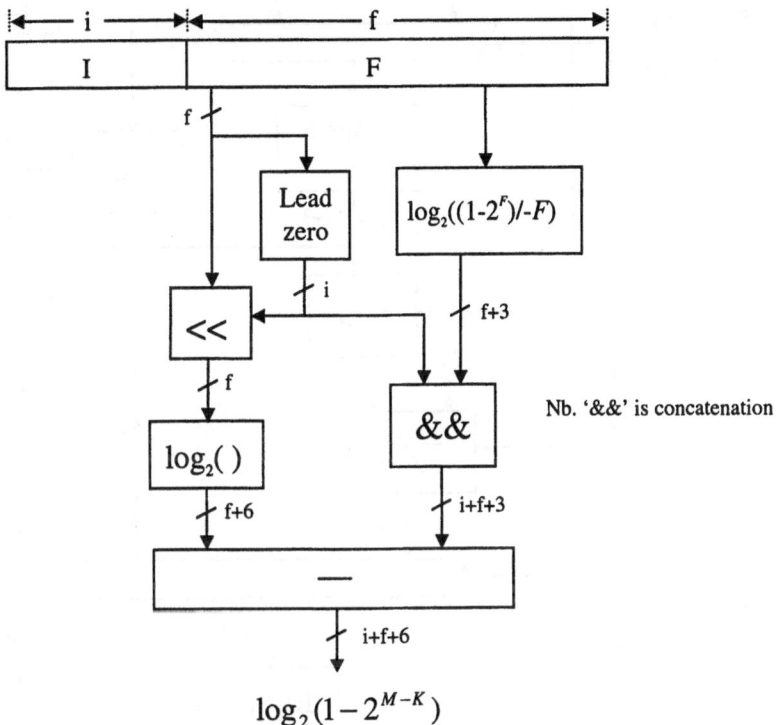

$$\log_2 (1 - 2^{M-K})$$

Fig. 6. A diagram of the hardware to calculate the subtraction function on the range (1,0)

To implement the logarithmic addition/subtraction function, transcendental functions need to be calculated either by using full table lookup or by function approximation. To guarantee correctly rounded results for function approximation methods the results need to be calculated with many extra bits of precision to overcome the 'table-makers-dilemma' problem [9]. In short the worst-case relative error is not achievable in any practical implementation so a compromise is taken.

The LNS addition/subtraction function will incur an error as a result of rounding, this is 2^{-f-1}. The addition/subtraction function approximation will also have an error ε_{dp}. The total error is $\varepsilon_{dp} + 2^{-f-1}$. The relative error for a realistic LNS implementation (i.e. one with approximation errors) is,

$$\varepsilon_{LNS} = 2^{(\varepsilon_{dp}+2^{-f-1})} -1 = \log_e(2) \times (\varepsilon_{dp} + 2^{-f-1}) \tag{10}$$

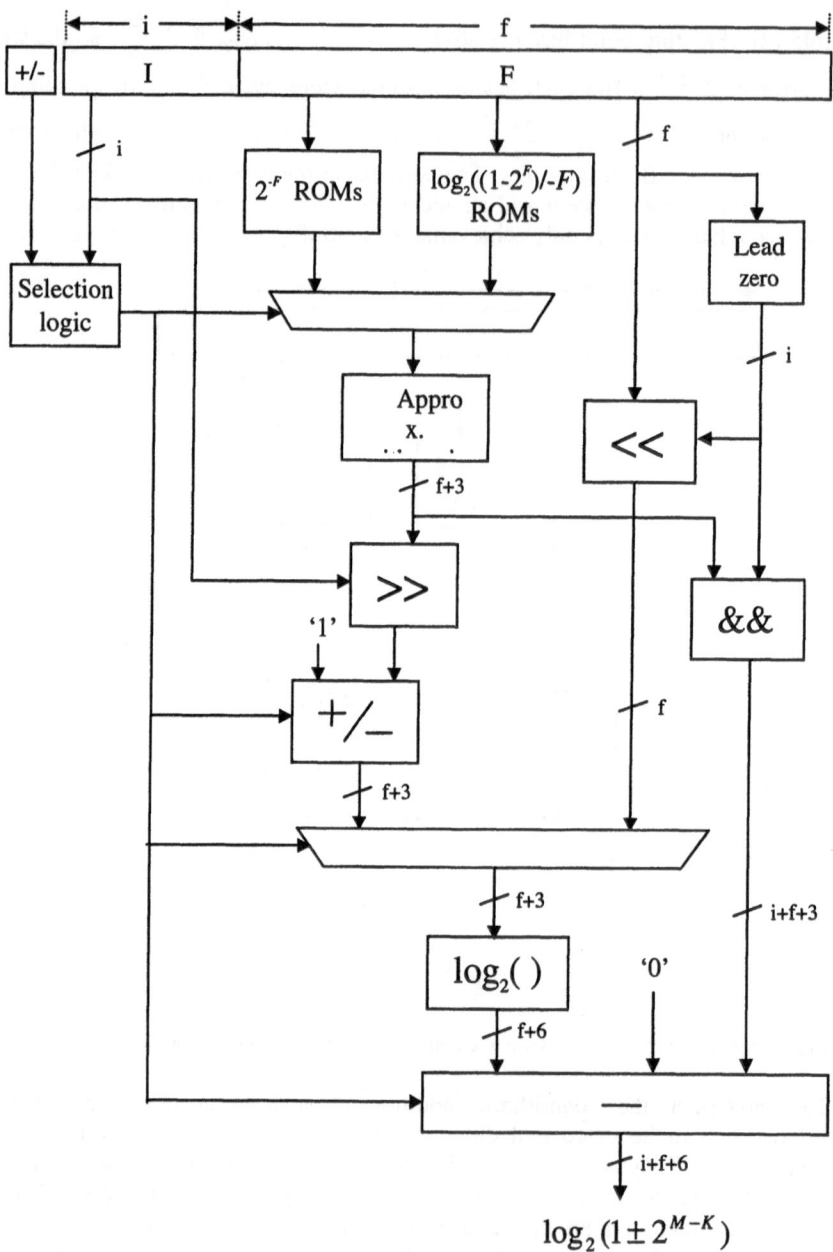

Fig. 7. The high-level hardware model to calculate the LNS addition/subtraction function

We would like the LNS to have an error that is equivalent or less than that of floating-point. The equation to calculate the permissible data path error that results in an equivalent accuracy of floating-point is,

$$\log_e(2) \times (\varepsilon_{dp} + 2^{-f-1}) \leq 2^{-f-1} \qquad (11)$$

Solving for $\mathcal{E}_{dp}$ gives,

$$\mathcal{E}_{dp} \le \left(\frac{1}{\log_e(2)} - 1\right) \times 2^{-f-1} \approx 1.771 \times 2^{-f-3} \approx 2^{-f-2.18} \tag{12}$$

Equation (12) requires the addition/subtraction function to be calculated correctly to three extra bits of precision to guarantee an error that is less than an equivalent floating-point system.

5   FPGA Implementation

A software model of the system has been developed using the MATLAB mathematical software package. The model allows the accuracy to be tested by enumeration to guarantee the BTFP accuracy criteria and provides a bit-true model to compare with the hardware model. The hardware model is written in VHDL with MATLAB being used to fill the coefficient ROMs needed for the function approximations.

5.1   Function Approximation

To calculate 2^{-f} and the correction function a simple 2nd order Lagrange polynomial approximation evaluated by Horner's method is used. This is not the optimum approximation technique but it is simple to create and suffices for this study. The basic approximation is rearranged to allow the coefficient ROMs to be addressed by the MSBs of the argument and the LSBs to be the 'x-terms' of the polynomial. The $\log_2(\ )$ function is evaluated by means of a third order Taylor approximation. The main choice for this is that such an approximation scheme has already been developed.

5.2   Shifter, Memory and Multiplier Structure

The shifters are constructed using layers of 4:1 and 2:1 multiplexers as done in [6] as these are the most efficient Virtex LUT-based multiplexer implementations. The shifters are binary encoded so each bit of the 'shift_amount' directly controls a multiplexer input. The multipliers are built based on the scheme given in [6], which uses one embedded multiplier and some extra logic to produce wider than 17X17 multipliers. The memory for the approximations is constructed using LUT-based memory. One LUT can be configured as a 4-bit in 1-bit out memory and many are multiplexed using dedicated embedded multiplexers to produce wider input memories.

6   Implementation Results

The addition/subtraction core has been implemented using Xilinx ISE 5.1.03i for an XC2V1000-4 part with results being generated for the two metrics of delay (ns) and area (slices). The delay is calculated by placing registers on the inputs and outputs and measuring the register-to-register delay as reported by the P&R tools. Table 1 shows the implementation results of an addition/subtraction core for an integer width of 8 and a fraction width of 23, which is the LNS format that is comparable to IEEE single precision floating-point. The results are given for implementations that use only the 'logic-elements'; use a combination of the 'logic-elements' and 4 embedded multipliers; and a theoretical implementation that uses 'logic-elements', 4 embedded multipliers and 2 blockRAMs.

Table 1. Implementation results of the LNS addition/subtraction function for an integer width of 8 and a fractional width of 23

	'Logic-elements' only	'Logic-elements' and 4 embedded multipliers	'Logic-elements', 4 embedded multipliers and 2 blockRAMs
Area (slices)	1832	1289	1001
Delay (ns)	99	101	X

For the theorectical implementation the blockRAMs replace two 128X36-bit LUT-based ROMs. A single 128X36-bit ROM uses 8*36=288 LUTs therefore two ROMs and two LUTs per slice means a saving of 288-slices.

Arnold [2] gives results for two FPGA implementations of the LNS addition function only. The unrestricted faithful rounding LNS adder uses 768-slices but has an accuracy that is worse than floating-point. Results for an unrestricted faithfully rounded subtraction function are not given. The estimated area for the BTFP LNS adder proposed by Lewis [1] is given in Arnold [2] as 2300-slices. Again, results for BTFP LNS subtraction are not given but would require a 10-fold increase in lookup ROM due to the singularity region [1]. Matousek et al. [8] implement a BTFP addition/subtraction function plus the logic to calculate the LNS sum using 1300-slices and 96 blockRAM cells. By using 96 blockRAMs the number of single precision LNS addition/subtraction components that can be fitted onto a single Virtex-II FPGA is restricted to one for all but the largest chip that could accommodate two.

7   Conclusion

A scheme has been presented that substantially reduces the lookup requirement and thus the area requirement of the implementation of an LNS addition/subtraction function on FPGA. Due to the area reductions it is now possible to accommodate 4 single-precision BTFP LNS addition/subtraction units on an XC2V1000 FPGA and potentially 45 on the largest 10M-gate chip [10]. A dual-path approach similar to that of floating-point addition has been adopted and this reduces the lookups to be of fractional length. BTFP accuracy has been used as this allows a fair comparison with floating-point [6], which is a future work. Pipelining, delay reduction and efficient fixed-point to LNS conversions and vice-versa are also future works.

Acknowledgement

A special thanks to Richard Walke and the RTSL team at QinetiQ, Malvern.

References

1. Lewis, D.M.: Interleaved Memory Function Interpolators with Application to an Accurate LNS Arithmetic Unit. IEEE Trans. on Comp. Aug. 1994. pp 974-982.
2. Arnold, M.G., Walter, C.: Unrestricted Faithful Rounding is Good Enough for Some LNS Applications. 15th symp. on Comp. Arith. 2001. pp 237-246.
3. Chen C., Chen R., Yang C.: Pipelined Computation of Very Large Word-length LNS Addition/Subtraction with Polynomial Hardware Cost. IEEE Trans. on Comp. Jul. 2000. pp 716-726.
4. Poliouris, V., Stouraitis, T.: A Novel Algorithm for Accurate Logarithmic Number System Subtraction. IEEE Symp. on Circuits and Sys. May. 1996. pp 268-271.
5. IEEE Standard for Binary Floating-Point Arithmetic, ANSI/IEEE std. 754. 1985.
6. Lee, B.R., Burgess, N.: Parameterisable Floating-Point operations on FPGA. 36th Asilomar Conf. on Signals, Systems and Comp. Nov. 2002.
7. Taylor, F.J., Gill, R., Joseph, J., Radke, J.: A 20-bit Logarithmic Number System Processor. IEEE Trans. on Comps. Feb. 1988. pp 190-199.
8. Matousek, R., Tichy, M., Pohl, M., Kadlec, J., Softley, C., Coleman, N.: Logarithmic Number System and Floating Point Arithmetics on FPGA. Field-Programmable Logic and Applications, LNCS 2438, Sept. 2002. pp 626-636.
9. Schulte, M.J., Swartzlander, E.E, Jr.: Hardware Designs for Exactly Rounded Elementary Functions. IEEE Trans. on Comp. Aug. 1994. pp 964-973.
10. Xilinx. Virtex-II Platform FPGA Handbook. Dec. 2000.

A Modular Reconfigurable Architecture for Efficient Fault Simulation in Digital Circuits

J. Soares Augusto, C. Beltrán Almeida, and H. C. Campos Neto

INESC/IST**, Aptd. 13069, 1000-029 Lisboa, Portugal,
jose.augusto@inesc-id.pt, Tel: +351 213100294, Fax: +351 213145843

Abstract. In this paper, a modular reconfigurable architecture for efficient stuck-at fault simulation in digital circuits is described. The architecture is based on a *Universal Faulty Gate Block*, which models each 2-input gate by a 4-input Look-Up Table (LUT) and a Shift-Register (SR) with 3 stages, and relies on collapsing the stuck-at fault list of the gates using equivalence and dominance relations between faults. An example is presented, the expected performance is estimated and the applicability and limitations of the architecture are discussed.

1 Introduction

Fault simulation is an important issue in the design and in the test of electronic circuits. It measures the Fault Coverage (FC) obtained with a set of Test Vectors (TVs) applied to the Circuit Under Test (CUT). It also discovers redundant TVs, that is, TVs which aren't unique in detecting some fault [1, 2]. It is a fundamental tool in the evaluation of the testability of a given design and, in fault-tolerant circuits, fault simulation measures the degree of tolerance.

However, fault simulation is a very time consuming task and it is important to do it fast. Software techniques to accomplish this goal have been developed: the more important are the parallel, deductive and concurrent techniques [1, 2].

Hardware fault simulation is also well known. It relies on using specialized computer architectures that exploit the parallelism and/or concurrency in many CAD algorithms. Most hardware simulators are directed towards logic simulation, but many can also be used for fault simulation. Two examples are the logic simulation machine in [3], which was organized as a distributed processing architecture composed of separate processing units dedicated to specific tasks of the simulation algorithm, and the 'Yorktown Simulation Engine' [4] which exploited the parallelism of 256 processors. See [2] for a review of hardware fault simulation.

Reconfigurable Hardware (RHw) can speed-up many algorithms. In the last years, it was applied successfully to important ones. Many test and simulation algorithms are NP-complete [2], which means RHw is an important tool in these areas.

** This work was developed under the EC MEDEA+ A503 ASSOCIATE project, with funding from the Portuguese Government Agency "Agência de Inovação".

There is already a substantial amount of literature on the implementation of test-related algorithms in RHw. In [5] a critical path tracing algorithm was implemented in RHw. The satisfiability (SAT) problem, which is NP-complete and is very important to CAD and Test, was also implemented in RHw. In [6] a method for the emulation of the PODEM algorithm formulated as a SAT problem was proposed and an implementation of a small SAT problem appears in [9]. In [7] each instance of a SAT problem is solved through the creation of a specialized circuit: simulations predicted it could solve a random 3-SAT problem with 400 variables in 20 minutes, using a clock of 1 MHz. This approach of synthesizing each instance of SAT is discussed in [8] and in other articles from the same group.

In [10] a novel approach for generating test vectors that detect faults in combinational CUTs was introduced. It automatically built a circuit which implemented the D-algorithm (an Automatic Test Pattern Generation algorithm) specialized for the combinational CUT. It was estimated that the approach would be two times faster than a software implementation.

In [11] instance-specific accelerators for minimum-cost covering problems based on a branch-&-bound algorithm were presented. An instance-specific hardware architecture implementing branch-&-bound in 3-valued logic and using reduction techniques borrowed from software solvers, resulted in significant speedups in small-sized covering problems.

In [12] a method that does serial fault emulation using FPGAs, which must be reconfigured for each fault, was proposed. To improve the efficiency due to the overhead of the reconfiguration time spent in the mapping of numerous faulty circuits, independent faults are injected simultaneously, as well as some sets of dependent faults. This capability needs extra supporting circuitry.

A more efficient approach is followed in [13], where only *partial* FPGA reconfiguration is used to inject the faults.

Reconfiguration models can be classified into Compile-Time Reconfiguration (CTR) and Run-Time Reconfiguration (RTR), which is much more slower in practice. Our approach falls into the 'faster' CTR model. Other important issue in RHw is data transfer between the host computer and the hardware: the data must be transferred in as few 'batches' as possible to avoid the large time overhead in data transfer operations. Our approach also minimizes this issue.

In summary, in this paper a reconfigurable hardware architecture that emulates faulty digital circuits and pursues efficient fault simulation is proposed. In the following sections we discuss briefly the stuck-at fault model, fault list reduction and fault simulation, we present the reconfigurable architecture and its basic building block, and apply it to an example. Performance in moderately large (1000 gates) circuits is predicted from 3 typical implementations. We conclude the paper with an appreciation of the merits and of the limitations of the approach.

2   Stuck-at Faults and Fault Simulation

2.1   The Fault Simulation Problem

The fault simulation problem is formulated as follows: given a circuit and a fault list, apply a set of test vectors to the primary inputs of both the nominal/good circuit and the faulty circuits (circuits whose functionality suffers from the effect of a single fault) and collect the test responses in the primary outputs. The *single stuck-at zero/one fault* model is usually used in logic fault simulation.

The test responses which are collected are used for fault detection, diagnosis and for the selection of TVs for production test. This selection delivers a (minimal) test set that detects (almost) all the faults.

The simulation of the faulty circuits with a large set of TVs can be lengthy. This paper proposes RHw as a means of emulating the faulty circuits and simulating efficiently the faults.

2.2   Stuck-at Faults, Fault Equivalence and Fault Dominance

Although logic single stuck-at faults (SAFs) don't model all the possible defects, they are the main fault model used in practice. The reasons of this ubiquity are: SAFs are a simple model; the number of SAFs is finite (twice the number of nodes); and TVs that catch a large percentage of the SAFs (i. e., which have a good fault coverage) usually detect most of the defects, single or multiple, in the circuit (the correlation between SAF fault coverage and defects fault coverage is good).

There are two SAFs: the stuck-at-0 (SA0) and the stuck-at-1 (SA1) faults. If a node holds a SA0 fault, it will always have a logic '0'. In case it holds a SA1 fault, it will always be at '1'. The SAFs are associated with a logical block. A block with q terminals (primary inputs and outputs) has $2q$ SAFs: one of each per terminal. Thus, a 2-input gate has 6 faults (4 in the inputs and 2 in the output).

As fault simulation is time consuming, in practice only *single-faults* are simulated. There is also the empirical observation that tests with good single faults coverage have good multiple faults coverage.

Faults in a gate can be collapsed with *fault equivalence and dominance* [1].

Fault equivalence: faults in a logic block are equivalent iff the response in the output is the same for them all when all the input logic combinations (2^{n_i}, for n_i inputs) are applied.

Fault dominance: Let T_g be the set of all TVs that detect a fault g. *Another fault f dominates the fault g iff f and g are functionally equivalent under T_g.* (The concept of fault dominance implies that all the TVs that detect the fault g also detect the dominant fault f, but not vice-versa.)

Let's reduce the fault list of a NOR gate $z = \overline{x + y}$. Table 1 shows the nominal output (z) and the output under the effect of the faults. It is clear that x-SA1, y-SA1 and z-SA0 are equivalent: the corresponding columns are equal for all inputs. From this equivalent set, only z-SA0 will be kept in the fault list.

Table 1. Faults in a 2-input NOR gate. The boxed values illustrate fault dominance.

(x,y)	$z = \overline{x+y}$	x-SA0	x-SA1	y-SA0	y-SA1	z-SA0	z-SA1
00	1	1	0	1	0	0	1
01	0	0	0	[1]	0	0	[1]
10	0	[1]	0	0	0	0	[1]
11	0	0	0	0	0	0	[1]

In table 1 the columns corresponding to x-SA0, y-SA0 and z-SA1 have boxes in the entries that differ from the nominal z. The only TV of the fault x-SA0 is $(xy) = 10$ and the only TV of y-SA0 is $(xy) = 01$. Both these TVs detect z-SA1, which means that *the fault z-SA1 dominates both x-SA0 and y-SA0.*

In case x-SA0 or y-SA0 are detected, it is guaranteed that the dominant fault z-SA1 also is detected. If none of the dominated faults is detected, there is no information about the detection of z-SA1. Thus, the removal of the dominant fault from the fault list leaves uncertainty about fault coverage. However, in practice FC is usually above 99% which means that almost all the dominated faults are detected. This fact justifies collapsing the dominant faults (z-SA1 in the NOR).

Fault equivalence and fault dominance reduce the fault list of the NOR gate to z-SA0, x-SA0 and y-SA0. This is shown in figure 1 where a black dot and a circle represent SA1 and SA0 faults, respectively. The same procedure can be applied to 2-input AND, NAND, OR and 'x-input-negated' gates (which are useful in expanding the 2-input XOR). The reduced fault lists are in table 2.

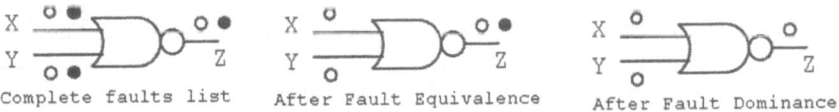

Complete faults list   After Fault Equivalence   After Fault Dominance

Fig. 1. Fault list reduction in the NOR with equivalence and dominance.

Table 2. Reduced fault lists in 2-input gates (inputs x, y, output z).

Gate	Reduced Fault List	Gate	Reduced Fault List
OR	z-SA1, x-SA0, y-SA0	OR/xNEG	z-SA1, x-SA1, y-SA0
NOR	z-SA0, x-SA0, y-SA0	NOR /xNEG	z-SA0, x-SA1, y-SA0
AND	z-SA0, x-SA1, y-SA1	AND/xNEG	z-SA0, x-SA0, y-SA1
NAND	z-SA1, x-SA1, y-SA1	NAND/xNEG	z-SA1, x-SA0, y-SA1

3   The Reconfigurable Architecture

The reconfigurable architecture we propose falls into the compile-time reconfiguration (CTR) model and is directed towards FPGA implementation. After the initial compilation into the FPGA, the nominal and all the faulty circuits are simulated in sequence with all the TVs the design/test-engineer wants to apply. The FPGA is re-programmed only when a different circuit is simulated.

3.1   The Universal Faulty Gate Block (UFGB)

The basic primitive block of the architecture is the *Universal Faulty Gate Block (UFGB)*. The UFGB models a 2-input AND, NAND, OR or NOR gate and its single SAFs. The gates with one input negated are also supported. However, *2-input NOR and 2-input NAND gates are sufficient* which means that any logic function can be implemented with only one of them. Thus, more complex gates are expanded, at compile time, to be represented by a combination of the aforementioned 2-input gates. In conclusion, the technique is applicable to any circuit and is limited only by the FPGA size and/or wiring and I/O limitations.

Fault equivalence and dominance collapse the 6 possible SAFs into 3 SAFs. Accordingly, the 'reconfigurable block' we propose implements 4 functions: the nominal function of the gate and the 3 stuck-at faults. This structure, that we call UFGB, replaces each gate in the CUT and, through 'soft' reconfiguration, permits to simulate the nominal gate and each single fault in the gate's fault list.

The UFGB consists of one Look-Up Table (LUT) with 4 inputs and 1 output, and of a 3 stage shift register (SR) wired as shown in figure 2. The UFGB structure is the same for all the supported 2-input gates: the differences between each gate are encapsulated in the LUT configuration.

As many FPGAs are based on LUTs with 4 (or more) inputs and 1 output, the LUT is a component already built into RHw. The SR is the serial connection of 3 flip-flops, or memory cells, which are also available in most reconfigurable devices.

Let's explain how the UFGB works using the NOR gate ($z = \overline{x + y}$) as an example. Figure 2 shows the original gate with the 3 faults to be simulated, and the UFGB that emulates the gate in the FPGA. 2 of the LUT's inputs are assigned to the original NOR inputs, x and y. The other 2 inputs are configuration signals named $c1$ and $c2$. The LUT is programmed as described in table 3 and works as a *multiplexer of logic functions*.

$c1$ and $c2$ configure the 'LUT gate' in nominal mode or in one of the 3 fault modes, which are simulated in sequence by applying a bit-stream '...0000110000...' into the $SRin$ input of the shift register (see the legend in figure 3). As the bit-stream is right-shifted, $c1$ and $c2$ span all the configuration options and the 3 faults are simulated in sequence. As soon as the 1's leave the nodes $c1$ and $c2$, the UFGB goes into nominal mode. The need of a 3-stage SR, instead of a 2-stage SR, is explained below.

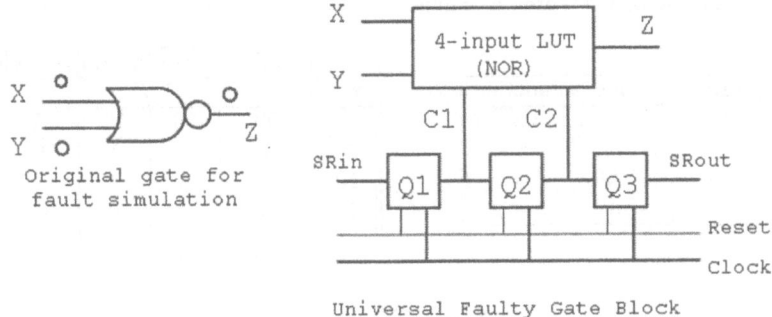

Fig. 2. Universal Faulty Gate Block (UFGB).

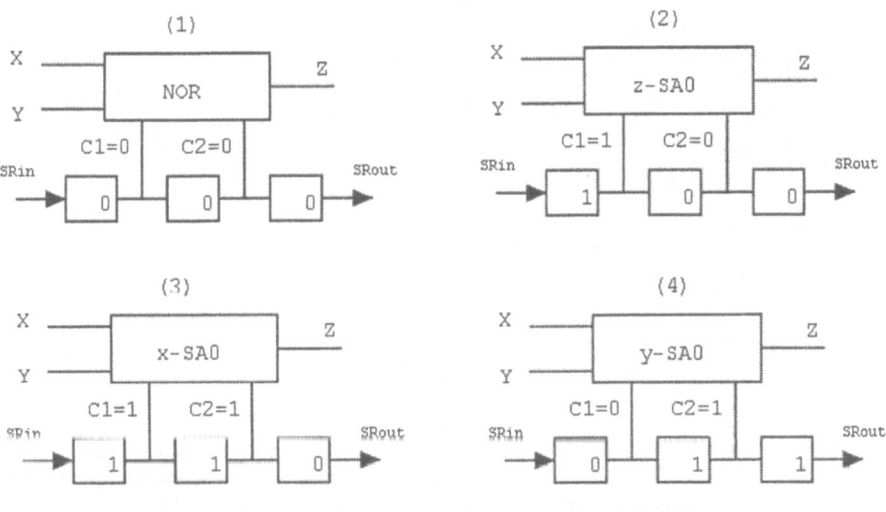

Fig. 3. UFGB configuration with a bit-stream '...00011000...' arriving at SRin: (1) with '000' in the SR, the LUT emulates a NOR; (2) with '100' in the SR the LUT emulates a z-SA0 fault; in (3) and (4) bit-stream shifting emulates the faults x-SA0 and y-SA0, respectively. With another shift, the SR is reset and the gate returns to nominal mode.

3.2   Fault Simulation in a Complete Circuit

Figure 4 shows a complete circuit built with one AND, one OR and one NOR, as well as the architecture which emulates it. The programming of the LUT for the NOR gate is shown in table 3. The other LUTs are programmed according to the gate nominal function and the fault list in table 2. In figures 3 and 4 the *Clock* and *Reset* lines shown in figure 2 are omitted in order to simplify the picture.

The flip-flops $m0a$ and $m0b$ are different from those in the UFGBs ($m1$ to $m9$): they have a *Set* input while the others have a *Reset* input as shown in

Table 3. LUT programming to replace the NOR gate $z = \overline{x + y}$.

(c1,c2)	Mode	Function
00	nominal	$z = \overline{x+y}$
10	z-SA0	$z = 0$
11	x-SA0	$z = \overline{y}$
01	y-SA0	$z = \overline{x}$

z	nominal	z-SA0	x-SA0	y-SA0
(x,y)	(c1,c2)			
	00	10	11	01
00	1	0	1	1
10	0	0	1	0
11	0	0	0	0
01	0	0	0	1

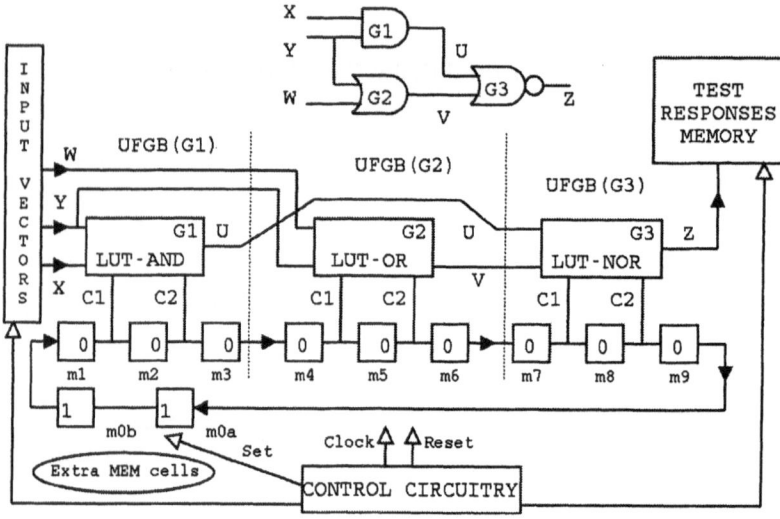

Fig. 4. Simulation of a circuit with 3 gates (top). The complete simulation architecture consists of 3 UFGBs (separated with dashed lines) with LUTs corresponding to each gate type, two extra memory cells $m0a$ and $m0b$, control circuitry, a block to apply the TVs and a block to save the test outputs ('Test Responses Memory'). In the beginning of the simulation (the logic values shown) the 'good' circuit is configured.

figure 2. In the beginning of the simulation $m0a$ and $m0b$ are set, while $m1$ to $m9$ are reset: a closed bit-stream with only two '1's is loaded in the flip-flops. In this stage, all the UFGBs are configured to the nominal function ($c1 = c2 = 0$ in all of them) and the nominal circuit is being simulated. After 1 clock cycle, the bit-stream rotates to the right and $c1 = 1, c2 = 0$ in the LUT-AND of G1: thus, we simulate the first fault in the fault list of G1 (according to table 2 it is the output-SA0 which in figure 4 is U-SA0). After a second clock cycle, $c1 = 1, c2 = 1$ in the LUT-AND of G1 and the second fault in this gate is simulated. And so on and so forth.

The third stage in each SR of the UFGBs can be explained with the help of figure 4. Consider, for instance, that $m3$ in the UFGB of G1 doesn't exist. When shifting the bit-stream there would be a state where $m1 = c1(G1) = 0$, $m2 = c2(G1) = 1$, $m4 = c1(G2) = 1$ and $m5 = c2(G2) = 0$ (remember $m3$ is supposedly removed). With this configuration *a double fault is being simulated*, as the logic conditions $(c1, c2)(G1) = 01$ and $(c1, c2)(G2) = 10$ configure simultaneously the two LUTs associated to G1 and G2 in a faulty mode. $m3$ is thus inserted to act as a 'waiting buffer' to the front-end bit '1' while the rear bit '1' is still configuring the LUT of G1 to simulate one fault. This reasoning applies also to $m6$ and $m9$, the last flip-flops in the SRs belonging to the UFGBs of the gates G2 and G3.

The good circuit is simulated twice until the bit-stream returns to the initial position: one time in the beginning, as shown in figure 4, and another time when $m9 = m0a = 1$. One of the flip-flops $m0a$ or $m0b$ can be discarded if the last flip-flop in the chain, $m9$, is initially 'set' instead of 'reset'. This, however, destroys the modularity.

The TVs can be applied through a memory or using a generator such as a Linear Feedback Shift Register (LFSR). The outputs can be saved in the 'Test Responses Memory' or applied to an LFSR configured as a signature analyzer.

The control circuitry in figure 4 controls the clocks and the simulation modes, the set and reset of the SR flip-flops and of $m0a$ and $m0b$, the addressing of the memory with the TVs, the saving of the results into the 'test responses memory', it signals the end of the simulation and does an eventual intermediate data transfer with the host computer. The number of *Clock* pulses applied to the SR chain can be programmed in the control circuitry or, otherwise, the end of a 'bit-stream round trip' can be detected by AND'ing $m0a$ and $m0b$. Note that the only control signals applied to the emulation hardware are *Clock*, *Reset* and *Set* which usually use dedicated lines in the FPGA: thus, there is an economy of routing resources, which is a key advantage of this simulation architecture.

4   Performance Estimation

To estimate fault simulation times in real circuits, 1000 UFGBs were compiled into a Xilinx Spartan-II-200 FPGA. This reconfigurable device has 2352 slices (i.e., reconfigurable units).

Three circuits, all with 1000 UFGBs, were synthesized: 1000 UFGBs in series (the z output of each UFGB is linked to the next x and y inputs); a rectangular array of 10×100 UFGBs; and a rectangular array of 20×50 UFGBs. In these arrays, the number of inputs is the same as the number of outputs and is equal to 10 and 20, respectively. The inputs of the intermediate stages in the arrays were cross-wired to the outputs of the previous stage. The number of slices used in the FPGA before and after placement and routing (P&R), and the input/output delay after P&R are shown in table 4.

In all the 3 cases almost all of the slices in the FPGA were used, which gives a consumption figure of 2.35 slices per UFGB. However, before P&R only an average of 1.6 slices per UFGB was used. As the FPGA is almost completely populated, it is acceptable this overhead due to the P&R constraints.

Each fault simulation will run in a time approximately equal to the input/output delay shown in table 4, which is approximately proportional to the depth (number of levels) of the circuit. Between 1ns and 2 ns per level are

Table 4. Results from implementing 3 circuits with 1000 UFGBs.

Circuit	Slices before P&R	Slices after P&R	in/out delay (ns)
1000 UFGBs in series	1607	2350	1290
10 × 100 UFGBs (10 in/out)	1606	2350	190
20 × 50 UFGBs (20 in/out)	1617	2350	108

measured in the example. The compilation time of the FPGA will have to be accounted for in more exact calculations, but it can be neglected when simulating many faults [12].

The delay allows the estimation of simulation times. For instance, in the 20 × 50 UFGBs array the simulation of all the faults (3000) with all the possible TVs ($2^{20} \approx 10^6$) would last $3 \times 10^3 \times 2^{20} \times 108 \times 10^{-9} \approx 340$ seconds (i.e., 5.7 minutes). In the 10 × 100 array it would be $3 \times 10^3 \times 2^{10} \times 190 \times 10^{-9} \approx 0.58$ seconds. In conclusion, when the number of inputs is less than about 20, all the TVs can be simulated in a reasonable amount of time!

The *fault simulation time per gate* (t_{fsgate}) allows us to compare our results with software fault simulation. The third circuit in the table could be a real example (i.e., 1000 gates, 50 levels). Thus $t_{fsgate} = 108$ ps. According to [13], where are reported *software fault simulation results* in the ISCAS'89 benchmark circuits s5378 and s13207, with a widely used commercial fault simulation tool in a Sun Ultra 10/440 workstation with 1 GB Ram, t_{fsgate} is 3.7 μs/gate and 13.2 μs/gate, respectively. This shows a potential of *more than a four order magnitude improvement in simulation time* of the approach here proposed.

5   Applicability and Limitations of the Architecture

All the faults (n gates will have $3n$ faults) can be simulated without reconfiguring the RHw. Fault simulation speed is limited by maximum clock speed in FPGA's flip-flops, paths' delays, and by the fill up of the test acquisition memory (a data transfer to the host computer must be done). Placement and routing can be a problem in case the maximum FPGA capacity is approached.

The architecture is directed towards combinational circuits. Unsupported gates can be synthesized with NANDs or NORs. Fault simulation of *sequential synchronous circuits* can be done by expanding them according to the combinational 'iterative array' model where each stage is a copy of the combinational block of the original sequential circuit. This approach is followed in the discovery of test vectors for faults in sequential circuits [1, 2]. Asynchronous circuits can eventually be tackled (their gates are replaced by UFGBs) but one must keep in mind that FPGAs' delays are different from ASICs' delays.

The validity of the emulation of the 'true' circuit by the proposed methodology is at logical level. Delays in FPGAs are different from ASICs' delays. Thus, delay-dependent issues such as timing verification, delay faults and stuck-open faults cannot be checked with our approach.

6   Conclusions and Further Work

We proposed a modular architecture suited to efficient stuck-at fault simulation in digital circuits. It is expected that in moderately sized circuits (e.g. with 1000 gates) one fault is simulated in less than 200 ns (5.000.000 faults/second simulation rate). The modularity of the architecture eases its generalization to simulate larger circuits using hardware 'virtualization'. The efficiency comes from the fact that *all the faults are simulated without performing total or partial reconfiguration*.

An automatic system is under development. It consists of tools that: convert unsupported gates (XORs and gates with more than 2 inputs) to 2-input gates; expand synchronous sequential circuits into the corresponding iterative array; extract the fault list from the circuit description; add control circuitry; define RAM partitions (where test vectors are kept, where the responses are saved); convert the circuit file into an HDL, compile it and download it into the RHw (FPGA); run the simulation, read the responses from the output memory, perform fault diagnosis and select test vectors.

References

1. M. Abramovici, M. Breuer and A. Friedman, "Digital Systems Testing and Testable Design" IEEE Press, 1990.
2. H. Fujiwara, "Logic Testing and Design for Testability", MIT Press, 1985.
3. M. Abramovici, Y. Levendel, P. Menon, "A logic simulation machine", Proc. 19th Design Automation Conference, pp. 65-73, 1982.
4. G. Pfister "The Yorktown Simulation Engine", Proc. 19th Design Automation Conference, pp. 51-54, 1982.
5. M. Abramovici, P. Menon, "Fault simulation on reconfigurable hardware", 5th IEEE Symposium on FPGA-Based Custom Computing Machines (FCCM'97), 1997.
6. M. Abramovici and D. G. Saab, "Satisfiability on Reconfigurable Hardware", 7th Int. Workshop on Field Programmable Logic and Applications, 1997.
7. T. Suyama, M. Yokoo, H. Sawada, "Solving Satisfiability Problems Using Logic Synthesis and Reconfigurable Hardware", 31st Hawaii Intl. Conf. on Sys. Sciences, 1998.
8. P. Zhong, M. Martonosi, P. Ashar, and S. Malik, "Accelerating Boolean Satisfiability with Configurable Hardware", Proc. IEEE Symp. on Field-Programmable Custom Computing Machines, April, 1998.
9. M. Abramovici, J. de Sousa, D. Saab, "A Massively-Parallel Easily-Scalable Satisfiability Solver Using Reconfigurable Hardware", Design Automation Conference (DAC'99), New Orleans, USA, 1999.
10. F. Kocan, D. G. Saab, "Concurrent D-Algorithm on Reconfigurable Hardware", Int. Conference on Computer Aided Design (ICCAD'99), 1999.
11. C. Plessl, M. Platzner, "Instance-Specific Accelerators for Minimum Covering", 1st Intl. Conf. on Eng. of Reconf. Systems and Algorithms, Las Vegas, USA, 2001.
12. K.-T. Cheng, S.-H Huang, W.-J. Dai, "Fault Emulation: a New Methodology for Fault Grading", IEEE Trans. CAD, vol. 18, no. 10, pp. 1487-95, 1999.
13. A. Parreira, J. Teixeira, M. Santos, "A Novel Approach to FPGA-Based Hardware Fault Modeling and Simulation", IEEE Int. Workshop on Design and Diag. of Elect. Circ. and Systems, Poland, April, 2003.

Evaluation of Testability of Path Delay Faults for User-Configured Programmable Devices*

Andrzej Krasniewski

Warsaw University of Technology, Institute of Telecommunications
Nowowiejska 15/19, 00-665 Warsaw, Poland
andrzej@tele.pw.edu.pl

Abstract. A model of the combinational section of a programmable device suitable for an analysis of testability of delay faults is proposed. All relevant factors that affect the evaluation of testability of path delay faults are identified and their impact on the outcome of the evaluation is discussed. A detailed analysis, supported by quantitative results, focuses on the selection of the set of target faults in terms of a class of logical paths and on the concept of defining testability measures for physical paths rather than for logical paths. Practical guidelines are formulated for the development of a procedure for the evaluation of testability of path delay faults.

1 Introduction

With the growing miniaturization, complexity and speed of today's ICs, increasingly more problems are associated with timing. These problems are caused not only by global and local disturbances of the fabrication process, but also by the impact of noise/interference effects, such as signal coupling (crosstalk), power supply noise and substrate noise, some of which are specific for the system environment [1] [2]. This has the following consequences:

- Timing-related problems cannot be adequately identified by testing an isolated chip by the manufacturer. This is especially true for programmable devices for which the operational (user-defined) configuration and clock frequency is not known at the time the device is fabricated. Therefore, although an application-independent test procedure aimed at delay faults, e.g. the one proposed in [3], might be useful, for highly dependable systems, it must be augmented with in-system test of a user-defined configuration (application-dependent test, application-oriented test [4]).
- It is extremely difficult, if not impossible, to generate deterministic patterns that would account for the above described timing-related effects [1]. Therefore, random testing might be a preferred solution.

Application-dependent random testing of delay faults in programmable devices can be implemented using the BIST techniques [5]. It should be emphasized that, for

---

* This work was supported by the State Committee for Scientific Research of Poland under grant no. 4 T11D 014 24

P.Y.K. Cheung et al. (Eds.): FPL 2003, LNCS 2778, pp. 828–838, 2003.

advanced FPGAs or CPLDs, by exploiting their reconfigurability or partial reconfigurability, BIST can be implemented at no circuitry or performance penalty [3] [5]-[7]. What is unclear about the BIST-based random testing is its quality.

The evaluation of efficiency of testing delay faults in FPGAs or CPLDs is difficult. As was shown in [8], the conventional methods for analysis of delay fault testability [9]-[11], developed for networks of simple gates, i.e. NOT, AND, NAND, OR, and NOR gates, are not directly applicable. This is because basic logic components of programmable devices (LUT-based FPGAs or similar) can implement arbitrary Boolean functions (and not only positive or negative unate functions) and, therefore, even such simple concepts like "a controlling value" do not apply.

Some specific aspects of the evaluation of quality of testing delay faults in a network of arbitrary logic components are presented in [8] [12]. In this paper, we discuss this problem in more detail.

2 Determining the Speed of a User-Programmed Device

The speed of a user-programmed FPGA/CPLD is determined by the maximum time it takes to propagate a transition along a path between two memory elements (we assume that all memory elements are controlled by the same clock). A path may include various types of logic components. We distinguish *active logic components* (ALCs), i.e. components that have at least two inputs which have not been fixed to constant values by programming the device. In our analysis, other (non-active) logic components are seen as part of the interconnection structure. This refers, in particular, to any component which - for a specific user-defined configuration - implements a single-argument function (an example being a multiplexer whose address is fixed or a 2-input gate whose one input is set to a non-controlling value). For a non-active component, a transition of a particular polarity (rising or falling) at its input always results in a transition of a specific polarity at its output. This may not be the case for an active component where for a transition of a particular polarity at its input, the occurrence of the output transition and its polarity depends on the state of the other inputs. In the case when a non-active logic component implements an inversion function, we "remove" this inversion and appropriately modify the function of ALCs driven by the considered component, so that not to change the network function.

The combinational part of a used-configured programmable device is represented as a network of single-output ALCs (a multiple-output component is represented as a set of single-output components) that implement arbitrary Boolean functions. Inputs and outputs of the network are special ALCs. Then, any path can be seen as an alternating sequence of ALCs and connections. This is illustrated in Fig. 1. It may be observed that the function implemented by ALC3 (which represents LUT2) has been modified to account for the "removed" inversion between the AND gate and LUT2.

The following assumptions are taken regarding propagation delays of components in the considered network of ALCs:
- delays are assigned to both ALCs and connections;
- ALC and connection delays may depend on the polarity of signal transitions;

- different ALC delays may be assigned to different ALC inputs;
- delays of a programmable ALC may depend on its specific user-defined function (e.g. different delays may be assigned to LUTs that implement different functions);
- for a multiple-destination connection (originating at some ALC and leading to several other ALCs), different delays may be assigned to its different branches.

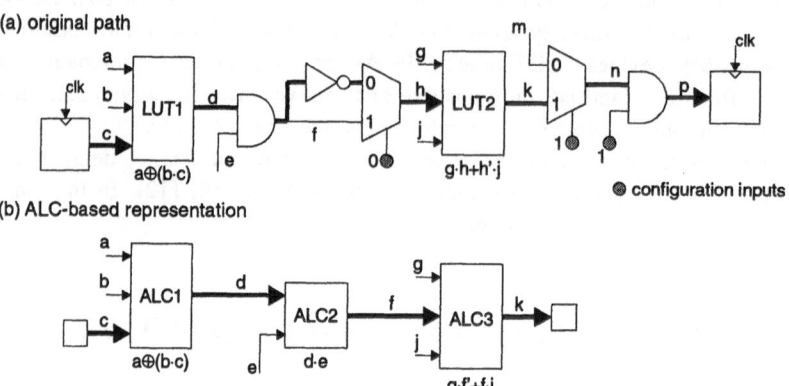

Fig. 1. Representation of a path in a network of ALCs

The propagation delay of a path depends on delays of its components - ALCs and connections. The delays of relevant portions of memory elements (flip-flops) at which the path originates and terminates are included in the delays of adjacent connections.

In the path-delay fault model, it is assumed that faults (excessive delays) are associated with logical paths [9] [10]. For a network of simple gates, a logical path is defined by a path (physical path, structural path) and the polarity of a transition that occurs at its input (or output); this is sufficient to determine the polarity of a transition at the input or output of each gate on the path. This may, however, not be the case for a network of ALCs. For example, for the path in Fig. 1(b), assuming that a rising transition at input c propagates through ALC1 (which requires b = 1), either a rising transition (for a = 0) or a falling transition (for a = 1) occurs at connection d.

For a network of ALCs, a *logical path* π(PTP) is defined by a path π and a *path transition pattern*, PTP, which specifies for each connection along the path, whether a rising (↑) or a falling (↓) transition occurs at the output of the ALC which feeds this connection. The path transition pattern must comply with the ALC functions, i.e. for each ALC along the path, transitions at the input and output of this ALC must have the same (different) polarity if the ALC function is positive (negative) unate in its on-path input [13]. Thus, the number of logical paths associated with a given (physical) path is 2^{K+1}, where K is the number of ALCs that implement functions which are binate in their on-path inputs. Path transitions patterns that do not comply with the ALC functions correspond to *pseudological paths*. For example, for path c-d-f-k in Fig. 1(b), the functions implemented by ALC1 and ALC3 are binate in their on-path inputs, whereas the function implemented by ALC2 is positive unate in its on-path input. Thus, the following 8 logical paths are associated with path c-d-f-k: c↑d↑f↑k↑, c↑d↑f↑k↓, c↑d↓f↓k↑, c↑d↓f↓k↓, c↓d↑f↑k↑, c↓d↑f↑k↓, c↓d↓f↓k↑, and c↓d↓f↓k↓.

Not all path transitions patterns that comply with the ALC functions can actually be produced in normal operation of the network, especially if the set of *functional input pairs*, i.e. vectors pairs that can occur at the input of the network in normal operation, does not contain all possible input pairs. A logical path π(PTP) is *feasible* if there exists a functional input pair which produces PTP.

For a feasible logical path π(PTP), there may or may not exist a functional input pair that - for a given *delay assignment* that specifies the propagation delay of each logical path in the network.- sensitizes π(PTP), i.e. propagates a transition along π, producing PTP (the sensitization of a logical path is formally defined in [12]). If such a functional input pair exists, π(PTP) is called *true*, otherwise π(PTP) is called *false*.

For a given delay assignment, a logical path π(PTP) *determines the speed of the network* if for all possible functional input pairs, the latest transition at the network output is an effect of the sensitization of π(PTP) (by some of these pairs). Thus, to determine the speed of the network, propagation delays of true paths should only be examined. However, for a "real" device, the delay assignment is unknown - due to imperfections of the manufacturing process, path delays can vary significantly. Therefore, to determine the speed of the network, *irredundant* logical paths, i.e. logical paths that might be true for *some* delay assignment must be considered.

It might appear that any irredundant logical path determines the speed of the network under some delay assignment. However, in [12] it was shown that such a statement is incorrect and that to determine the speed of the network it is sufficient to examine a subset of irredundant paths, called *delay-essential* paths.

The following remarks should be made regarding the above defined concepts:
- Although "irredundant logical path" and "delay-essential logical path" are clearly timing-related concepts, no knowledge of network timing is necessary to decide whether or not a particular logical path is irredundant or delay-essential [12].
- The set of delay faults associated with delay-essential logical paths is an "ideal" set of target faults for any timing-oriented test procedure; this set contains all those and only those logical paths that determine the speed (maximum propagation delay) of the network under the unknown delay assignment.

3   Evaluation of Testability of Path Delay Faults

When evaluating the testability of path delay faults for a user-configured device, several factors should be considered, as shown in Fig. 2. We focus on the factors that are unique or particularly relevant for programmable devices, i.e. which differentiate networks of ALCs from networks of simple gates. These differences stem from that in a network of ALCs, a possibly large number of logical paths of possibly different testability characteristics are associated with each physical path, whereas in a network of simple gates, each physical path has exactly two corresponding logical paths.

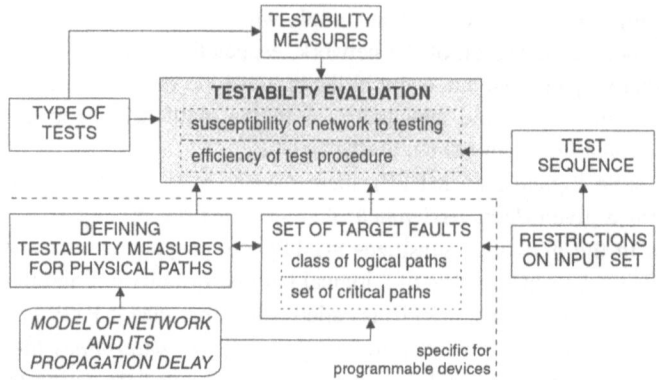

Fig. 2. Evaluation of testability of path delay faults for a user-configured programmable device

3.1 Testability Measures

There are two facets of the evaluation of testability: the evaluation of susceptibility of a network to testing and the evaluation of efficiency of a specific test procedure. The susceptibility of a network to testing is usually measured by *fault testability*, i.e. the percentage of faults that are detectable. The efficiency of a specific test procedure is usually measured by *fault coverage*, i.e. the percentage of faults that are detected by a test sequence produced by the considered procedure.

3.2 Type of Tests

For path delay faults, various types of tests are defined, including weak non-robust tests (WNR-tests), often referred to as non-robust tests, strong non-robust tests (SNR-tests), also referred to as restricted non-robust tests, and robust tests (R-tests). For networks of arbitrary logic components, the requirements for these types of tests can be found in [14]. The values of testability measures are calculated assuming a specific type of tests. Thus, we can report, for example, fault testability for robust tests (R-testability) or fault coverage by WNR tests (WNR-coverage).

3.3 Set of Target Faults

The value of a particular testability measure can only be calculated for a given *set of target faults* (fault model). As was stated earlier, in the considered network of ALCs, there is a one-to-one correspondence between logical paths and path delay faults. Therefore, the set of target faults is defined by a class of logical paths (a set of logical paths having a certain attribute) or by its appropriately specified subset.

a)   classes of logical paths

A straightforward approach is to define the set of target faults (target logical paths) based on some specific type of testability. This way, the set of target faults can be defined as the set of WNR-testable logical paths, the set of SNR-testable logical paths, or the set of R-testable logical paths. Such an approach, taken in many publications on testing delay faults in gate networks, neglects, however, the key question of "which faults actually affect the speed of the network".

As was stated in Section 2, the "ideal" set of target path delay faults in a network of ALCs is defined by the class of delay-essential logical paths. Although this class can be identified based exclusively on the network structure and functions implemented by ALCs, an examination of logical *multipaths* is required [12]. This might be too complex even for medium-size networks. Therefore, a practical solution would be to rely on a "reasonable approximation" of the set of delay-essential logical paths. The proposed solution is based on the relationship between the various classes of logical paths. As shown in Fig. 3, the set of delay-essential paths is a superset of the set of SNR-testable paths and is a subset of the set of irredundant paths. Thus, these two sets, whose identification does not involve an examination of multipaths, can be used to obtain lower- and upper-bound estimates on the "exact" testability measures. Other, less exact, estimates would rely on the set of feasible paths and all logical paths (lower-bound estimates) and on the set of R-testable paths (upper-bound estimates).

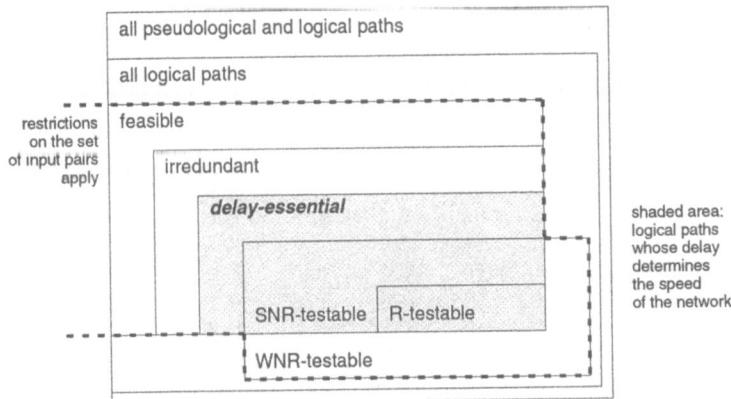

Fig. 3. Relationship between various classes of logical paths in a network of ALCs

The impact of various possible selections of the set of target faults on the values of testability measures is illustrated for the example network of ALCs shown in Fig. 4. For this network, the values of fault testability and fault coverage for the various types of testability are given in Table 1. The values of fault coverage are calculated for the specific test sequence composed of only 8 input vectors (7 input pairs) that has been developed so that to maximize the efficiency of testing; this sequence was selected out of $(2^6)^8 = 2.81 \cdot 10^{14}$ possible sequences using a combination of automatic search and manual optimization.

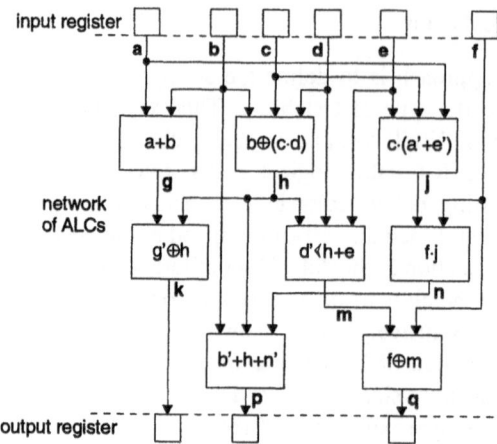

Fig. 4. Example network of ALCs

The data in Table 1 illustrate a strong impact of the class of logical paths that defines the set of target faults on the values of testability measures. For example, depending on the class of logical paths, the value of R-testability may change from 55.6% (or even from 26.0% if pseudological paths are also considered) to 100.0%, with the "exact" value (corresponding to the set of delay-essential paths) of 86.2%. Similarly, the WNR-coverage may change from 25.6% (or even from 12.0%) to 42.0%, with the "exact" value of 37.9%.

Table 1. Path delay fault testability and fault coverage for the network of Fig. 4

target logical paths	no. of paths	fault testability [%]			fault coverage [%]		
		WNR	SNR	R	WNR	SNR	R
all pseudological and logical	192	35.9	29.2	26.0	12.0	8.3	6.3
all logical	90	76.6	62.2	55.6	25.6	17.8	13.3
feasible	74	75.7	75.7	67.6	29.7	21.6	16.2
irredundant	70	80.0	80.0	71.4	31.4	22.9	17.1
delay-essential	58	96.6	96.6	86.2	37.9	27.6	20.7
WNR-testable	69	100.0	81.2	72.5	33.3	23.2	17.4
SNR-testable	56	100.0	100.0	89.3	39.3	28.6	21.4
R-testable	50	100.0	100.0	100.0	42.0	30.0	24.0

b) critical paths

An alternative to defining the set of target faults by the set of all logical paths of a certain class is to consider only critical paths in such a class, i.e. the "longest" paths, determined based on structural properties (number of components), timing simulation or some other method [15].

Most frequently, the attribute "critical" is associated with a physical path. Then, on average more logical paths are associated with a critical physical path than with a non-critical one (long paths include more ALCs than short ones). This, however, does not imply that the same relationship holds for delay-essential logical paths or other classes of logical paths that can be used to define the set of target faults. For example, for the network of Fig. 4, under assumption that the set of critical paths includes all paths that contain 3 ALCs (4 connections), an average number of logical paths associated with a physical path is 5.0 for critical paths and 4.62 for non-critical paths, whereas an average number of delay-essential logical paths associated with a physical path is 2.33 for critical paths and 3.38 for non-critical paths. In general, if the set of target faults is restricted to critical paths, more-difficult-to-satisfy sensitization requirements lead to lower values of the testability measures.

3.4 Defining Testability Measures for Physical Paths

So far, we have assumed that all logical paths of a certain class (and the associated delay faults) equally contribute to the values of testability measures. Under such an assumption, a physical path having a large number of logical paths of a certain class has a significantly higher impact on the values of testability measures than a physical path having a small number of logical paths. If this is undesirable, an idea of defining testability measures for physical paths rather than logical paths can be considered.

The simplest way of implementing this idea is to assume that a physical path π has a certain testability-oriented feature, e.g. is delay-essential or is R-covered by some test sequence, if at least one logical path associated with π has this feature. The impact of such an assumption is illustrated in Table 2, which gives the values of fault testability and fault coverage for the network of Fig. 4 for the selected classes of physical paths. The fault coverage in Table 2 is calculated for the same test sequence as the fault coverage in Table 1. It is clearly seen that the values of testability measures calculated for physical paths (Table 2) are significantly higher than the corresponding values calculated for logical paths (Table 1).

Table 2. Testability measures for physical paths in the network of Fig. 4: simple approach

target physical paths	no. of paths	fault testability [%]			fault coverage [%]		
		WNR	SNR	R	WNR	SNR	R
all physical	19	94.7	89.5	84.2	73.7	63.2	42.1
irredundant	18	94.4	94.4	88.9	77.8	66.7	44.4
delay-essential	17	100.0	100.0	94.1	82.4	70.6	47.1
WNR-testable	18	100.0	94.4	88.9	77.8	66.7	44.4
SNR-testable	17	100.0	100.0	94.1	82.4	70.6	47.1
R-testable	16	100.0	100.0	100.0	81.3	68.8	50.0

A more sophisticated way of implementing the idea of defining testability measures for physical paths is to assume that a physical path π has a certain testability-oriented feature if all logical paths in a specific subset of LP(π) (LP(π) is the set of all logical paths of a certain class associated with π) have this feature. Such a subset of LP(π), SLP(π), must satisfy the following property: if LP(π) includes a logical path that has a transition of a certain polarity associated with some connection c, then SLP(π) also includes such a path. To illustrate this idea, consider path a-g-k in the network of Fig. 4. The set of delay-essential logical paths for this path is LP(a-g-k) = {a↑g↑k↑, a↑g↑k↓, a↓g↓k↑, a↓g↓k↓}. Thus, to decide whether or not a-g-k has a certain testability-oriented feature, it is sufficient to examine two logical paths: either those in

SLP'(a-g-k) = {a↑g↑k↑, a↓g↓k↓} or those in SLP"(a-g-k) = {a↑g↑k↓, a↓g↓k↑}. The values of the testability measures for the network of Fig. 4 calculated based on the above described concept are given in Table 3 (the fault coverage is calculated for the same test sequence as the fault coverage in Table 1). It can be seen that, as expected, the values in Table 3 are significantly lower than those in Table 2.

Table 3. Testability measures for physical paths in the network of Fig. 4: more sophisticated approach

target physical paths	no. of paths	fault testability [%]			fault coverage [%]		
		WNR	SNR	R	WNR	SNR	R
all physical	19	94.7	89.5	84.2	36.8	21.1	15.8
irredundant	18	94.4	94.4	88.9	38.9	22.2	16.7
delay-essential	17	100.0	100.0	94.1	41.2	23.5	17.6
WNR-testable	18	100.0	94.4	88.9	38.9	22.2	16.7
SNR-testable	17	100.0	100.0	94.1	41.2	23.5	17.6
R-testable	16	100.0	100.0	100.0	43.8	25.0	18.8

3.5 Accounting for Restrictions on the Set of Input Pairs

The set of vector pairs that occur at the input of a given network of ALCs in normal operation is usually restricted. Such restrictions should be considered when defining the set of target faults and calculating the corresponding testability measures. This problem is discussed in more detail in [16]. Generally, it can be stated that the restrictions imposed on the set of input pairs affect the number of logical paths in most classes shown in Fig. 3. This obviously has an impact on the values of testability measures. This impact is particularly significant when the fault coverage is calculated for a test sequence that has been developed taking into account the restrictions on the input set.

4   Observations and Practical Guidelines

The results in Tables 1-3 show that for the network of Fig. 4, the values of path delay fault testability measures strongly depend on assumptions taken when calculating these values. For example, the coverage of the path delay faults, calculated for the considered test sequence under different assumptions, varies from 6.3% to 82.4%.

Our analysis of the impact of various factors on the outcome of the evaluation of path delay fault testability leads to the following observations and guidelines:
- The selection of the class of logical paths that defines the set of target faults poses an accuracy-complexity trade-off. The exact values of testability measures, corresponding to the set of delay-essential logical paths, are very difficult, if not impossible, to calculate. On the other hand, their easy-to-calculate approximations may be unacceptably inaccurate. As a compromise, the sets of target faults corresponding to irredundant and SNR-testable logical paths are recommended to provide reasonably accurate estimates of the exact values of testability measures.

- To reduce the computational load associated with the testability evaluation, one can decide to consider only critical paths. Such a decision may also be justified by a higher likelihood of affecting the speed of a device by faults associated with the longest paths. This approach would, however, lead to substantial savings only if the set of critical paths is a relatively small subset of the set of all paths, which is usually not the case for most speed-optimized circuits, for which the detection of timing-related problems is particularly important.
- As the number of physical paths is significantly lower than the number of logical paths, the idea of defining testability measures for physical paths looks attractive, even in the case when the calculation of such measures involves an examination of logical paths. Especially, the straightforward approach of assigning a testability-oriented feature to a physical path if some associated logical path has this feature would be recommended in the case when the rising and falling delays of individual components can be assumed equal or, at least, not significantly different.
- If the set of vector pairs that occur at the input of the network in normal operation is restricted, such restrictions should be considered when defining the set of target faults and evaluating the testability. This is of critical importance when the fault coverage is calculated for a test sequence that has been developed taking into account the restrictions on the input set.

5 Conclusion

We have shown that the evaluation of testability of delay faults for a user-configured programmable device is difficult both conceptually and computationally. To deal with this problem, we have developed a model of the combinational section of a programmable device (referred to as a network of ALCs) which contains a minimal number of components necessary for an analysis of path delay fault testability. We have identified all relevant factors that affect the evaluation of testability of path delay faults in a network of ALCs. An examination of the impact of these factors on the values of testability measures has resulted in several practical guidelines for the development of a procedure for the evaluation of testability of path delay faults.

The key conclusion of our study can be formulated as follows: Whenever the value of any measure of path delay fault testability for a programmable device is reported, a detailed explanation on what is reported should be given. This appears obvious, but is not. In many publications, the presented values are claimed to represent "the coverage of path delay faults by robust tests" or simply "the coverage of path delay faults". This leads to uncertainty about the meaning of the reported results. For example, if some test procedure is claimed to cover 50% of path delay faults, this would probably mean an excellent test quality in the case when this value represents the robust coverage of delay-essential logical paths, but a rather poor test effort if it represents the coverage of robustly testable logical paths by non-robust tests.

References

1. Krstic, A., Liou, J.-J.: Delay Testing Considering Crosstalk-Induced Effects. In: Proc. IEEE Int. Test Conf. (2001) 558-567
2. Metra, C., Pagano, A., Ricco, B.: On-Line Testing of Transient and Crosstalk Faults Affecting Interconnections of FPGA-Implemented Systems. In: Proc. IEEE Int. Test Conf. (2001) 939-947
3. Abramovici, M., Stroud, C.: BIST-Based Delay-Fault Testing in FPGAs. In: Proc. IEEE Int. On-Line Testing Workshop (2002)
4. Renovell, M., Faure, P., Portal, J.M., Figueras, J., Zorian, Y.: IS-FPGA: A New Symmetric FPGA Architecture with Implicit SCAN. In: Proc. IEEE Int. Test Conf. (2001) 924-931
5. Krasniewski, A.: Application-Dependent Testing of FPGA Delay Faults. In: Proc. 25th EUROMICRO Conf., vol. 1 (1999) 260-267
6. Stroud, C., Konala, S., Chen, P., Abramovici, M.: Built-In Self-Test of Logic Blocks in FPGAs (Finally, a Free Lunch: BIST Without Overhead!). In: Proc. 14th VLSI Test Symp. (1996), 387-392
7. Harris, I.G., Menon, P.R., Tessier, R.: BIST-Based Delay Path Testing in FPGA Architectures. In: Proc. IEEE Int. Test Conf. (2001) 932-938
8. Krasniewski, A.: Testing FPGA Delay Faults: It Is Much More Complicated Than It Appears. In: Ciazynski W. et al. (eds.), Programmable Devices and Systems, Pergamon - Elsevier Science (2002) 281-286
9. Sparmann, U. et al.: Fast Identification of Robust Dependent Path Delay Faults. In: Proc. 32nd ACM/IEEE Design Automation Conf. (1995) 119-125
10. Cheng, K.-T., Chen, H.-C.: Classification and Identification of Nonrobust Untestable Path Delay Faults. IEEE Trans. on CAD, **8** (1996) 845-853
11. Krstic, A., Cheng, K.-T., Chakradhar, S.T.: Primitive Delay Faults: Identification, Testing, and Design for Testability. IEEE Trans. on CAD, **11** (1999) 669-684
12. Krasniewski, A.: Sensitization of Logical Paths in a Network of Arbitrary Logic Components: Theory and Application to Delay Fault Testing. In: Proc. IEEE Design and Diagnostics of Electronics Circuits and Systems Workshop (2003) 143-150
13. Krasniewski, A.: On the Set of Target Path Delay Faults in Sequential Subcircuits of LUT-Based FPGAs. In: Glesner, M., Zipf, P., Renovell, M. (eds.), Proc. FPL 2002, LNCS, vol. 2438, Springer Verlag (2002) 616-626
14. Underwood, B., Law, W.-O., Kang, S., Konuk, H.: Fastpath: A Path-Delay Test Generator for Standard Scan Designs. In: Proc. IEEE Int'l Test Conf. (1994) 154-163
15. Majhi, A.K., Agrawal, V.D., Jacob, J., Patnaik, L.M.: Line Coverage of Path Delay Faults. IEEE Trans. on VLSI Systems, **8**, (2000) 610-613
16. Krasniewski, A.: Evaluation of the Quality of Testing Path Delay Faults Under Restricted Input Assumption. In: Proc. IEEE Int. On-Line Testing Symp. (2003) (in printing)

Fault Simulation
Using Partially Reconfigurable Hardware

A. Parreira, J.P. Teixeira, A. Pantelimon, M.B. Santos, and J.T. de Sousa

IST / INESC-ID, R. Alves Redol, 9, 1000-029 Lisboa, Portugal[1]
marcelino.santos@inesc.pt

Abstract. This paper presents a fault simulation algorithm and that uses efficient partial reconfiguration of FPGAs. The methodology is particularly useful for evaluation of BIST effectiveness, and for applications in which multiple fault injection is mandatory, such as safety-critical applications. A novel fault collapsing methodology is proposed, which efficiently leads to the minimal stuck-at fault list at the look-up-tables' terminals. Fault injection is performed using local partial reconfiguration with small binary files. Our results on the ISCAS'89 sequential circuit benchmarks show that our methodology can be orders of magnitude faster than software or fully reconfigurable hardware fault simulation..

Introduction

Design for Testability (DfT) techniques are mandatory for cost-effective product development. Full scan based DfT techniques are very popular for sequential circuits, but these techniques impact device performance, area overhead, test application time and prevent at-speed testing, which is crucial to detecting dynamic faults [1]. To solve these problems partial scan or self-test strategies have been used, which requires testability analysis of the sequential blocks.

During the design phase, test quality may be assessed by Fault Simulation (FS). For complex devices FS may be a very costly process, especially for sequential circuits. In fact, circuit complexity, test pattern length and fault list size may lead to a large computational effort. Although many efficient algorithms have been proposed for software fault simulation (see for example [2] and [3]), for complex circuits it is still a very time-consuming task and can significantly lengthen the time-to-market.

FS can be implemented in software or hardware [1]. The ease of developing software tools for FS (taking advantage of the flexibility of software programming) made software simulation the most current approach. However, the recent advent of very complex programmable logic devices, namely Field Programmable Gate Arrays (FPGAs), created an opportunity for Hardware Fault Simulation (HFS), which may be an attractive solution for at least a subset of practical situations. In fact, evaluating Built-In Self-Test (BIST) effectiveness may require a heavy computational effort in fault simulation. Long test sessions are needed to evaluate systems composed of several modules that have to be tested simultaneously in order to evaluate aliasing.

---

[1] This work has been partially funded by FCT (Portugal), POCTI/ESE41788/2001.

P.Y.K. Cheung et al. (Eds.): FPL 2003, LNCS 2778, pp. 839–848, 2003.

One example is functional BIST in a System on a Chip (SoC) with data compression cores being used as part of the signature compression logic. Another fault simulation task not efficiently performed by software tools is *multiple* fault simulation, mainly because of the enormous number of possible fault combinations. Unfortunately, multiple fault simulation may become mandatory, namely in the certification process of safety-critical applications [4].

Different HFS approaches using FPGAs have been proposed in the literature to overcome the difficulties with Software Fault Simulation (SFS) for sequential circuits. *Dynamic* fault injection using dedicated extra hardware, and allowing the injection of different faults without reconfiguring the FPGA, was proposed in [5-8]. The additional hardware proposed for implementing dynamic fault injection uses a Shift Register (SR) whose length corresponds to the size of the fault list. Each fault is injected when the corresponding position in the SR has logic "1", while all other positions have logic "0". Upon initialization, only the first position of the SR is set to "1". Then the "1" is shifted along the SR, activating one fault at a time. This technique was further optimized in [9]. However, a major limitation of this technique is the fact that the added hardware increases with the number of faults to inject, which limits the size of the circuits that can be simulated. In [6], it is also reported that some parallelism is possible by injecting independent faults at the same time. Unfortunately, the speedup reported was of only 1.36 times faster than serial fault simulation. In [10], a new approach that included a backward propagation network to allow critical path tracing [11] is proposed. This information allows multiple faults to be simulated for each test vector, but also requires heavy extra hardware and only combinational circuits have been analyzed.

A serial fault simulation technique that required only partial reconfiguration during logic emulation was proposed in [12], showing that no extra logic need be added for fault injection purposes, and that HFS can be about two orders of magnitude faster than SFS for designs over 100,000 gates. More recently, other hardware fault injection approaches were proposed in [13-15] using the JBITS [16] interface for partial FPGA reconfiguration. The JBITS software can achieve effective injection of faults on Look-Up-Tables (LUTs) [13] [14] or erroneous register values [15]. These works do not report any results on partial FPGA reconfiguration times for fault injection - only some predictions are given. Moreover, the approach that uses JBITS requires the Java SDK 1.2.2 [17] platform and the XHWIF [18] hardware interface.

The purpose of this paper is to present a Reconfigurable Hardware Fault Simulation (RHFS) methodology that uses partial reconfiguration to implement a novel, highly efficient hardware fault simulation tool, which we called f^2s. We show that partial FPGA reconfiguration can be accomplished very efficiently by deriving very small bit files for fault injection. These bit files are obtained from the full configuration bit file by means of its direct manipulation, without requiring any additional software tools. Moreover, the new RHFS tool does not require the use any specific hardware interface with the FPGA.

This paper is organized as follows. In section 2 the LUT extraction from the bit file is explained and the fault collapsing technique is proposed. In section 3 the new fault simulation tool, f^2s, is presented. Section 4 presents experimental results that compare SFS and HFS to the new RHFS approach using two ISCAS'89 benchmark circuits [19]. Finally, section 5 concludes the paper.

2 LUT Extraction, Fault Injection and Collapsing

Virtex FPGAs [20] are essentially composed of an array of Configurable Logic Blocks (CLBs) surrounded by a ring of I/O blocks. The CLBs are the building blocks for implementing custom logic. Each CLB contains two slices. Each slice contains two 4 input Look-Up-Tables (LUTs), 2 flip-flops and some carry logic. For the fault model used in the next section, only LUT extraction is required. Each LUT can be used to implement a function of 0, 1, 2, 3 or 4 inputs. It is important to identify the number of inputs relevant for each used LUT, in order to include the corresponding faults in the fault list. Since each LUT has 4 inputs, the LUT contents consist of a 16-bit vector, one vector for each combination of the inputs. Let y_{abcd} be the output value for the combination of input values $abcd$. The LUT configuration 16-bit vector can be denoted $y_{0000}, y_{0001}, y_{0010}, \ldots, y_{1111}$, where each bit position corresponds to the LUT output value for the respective combination of the inputs i_3, i_2, i_1 and i_0. If one input has a fixed value, then an 8 bit vector is obtained. For instance, if we have always $i_3=0$, then the relevant 8 bit vector is $y_{0000}, y_{0001}, y_{0010}, y_{0011}, y_{0100}, y_{0101}, y_{0110}, y_{0111}$.

The only circuit information required for our RHFS method is the configuration bit file of the FPGA. After retrieving the information of each LUT from the bit file [21], the relevance of each input i_x (x=0,1, 2 and 3) is evaluated comparing the 8-bit vectors corresponding to $i_x=0$ and $i_x=1$. If these two vectors are identical then the input i_x is not active. This is how we extract the LUT types, LUT2, LUT3, LUT4, etc., according to their number of relevant inputs.

2.2 Line Stuck-AT Fault Injection and Collapsing

The most popular fault model for digital circuit fault simulation is the single Line-Stuck-At (LSA) fault model. For the Virtex FPGAs, RHFS injects LSA faults in the outputs and inputs of LUTs. After extracting the number of active inputs in each LUT, the LSA fault list is created. Each LUT has two LSA faults (LSA-0 and LSA-1) per input or output. Thus, the total number of faults is 2, 4, 6, 8 and 10 for LUTs with 0, 1, 2, 3 or 4 active inputs, respectively. The required data for each fault injection consists of the 16-bit LUT configuration vector, modified in order to inject the fault behavior.

In order to inject a LSA-v fault ($v = 0$ or $v =1$) at the output of a LUT, all 16 bit positions of the LUT reconfiguration vector must be set to the logic value v. The reconfiguration vector that corresponds to the LUT input a stuck at value v is obtained by copying the values y_{vbcd} to $y_{\neg vbcd}$ for each bcd combination. For instance, the vector for the fault input a LSA-0 is obtained by copying y_{0000} to y_{1000}, y_{0001} to y_{1001}, y_{0110} to y_{1110}, $\ldots$, y_{0111} to y_{1111}.

After computing the 16-bit reconfiguration vectors for each fault, fault collapsing can be performed efficiently by grouping the faults with identical reconfiguration vectors. To illustrate this, consider the LUT functionality $F=a.b+c.d$. Table 1 shows the configuration vectors for the fault free LUT (F) and the configuration vectors that model all LSA faults in the LUT (i/v means line i stuck-at value v). The LUT input combination that corresponds to each vector position (row) is given in the four left most columns of Table 1. Analyzing this table, we notice that two fault pairs can be collapsed using the methodology presented: ($a/0$, $b/0$) and ($c/0$, $d/0$). The shaded cells in the table indicate the LUT input combinations (or LUT test vectors) that activate the fault given in the respective column.

Table 1. Fault reconfiguration vectors for LSA faults of a LUT implementing $F=a.b+c.d$.

a	b	C	d	F	a/0	b/0	c/0	d/0	a/1	b/1	c/1	d/1	F/1	F/0
0	0	0	0	0	0	0	0	0	0	0	0	0	1	0
0	0	0	1	0	0	0	0	0	0	0	1	0	1	0
0	0	1	0	0	0	0	0	0	0	0	0	1	1	0
0	0	1	1	1	1	1	0	0	1	1	1	1	1	0
0	1	0	0	0	0	0	0	0	1	0	0	0	1	0
0	1	0	1	0	0	0	0	0	1	0	1	0	1	0
0	1	1	0	0	0	0	0	0	1	0	0	1	1	0
0	1	1	1	1	1	1	0	0	1	1	1	1	1	0
1	0	0	0	0	0	0	0	0	0	1	0	0	1	0
1	0	0	1	0	0	0	0	0	0	1	1	0	1	0
1	0	1	0	0	0	0	0	0	0	1	0	1	1	0
1	0	1	1	1	1	1	0	0	1	1	1	1	1	0
1	1	0	0	1	0	0	1	1	1	1	1	1	1	0
1	1	0	1	1	0	0	1	1	1	1	1	1	1	0
1	1	1	0	1	0	0	1	1	1	1	1	1	1	0
1	1	1	1	1	1	1	1	1	1	1	1	1	1	0

3 Reconfigurable Hardware Fault Simulation

A new tool for FPGA based fault simulation (f^2s) has been developed. One great advantage of f^2s is the fact that it takes as input solely the binary file with the complete FPGA configuration. From this file, f^2s identifies the target device for which the mapping was performed. This identification is mandatory in order to know the number of rows and columns for LUT bit positioning. LUT extraction is carried out as described in section 2.1 and the fault list is collapsed as described in section 2.2. After fault collapsing, f^2s creates a report with all the LUTs in use, the original fault list, the collapsed fault list and the estimated reconfiguration times for different hardware interfaces. Fig. 1 shows this report for the s5378 ISCAS'89 benchmark circuit. Next, all partial reconfiguration bit files are generated in sequence for each fault. In this way, each bit file reprograms only the minimum number of frames (smallest amount of data that can be read or written with a single command) required to eject the previous fault and inject the next one. Both single and multiple fault injection are supported.

Debugging and validation of the tool was carried out using a JTAG 200 Kb parallel III port. The experimental values obtained for the reconfiguration times matched the estimated values. Since the current version of the tool does not support any type of external test vector application, it is more suitable for BIST quality evaluation purposes. However, the methodology can be easily extended to apply external vectors supplied from an internal (FPGAs have internal RAM blocks) or external memory.

```
BIT_FILE_NAME:  = s5378.ncd
BIT_PART_STRING:  = v2000ebg560
BIT_DATE_STRING:  = 2003/01/16
BIT_TIME_STRING:  = 18:01:59
BIT_IMAGE:  size 1269956


FAULTS STACK AT ZERO/ONE at LUT's INPUTS / OUTPUTS / SELECT RAM
BITS
LUT -- LOCATION , TYPE , CONTENTS AND FAULT TYPE TO INJECT (
equivalent faults , between – and separated by / )
row=49 col=19  slice=1  lut=G    type=2  vector=0 0 0 0 0 0 1 1 0 0 0 0 0 0 1 1
       faults= -- O_S_1/I1_S_0/I2_S_0 – O_S_0 – I2_S_1 – I1_S_1

...
...
row=55 col=73  slice=0  lut=G    type=4  vector=0 0 0 0 0 0 0 0 0 0 0 0 0 0 1 0
       faults= -- O_S_1/I0_S_0/I1_S_0/I2_S_0/I3_S_1 – O_S_0 – I1_S_1 – I2_S_1
– I0_S_1 – I3_S_0
```

lut capacity	38400	
lut not used	37996	98.9479166666667
fauls to detect	3610	
fauls to inject	2707	0.749861495844875
lut type 1	0	0
lut type 2	66	0.171875
lut type 3	83	0.216145833333333
lut type 4	255	0.6640625

TIME (hour:min:sec) to program and inject 0:6:51   (based on 200 Kb parallel III port speed – jtag)
TIME (hour:min:sec) to program and inject 0:0:16   (based on 5 Mb parallel IV port speed – jtag)
TIME (hour:min:sec) to program and inject 0:0:1   (based on 66 Mb usb port speed – serial)

Fig. 1. Fault list report from f^2s

844   A. Parreira et al.

4 Experimental Results

Software, Hardware and Reconfigurable Hardware Fault Simulation are compared in this section. The focus is to determine the trade-offs associated with the fundamentals of the three approaches: SFS costs increase steadily with circuit complexity, test length and fault list size, especially for sequential circuits. HFS has a significant initial cost, associated with FPGA synthesis and implementation, and a low cost in the hardware emulation process itself. Finally, RHFS represents a trade-off between the previous two approaches. Because reconfiguration is used in fault injection, hardware mechanisms for fault injection need not be added to the original circuit. This gives rise to a much simpler circuit that can be compiled faster. On the other the fault replacements by means of partial reconfiguration will take more time than simply shifting the fault register as in HFS. Thus the RHFS fault simulation times increase faster with the number of vectors when compared to HFS, but slower when compared to SFS.

In order to compare the three approaches, two ISCAS'89 circuits, s5378 and s13207, have been used as test vehicles. The complexity of these circuits and their fault list sizes before and after collapsing are shown in Table 2. From the analysis of this table, it is clear that the original fault list of LSAs at LUT terminals is efficiently collapsed by the tool. Table 3 shows the LSA fault collapsing results for the s5378 circuit obtained with the f's tool in terms of the distribution of the faults among the various types of extracted LUTs. Pseudo-random test vectors have been used in the experiments. The vectors are generated internally in the FPGA using an LFSR, as it is common for BIST. Since we are mainly interested in measuring the fault simulation time, we have not obtained fault coverage values. However fault coverage can be obtained by emulating the fault free circuit, saving the response sequences and later comparing them to the output response sequences of the faulty circuits. Work in this direction is in progress.

Table 2. LUT usage for a Virtex XCV2000E FPGA and corresponding fault list sizes for the two ISCAS'89 sequential circuit benchmarks.

Benchmark Circuit	#LUTs used	#LUTs unused	#LSA faults before collapsing	#LSA faults after collapsing
s5378	404	37996	3610	2707
s13207	795	37605	7126	5434

Table 3. Collapsed fault distribution over LUT types for the s5378 benchmark circuit.

LUT Type	# faults						
	4	5	6	7	8	9	10
LUT2	63		3				
LUT3		40	24	7	14		
LUT4			54	47	118	11	24

Software fault simulation has been carried out using a widely used commercial fault simulation tool running on a Sun ultra10/440 workstation of 1 GB of RAM capacity. The SFS time, t_{SFS}, is computed by the commercial EDA tool. The fault simulation time, t_{HFS}, for the HFS or RHFS approaches is computed by the following expression:

$$t_{HFS} = t_{comp} + t_{reconf} + \frac{\# faults \times \# vectors}{f_{HFS}}$$

where t_{comp} is the sum of the FPGA synthesis and configuration times, t_{reconf} is the time required for reconfiguration (which is zero in the case of HFS) and f_{HFS} is the frequency of vector hardware application. The value of t_{comp} is obtained after synthesis and mapping of the circuit using the Xilinx Synthesis Tool (XST). The f's tool analyses the bit stream and computes all reconfiguration frames for each fault. According to the number of generated reconfiguration frames, it reports the value of t_{reconf} for different hardware interfaces. Figs. 3 and 4 show the fault simulation times for the two ISCAS'89 benchmark circuits, s5378 and s13207, respectively, for $f_{HFS} =$ 50MHz, using a Celoxica RC1000-PP board containing a Xilinx Virtex 2000E FPGA. Note that, unlike in SFS, in HFS or RHFS no fault dropping is available since all faults are simulated in sequence. The curve marked "RHFS (200Kb)" refers to partial reconfiguration using a JTAG 200 Kb parallel III port. The curve marked "RHFS (5Mb)" refers to partial reconfiguration using a JTAG 5 Mb parallel IV port.

From these results it is clear that HFS (hardware fault simulation with a unique configuration file) is advantageous over SFS only when the number of test vectors exceeds a certain value that mainly depends on the FPGA compilation time t_{comp} [9]. Once the FPGA is compiled, the cost of applying test vectors is very small compared to SFS, despite the fact that the slope of t_{SFS} depends on the computational resources of the machine that is running it. On the other hand, RHFS (hardware fault simulation using partial reconfiguration files for fault replacement) is faster than SFS even for a small number of test vectors. This results from the fact that $t_{comp} + t_{recomp}$ for RHFS is much lower than t_{comp} for the HFS approach. In the final version of this paper, results comparing RHFS, HFS and SFS times for a larger set of benchmark circuits will be presented, which should also include larger circuits like the ones reported in [22].

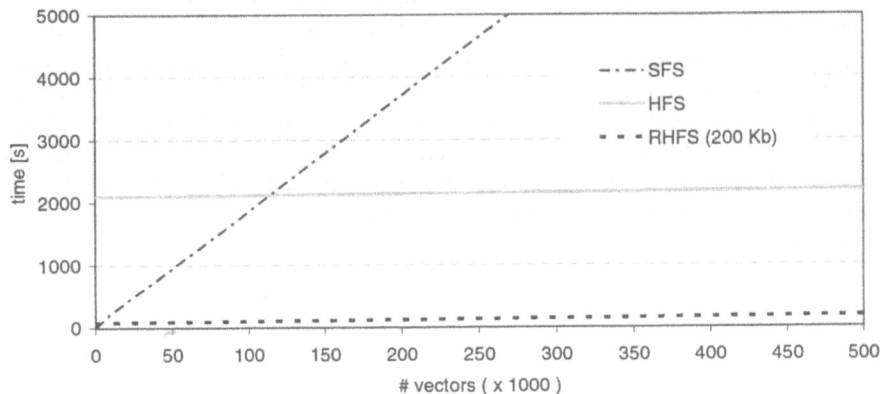

Fig. 2. Reconfigurable HW, fixed HW and SW fault simulation times for the s5378 benchmark circuit.

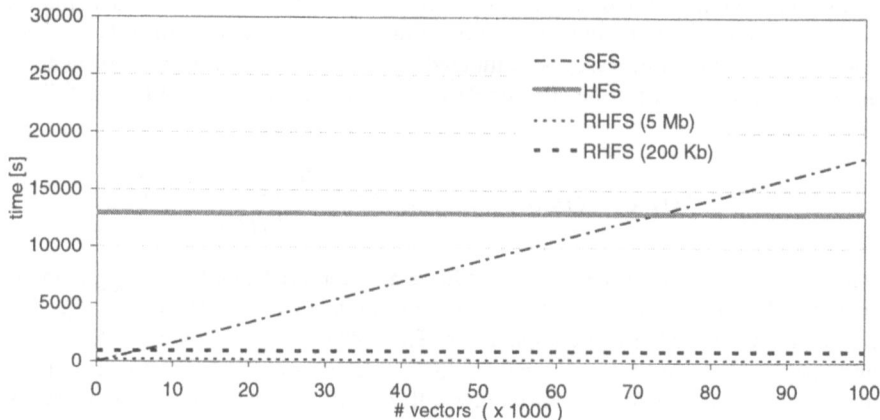

Fig. 3. Reconfigurable HW, fixed HW and SW fault simulation times for the s13207 benchmark circuit

Conclusions

In this work we have presented a novel approach to fault simulation using partial reconfiguration of FPGAs for fault replacement (RHFS). The approach has been compared to traditional software fault simulation (SFS), and to previous fixed hardware fault simulation (HFS). The superiority of the new approach has been demonstrated on two ISCAS'89 benchmark circuits.

Compared to software fault simulation, we found that RHFS is faster than SFS, even for a limited number of test vectors. After performing FPGA compilation during some tens of minutes, the simulation time for RHFS increases linearly with the number of test vectors at a rate that, for a 50MHz clock frequency, is two orders of magnitude lower than the increase rate observed for SFS.

Our fault simulation methodology uses the LSA fault model at LUT terminals. We presented a fault collapsing technique based on comparing the LUT reconfiguration vectors for each fault that leads to an efficiently optimized fault list.

Our RHFS methodology was embodied in a software tool, f²s, which takes as unique input the binary file for FPGA configuration. This is a significant advantage over other partial reconfiguration techniques that depend on a variety of file formats and tools. When processing the bit file, f²s, identifies the target FPGA device, extracts the LUTs in use and their types, and generates minimal reconfiguration bit files for fault injection. Current work is oriented towards supporting external vectors, automation of vector formatting and fault coverage evaluation. The goal is to make f²s a competitive fault simulation tool.

References

[1] M.L. Bushnel, V.D. Agrawal, "Essentials of Electronic Testing for Digital Memory and Mixed-Signal VLSI Circuits", Kluwer Academic Pubs., 2000.

[2] T.M. Niermann, W. T. Cheng, J. H. Patel, "PROFS: A fast, memory-efficient sequential circuit fault simulator", IEEE Trans. Computer-Aided Design, pp.198-207, 1992.

[3] E.M. Rudnick, J.H. Patel, "Overcoming the Serial Logic Simulation Bottleneck in Parallel Fault Simulation", Proc. of the IEEE International Test Conference (ITC), pp. 495-501, 1997.

[4] F.M. Gonçalves, M.B. Santos, I.C. Teixeira and J.P. Teixeira, "Design and Test of Certifiable ASICs for Safety-critical Gas Burners Control", Proc. of the 7th. IEEE Int. On-Line Testing Workshop (IOLTW), pp. 197-201, July, 2001.

[5] R.W.Wieler, Z. Zhang, R. D. McLeod, "Simulating static and dynamic faults in BIST structures with a FPGA based emulator", Proc. of IEEE Int. Workshop of Field-Programmable Logic and Application, pp. 240-250, 1994.

[6] K. Cheng, S. Huang, W. Dai, "Fault Emulation: A New Methodology for Fault Grading", IEEE Trans. On Computer-Aided Design of Integrated Circuits and Systems, vol. 18, no. 10, pp1487-1495, October 1999.

[7] Shih-Arn Hwang, Jin-Hua Hong and Cheng-Wen Wu, "Sequential Circuit Fault Simulation Using Logic Emulation", IEEE Transations on Computer-Aided Design of Integrated Circuits and Systems, vol. 17, no. 8, pp. 724-736, August 1998.

[8] P.Civera, L.Macchiarulo, M.Rebaudengo, M.Reorda, M.Violante, "An FPGA-based approach for speeding-up Fault Injection campains on safety-critical circuits", IEEE Journal of Electronic Testing Theory and Applications, vol. 18, no.3, pp. 261-271, June 2002.

[9] M.B. Santos, J. Braga, I. M. Teixeira, J. P. Teixeira, "Dynamic Fault Injection Optimization for FPGA-Based Harware Fault Simulation", Proc. of the Design and Diagnostics of Electronic Circuits and Systems Workshop (DDECS), pp. 370-373, April, 2002.

[10] Miron Abramovici, Prem Menon, "Fault Simulation on Reconfigurable Hardware" , *IEEE Symposium on FPGAs for Custom Computing Machines*, pp. 182 190, 1997.

[11] M. Abramovici, P. R. Menon, D. T. Miller, "Critical Path Tracing: An Alternative to Fault Simulation", IEEE Design Automation Conference, pp. 468 - 474 , 1984.

[12] L. Burgun, F. Reblewski, G. Fenelon, J. Barbier, O. Lepape, "Serial fault simulation", Proc. Design Auomation Conference, pp. 801-806, 1996.

[13] L.Antoni, R. Leveugle, B. Fehér, "Using Run-Time Reconfiguration for Fault Injection in Hardware Prototypes", IEEE International Symposium on Defect and Fault Tolerance in VLSI Systems, pp.405-413, October 2000.

[14] L.Antoni, R. Leveugle, B. Fehér, "Using Run-Time Reconfiguration for Fault Injection Applications", IEEE Instrumentation and Measurement Technology Conference, vol. 3, pp.1773-1777, May 2001.

[15] L.Antoni, R. Leveugle, B. Fehér, "Using Run-Time Reconfiguration for Fault Injection in Hardware Prototypes", IEEE International Symposium on Defect and Fault Tolerance in VLSI Systems, pp. 245-253, 2002.

[16] S. Guccione, D. Levi, P.Sundararajan, "Jbits: A Java-based Interface for Reconfigurable Computing", Proc. of the 2nd Military and Aerospace Applications of Programmable Devices and Technologies Conference (MAPLD), pp. 27, 1999.

[17] E. Lechner, S. Guccione, "The Java Environment for Reconfigurable Computing", Proc. of the 7th International Workshop on Field-Programmable Logic and Applications (FPL), Lecture Notes in Computer Science 1304, pp.284-293, September 1997.

[18] P.Sundararajan, S.Guccione, D.Levi, "XHWIF: A portable hardware interface for reconfigurable computing", Proc. of Reconfigurable Technology: FPGAs and Reconfigurable Processors for Computing and Communications, SPIE 4525, pp.97-102, August 2001.

[19] F. Brglez, D. Bryan, K. Kominski, "Combinational Profiles of Sequential Benchmark Circuits", Proc. Int. Symp. on Circuits and Systems (ISCAS), pp. 1229-34, 1989.

[20] Xilinx Inc., "Virtex-E 1.8V Field Programmable Gate Arrays", Xilinx DS022, 2001.

[21] Xilinx Inc., "Virtex Series Configuration Architecture User Guide", Application Note: Virtex Series, XAPP151 (v1.5), September 27, 2000.

[22] H. Cha, E. Rudnick, J.Patel, R. Iyer, G.Shoi, "A Gate-Level Simulation Environment for Alpha-Particle-Induced Transient Faults", IEEE Transactions on Computers, vol. 45, no.11, pp1248-1256, November 1996.

Switch Level Fault Emulation

Seyed Ghassem Miremadi and Alireza Ejlali

Sharif University of Technology, Department of Computer Engineering
Azadi Ave., Tehran, Iran
miremadi@sharif.edu
ejlali@ce.sharif.edu

Abstract. The switch level is an abstraction level between the gate level and the electrical level, offers many advantages. Switch level simulators can reliably model many important phenomena in CMOS circuits, such as bi-directional signal propagation, charge sharing and variations in driving strength. However, the fault simulation of switch level models is more time-consuming than gate level models. This paper presents a method for fast fault emulation of switch level circuits using FPGA chips. In this method, gates model switch level circuits and we can emulate mixed gate-switch level models. By the use of this method, FPGA chips can be used to accelerate the fault injection campaigns into switch level models.

1 Introduction

Logic emulation systems [1], [2] are now commercially available for fast prototyping, real-time operation, and logic verification. In many cases emulation is about 10^3 to 10^6 faster than simulation [3]. A logic emulator can implement a synthesizable model on a board composed of dozens of field programmable gate array (FPGA) chips. It cannot be used to emulate switch level models. However, in this paper, we present a method for modeling switch level circuits by gate level models. The resulted gate level model may be used as a starting point for synthesis into an interconnection of logic cells or FPGA layouts. In contrast to most cases for which transistors are grouped together to form gates, in this method, gates are grouped together to form switch models of transistors.

The advantages of this method are:

1. Use of FPGAs for fast emulation of CMOS switch level or mixed switch-gate level circuits.
2. Fault emulation [4] of switch level fault models such as transistor-stuck-on and transistor-stuck-off faults.
3. Emulation of three-state CMOS gates by the use of FPGAs, which do not support three-state logic directly.

Section 2 presents a brief review of switch level models. Section 3 presents the method for modeling switch level circuits by logic gates. Section 4 describes how the presented model can be used in switch level fault injection. Section 5 describes the mixed-mode approach, which is used for emulating CMOS circuits. In section 6, the

P.Y.K. Cheung et al. (Eds.): FPL 2003, LNCS 2778, pp. 849–858, 2003.

experimental evaluation of the presented method is discussed. Finally, the conclusion is in Section 7.

2   A Review of Switch Level Models

The switch level is an abstraction level between the gate level and the electrical level, offers many advantages. By operating directly on the transistor network, switch level simulators can reliably model many important phenomena in CMOS circuits, such as bi-directional signal propagation, charge sharing and variations in driving strength [5][6].

On the switch level, transistors are viewed as switches and the circuit state is described by discrete values, consisting of a logic state (e.g. 0, 1, U - unknown, Z - high impedance) and a strength (e.g. weak, strong, medium, ...) [7][8].

Different number of strength levels may be used in switch level models. In some methods, only two levels of strengths are used (e.g. weak and strong) [7], while others use more levels. For example, Verilog is a hardware description language, which supports switch level modeling with eight different levels of strength [8].

Each switch is a bi-directional element, which is controlled by the gate terminal G. A point at which two or more transistors have a common connection is called a node. A node may be driven by signals, which have different strengths. If a node is driven with the driver signals S_1, S_2, ..., S_n, the value (i.e. the logic state and the strength) of the node is equal to the value of the strongest driver signal. If two driver signals, with same strengths and different logic states, have the highest strength, the logic state of the node will be U (unknown).

In some methods, resistive switches are used. These switches reduce the strength of signals that propagate through them. For instance, in Verilog HDL both resistive and non-resistive switches can be used [8].

3   Switch Level Modeling Using Gate Level Circuits

We present in this section a method for modeling switch level circuits by logic gates. In this method, each transistor switch and each node is replaced by a corresponding gate level circuit. These circuits model the behavior of transistor switches and nodes. Figure 1b shows the gate level circuit, which models an NMOS transistor, and Figure 1d shows another gate level circuit, which models a node. We will refer to these circuits as *pseudo transistor* and *pseudo node* respectively.

Drain and source terminals of a MOS transistor switch are bi-directional connections, but all wires in a gate level circuit are uni-directional, so two connections with different directions are used for each of the drain and source terminals as shown in Figures 1b and 1d.

As an illustration, we will show how pseudo transistors and pseudo nodes can be used for modeling a CMOS NAND gate. Figure 2 shows a CMOS NAND gate and its counterpart gate level circuit, which is constructed with pseudo transistors and pseudo nodes.

Each terminal of a pseudo transistor or a pseudo node consists of K wires, which transfer K-bit codes between pseudo transistors and pseudo nodes. For example, S1 terminal of the pseudo transistor shown in Figure 1b consists of K wires.

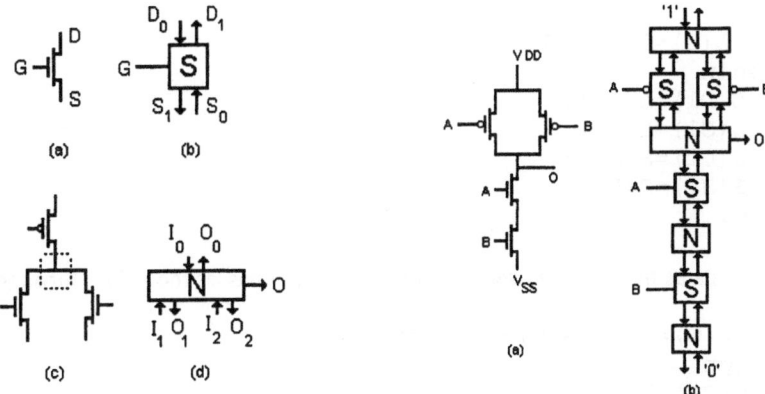

Fig. 1. (a) An NMOS switch, (b) The pseudo transistor,(c) A node, (d) The psudo node.

Fig. 2. A CMOS NAND gate (a) The CMOS Circuit, (b) The gate level model

The K-bit code is used for coding different logic states and strengths. This K-bit code is divided into two parts: state code and strength code. State code is a 2-bit code, which codes the logic state of the terminal, while strength code is an n-bit code, which codes the strength level of it. The width of the strength code depends on the number of strength levels.

In this paper, strengths are denoted by L_i, where i identifies the strength level. For example, if four different strength levels are used, the possible strengths will be L0, L1, L2, and L3, where L0 is the weakest one and L3 is the strongest one.

The logic states, which are commonly used in switch level simulations, are '1', '0', 'U' and 'Z' [7]. Here, only '1', '0' and 'U' states are coded by state codes and 'Z' state is indicated by the strength L0.

3.1 The Operation of the Pseudo Transistors

Let L(X) be the logic state of the terminal or wire X and let S(X) be its strength. The pseudo transistor, shown in Figure 1b, is a combinational circuit, which operates according to the following rules:

```
1.  If L(G)=0 and S(G)>L0 then
    begin
        S(D1)=L0;
        S(S1)=L0;
        L(D1)= don't care;
        L(S1)= don't care;
    end;
```

Rule 1 states that when a pseudo transistor is off, it passes the binary code of 'Z' value to pseudo node circuits which are connected to it indicating that the transistor does not take part in resolving the value of the nodes. Note that 'Z' value is represented by strength L0

```
2. If L(G)=1 and S(G)>L0 then
   begin
      L(D1)=L(S0);
      L(S1)=L(D0);
      S(D1)=S(S0);
      S(S1)=S(D0);
   end;
```

Rule 2 states that when a pseudo transistor is on, it passes the values through itself.

```
3. If L(G)=U or S(G)=L0 then
   beign
      L(D1)=U;
      L(S1)=U;
      S(D1)=S(S0);
      S(S1)=S(D0);
   end;
```

Rule 3 states that when the gate value is unknown or high impedance, there is no way to determine whether the transistor switch is on or off. Therefore, drain and source values will be unknown. Since the transistor switch may be on or off, it may passes strengths through itself or not. In this case, by considering the worst case it is supposed that strengths pass through the transistor switch.

These three rules model the behavior of NMOS switches. The corresponding rules for PMOS switches can be easily obtained from the above rules by interchanging 'L(G)=1' and 'L(G)=0'.

3.2 The Operation of the Pseudo Nodes

When all of the MOS transistors, which are connected to a node are off, the node retains its previous value. This is caused by the charge stored on the stray capacitance C associated with the node. For this reason, a sequential circuit is used as a pseudo node. Memory elements (flip-flops) are used for modeling gate capacitance of MOS transistors or capacitance of wires, which can store signal values.

It should be noted that the stored values are weak values because they result from small charges, which have not the current drive capability [8]. Therefore, in this paper, the strength L1 is assigned to each stored value. The strength L0 can not be used for this purpose, because it indicates the 'Z' state.

Figure 3 shows the internal structure of a pseudo node. In this figure, the output O is called the main output of the node and the outputs O_i are called returned outputs.

The boxes labeled "Resolution circuit" and "Attenuator" in Figure 3 are combinational circuits, while box labeled "Pseudo capacitor" is a sequential circuit.

As mentioned previously, if a node is driven with the driver signals I_1, I_2, ..., I_n, the value of the node is equal to the value of the strongest driver signal, and if two driver signals, with same strengths and different logic states, have the highest strength, the logic state of the node will be U (unknown). This is the way a resolution circuit resolves a single value from multiple driving values.

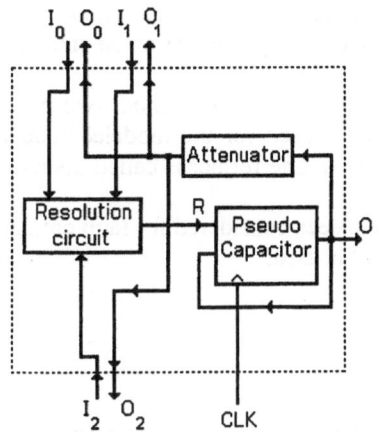

Fig. 3. The internal structure of a pseudo node

Table 1. The state table of the pseudo capacitor circuit

Present State	Next State						
	$R=U_n$	$R=1_n$	$R=0_n$	$R=U_l$	$R=1_l$	$R=0_l$	$R=Z$
U_m	U_n	1_n	0_n	U_l	U_l	U_l	U_l
1_m	U_n	1_n	0_n	U_l	1_l	U_l	1_l
0_m	U_n	1_n	0_n	U_l	U_l	0_l	0_l
U_l	U_n	1_n	0_n	U_l	U_l	U_l	U_l
1_l	U_n	1_n	0_n	U_l	1_l	U_l	1_l
0_l	U_n	1_n	0_n	U_l	U_l	0_l	0_l
Z	U_n	1_n	0_n	U_l	1_l	0_l	Z

$n>1, m>1$

A pseudo capacitor models the behavior of the parasitic capacitance associated with the node. If we consider the state of this sequential circuit to be its output, its operation can be summarized in Table 1 as a state table.

By considering the operation of a node capacitance in a switch level circuit, the derivation of this state table is carried out most easily. For example, according to the first three columns of the table (columns labeld $R=U_n$, $R=1_n$ and $R=0_n$), when the value of the input R have the strength greater than L1, the output O will be equal to input R. This is because the values stored in a node capacitance can be overridden by stronger values supplied by turned-on transistor switches. As another example, consider the row 'Present State=1_m' and the column '$R=0_l$', in this case the next state will be 'U_l'. This is because a '1_l' value is stored in the stray capacitance because of the previous '1_m' value of the node. When the transistor switches connected to the node supply the weak value '$R=0_l$' to the node, the result is 'U_l'. Note that the input R of the pseudo capacitor circuit is the output of the resolution circuit, therefore its value is determined by all the pseudo transistors, which are connected to the node.

The combinational circuit labeled "Attenuator", reduces the strength of the output O, and transfers the result to all the transistor switches, which are connected to the node. It should be noted that this circuit does not change the logic state of the output O. Also it does not change the strength of the output O if its strength is L0 (i.e. 'Z' state). The benefits of the attenuator circuit are as follows:

- This circuit ensures that a value with the strength greater than L1 can not be stored in pseudo nodes. Note that when two pseudo nodes are connected with a turned on pseudo transistor switch, they send their values for each other at each clock pulse. In this case the attenuator circuits of these nodes ensure that strong values can not loop between these two pseudo nodes. They attenuate such looping values, therefore only weak values with strength L0 can be stored in such loops.

- By the use of this circuit, resistive transistor switches can be modeled, because the strength of a signal is reduced by 1, whenever it propagates from a node to another node through a transistor switch.

Since pseudo nodes are sequential circuits, when a combinational circuit, such as the NAND gate shown in Figure 2a, is modeled using the presented model (See Figure 2b), a sequential circuit is resulted.

If one wants to model a CMOS sequential circuit by this method, two different clock signals will be needed. One clock signal is used for the modeled sequential circuit itself and the other is used for pseudo nodes. The former is called main clock, while the latter is called emulation clock.

In this paper, only combinational circuits have been considered for fault emulation, therefore only one clock signal (emulation clock) is used.

4   Fault Injection at the Switch Level

Some of the advantages of the fault injection at the switch level over the fault injection at the gate level are as follows:
- Many regions are not reachable for gate level fault injection, as they are internal to the gates. But real defects may occur inside a gate.
- There are some defects, which can not be modeled by gate level stuck-at faults.

This section describes how the presented model can be used for injecting transistor-stuck fault models into switch level circuits.

A transistor is stuck-on if it remains on regardless of its gate value. In a similar manner, a transistor is stuck-off if it remains off regardless of its gate value.

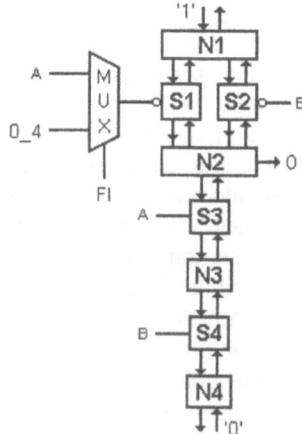

Fig. 4. Injection of transistor-stuck-on fault

Figure 4 shows how a transistor-stuck fault can be injected into the switch level model of a CMOS NAND gate (Shown in figure 2) using the presented gate level model. The signal "FI" in the Figure 4 is a fault injector signal, which is used for activating a fault. This signal is the selection line of a 2-line to 1-line multiplexer. When the input A is selected, the circuit is fault free, and when the input 0_4 (with the logic state 0 and the strength L4) is selected a transistor-stuck-on fault is injected because, in this case, the pseudo transistor "S1" is on regardless of the value of A.

It is not necessary to choose the value 0_4 in order to turn on the pseudo transistor "S1". In fact, any value with the logic state 0 and the strength greater than L0 can be used to turn on the pseudo transistor. Note that each input and output of the multiplexer consists of K wires, which transfer K-bit codes.

5 Mixed-Mode Emulation

If the presented model is used for emulating a switch level circuit using FPGAs, a large number of FPGA resources may be consumed. In order to attain a better utilization of FPGA resources a mixed-mode emulation approach has been used. In this approach parts of the circuit are emulated at the gate level while the rest of the circuit (such as faulty portions of the circuit) is emulated at the switch level. This method is similar to the method, which is used in mixed-mode fault simulators [9]. The difference is that the presented mixed-mode method has been used to address the FPGA resources utilization problem while the mixed-mode simulators have been used to address the time explosion problem.

[4] discusses a dual-railed logic which is used for emulating gate level circuits. The idea is to use two wires to represent a three-valued logic signal. In a similar manner, in this paper a dual-railed logic with three-valued logic signals (0, 1 and U) is used for gate level parts of the mixed-model. This is because switch level parts have logic values such as "U" and "Z" which cannot propagate through two-valued logic gates.

Whenever a signal propagates form the switch level parts to the gate level parts of the circuit only its state code is transferred and its strength code is omitted. As mentioned in Section 3, state code is a 2-bit code, which codes the logic state of a signal. Therefore, it can be applied to dual-railed logic gates.

Figure 5 shows how a signal propagates from the switch level portion of a mixed-model to its gate level portion.

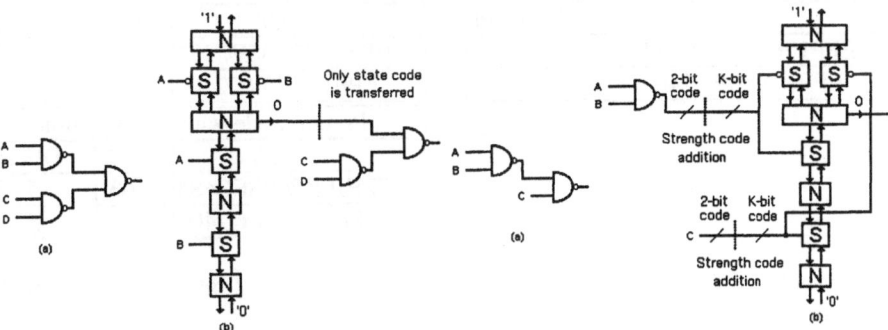

Fig. 5. mixed-mode switch-gate level emulation (a) a gate level circuit (b) its mixed-mode model for emulating

Fig. 6. mixed-mode switch-gate level emulation (a) a gate level circuit (b) its mixed-mode model for emulating

When a signal propagates from the gate level portion of a circuit to its switch level portion, only a strength code with the strength greater than L0 is added to the propagated signal.

Figure 6 shows how a signal propagates from the gate level portion of a mixed-model to its switch level portion. In most cases, signals, which propagate from gate level parts to the switch level parts, are connected to the gates of the transistor switches. In such cases, the value of the added strength code is insignificant (See rules 1,2 and 3 in Section 3). It only has to be greater than L0.

6 Experimental Results

This section presents the results, which are obtained from the experimental evaluation of the method by the use of two Altera FPGAs (FLEX10K200SFC484 with 9984 LCs and 10160 DFFs, and FLEX10K50RC240 with 2880 LCs and 3184 DFFs). The experimented model has 8 different levels of strength.

Table 2 shows the resources, which are used by each pseudo transistor. As shown in Table 2, pseudo transistors do not consume DFFs (these circuits are combinational). Table 3 shows the number of resources, which are used by pseudo nodes. This table is made up of three parts based on the number of the pseudo transistors, which are connected to the pseudo node.

As shown in Table 3, the number of DFFs, which the pseudo node uses is 5 regardless of the number of the pseudo transistors, which are connected to it. This is because only the pseudo capacitor of a node needs DFFs.

Table 2. Resources used by pseudo transistors (with 8 levels of strength)

P type pseudo transistor		
FPGA	LCs	DFFs
EPF10K200SFC484	11	0
EPF10K50RC240	11	0
N type pseudo transistor		
FPGA	LCs	DFFs
EPF10K200SFC484	11	0
EPF10K50RC240	11	0

Table 3. Resources used by pseudo nodes(with 8 levels of strength)

Pseudo node with 2 connections		
FPGA	LCs	DFFs
EPF10K200SFC484	30	5
EPF10K50RC240	30	5
Pseudo node with 3 connections		
FPGA	LCs	DFFs
EPF10K200SFC484	57	5
EPF10K50RC240	56	5
Pseudo node with 4 connections		
FPGA	LCs	DFFs
EPF10K200SFC484	63	5
EPF10K50RC240	63	5

The width of the strength code is 3 and the state code is a 2-bit code. Therefore, 5 DFFs are used.

Table 4 shows the resources, which are used by dual-railed NAND gates. These results show the importance of the mixed-mode switch-gate level emulation presented in the previous section.

Our experiments show that the speedup grows with the circuit size and the emulation of a switch level circuit can be about 27 to 532 times faster than its simulation using Verilog switch level models. In these experiments a ModelSim simulator (Version 5.4a) is used which is run on a Pentium II system (333MHz, RAM=192 MB). For example for an 8-bit Magnitude comparator circuit, which its gate level model is shown in Figure 7, the resulted speedup is 417 when 10000 different inputs are applied to the circuit. In this experiment mixed-mode emulation is not used and circuit is emulated using a pure switch level model. In fact, each gate of the circuit shown in Figure 7 is replaced with its corresponding switch level model.

Table 4. Resources used by dual-railed NAND gates

NAND gate with 2 inputs		
FPGA	LCs	DFFs
EPF10K200SFC484	2	0
EPF10K50RC240	2	0
NAND gate with 3 inputs		
FPGA	LCs	DFFs
EPF10K200SFC484	3	0
EPF10K50RC240	3	0
NAND gate with 4 inputs		
FPGA	LCs	DFFs
EPF10K200SFC484	4	0
EPF10K50RC240	4	0

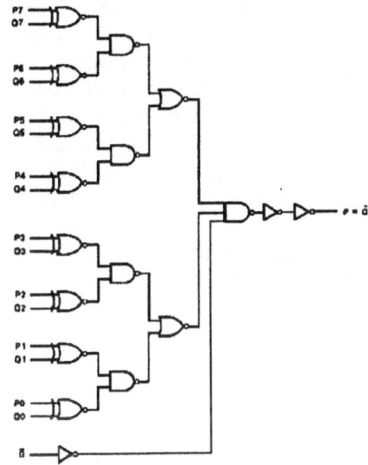

Fig. 7. An 8-bit Magnitude Comparator used in the experiments

7 Conclusions

In this paper a novel method is presented in order to emulate switch level circuits and switch level faults. The advantages of this method are:
1. Use of FPGAs for fast emulation of CMOS switch level or mixed switch-gate level circuits.
2. Fault emulation of switch level fault models such as transistor-stuck-on and transistor-stuck-off faults.
3. Emulation of three-state CMOS gates by the use of FPGAs, which do not support three-state logic directly.

The experiments show that the emulation of a switch level circuit can be very faster than its simulation

It is shown that a mixed-mode emulation approach can be used in order to optimize the utilization of FPGA resources when switch level emulation is used.

References

1. J. Varghese, M. Butts, and J. Batcheller, "An efficient logic emulation system," IEEE Trans. VLSI Syst., vol. 1, pp. 171-174, June 1993.
2. S. Walters, "Computer-aided prototyping for ASIC-based system," IEEE Design Test Comput., pp. 4-10, June 1991.
3. U. R. Khan, H. L. Owen, J. L. Hughes, "FPGA Architecture for ASIC Hardware Emulator", Proc. 6, IEEE ASIC Conference, p.336, 1993.
4. K.T. Cheng, S.Y. Huang, and W.J. Dai, "Fault Emulation: A New Methodology for Fault Grading," IEEE Trans. Computer-Aided Design of Integrated Circuits and Systems, vol. 18, no. 10, pp. 1487-1495, October 1999.
5. P. Dahlgren, P. Liden, "Efficient Modeling of Switch-Level Networks Containing Undetermined Logic Node States", Proc. IEEE/ACM Int. Conf. on CAD, pp. 746-752, 1993.
6. R.E. Bryant, "A Switch-Level Model and Simulator for MOS Digital Systems", IEEE Trans. Computers, Vol. C-33, No. 2, pp. 160-177, Feb 1984.
7. M. Abramovici, M. A. Breuer, and A. D. Friedman, "Digital Systems Testing and Testable Design", Revised edition, IEEE Press, 1995.
8. Verilog Hardware Descriptor Language Reference Manual (LRM) DRAFT, IEEE 1364, April, 1995.
9. G. S. Choi, and R. K. Iyer, "FOCUS: An Experimental Environment for Fault Sensitivity Analysis", IEEE Trans. Computers, vol. 41, no. 12, pp.1515-1526, Dec. 1992.

An Extensible, System-On-Programmable-Chip, Content-Aware Internet Firewall*

John W. Lockwood, Christopher Neely, Christopher Zuver,
James Moscola, Sarang Dharmapurikar, and David Lim

Applied Research Laboratory
Washington University in Saint Louis
1 Brookings Drive, Campus Box 1045
Saint Louis, MO 63130 USA
{lockwood,cen1,cz2,jmm5,sarang,dlim}@arl.wustl.edu
http://www.arl.wustl.edu/arl/projects/fpx/

Abstract. An extensible firewall has been implemented that performs packet filtering, content scanning, and per-flow queuing of Internet packets at Gigabit/second rates. The firewall uses layered protocol wrappers to parse the content of Internet data. Packet payloads are scanned for keywords using parallel regular expression matching circuits. Packet headers are compared to rules specified in Ternary Content Addressable Memories (TCAMs). Per-flow queuing is performed to mitigate the effect of Denial of Service attacks. All packet processing operations were implemented with reconfigurable hardware and fit within a single Xilinx Virtex XCV2000E Field Programmable Gate Array (FPGA). The single-chip firewall has been used to filter Internet SPAM and to guard against several types of network intrusion. Additional features were implemented in extensible hardware modules deployed using run-time reconfiguration.

1   Introduction

Recently, demand for Internet security has significantly increased. Internet connected hosts are now frequently attacked by malicious machines located around the world. Hosts can be protected from remote machines by filtering traffic through a firewall. By actively dropping harmful packets and rate-limiting unwanted traffic flows, the harm caused by attacks can be reduced.

While some types of attacks can be thwarted solely by examination of packet headers, other types of attacks – such as network intrusion, Internet worm propagation, and SPAM proliferation – require that firewalls process entire packet payloads [1]. Few existing firewalls have the capability to scan entire packet payloads. Of those that do, most are software-based and cannot process packets at the high-speed rates used by modern networks. Hardware-accelerated firewalls are needed to process entire packet payloads at high speeds.

* This research was supported in part by a grant from Global Velocity, the National Science Foundation (ANI-0096052), and a gift from Xilinx

P.Y.K. Cheung et al. (Eds.): FPL 2003, LNCS 2778, pp. 859–868, 2003.

Application Specific Integrated Circuits (ASICs) have been used in firewalls to implement some packet filtering functions. ASICs allow firewalls to achieve high throughput by processing packets in deep pipelines and parallel circuits. But ASICs can only protect networks from threats known to the designer when the chip was fabricated. Once an ASIC is fabricated, its function is static and cannot be altered. The ability to protect networks against both the present and future threats is the real challenge in building modern firewalls [2].

2   System-On-Programmable-Chip (SOPC) Firewall

A System-On-Programmable-Chip (SOPC) Internet firewall has been implemented that protects high-speed networks from present and future threats. In order to protect networks against current threats, the SOPC firewall parses Internet protocol headers, scans packet payloads, filters traffic, and performs per-flow queuing. In order to protect against future threats, the SOPC is highly extensible. This paper details the implementation of that firewall in a single Xilinx Virtex XCV2000E FPGA.

The top-level architecture of the SOPC firewall is shown in Figure 1. When a packet first enters the SOPC firewall, it is processed by a set of *layered protocol wrappers*. These wrappers segment and reassemble frames; verify and compute checksums; and read and write the headers of the Internet packet. Once the packet has been parsed by the wrappers, a *payload scanner* searches the entire content of the packet for keywords and regular expressions. Next, a *Ternary Content Addressable Memory (TCAM) filter* classifies packets based on a set of reconfigurable rules. The packet then passes through one or more *extensible modules*, where additional packet processing or packet filtering is implemented. Next, the data and flow ID are passed to a *queue manager*, which schedules the packets for transmission from a *flow buffer*. Finally, the packet is transmitted to a switch or to a Gigabit Ethernet or SONET line card. Details of these components are given in the sections that follow.

2.1   Protocol Processing

To process packets, a set of layered protocol wrappers was implemented to parse protocols at multiple layers [3]. At the lowest layer, data is segmented and reassembled from short cells into complete frames. At the network layer of the protocol stack, the fields of the Internet Protocol (IP) packets are verified and computed. At the transport layer, the port numbers and packet lengths are used to extract application-level data.

2.2   Payload Processing with Regular Expressions

Many types of Internet traffic cannot be classified by header inspection [4]. For example, junk email (SPAM) typically contains perfectly valid network and transport-layer data. In order to determine that a message is SPAM, a filtering

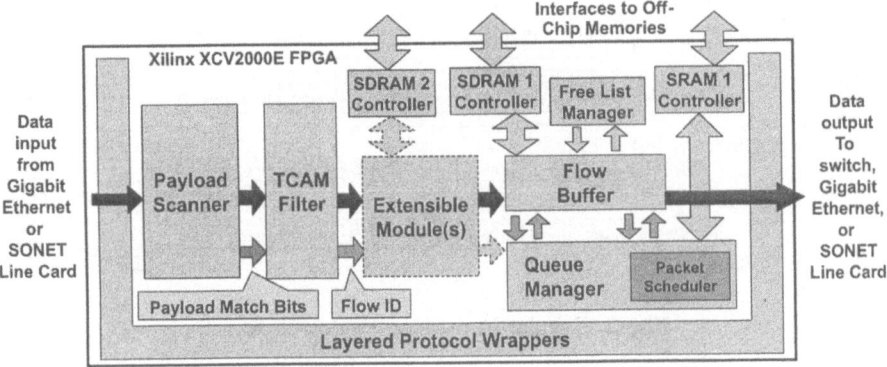

Fig. 1. Block Diagram of System-On-Chip Firewall

device must be able to classify packets based on the content rather than just the values that appear in the packet headers.

Dynamically reconfigurable hardware performs content classification functions effectively. Unlike an ASIC, new finite automata can be programmed into into hardware to scan for specific content. In order to scan packet payloads, a regular expression matching circuit was implemented. Regular expressions provide a shorthand means to specify the value of a string, a substring (specified by '( )'), alterative values (separated with '|'), any one character (specified by '.'), zero or one characters (specified by '?'), or zero or more characters (specified by '*'). For example, to match all eight case variations of the phrase "make money fast" and to allow any number of characters to reside between keywords, the expression: "(M|m)ake.*(M|m)oney.*(F|f)ast" would be specified.

The architecture of a payload scanner with four parallel content scanners searching for eight Regular Expressions, *RE[1]-RE[8]*, is illustrated in Figure 2. A match is detected when a sequence of incoming data causes a state machine to reach an accepting state. In order to differentiate between multiple groups of regular expressions, the scanner instantiates a sequence of parallel machines that each search for different expressions. In order to maximize throughput, multiple sets of parallel content matching circuits are instantiated. When data arrives, a *flow dispatcher* sends it to an available buffer which then streams data though the sequence of regular expression search engines. Finally, a *flow collector* combines the traffic into a single outgoing stream of data. The final result of the circuit is a set of *payload match bits* that identify which regular expressions were present in each data flow. To implement other high-speed payload scanning circuits, an automated design process was created to generate circuits from a specification of regular expressions [5].

2.3   Rule Processing

A diagram of the rule matching circuit used on the SOPC firewall is shown in Figure 3. An on-chip, TCAM is used to classify packets as belonging to a specific

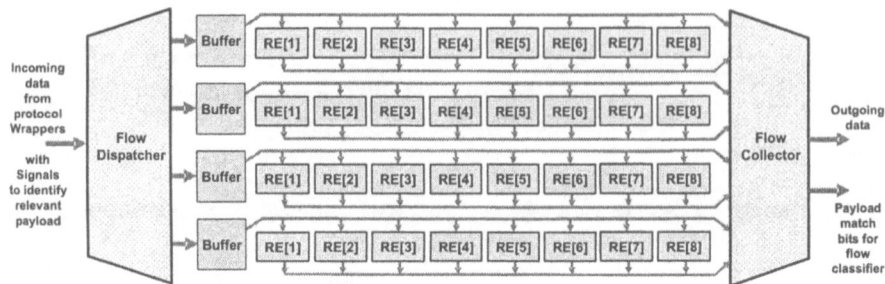

Fig. 2. Regular Expression Payload Scanner

flow. As an Internet Protocol (IP) packet arrives, *bits in the IP header*, which include the *source address, destination address, source port, destination port,* and *protocol* as well as the *payload match bits* are simultaneously compared to the *CAM value* fields in all rows of the TCAM table. Once all of the values have been compared, a *CAM mask* is applied to select which bits of each row must match and which bits can be ignored. If all of the values match in all of the bit locations that are unmasked, then that row of the TCAM is considered to be a match. The *flow identifier* associated with the rule in the highest-priority matching TCAM is then assigned to the flow. Rules can be added or modfied by sending control packets to the firewall to change values in the *CAM mask*, *CAM value*, and *flow ID* fields. Large CAMs can be implemented on the FPGA by using block RAMs [6].

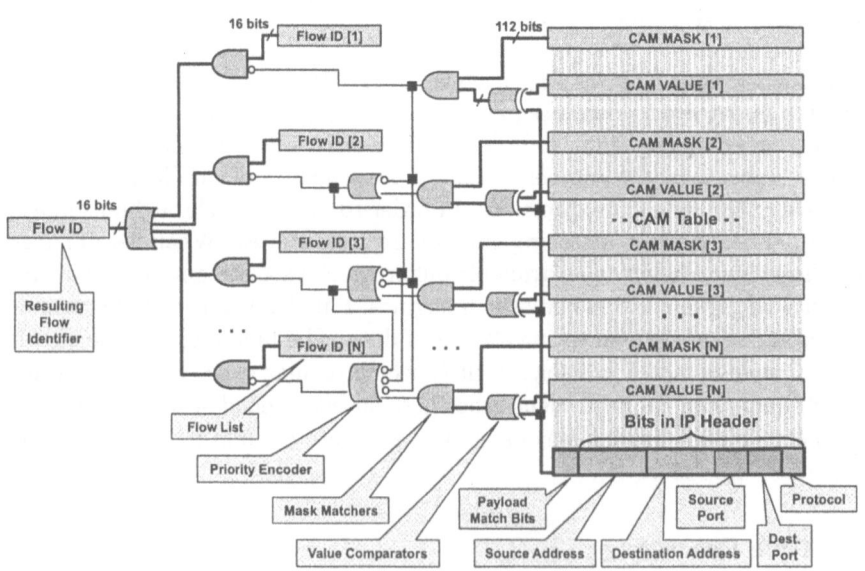

Fig. 3. Header Processing Module

2.4   Flow Buffering

To provide different levels of Quality of Service (QoS) for different traffic passing through the network, the SOPC firewall implements both class-based and per-flow queuing. Class-based queuing allows certain types of traffic to receive better service than other traffic. Per-flow queuing ensures that no single traffic flow consumes all of the network's available bandwidth [7].

To support multiple classes of service, traffic flows are organized by the firewall into four priority classes. Within each class, multiple linked lists of packets are maintained. Management of queues and tracking of free memory space is performed by the FPGA hardware in constant-time using linked-list data structures.

A diagram of the flow buffer and queue manager is shown in Figure 4. The queue manager includes circuits to enqueue traffic flows, dequeue traffic flows, and to schedule flows for transmission. Within the scheduler, four separate queues of flows are maintained (one for each class).

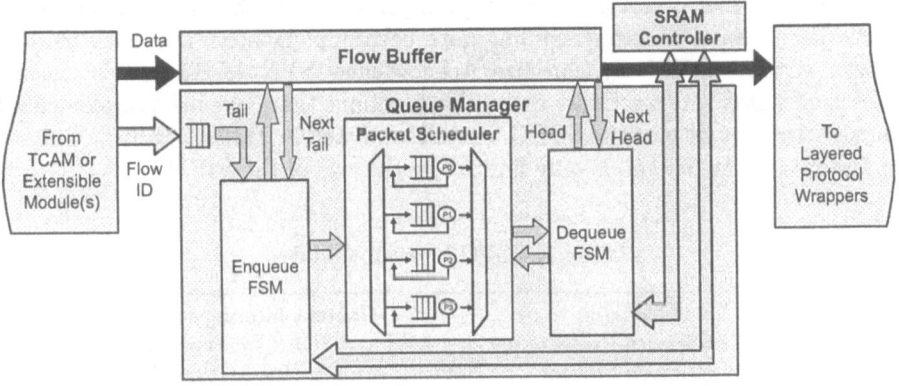

Fig. 4. Queue manager for organizing traffic flows

When a packet arrives, the packet's data is delivered to the *flow buffer* and the packet's flow identifier is passed to the *enqueue FSM* in the *Queue Manager*. Using the *flow ID*, the *enqueue FSM* reads SRAM to retrieve the flow's state. Each entry of the flow's state table contains a pointer to the *head* and *tail* of a linked list of stored packets in SDRAM as well as counters to track the number of packets read and written to that flow.

Meanwhile, the flow buffer stores the packet into memory at the location specified by the enqueue FSM's tail pointer. The flow buffer includes a controller to store the packet in Synchronous Dynamic Random Access Memory (SDRAM) [8]. After writing a packet to the next available memory location, the value of the tail pointer is passed to the queue manager to identify the next free memory location.

Within each class of traffic, the queue manager performs round-robin queuing of individual traffic flows. When the first packet of a flow arrives that has no packets already buffered, the flow identifier is inserted into a scheduling queue for that packet's class of service and the flow state table is updated. When another packet of a flow that is already scheduled arrives, the packet is simply appended to the linked list and the packet write count is incremented.

To transmit packets, the *dequeue FSM* reads a flow ID from the scheduler. The scheduler dequeues the flow from the next available flow in the highest priority class of traffic. The dequeue FSM then reads the flow state to obtain a pointer to the data in the flow buffer. The flow identifier is then removed from the head of the scheduler's queue and re-enters at the tail of the same queue if that flow has additional packets to transmit.

2.5   Extensible Features and Design Tools

Custom functionality is programmed into the SOPC firewall to implement specific functions demanded for a particular network application or service. Extensible modules that perform data encryption, enhanced packet filtering functions, and other types of packet scheduling have been implemented. One such module blocks patterns of TCP Synchronize/Acknowledge (SYN/ACK) packets characteristic of a DoS attacks. Other extensible functions that have been implemented as modules that fit into the SOPC firewall are listed in Table 1 [9] [10]. The set of features in the firewall is only limited by the size of the FPGA.

Table 1. SOPC Firewall Modules

Virus blocking	Content filtering
Denial of Service Protection	AES and 3DES Decryption
Bitmap image filtering	Network Address Translation (NAT)
Internet route lookup	IP Version 6 (IPV6) tunneling
Resource Reservation (RSVP)	Domain Name Service (DNS) caching

To facilitate development of extensible modules, an integration server was created that allows modules to be automatically configured using an application developer's information submitted via the web [11]. The tool transforms time-consuming and mistake-prone task of individually wiring ports on newly created modules into the relatively simple task of selecting where a module fits within the system interface. Developers upload their extensible modules to the integration server, which registers them in a database. The server parses the uploaded VHDL files and prompts the user to supply the port mapping to one or more of the standard, modular interfaces on the SOPC firewall. Then a bitfile is generated that contains the new module integrated into the baseline firewall, which is then returned to the user for testing. The server can also be used to create a bitfile for a SOPC firewall with any subset of previously uploaded extensible modules.

The integration server tracks and displays the estimated device utilization for each module to aid the selection process.

To further ease the development of extensible modules, a networked test server was developed to allow new extensible modules to be tested over the Internet [12]. After developing an extensible module on the integration server, a developer submits the resulting placed and routed design via a web interface to the test server. The test server: (1) remotely programs the FPGA platform, (2) generates test traffic in the form of network packets, and (3) displays the resulting packets. The traffic generated by the test server is processed by the firewall and returned to the test server. The developer then compares the contents of the resulting packets to expected values to verify correctness. Automation simplifies the process of migrating designs from concept to implementation.

3   Results

The results after place and route for the synthesized SOPC Firewall on the Xilinx Virtex XCV2000E are listed in Table 2. The core logic occupied 43% of the logic and 39% of the block RAMs. Placement of the SOPC circuitry was constrained using Synplicity's Amplify tool to lock the location of modules into specific regions of the FPGA. A view of the placed and routed Xilinx Virtex XCV2000E is shown in Figure 5. Note that the center region of the chip was left available for insertion of extensible modules.

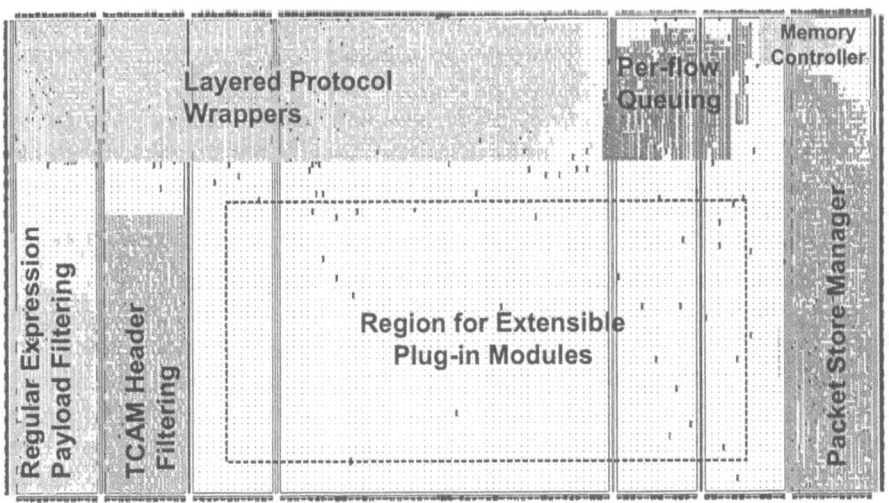

Fig. 5. FPGA Layout of the SOPC Firewall

Table 2. SOPC Firewall Implementation Statistics

Resource	Virtex XCV2000E Utilization Device Utilization	Utilization Percentage
Logic Slices	8342 out of 19200	43%
BlockRAMs	63 out of 160	39%
External IOBs	286 out of 512	55%

3.1 Throughput

The components of the SOPC firewall that implement the protocol wrappers, CAM filter, flow buffer, and queue manager were synthesized and operated at 62.5 MHz. Each of these components process 32 bits of data in every cycle, thus giving the SOPC firewall a throughput of 32*62.5MHz = 2 Gigabits/second. The regular expression scanning circuit for the SPAM pipeline synthesized at 37 MHz. Given that each pipeline processes 8 bits of data per cycle, the throughput of the SPAM filter is 8*37MHz=296 Megabits/second per pipeline. By running 8 SPAM pipelines in parallel, the payload matching circuit achieves a throughput of 8*8*37MHz=2.368 Gigabits/second.

3.2 Testing the SOPC Firewall on the FPX Platform

The Field Programmable Port Extender (FPX) platform was used to evaluate the performance of the SOPC firewall with real Internet traffic. The FPX is an open hardware platform that supports partial run-time reconfiguration [13] [14]. As with other Internet-based reconfigurable systems, the FPX allows some or all of the hardware to be dynamically reprogrammed over the network [15] [16]. On the FPX, however, all reconfiguration operations are performed in hardware and the network interfaces run at multi-gigabit/second rates [17].

The core components of the SOPC firewall include the layered protocol wrappers, TCAM packet filters, the payload processing circuit, and the per-flow packet buffer with the SRAM and SDRAM controllers. All of these components were synthesized into the Virtex XCV2000E device that implements the Reprogrammable Application Device (RAD) on the FPX. The resulting bitfile was uploaded into the RAD and in-system testing was performed.

Using the features described above, a SPAM filter was implemented on the SOPC firewall to detect and filter unwanted junk email, commonly referred to as SPAM. By scanning the payloads of packets passing through our network, potentially unwanted messages were filtered before they reached their targeted endpoint. To implement our SPAM filter, we observed that within our network, emails could be classified into one of eight general categories. Within these categories, we identified several regular expressions that were commonly present or absent in SPAM-infested email. Our listed included the terms: "MAKE MONEY FAST", "Limited Time Offer", as well as 32 other terms. These terms were programmed into hardware as regular expressions in a way that allowed case-insensitive matching where needed. Meanwhile, the TCAM was programmed with rules to assign SPAM-infested packets to put SPAM-infested email in lower-priority traffic flows [10].

Actual network traffic was sent to the hardware over the network from remote hosts. The SOPC firewall filtered traffic passing between a local area network and a wide area network, as shown in Figure 6. Malicious packets were dropped, the SPAM was rate-limited, and all other flows received a fair share of bandwidth.

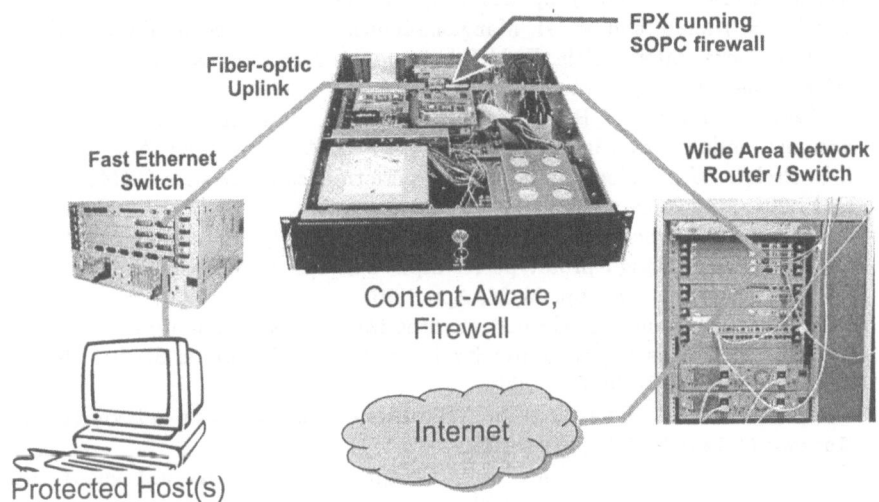

Fig. 6. Firewall Configuration

4   Conclusions

An extensible firewall has been implemented as a reconfigurable System-On-Programmable-Chip (SOPC). In addition to the standard features implemented by other Internet firewalls, the SOPC firewall performs payload scanning and per-flow queuing. The circuit was implemented on a Xilinx XCV-2000E FPGA. The resulting bitfile was tested on the Field Programmable Port Extender (FPX) network platform.

By using parallel hardware and deeply pipelined circuits, the SOPC firewall can process protocol headers with TCAMS and search the entire payload using regular expression matching at rates over 2 Gigabits/second. Attacks are mitigated by sorting each traffic flow into its own queue and scheduling traffic from all flows. These features allow the firewall to filter SPAM and protect networks from intrusion. A region of gates in the FPGA was left available to be used for extensible plugins. By using reprogrammable this hardware, new modules can be added and the firewall can protect a network against future threats.

References

1. R. Franklin, D. Carver, and B. L. Hutchings, "Assisting network intrusion detection with reconfigurable hardware," in *FCCM*, (Napa, CA), Apr. 2002.
2. J. W. Lockwood, "Evolvable internet hardware platforms," in *The Third NASA/DoD Workshop on Evolvable Hardware (EH'2001)*, pp. 271–279, July 2001.
3. F. Braun, J. Lockwood, and M. Waldvogel, "Reconfigurable router modules using network protocol wrappers," in *Field-Programmable Logic and Applications (FPL)*, (Belfast, Northern Ireland), pp. 254–263, Aug. 2001.
4. Y. Cho, S. Nahab, and W. H. Mangione-Smith, "Specialized hardware for deep network packet filtering," in *Field Programmable Logic and Applications (FPL)*, (Montpellier, France), Sept. 2002.
5. J. Moscola, J. Lockwood, R. P. Loui, and M. Pachos, "Implementation of a content-scanning module for an Internet firewall," in *FCCM*, (Napa, CA), Apr. 2003.
6. J.-L. Brelet, "Using block RAM for high performance read/write CAMs." Xilinx XAPP204, May 2002.
7. H. Duan, J. W. Lockwood, S. M. Kang, and J. Will, "High-performance OC-12/OC-48 queue design prototype for input-buffered ATM switches," in *INFO-COM'97*, (Kobe, Japan), pp. 20–28, Apr. 1997.
8. S. Dharmapurikar and J. Lockwood, "Synthesizable design of a multi-module memory controller." Washington University, Department of Computer Science, Technical Report WUCS-01-26, Oct. 2001.
9. "Acceleration of Algorithms in Hardware." http://www.arl.wustl.edu/~lockwood/class/cs535/, Sept. 2001.
10. "Reconfigurable System-On-Chip Design." http://www.arl.wustl.edu/~lockwood/class/cs536/, Dec. 2002.
11. D. Lim, C. E. Neely, C. K. Zuver, and J. W. Lockwood, "Internet-based tool for system-on-chip integration," in *International Conference on Microelectronic Systems Education (MSE)*, (Anaheim, CA), June 2003.
12. C. E. Neely, C. K. Zuver, and J. W. Lockwood, "Internet-based tool for system-on-chip project testing and grading," in *International Conference on Microelectronic Systems Education (MSE)*, (Anaheim, CA), June 2003.
13. E. L. Horta, J. W. Lockwood, D. E. Taylor, and D. Parlour, "Dynamic hardware plugins in an FPGA with partial run-time reconfiguration," in *Design Automation Conference (DAC)*, (New Orleans, LA), June 2002.
14. T. Sproull, J. W. Lockwood, and D. E. Taylor, "Control and configuration software for a reconfigurable networking hardware platform," in *IEEE Symposium on Field-Programmable Custom Computing Machines, (FCCM)*, (Napa, CA), Apr. 2002.
15. S. McMillan and S. Guccione, "Partial run-time reconfiguration using JRTR," in *Field-Programmable Logic and Applications (FPL)*, (Villach, Austria), pp. 352–360, Aug. 2000.
16. H. Fallside and M. J. S. Smith, "Internet connected FPL," in *Field-Programmable Logic and Applications (FPL)*, (Villach, Austria), pp. 48–57, Aug. 2000.
17. J. W. Lockwood, J. S. Turner, and D. E. Taylor, "Field programmable port extender (FPX) for distributed routing and queuing," in *ACM International Symposium on Field Programmable Gate Arrays (FPGA'2000)*, (Monterey, CA, USA), pp. 137–144, Feb. 2000.

IPsec-Protected Transport of HDTV over IP*

Peter Bellows[1], Jaroslav Flidr[1], Ladan Gharai[1], Colin Perkins[1], Pawel Chodowiec[2], and Kris Gaj[2]

[1] USC Information Sciences Institute, 3811 N. Fairfax Dr. #200, Arlington VA 22203, USA; {pbellows|jflidr|ladan|csp}@isi.edu
[2] Dept. of Electrical and Computer Engineering, George Mason University, 4400 University Drive, Fairfax VA 22030, USA; {pchodow1|kgaj}@gmu.edu

Abstract. Bandwidth-intensive applications compete directly with the operating system's network stack for CPU cycles. This is particularly true when the stack performs security protocols such as IPsec; the additional load of complex cryptographic transforms overwhelms modern CPUs when data rates exceed 100 Mbps. This paper describes a network-processing accelerator which overcomes these bottlenecks by offloading packet processing and cryptographic transforms to an intelligent interface card. The system achieves sustained 1 Gbps host-to-host bandwidth of encrypted IPsec traffic on commodity CPUs and networks. It appears to the application developer as a normal network interface, because the hardware acceleration is transparent to the user. The system is highly programmable and can support a variety of offload functions. A sample application is described, wherein production-quality HDTV is transported over IP at nearly 900 Mbps, fully secured using IPsec with AES encryption.

1 Introduction

As available network bandwidth scales faster than CPU power[1], the overhead of network protocol processing is becoming increasingly dominant. This means that high-bandwidth applications receive diminishing marginal returns from increases in network performance. The problem is greatly compounded when security is added to the protocol stack. For example, the IP security protocol (IPsec) [2] requires complex cryptographic transforms which overwhelm modern CPUs. IPsec benchmarks on current CPUs show maximum throughput of 40-90 Mbps, depending on the encryption used [3]. With 1 Gbps networks now standard and 10 Gbps networks well on their way, the sequential CPU clearly cannot keep up with the load of protocol and security processing. By constrast, application-specific parallel computers such as FPGAs are much better suited to cryptography and other streaming operations. This naturally leads us to consider using dedicated hardware to offload network processing (especially cryptography), so more CPU cycles can be dedicated to the applications which use the data.

This paper describes a prototype of such an offload system, known as "GRIP" (Gigabit-Rate IPsec). The system is a network-processing accelerator card based on

---

* This work is supported by the DARPA Information Technology Office (ITO) as part of the Next Generation Internet program under Grants F30602-00-1-0541 and MDA972-99-C-0022, and by the National Science Foundation under grant 0230738.

P.Y.K. Cheung et al. (Eds.): FPL 2003, LNCS 2778, pp. 869–879, 2003.

Xilinx Virtex FPGAs. GRIP integrates seamlessly into a standard Linux implementation of the TCP/IP/IPsec protocols. It provides full-duplex gigabit-rate acceleration of a variety of operations such as AES, 3DES, SHA-1, SHA-512, and application-specific kernels. To the application developer, all acceleration is completely transparent, and GRIP appears as just another network interface. The hardware is very open and programmable, and can offload processing from various levels of the network stack, while still requiring only a single transfer across the PCI bus. This paper focuses primarily on our efforts to offload the complex cryptographic transforms of IPsec, which, when utilized, are the dominant performance bottleneck of the stack.

As a demonstration of the power of hardware offloading, we have successfully transmitted an encrypted stream of live, production-quality HDTV across a commodity IP network. Video is captured in an HDTV frame-grabber at 850 Mbps, packetized and sent AES-encrypted across the network via a GRIP card. A GRIP card on a receiving machine decrypts the incoming stream, and the video frames are displayed on an HDTV monitor. All video processing is done on the GRIP-enabled machines. In other words, the offloading of the cryptographic transforms frees enough CPU time for substantial video processing with no packet loss on ordinary CPUs (1.3 GHz Pentium III).

This paper describes the hardware, device driver and operating system issues for building the GRIP system and HDTV testbed. We analyze the processing bottlenecks in the accelerated system, and propose enhancements to both the hardware and protocol layers to take the system to the next levels of performance (10 Gbps and beyond).

2   GRIP System Architecture

The overall GRIP system is diagrammed in figure 1. It is a combination of an accelerated network interface card, a high-performance device driver, and special interactions with the operating system. The interface card is the SLAAC-1V FPGA coprocessor board [4] combined with a custom Gigabit Ethernet mezzanine card. The card has a total of four FPGAs which are programmed with network processing functions as follows. One device (X0) acts as a dedicated packet mover / PCI interface, while another (GRIP) provides the interface to the Gigabit Ethernet chipset and common offload functions such as IP checksumming. The remaining two devices (X1 and X2) act as independent transmit and receive processing pipelines, and are fully programmable with any acceleration function. For the HDTV demonstration, X1 and X2 are programmed with AES-128 encryption cores.

The GRIP card interfaces with a normal network stack. The device driver indicates its offload capabilities to the stack, based on the modules that are loaded into X1 and X2. For example in the HDTV application, the driver tells the IPsec layer that accelerated AES encryption is available. This causes IPsec to defer the complex cryptographic transforms to the hardware, passing raw IP/IPsec packets down to the driver with all the appropriate header information but no encryption. The GRIP driver looks up security parameters (key, IV, algorithm, etc.) for the corresponding IPsec session, and prefixes these parameters to each packet before handing it off to the hardware. The X0 device fetches the packet across the PCI bus and passes it to the transmit pipeline (X1). X1 analyzes the packet headers and security prefix, encrypting or providing other security

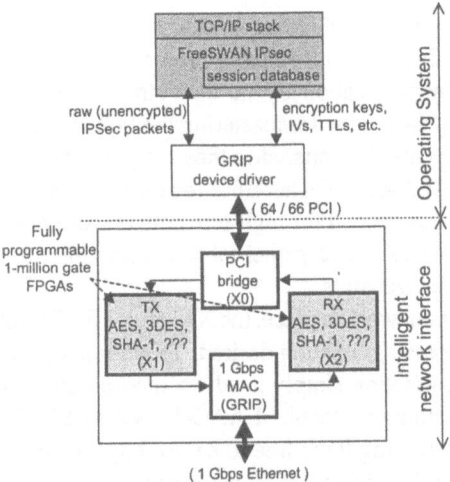

Fig. 1. GRIP system architecture

services as specified by the driver. The packet, now completed, is sent to the Ethernet interface on the daughter card. The receive pipeline is just the inverse, passing through the X2 FPGA for decryption. Bottlenecks in other layers of the stack can also be offloaded with this "deferred processing" approach.

3   GRIP Hardware

3.1   Basic Platform

The GRIP hardware platform provides an open, extensible development environment for experimenting with 1 Gbps hardware offload functions. It is based on the SLAAC-1V FPGA board, which was designed for use in a variety of military signal processing applications. SLAAC-1V has three user-programmable Xilinx Virtex 1000 FPGAs (named X0, X1 and X2) connected by separate 72-bit systolic and shared busses. Each FPGA has an estimated 1 million equivalent programmable gates with 32 embedded SRAM banks, and is capable of clock speeds of up to 150 MHz. The FPGAs are connected to 10 independent banks of 1 MB ZBT SRAM, which are independently accessible by the host through passive bus switches. SLAAC-1V also has an on-board flash/SRAM cache for storing FPGA bitstreams, allowing for rapid run-time reconfiguration of the devices. For the GRIP project, we have added a custom 1 Gigabit Ethernet mezzanine card to SLAAC-1V. It has a Vitesse 8840 Media Access Controller (MAC), and a Xilinx Virtex 300 FPGA which interfaces to the X0 chip through a 72-bit connector. The Virtex 300 uses 1 MB of external ZBT-SRAM for packet buffering, and performs common offload functions such as filtering and checksumming.

The GRIP platform defines a standard partitioning for packet processing, as described in section 2. As described, the X0 and GRIP FPGAs provide a static framework that manages basic packet movement, including the MAC and PCI interfaces. The X0 FPGA contains a packet switch for shuttling packets back and forth between the other FP-GAs on the card, and uses a 2-bit framing protocol ("start-of-frame" / "end-of-frame") to ensure robust synchronization of the data streams. By default, SLAAC-1V has a high-performance DMA engine for mastering the PCI bus. However, PCI transfers for a network interface are small compared to those required for the signal processing applications targeted by SLAAC-1V. Therefore for the GRIP system, the DMA engine was tuned with key features needed for high-rate network-oriented traffic, such as dynamic load balancing, 255-deep scatter-gather tables, programmable interrupt mitigation, and support for misaligned transfers.

With this static framework in place, the X1 and X2 FPGAs are free to be programmed with any packet-processing function desired. To interoperate with the static framework, a packet-processing function simply needs to incorporate a common I/O module and adhere to the 2-bit framing protocol. SLAAC-1V's ZBT SRAMs are not required by the GRIP infrastructure, leaving them free to be used by packet-processing modules. Note that this partitioning scheme is not ideal in terms of conserving resources - less than half of the circuit resources in X0 and GRIP are currently used. This scheme was chosen because it provides a clean and easily programmable platform for network research. The basic GRIP hardware platform is further documented in [5].

3.2   X1/X2 IPsec Accelerator Cores

A number of packet-processing cores have been developed on the SLAAC-1V / GRIP platform, including AES (Rijndael), 3DES, SHA-1, SHA-512, SNORT-based traffic analysis, rules-based packet filtering (firewall), and intrusion detection [6–8]. For the secure HDTV application, X1 and X2 were loaded with 1 Gb/s AES encryption cores. We chose a space-efficient AES design, which uses a single-stage iterative datapath with inner-round pipelining. The cores support all defined key sizes (128, 192 and 256-bit) and operate in either CBC or counter mode. Because of the non-cyclic nature of counter mode, the counter-mode circuit can maintain maximum throughput for a single stream of data, whereas the CBC-mode circuit requires two interleaved streams for full throughput. For this reason, counter mode was used in the demonstration system.

The AES cores are encapsulated by state machines that read each packet header and any tags prefixed by the device driver, and separate the headers from the payload to be encrypted / decrypted. The details of our AES designs are given in [6]. We present FPGA implementation results for the GRIP system in section 7.

4   Integrating GRIP with the Operating System

The integration of the hardware presented in the section 3 is a fairly complex task because, unlike ordinary network cards or crypto-accelerators, GRIP offers services to three layers of the OSI architecture: the physical, link and network layers. To make a complex matter worse, the IPsec stack - the main focus of current GRIP research - is located in neither the network nor link layers. Rather, it could be described as a link-layer component "wrapped" in the IP stack (figure 2). Thus care must be taken to provide a continuation of services even though parts of higher layers have been offloaded.

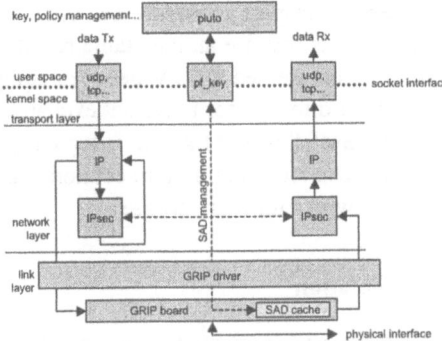

Fig. 2. The IPsec (FreeSWAN) stack in the kernel achitecture.

For this study we used FreeSWAN, a standard implementation of IPsec for Linux [9]. FreeSWAN consists of two main parts: KLIPS and Pluto. KLIPS (KerneL IP Security) contains the Linux kernel patches that implement the IPsec protocols for encryption and authentication. Pluto negotiates the Security Association (SA) parameters for IPsec-protected sockets. Figure 2 illustrates the integration of FreeSWAN into the system architecture. Pluto negotiates new security associations (SA's) using the ISAKMP protocol. When a new SA is negotiated, it is sent to the IPsec stack via the pf_key socket, where it is stored in the Security Association Database (SAD). At this point, the secure channel is open and ready to go. Any time a packet is sent to an IPsec-protected socket, the IPsec transmit function finds the appropriate SA in the database and performs the required cryptographic transforms. After this processing, the packet is handed back to IP which passes it to the physical interface. The receive mode is the inverse but somewhat less complex. When there are recursive IPsec tunnels or multiple IPsec interfaces, the above process can repeat many times.

In order to accommodate GRIP acceleration we made three modifications. First, we modified Pluto so that AES Counter mode is the preferred encryption algorithm for negiotiating new SA's. Second, we altered the actual IPsec stack so that new SA's are communicated to the GRIP device driver using the driver's private space. The driver then caches the security parameters (encryption keys, etc.) on the GRIP card for use by the accelerator circuits. Finally, the IPsec transmit and receive functions were slightly modified to produce proper initialization vectors for AES counter mode. Any packet associated with an AES SA gets processed as usual - IPsec headers inserted, initialization vectors generated, etc. The only difference is that the packet is passed back to the stack without encryption. The GRIP driver recognizes these partially-processed packets and tags them with a special prefix that instructs the card to perform the encryption.

5   Example Application: Encrypted Transport of HDTV over IP

5.1   Background

To demonstrate the performance of the GRIP system, we chose a demanding real-time multimedia application: transport of High Definition Television (HDTV) over IP. Studios and production houses need to transport uncompressed video through various cycles of production, avoiding the artifacts that are an inevitable result of multiple compression cycles. Local transport of uncompressed HDTV between equipment is typically done with the SMPTE-292M standard format for universal exchange [10]. When production facilities are distributed, the SMPTE-292M signal is typically transported across dedicated fiber connections between sites, but a more economical alternative is desirable. We consider the use of IP networks for this purpose.

5.2   Design and Implementation

In previous work [11] we have implemented a system that delivers HDTV over IP networks. The Real-time Transport Protocol (RTP) [12] was chosen as the delivery service. RTP provides media framing, timing recovery and loss detection, to compensate for the inherent unreliability of UDP transport. HDTV capture and playout was via DVS HDstation cards [13], which are connected via SMPTE-292M links to an HDTV camera on the transmitter and an HDTV plasma display on the receiver. These cards were inserted into dual-processor Dell PowerEdge 2500 servers with standard Gigabit Ethernet cards and dual PCI busses (to reduce contention with the capture/display cards). Since the GRIP card appears to the system as a standard Ethernet card, it was possible to substitute a GRIP card in place of the normal Ethernet, and run the HDTV application unmodified.

The transmitter captures the video data, fragments it to match the network MTU, and adds RTP protocol headers. The native data rate of the video capture is slightly above that of gigabit Ethernet, so the video capture hardware is programmed to perform color sub-sampling from 10 to 8 bits per component, for a video rate of 850 Mbps. The receiver code takes packets from the network, reassembles video frames, corrects for the effects of network timing jitter, conceals lost packets, and renders the video.

5.3   Performance Requirements

As noted previously, the video data rate (after colour subsampling) is 850 Mbps. Each video frame is 1.8 million octets in size. To fit within the 9000 octet gigabit Ethernet MTU, frames are fragmented into approximately 200 RTP packets for transmission. The high packet rates are such that a naive implementation can saturate the memory bandwidth; accordingly, a key design goal is to avoid data copies. We implement scatter send and receive (implemented using the recvfrom() system call with MSG_PEEK to read the RTP header, followed by a second call to recvfrom() to read the data) to eliminate data marshalling overheads. Throughput of the system is limited by the interrupt processing and DMA overheads. We observe a linear increase in throughput as the MTU is increased, and require larger than normal MTU to successfully support the full data rate. It is clear that the system is operating close to the limit, and that adding IPsec encryption will not be feasible without hardware offload.

6   Related Work

Two common commercial implementations of cryptographic acceleration are *VPN gateways* and *crypto-accelerators*. The former approach is limited in that it only provides security between LANs with matching hardware (datalink layer security), not end-to-end (network layer) security. The host-based crypto-accelerator reduces the CPU overhead by offloading cryptography, but overwhelms the PCI bus at high data rates. GRIP differs from these approaches in that it is a reprogrammable, full system solution, integrating accelerator hardware into the core operation of the TCP/IP network stack.

A number of other efforts have demonstrated the usefulness of dedicated network processing for accelerating protocol processing or distributed algorithms. Examples of these efforts include HARP[14], Typhoon[15], RWCP's GigaE PM project[16], and EMP [17]. These efforts rely on embedded processor(s) which do not have sufficient processing power for full-rate offload of complex operations such as AES, and are primarily primarily focused on unidirectional traffic. Other research efforts have integrated FPGAs onto NICs for specific applications such as routing [18], ATM firewall [19], and distributed FFT [20]. These systems accelerate end applications instead of the network stack, and often lacked the processing power of the GRIP card.

7   Results

7.1   System Performance

The HDTV demonstration system was built with symmetric multiprocessor (SMP) Dell PowerEdge 2500 servers (2x1.3 GHz) and Linux 2.4.18 kernels, as described in section 5.2, substituting a GRIP card in place of the standard Ethernet. The full, 850 Mbps HDTV stream was sent with GRIP-accelerated AES encryption and no compression. In addition, we tested for maximum encrypted bandwidth using iperf [21]. Application and operating system bottlenecks were analyzed by running precision profiling tools for 120 second intervals on both the transmitter and receiver. Transmitter and receiver profiling results are comparable, therefore only the transmitter results are presented for brevity. The profiling results are given in figure 7.1.

Library/Function	bandwidth	idle	kernel	IPsec	grip driver	appplication	libc
HDTV-SMP	893 Mbps	62%	28%	4%	3%	<1%	< 1%
iperf-SMP	989 Mbps	47%	35%	4%	4%	2%	8%
iperf-UP	989 Mbps	0%	70%	9%	4%	3%	12%

Fig. 3. Transmitter profiling results running the HTDV and iperf applications, showing percentage of CPU time spent in various functions.

The HDTV application achieved full-rate transmission with no packets dropped. Even though the CPU was clearly not overloaded (idle time > 60%!), stress tests such as running other applications showed that the system was at the limits of its capabilities. Comparing the SMP and UP cases under iperf, we can see that the only change (after taking into account the 2X factor of available CPU time under SMP) is the amount of idle time. Yet in essence, the performance of the system was unchanged.

To explain these observations, we consider system memory bandwidth. We measured the peak main memory bandwidth of the test system to be 8 Gbps with standard bench-marking tools. This means that in order to sustain gigabit network traffic, each packet can be transfered at most 8 times to/from main memory. We estimate that standard packet-processing will require three memory copies per packet: from the video driver's buffer to the hdtv application buffer, from the application buffer to the network stack, and a copy within the stack to allow IPsec headers to be inserted. The large size of the video buffer inhibits effective caching of the first copy and the read-access of the second copy; this means these copies consume 3 Gbps of main memory bandwidth for 1 Gbps network streams. Three more main memory transfers occur in writing the video frame from the capture card to the system buffer, flushing ready-to-transmit packets from the cache, and reading packets from memory to the GRIP card. In all, we estimate that a 1 Gbps net-work stream consumes 6 Gbps of main memory bandwidth on this system. Considering that other system processes are also executing and consuming bandwidth, and that the random nature of network streams likely reduces memory efficiency from the ideal peak performance, we conclude that main memory is indeed the system bottleneck.

7.2   Evaluating Hardware Implementations

Results from FPGA circuit implementations are shown in figure 7.2. As shown in the figure, the static packet-processing infrastructure easily achieves 1 Gbps throughput. Only the AES and SHA cores have low timing margins. Note that there are more than enough resources on SLAAC-1V to combine both AES encryption and a secure hash function at gigabit speeds. Also note that the target technology, the Virtex FPGA family, is five years old; much higher performance could be realized with today's technology.

Design	CLB Util.	BRAM Util	Pred. Perf. (MHz / Gbps)	Measured Perf. (MHz / Gbps)
X0	47%	30%	PCI: 35 / 2.24	33 / 2.11
			I/O: 54 / 1.73	33 / 1.06
X1 / X2 (AES)	17%	65%	CORE: 90 / 1.06	90 / 1.06
			I/O: 47 / 1.50	33 / 1.06
GRIP	35%	43%	41 / 1.33	33 / 1.06
Other modules:				
3DES	31%	0%	77 / 1.57	83 / 1.69
SHA-1	16%	0%	64 / 1.00	75 / 1.14
SHA-512	23%	6%	50 / 0.62	56 / 0.67

Fig. 4. Summary of FPGA performance and utilization on Virtex 1000 FPGAs

8   Conclusions and Future Work

Network performance is currently doubling every eight months [1]. Modern CPUs, advancing at the relatively sluggish pace of Moore's Law, are fully consumed by full-rate data at modern line speeds, and completely overwhelmed by full-rate cryptography. This disparity between network bandwidth and CPU power will only worsen as these trends continue. In this paper we have proposed an accelerator architecture that attempts to resolve these bottlenecks now and can scale to higher performance in the future. The unique contributions of this work are not the individual processing modules themselves; for example, 1 Gbps AES encryption has been demonstrated by many others. Rather, we believe the key result is the full system approach to integrating accelerator hardware directly to the network stack itself. The GRIP card is capable of completing packet processing for multiple layers of the stack. This gives a highly efficient coupling to the operating system, with only one pass across the system bus per packet. We have demonstrated this system running at full 1 Gbps line speed with end-to-end encryption on commodity PCs. This provides significant performance improvements over existing implementations of end-to-end IPsec security.

As demonstrated by the HDTV system, this technology is very applicable to signal processing and rich multimedia applications. It could be applied to several new domains of secure applications, such as immersive media (e.g. the collaborative virtual operating room), commercial media distribution, distributed military signal processing, or basic VPNs for high-bandwidth networks.

We would like to investigate other general-purpose offload capabilities on the current platform. A 1 Gbps secure hash core could easily be added to the processing pipelines to give accelerated encryption and authentication simultaneously. More functions could be combined by using the rapid reconfiguration capabilities of SLAAC-1V to switch between a large number of accelerator functions on-demand. Packet sizes obviously make a big difference - larger packets mean less-frequent interrupts. The GRIP system could leverage this by incorporating TCP/IP fragmentation and reassembly, such that PCI bus transfers are larger than what is supported by the physical medium. Finally, several application-specific kernels could be made specifically for accelerating the HDTV system, such as RTP processing and video codecs.

Our results suggest that as we look towards the future and consider ways to scale this technology to multi-gigabit speeds, we must address the limitations of system memory bandwidth. At these speeds, CPU-level caches are of limited use because of the large and random nature of the data streams. While chipset technology improvements help by increasing available bandwidth, performance can also greatly improve by reducing the number of memory copies in the network stack. For a system such as GRIP, three significant improvements are readily available. The first and most beneficial is a direct DMA transfer between the grabber/display card and the GRIP board. The second is the elimination of the extra copy induced by IPsec, by modifying the kernel's network buffer allocation function so that the IPsec headers are accomodated. The third approach is to implement the zero-copy socket interface.

FPGA technology is already capable of multi-gigabit network acceleration. 10-Gbps AES counter mode implementations are straightforward using loop-unrolling [22]. Cyclic transforms such as AES CBC mode and SHA will require more aggressive techniques such as more inner-round pipelining, interleaving of data streams, or even multiple units in parallel. We believe that 10 Gbps end-to-end security is possible with emerging commodity system bus (e.g. PCI Express), CPU, and network technologies, using the offload techniques discussed.

References

1. Calvin, J.: Digital convergence. In: Proceedings of the Workshop on New Visions ofr Large-Scale Networks: Research and Applications, Vienna, Virginia (2001)
2. IP Security Protocol (IPsec) Charter: Latest RFCs and Internet Drafts for IPsec, http://ietf.org/-html.charters/ipsec-charter.html. (2003)
3. FreeS/WAN: IPsec Performance Benchmarking, http://www.freeswan.org/freeswan_trees/-freeswan-1.99/doc/performance.html. (2002)
4. Schott, B., Bellows, P., French, M., Parker, R.: Applications of adaptive computing systems for signal processing challenges. In: Proceedings of the Asia South Pacific Design Automation Conference, Kitakyushu, Japan (2003)
5. Bellows, P., Flidr, J., Lehman, T., Schott, B., Underwood, K.D.: GRIP: A reconfigurable architecture for host-based gigabit-rate packet processing. In: Proc. of the IEEE Symposium on Field-Programmable Custom Computing Machines, Napa Valley, CA (2002)
6. Chodowiec, P., Gaj, K., Bellows, P., Schott, B.: Experimental testing of the gigabit IPsec-compliant implementations of Rijndael and Triple-DES using SLAAC-1V FPGA accelerator board. In: Proc. of the 4th Int'l Information Security Conf., Malaga, Spain (2001)
7. Grembowski, T., Lien, R., Gaj, K., Nguyen, N., Bellows, P., Flidr, J., Lehman, T., Schott, B.: Comparative analysis of the hardware implementations of hash functions SHA-1 and SHA-512. In: Proc. of the 5th Int'l Information Security Conf., Sao Paulo, Brazil (2002)
8. Hutchings, B.L., Franklin, R., Carver, D.: Assisting network intrusion detection with re-configurable hardware. In: Proc. of the IEEE Symposium on Field-Programmable Custom Computing Machines, Napa Valley, CA (2002)
9. FreeS/Wan: http://www.freeswan.org/. (2003)
10. Society of Motion Picture and Television Engineers: Bit-serial digital interface for high-definition television systems (1998) SMPTE-292M.
11. Perkins, C.S., Gharai, L., Lehman, T., Mankin, A.: Experiments with delivery of HDTV over IP networks. Proc. of the 12th International Packet Video Workshop (2002)
12. Schulzrinne, H., Casner, S., Frederick, R., Jacobson, V.: RTP: A transport protocol for real-time applications (1996) RFC 1889.
13. DVS Digital Video Systems: http://www.dvs.de/. (2003)
14. Mummert, T., Kosak, C., Steenkiste, P., Fisher, A.: Fine grain parallel communication on general purpose LANs. In: In Proceedings of 1996 International Conference on Supercomputing (ICS96), Philadelphia, PA, USA (1996) 341–349
15. Reinhardt, S.K., Larus, J.R., Wood, D.A.: Tempest and typhoon: User-level shared memory. In: International Conference on Computer Architecture, Chicago, Illinois, USA (1994)
16. Sumimoto, S., Tezuka, H., Hori, A., Harada, H., Takahashi, T., Ishikawa, Y.: The design and evaluation of high performance communication using a Gigabit Ethernet. In: International Conference on Supercomputing, Rhodes, Greece (1999)
17. Shivam, P., Wyckoff, P., Panda, D.: EMP: Zero-copy OS-bypass NIC-driven Gigabit Ethernet message passing. In: Proc. of the 2001 Conference on Supercomputing. (2001)
18. Lockwood, J.W., Turner, J.S., Taylor, D.E.: Field programmable port extender (FPX) for distributed routing and queueing. In: Proc. of the ACM International Symposium on Field Programmable Gate Arrays, Napa Valley, CA (1997) 30–39

19. McHenry, J.T., Dowd, P.W., Pellegrino, F.A., Carrozzi, T.M., Cocks, W.B.: An FPGA-based coprocessor for ATM firewalls. In: Proc. of the IEEE Symposium on FPGAs for Custom Computing Machines, Napa Valley, CA (1997) 30–39
20. Underwood, K.D., Sass, R.R., Ligon, W.B.: Analysis of a prototype intelligent network interface. Concurrency and Computing: Practice and Experience (2002)
21. National Laboratory for Applied Network Research: Network performance measuring tool, http://dast.nlanr.net/Projects/Iperf/. (2003)
22. Jarvinen, K., Tommiska, M., Skytta, J.: Fully pipelined memoryless 17.8 Gbps AES-128 encryptor. In: Eleventh ACM International Symposium on Field- Programmable Gate Arrays (FPGA 2003), Monterey, California (2003)

Fast, Large-Scale String Match for a 10Gbps FPGA-Based Network Intrusion Detection System

Ioannis Sourdis and Dionisios Pnevmatikatos

Microprocessor and Hardware Laboratory,
Electronic and Computer Engineering Department,
Technical University of Crete, Chania, GR 73 100, Greece
{sourdis,pnevmati}@mhl.tuc.gr

Institute of Computer Science (ICS),
Foundation for Research and Technology-Hellas (FORTH),
Vasilika Vouton, Heraklion, GR 71110, Greece
pnevmati@ics.forth.gr

Abstract. Intrusion Detection Systems such as Snort scan incoming packets for evidence of security threats. The most computation-intensive part of these systems is a text search against hundreds of patterns, and must be performed at wire-speed. FPGAs are particularly well suited for this task and several such systems have been proposed. In this paper we expand on previous work, in order to achieve and exceed a processing bandwidth of 11Gbps. We employ a scalable, low-latency architecture, and use extensive fine-grain pipelining to tackle the fan-out, match, and encode bottlenecks and achieve operating frequencies in excess of 340MHz for fast Virtex devices. To increase throughput, we use multiple comparators and allow for parallel matching of multiple search strings. We evaluate the area and latency cost of our approach and find that the match cost per search pattern character is between 4 and 5 logic cells.

1 Introduction

The proliferation of Internet and networking applications, coupled with the widespread availability of system hacks and viruses have increased the need for network security. Firewalls have been used extensively to prevent access to systems from all but a few, well defined access points (ports), but firewalls cannot eliminate all security threats, nor can they detect attacks when they happen.

Network Intrusion Detection Systems (NIDS) attempt to detect such attempts by monitoring incoming traffic for suspicious contents. They use simple rules (or search patterns) to identify possible security threats, much like virus detection software, and report offending packets to the administrators for further actions. Snort is an open source NIDS that has been extensively used and studied in the literature [1–4]. Based on a rule database, Snort monitors network traffic and detect intrusion events. An example of a Snort rule is:

P.Y.K. Cheung et al. (Eds.): FPL 2003, LNCS 2778, pp. 880–889, 2003.

```
alert tcp any any ->192.168.1.0/24 111(content: "idc|3a3a|"; msg:
"mountd access";)
```

A rule contain fields that can specify a suspicious packet's protocol, IP address, Port, content and others. The "content" (idc|3a3a|) field contains the pattern that is to be matched, written in ascii, hex or mixed format, where hex parts are between vertical bar symbols "|". Patterns of Snort V1.9.x distribution contain between one and 107 characters.

NIDS rules may refer to the header as well as to the payload of a packet. Header rules check for equality (or range) in numerical fields and are straightforward to implement. More computation-intensive is the text search of the packet payload against hundreds of patterns that must be performed at wire-speed [3, 4]. FPGAs are very well suited for this task and many such systems have been proposed [5–7]. In this paper we expand on previous work, in order to achieve and exceed a processing bandwidth of 10Gbps, focusing on the string-matching module.

Most proposed FPGA-based NIDS systems use finite automata (either deterministic or non-deterministic) [8, 9, 6] to perform the text search. These approaches are employed mainly for their low cost, which is reported to be is between 1 and 1.5 logic elements per search pattern character. However, this cost increases when other system components are included. Also, the operation of finite automata is limited to one character per cycle operation. To achieve higher bandwidth researchers have proposed the use of packet-level parallelism [6], whereby multiple copies of the automata work on different packets at lower rate. This approach however may not work very well for IP networks due to variability of the packet size and the additional storage to hold these (possibly large) packets. Instead, in this work we employ full-width comparators for the search. Since all comparators work on the same input (one packet), it is straightforward to increase the processing bandwidth of the system by adding more resources (comparators) that operate in parallel. We evaluate the implementation cost of this approach and suggest ways to remedy the higher cost as compared to (N)DFA. We use extensive pipelining to achieve higher operating frequencies, and we address directly the fan-out of the packet to the multiple search engines, one of limiting factors reported in related work [2]. We employ a pipelined fan-out tree and achieve operating frequencies exceeding 245 MHz for VirtexE and 340 MHz for Virtex2 devices (post place & route results).

In the following section we present the architecture of our FPGA implementation, and in section 3 we evaluate the performance and cost of our proposed architecture. In section 4 we give an overview of FPGA-based string matching and compare our architecture against other proposed designs, and in section 5 we summarize our findings and present our conclusions.

2   Architecture of Pattern Matching Subsystem

The architecture of an FPGA-based NIDS system includes blocks that match header fields rules, and blocks that perform text match against the entire packet payload. Of the two, the computationally expensive module is the text match. In this work we assume that it it relatively straightforward to implement the first module(s) at high speed since they involve a comparison of a few numerical fields only, and focus in making the pattern match module as fast as possible.

If the text match operates at one (input) character per cycle, the total throughput is limited by the operating frequency. To alleviate this bottleneck suggested using packet parallelism where multiple copies of the match module scan concurrently different packet data. However, due to the variable size of the IP packets, this approach may not offer the guaranteed processing bandwidth. Instead, we use discrete comparators to implement a CAM-like functionality. Since each of these comparators is independent, we can use multiple instances to search for a pattern in a wider datapath. A similar approach has been used in [7].

The results of the system are (i) an indication that there was indeed a match, and (ii) the number of the rule that did match. Our architecture uses fine grain pipeline for all sub-modules: fan-out of packet data to comparators, the comparators themselves, and for the encoder of the matching rule. Furthermore to achieve higher processing throughput, we utilize N parallel comparators per search rule, so as to process N packet bytes at the same time. In the rest of this section we expand on our design in each of these sub-modules. The overall architecture we assume is depicted in Figure 1. In the rest of the paper we concentrate on the text match portion of the architecture, and omit the shaded part that performs the header numerical field matching. We believe that previous work in the literature have fully covered the efficient implementation of such functions [6, 7]. Next we describe the details of the three main sub-systems: the comparators, the encoder and the fan-out tree.

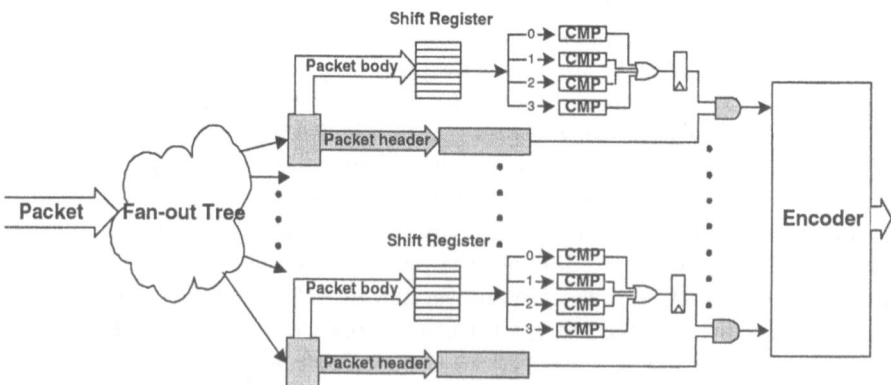

Fig. 1. Envisioned FPGA NIDS system: Packets arrive and are fan-out to the matching engines. N parallel comparators process N characters per cycle (four in this case), and the matching results are encoded to determine the action for this packet. Shaded is the header matching logic that involves numerical field matching.

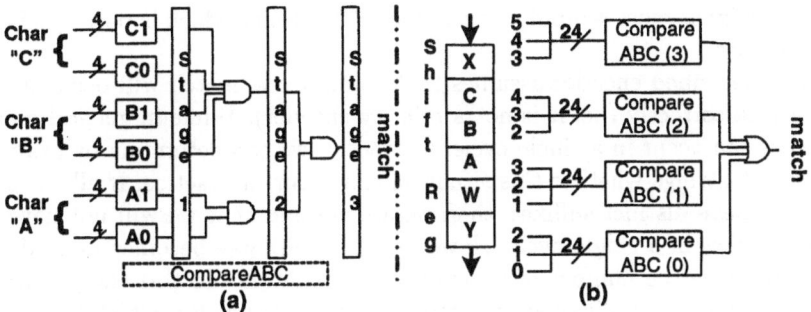

Fig. 2. (a) Pipelined comparator, which matches pattern "ABC". (b) Pipelined comparator, which matches pattern "ABC" starting at four different offsets.

2.1   Pipelined Comparator

Our pipelined comparator is based on the observation that the minimum amount of logic in each pipeline stage can fit in a 4-input LUT and its corresponding register. This decision was made based on the structure of Xilinx CLBs, but the structure of recent Altera devices is very similar so our design should be applicable to Altera devices as well. In the resulting pipeline, the clock period is the sum of wire delay (routing) plus the delay of a single logic cell (one 4-input LUT + 1 flip-flop). The area overhead cost of this pipeline is zero since each logic cell used for combinational logic also includes a flip-flop. The only drawback of this deep pipeline is a longer total delay (in clock cycles) of the result. However, since the correct operation of NIDS systems does not depend heavily on the actual latency of the results, this is not a crucial restriction for our system architecture. In section 3 we evaluate the latency of our pipelines to show that indeed they are within reasonable limits.

Figure 2(a) shows a pipelined comparator that matches the pattern "ABC". In the first stage comparator matches the 6 half bytes of the incoming packet data, using six 4-input LUTs. In the following two stages the partial matches are AND-ed to produce the overall match signal. Figure 2(b) depicts the connection of four comparators that match the same pattern shifted by zero, one, two and three characters (indicated by the numerical suffix in the comparator label). Comparator comparator_ABC(0) checks bytes 0 to 2, comparator_ABC(1) checks bytes 1 to 3 and so on. Notice that the choice of four comparators is only indicative; in general we can use N comparators, allowing the processing of N bytes per cycle.

2.2   Pipelined Encoder

After the individual matches have been determined, the matching rule has to be encoded and reported to the rest of the system (most likely software). We use a hierarchical pipelined encoder. In every stage, the combinational logic is described by at most 4-input, 1-output logic functions, which is permitted in our architecture.

The described encoder assumes that at most one match will occur in order to operate correctly (i.e. it is not a priority encoder). While in general multiple matches can occur in a single cycle, in practice we can determine by examining the search strings whether this situation can occur in practice. If all the search patterns have distinct suffixes, then we are ensured that we will not have multiple matches in a single cycle. However, this guarantee becomes more difficult as we increase the number of concurrent comparators. To this end we are currently working on a pipelined version of a priority encoder, which will be able to correctly handle any search string combination.

2.3   Packet Data Fan-Out

The fan-out delay is major slow-down factor that designers must take into account. While it involves no logic, signals must traverse long distances and potentially suffer significant latencies. To address this bottleneck we created a register tree to "feed" the comparators with the incoming data. The leaves of this tree are the shift registers that feed the comparators, while the intermediate nodes of the tree serve as buffers and pipeline registers at the same time. To determine the best fan-out factor for the tree, we experimented with the Xilinx tools, and we determined that for best results, the optimal fan-out factor changes from level to level. In our design we used small fan-out for the first tree levels and increase the fan-out in the later levels of the tree up to 15 in the last tree level. Intuitively, that is because the first levels of the tree feed large blocks and the distance between the fed nodes is much larger than in last levels. We also experimented and found that the optimal fan-out from the shift-registers is 16 (15 wires to feed comparators and 1 to the next register of shift register).

2.4   VHDL Generator

Deriving a VHDL representation starting from a Snort rule is very tedious; to handle tens of hundred of rules is not only tedious but extremely error prone. Since the architecture of our system is very regular, we developed a C program that automatically generates the desired VHDL representation directly from Snort pattern matching expressions, and we used a simple PERL script to extract all the patterns from a Snort rule file.

3   Performance Evaluation

The quality of an FPGA-based NIDS can be measured mainly using performance and area metrics. We measure performance in terms of operating frequency (to indicate the efficiency of our fine grain pipelining) and total throughput that can be serviced, and we measure total area, as well as area cost per search pattern character.

We evaluate our proposed architecture we used three sets of rules. The first is an artificial set that cannot be optimized (i.e. at every position all search characters are distinct), that contains 10 rules matching 10 characters each. We also used the "web-attacks.rules" from the Snort distribution, a set of 47 rules to show performance and cost for a medium size rule set, and we used the entire set of web rules (a total of 210 rules) to test the scalability of our approach for large rule sets. The average search pattern length for these sets was 10.4 and 11.7 characters for the Web-attack and all the Web rules respectively.

We synthesized each of these rule sets using the Xilinx tools (ISE 4.2i) for several devices (the $-N$ suffix indicates speed grade): Virtex 1000-6, VirtexE 1000-8, Virtex2 1000-5, VirtexE 2600-8 and Virtex2 6000-5. The structure of these devices is similar and the area cost of our design is expected (and turns out) to be almost identical for all devices, with the main difference in the performance.

The top portion of Table 1 summarizes our performance results. It lists the number of bits processed per cycle, the device, the achieved frequency and the corresponding throughput (in Gbps). It also lists the total area and the area required per search pattern character (in logic cells) of rules, the corresponding device utilization, as well as the dimensions of the rule set (number of rules and average size of the search patterns). For brevity we only list results for four parallel comparators, i.e. for processing 32 bits of data per cycle. The reported operating frequency gives a lower bound on the performance using a single (or fewer) comparators.

In the top portion of the table we can see that for our synthetic rule set (labeled 10x10) we are able to achieve throughput in excess of 6 Gbps for the simplest devices and over 12 Gbps for the advanced devices. For the actual Web attack rule set (labeled 47x10.4), we are able to sustain over 5 Gbps for the simplest Virtex 1000 device (at 171 MHz), and about 11 Gbps for a Virtex2 device (at 345 MHz). The performance with a VirtexE device is almost 8 Gbps at 245 MHz. Since the architecture allows a single logic cell at each pipeling stage, and the percentage of the wire delay in the critical path is around 50%, it is unlikely that these results can be improved significantly.

However the results for larger rule sets are more conservative. The complete set of web rules (labeled 210x11.7) operates at 204MHz and achieve a throughput of 6.5 Gbps on a VirtexE, and at 252MHz having 8 Gbps throughput on a Virtex2 device. Since the entire design is larger, the wiring latency contribution to the critical path has increased to 70% of the cycle time. The total throughput is still substantial, and can be improved by using more parallel comparators, or possibly by splitting the design in sub-modules that can be placed and routed in smaller area, minimizing the wire distances and hence latency.

In terms of implementation cost of our proposed architecture, we see that each of the search pattern characters costs between 15 and 20 logic cells depending on the rule set. However, this cost includes the four parallel comparators, so the actual cost of each search pattern character is roughly 4-5 logic cells multiplied by N for N times larger throughput.

Table 1. Detailed comparison of string matching FPGA designs

Description	Input Bits/ c.c.	Device	Freq. MHz	Throu- ghput (Gbps)	Logic Cells[1]	Logic Cells /char	Utili- zation	#Patterns × #Characters
Sourdis- Pnevmatikatos Discrete Comparators	32	Virtex 1000	193	6.176	1,728	17.28	7%	10 × 10
			171	5.472	8,132	16.64	33%	47 × 10.4
		VirtexE 1000	272	8.707	1,728	17.28	7%	10 × 10
			245	7.840	7,982	16.33	33%	47 × 10.4
		Virtex2 1000	396	12.672	1,728	16.86	16%	10 × 10
			344	11.008	8,132	16.64	80%	47 × 10.4
		VirtexE 2600	204	6.524	47,686	19.40	94%	210 × 11.7
		Virtex2 6000	252	8.064	47,686	19.40	71%	210 × 11.7
Sidhu et al.[9] NFAs/Reg. Expression	8	Virtex 100	93.5	0.748	280	~31	11%	(1×) 9^4
			57.5	0.460	1,920	~66	80%	(1×) 29^4
Franklin et al.[8] Regular Expressions	8	Virtex 1000	31	0.248	20,618	2.57	83%	8,003[5]
			99	0.792	314	3.17	1%	99
			63.5	0.508	1726	3.41	7%	506
		VirtexE 2000	50	0.400	20,618	2.57	53%	8,003
			127	1.008	314	3.17	1%	99
			86	0.686	1726	3.41	4%	506
Lockwood[6] DFAs 4 Parallel FSMs on different Packets	32	VirtexE 2000	37	1.184	4,067[2]	16.27	22%[2]	34 × 8[3]
Lockwood[10] FSM+counter	32	VirtexE 1000	119	3.808	98	8.9	0.4%	1 x 11
Gokhale et al.[5] Discrete Comparators	32	VirtexE 1000	68	2.176	9,722	15.2	39%	32 × 20
Young Cho et al.[7] Discrete Comparators	32	Altera EP20K	90	2.880	N/A	10	N/A	N/A

[1] Two *Logic Cells* form one *Slice*, and two Slices form one *CLB* (4 Logic Cells).
[2] These results does not includes the cost/area of infrastructure and protocol wrappers
[3] 34 regular expressions,with 8 characters on average, (about 250 character)
[4] One regular Expression of the form (a | b)*a(a | b)^k for k = 8 and 28. Because of the * operator the regular expression can match more than 9 or 29 characters.
[5] Sizes refer to Non-meta characters and are roughly equivalent to 800, 10, and 50 patterns of 10 characters each.

We compute the latency of our design taking into account the three compo- nents of our pipeline: fan-out, match, encode. Since the branching factor is not fixed in the fan-out tree, we cannot offer a closed form for the number of stages. The pipeline depths for the designs we have implemented are: $3 + 5 + 4 = 12$ for the Synth10 rule set, $3 + 6 + 5 = 14$ for the Web Attacks rule set, and $5 + 7 + 7 = 19$ for the Web-all rule set. For 1,000 patterns and pattern lengths of 128 characters, we estimate the total delay of the system to be between 20 and 25 clock cycles.

We also evaluated resource sharing to reduce the implementation cost. We sorted the 47 web attack rules, and we allowed two adjacent patterns to share comparator i if their i^{th} characters were the same, and found that the number of logic cells required to implement the system was reduced by about 30%. Due to space limitations, we do not expand on this option in detail. However, it is a very promising approach to reduce the implementation cost, and allow even more rules to be packed in a given device.

4   Comparison with Previous Work

In this section we attempt a fair comparison with previous reported research. While we have done our best to report these results with the most objective way, we caution the reader that this task is difficult since each system has its own assumptions and parameters, occasionally in ways that are hard to quantify.

One of the first attempts in string matching using FPGAs, presented in 1993 by Pryor, Thistle and Shirazi [11]. Their algorithm, implemented on Splash 2, succeeded to perform a dictionary search, without case sensitivity patterns, that consisted of English alphabet characters (26 characters). Pryor et al managed to achieve great performance and perform a low overhead AND-reduction of the match indicators using hashing.

Sidhu and Prassanna [9] used Regular Expressions and Nondeterministic Finite Automata (NFAs) for finding matches to a given regular expression. They focused in minimizing the space -$O(n^2)$- required to perform the matching, and their automata matched 1 text character per clock cycle. For a single regular expression, the constructed NFAs and FPGA circuit was able to process each text character in 17.42-10.70ns (57.5-93.5 MHz) using a Virtex XCV100 FPGA.

Franklin, Carver and Hutchings [8] also used regular expressions to describe patterns. The operating frequency of the synthesized modules was about 30-100 MHz on a Virtex XCV1000 and 50-127 MHz on a Virtex XCV2000E, and in the order of 63.5 MHz and 86 MHz respectively on XCV1000 and XCV2000E for a few tens of rules.

Lockwood used the Field Programmable Port Extender (FPX) platform, to perform string matching. They used regular expressions (DFAs) and were able to achieve operation at 37 MHz on a Virtex XCV2000E [6]. Lockwood also implemented a sample application on FPX using a single regular expression and were able to achieve operation at 119 MHz on a Virtex V1000E-7 device [10].

Gokhale, et al [5] using CAM to implement Snort rules NIDS on a Virtex XCV1000E. Their hardware runs at 68MHz with 32-bit data every clock cycle, giving a throughput of 2.2 Gbps, and reported a 25-fold improvement on the speed of Snort v1.8 on a 733MHz PIII and an almost 9-fold improvement on a 1 GHz PowerPC G4.

Closer to our work is the recent work by Cho, Navab and Mangione-Smith [7]. They designed a deep packet filtering firewall on a FPGA and automatically translated each pattern-matching component into structural VHDL. The content pattern match unit micro-architecture used 4 parallel comparators for every pattern so that the system advances 4 bytes of input packet every clock cycle. The design implemented in an Altera EP20K device runs at 90MHz, achieving 2.88 Gbps throughput. They require about 10 logic cells per search pattern character. However, they do not include the fan-out logic that we have, and do not encode the matching rule. Instead they just OR all the match signals to indicate that some rule matched.

The results of these works are summarized in the bottom portion of Table 1, and we can see that most previous works implement a few tens of rules at most, and achieve throughput less than 4 Gbps. Our architecture on the same or equivalent devices achieves roughly twice the operating frequency and throughput. In terms of best performance, we achieve 3.3 times better processing throughput compared with the fastest published design which implements a single search pattern. Our 210-rule implementation achieves at least a 70% improvement in throughput compared to the fastest existing implementation.

5   Conclusions and Future Work

We have presented an architecture for Snort rule match in FPGAs. We propose the use of extensive fine grain pipelining in order to achieve high operating frequencies, and parallel comparators to increase the processing throughput. This combination proves very successful, and the throughput of our design exceeded 11 Gbps for about 50 Snort rules. These results offer a distinct step forward compared to previously published research. If latency is not critical to the application, fine grain pipelining is very attractive in FPGA-based designs: every logic cell contains one LUT and one Flip-Flop, hence the pipeline area overhead is zero. The current collection of Snort rules contains less than 1500 patterns, with an average size of 12.6 characters. Using the area cost as computed earlier, we need about 3 devices of 120,000 logic cells to include the entire Snort pattern matching, and about 4 devices to include the entire snort rule set including header matching. These calculations do not include area optimizations, which can lead to further significant improvements.

Throughout this paper we used four parallel comparators. However, a different level of parallelism can also be used depending on the bandwidth demands. Reducing the processing width leads to a smaller, possibly higher frequency design, while increasing the processing width leads to a bigger and probably lower frequency design. Throughput depends on both frequency and processing width, so we need to seek for the cost effective tradeoff of these two factors.

Despite the significant body of research in this area, there are still improvements that we can use to seek better solutions. In our immediate goals is to use the hierarchical decomposition of large rule set designs, and attempt to use the multiple clock domains. The idea is to use a slow clock to drive long wide busses to distribute data and a fast clock for local processing that only uses local wiring. The target would be to retain the frequency advantage of our medium-sized design (47 rules) for a much larger rule set. All the devices we used already support multiple clock domains, and with proper placement and routing tool support this approach will also be quicker to implement: each of the modules can be placed and routed locally one after the other, reducing the memory and processing requirements for placement and routing.

Furthermore, future devices offer significant speed improvements, and it would be interesting to see whether the fine grain pipelining will be as effective for these devices (such as the Virtex 2 Pro) as it was for the devices we used in this work.

Acknowledgments

This work was supported in part by the IST project SCAMPI (IST-2001-32404) funded by the European Union.

References

1. SNORT official web site: ⟨http://www.snort.org⟩
2. Roesch, M.: Snort - lightweight intrusion detection for networks. In: Proceedings of LISA'99: 13th Administration Conference. (1999) Seattle Washington, USA.
3. Desai, N.: Increasing performance in high speed NIDS. In: www.linuxsecurity.com. (2002)
4. Coit, C.J., Staniford, S., McAlerney, J.: Towards faster string matching for intrusion detection or exceeding the speed of snort. In: DISCEXII, DAPRA Information Survivability conference and Exposition. (2001) Anaheim, California, USA.
5. Gokhale, M., Dubois, D., Dubois, A., Boorman, M., Poole, S., Hogsett, V.: Granidt: Towards gigabit rate network intrusion detection technology. In: Proceedings of 12th International Conference on Field Programmable Logic and Applications. (2002) France.
6. Moscola, J., Lockwood, J., Loui, R.P., Pachos, M.: Implementation of a content-scanning module for an internet firewall. In: Proceedings of IEEE Workshop on FPGAs for Custom Computing Machines. (2003) Napa, CA, USA.
7. Young H. Cho, S.N., Mangione-Smith, W.: Specialized hardware for deep network packet filtering. In: Proceedings of 12th International Conference on Field Programmable Logic and Applications. (2002) France.
8. Franklin, R., Carver, D., Hutchings, B.: Assisting network intrusion detection with reconfigurable hardware. In: IEEE Symposium on Field-Programmable Custom Computing Machines. (2002)
9. Sidhu, R., Prasanna, V.K.: Fast regular expression matching using fpgas. In: IEEE Symposium on Field-Programmable Custom Computing Machines. (2001) Rohnert Park, CA, USA.
10. Lockwood, J.W.: An open platform for development of network processing modules in reconfigurable hardware. In: IEC DesignCon '01. (2001) Santa Clara, CA, USA.
11. Pryor, D.V., Thistle, M.R., Shirazi, N.: Text searching on splash 2. In: Proceedings of IEEE Workshop on FPGAs for Custom Computing Machines. (1993) 172–177

Irregular Reconfigurable CAM Structures for Firewall Applications

T.K. Lee, S. Yusuf, W. Luk, M. Sloman, E. Lupu, and N. Dulay

Department of Computing,
Imperial College, 180 Queen's Gate, London SW7 2BZ, UK
{tk197,sy99,w.luk,m.sloman,e.c.lupu,n.dulay}@doc.ic.ac.uk

Abstract. Hardware packet-filters for firewalls, based on content-addressable memory (CAM), allow packet matching processes to keep in pace with network throughputs. However, the size of an FPGA chip may limit the size of a firewall rule set that can be implemented in hardware. We develop two irregular CAM structures for packet-filtering that employ resource sharing methods, with various trade-offs between size and speed. Experiments show that the use of these two structures are capable of reduction, up to 90%, of hardware resources without losing performance.

1 Introduction

FPGA-based firewall processors have been developed for high-throughput networks [3, 6, 7]. Such firewall processors must be able to carry out packet matching effectively based on filter rules. Each filter rule consists of a set of logical operations on different fields of an input packet header. A 'don't care' condition indicates that a field can match any value.

Content-Addressable Memory (CAM) is a searching device that consists of an array of storage locations. A search result can be obtained in constant time through parallel matching of the input with the data in the memory array. CAM based hardware packet-filters [4] are fast and support various data widths [3]. However, the size of an FPGA may limit the number of filter rules that can be implemented in hardware [7]. We describe two hardware structures that employ resource sharing methods to reduce hardware resource usage for packet-filtering firewalls. Resource usage reduces approximately linearly with the degree of grouping of the filter rules in a rule set. These two structures, when applied to CAM based packet-filters, offer various trade-offs between speed and size under different situations involving parallel and pipelined implementations. The contributions described in this paper include:

1. two hardware irregular CAM structures for implementing filter rules;
2. a strategy to generate hardware firewall processors;
3. an evaluation of the effectiveness of the irregular CAM structures, comparing them against regular CAMs.

P.Y.K. Cheung et al. (Eds.): FPL 2003, LNCS 2778, pp. 890–899, 2003.

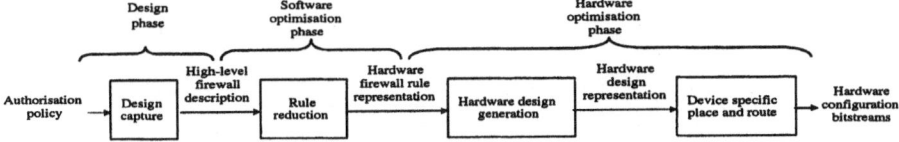

Fig. 1. An overview of our development framework for reconfigurable firewall processors. There are three main phases: the *design phase*, the *software optimisation phase*, and the *hardware optimisation phase*.

The rest of the paper is organised as follows. Section 2 gives an overview of our design framework. Section 3 describes our hardware structures for filter rules. Section 4 outlines the design generation. Section 5 evaluates the performance of our approach in terms of speed and size, and Section 6 provides a summary of current and future work.

2 Framework Overview

As shown in Figure 1, our framework for developing reconfigurable-hardware packet filtering firewalls consists of three main phases: the design phase, the software optimisation phase, and the hardware optimisation phase.

During the *design phase*, the requirements of a firewall are captured as a high-level description. We use a subset of Ponder [2], a policy specification language, to create our firewall description language [5]. This firewall description uses Ponder's parameterised types and the concept of domains. Our high-level firewall description supports abstraction from details of the hardware implementation. It uses constraints to specify low-level hardware requirements such as placement and partitioning, run-time reconfiguration, timing and size requirements, and hardware software co-operation.

During the *software optimisation phase*, high-level firewall rules are reduced and converted to a hardware firewall rule representation, using parameterised library specifications. We have developed a series of rule reduction steps to reduce hardware usage by employing rule elimination and rule sharing methods [5]. A rule set is divided into a number of groups of hardware filter rules. Sequencing is performed to preserve the ordering and the semantics of a rule set. Reordering and partitioning is conducted to facilitate the grouping process. Each group consists of either a list of rules related by common attributes, or a singleton rule if no sharing with other rules can be found within the same partition.

During the *hardware optimisation phase*, hardware firewall rule representations are converted to a hardware design which is then used to produce the hardware configuration bitstreams for downloading onto an FPGA. The next section describes the irregular CAM structures that we develop, and their use in implementing a rule set in hardware and in facilitating resource sharing.

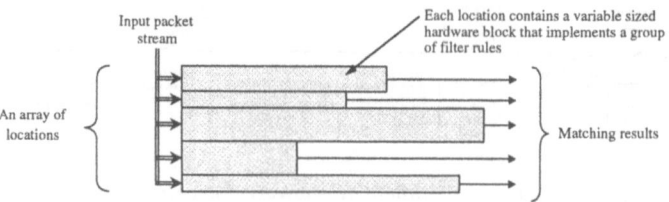

Fig. 2. A rule set in hardware implemented as an irregular CAM structure. Each location contains a *variable sized* hardware block that implements a *group* of filter rules. An array of hardware blocks together form a CAM structure that supports the whole rule set.

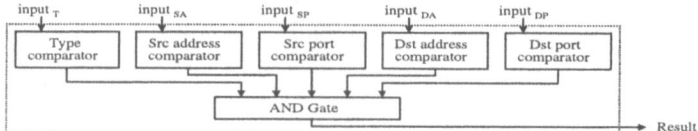

Fig. 3. A hardware block for a group of singleton filter rule. Instead of having a single matching processor as in a regular CAM, our hardware blocks for filter rules contain several field-level hardware comparators that correspond to different fields of an input packet. When a hardware block is instantiated, filter rules that contain 'don't care' fields will have the corresponding hardware comparators eliminated. This example shows the situation when no 'don't care' fields are involved in a filter rule.

3   Architecture of Irregular CAM for Firewall Rules

Our approach for implementing a firewall rule set in hardware is to construct a specialised CAM structure, as shown in Figure 2, to perform packet-filtering. Conventional regular CAM structures store a *single* matching criterion in each memory location. However, instead of having a one-to-one mapping of a filter rule to a CAM location, we construct each CAM location as a hardware block that implements a *group* of filter rules.

A rule set is divided into several groups of filter rules as described in Section 2. Each of these groups is implemented as a hardware block that corresponds to a CAM location. These *variable sized* hardware blocks together then produce an irregular CAM structure that represents the whole filter rule set.

Hardware blocks are instantiated according to the types of grouping and the combinations of field attributes. Figure 3 shows a hardware block for a group of singleton filter rule. It contains several field-level hardware comparators that correspond to different fields of the input. Matching results are obtained as the unified result from all the individual comparators. This design is functionally the same as a regular CAM, except that a regular CAM design will normally use only one comparator for the input data and does not need the AND gate. Separating the matching processor into field-level comparators, however, allows

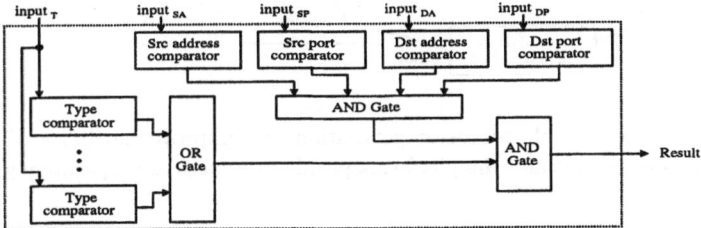

Fig. 4. A hardware block of a group of shared filter rules using the *Siamese Twins* structure. Individual fields of the filter rules having identical data values are shared by using the same hardware comparators. Fields that cannot be shared have their corresponding parts OR-ed together. This example shows that a block is instantiated with all but the *Type*-field comparator being shared.

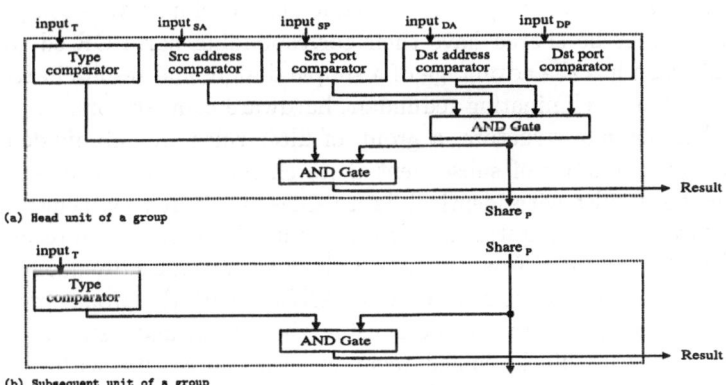

Fig. 5. Hardware blocks of a group of shared filter rules using the *Propaganda* structure. A group of filter rules are sub-divided into a head unit and a number of subsequent units chained to the head. Individual fields of the filter rules having identical data values are shared by using the same hardware comparators in the head unit. The comparison result of the shared fields is propagated from the head unit to each of the subsequent units in the group. Fields that cannot be shared are AND-ed with the propagating result individually. This example shows that the blocks are instantiated with all but the *Type*-field being shared.

us to achieve reduction in resource usage at the expense of introducing a multi-input AND-gate. When a hardware block is instantiated, filter rules that contain 'don't care' fields will have the corresponding hardware comparators eliminated.

Within a group of rules, hardware resources are shared by attributes that are common. There are two levels of sharing: field-level sharing and bit-level

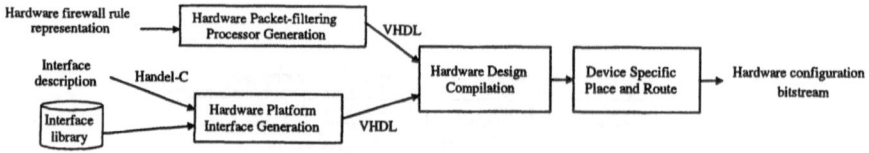

Fig. 6. Hardware firewall processor generation. To improve flexibility, the packet-filtering processing unit and the platform-specific interfaces are separately generated.

sharing [5]. We develop two irregular CAM structures: the *Siamese Twins* and the *Propaganda.* They both provide field-level sharing but have different trade-offs between size and speed. A group of shared filter rules are implemented as hardware blocks using either the Siamese Twins structure, which is optimised for area, or the Propaganda structure, which is optimised for speed. Figure 4 and Figure 5 show some examples of hardware blocks for a group of filter rules using the two structures.

In the Siamese Twins structure, the fields of a group of filter rules with identical data values are shared by using the same field-level hardware comparators. Fields that cannot be shared have their corresponding parts OR-ed together. This organisation has the advantage of a simple design, and results in reduction of resource usage by eliminating redundant hardware comparators.

In the Propaganda structure, a group of filter rules are sub-divided into a head unit and a number of subsequent units chained to the head. Individual fields of the filter rules with identical data values are shared by using the same field-level hardware comparators in the head unit. The comparison result of the shared fields is propagated from the head unit to each of the subsequent units in the group. Fields that cannot be shared are AND-ed with the propagating result individually. Filter rules implemented using the Propaganda structure result in a list of hardware blocks joined together. Each filter rule within a group corresponds to a hardware block. The length of the list varies with the number of rules in a group. This is in contrast to the hardware blocks of Siamese Twins or singleton filter rules, where there is only one unit.

4   Hardware Firewall Processor Generation

To generate our hardware firewall processors (Figure 6), we separate the generation of the platform-specific interfaces from the generation of the filtering processor unit. The interfaces are written in the Handel-C language [1], which facilitates porting the design to various hardware platforms.

We design the hardware code generator that takes the hardware firewall rule representation as input, and generate the hardware packet-filtering processor (Figure 6) in VHDL. During the implementation of a CAM location, a hardware block is instantiated according to the attributes of the field in a group of filter rules. These include the combinations of fields that are shared and not shared, and the number of rules in a group. Furthermore, there are structures for replicating and connecting the non-shared fields for a group of rules. All our hardware blocks can be used in both parallel and pipelined mode.

Our implementations target the Xilinx Virtex series FPGAs. We follow the vendor's recommendation [8] of reprogramming lookup tables as Shift Registers (SRL 16). The JBits tool is then used to reconfigure the SRL 16 blocks to desired matching values, for various locations in our irregular CAMs.

5 Performance Evaluation

To analyse the performance of our irregular CAM, we compare implementations that employ the Siamese Twins and the Propaganda structures against those based on regular CAMs. We evaluate the implementations in terms of clock speed and hardware resource consumption.

In addition to using rule sets from network sites, we also generate artificial rule sets. Our rule set generator is based on real filter rule sets and covers a wider spectrum of possible real situations as well as some worst-case scenarios. The test data include the effects of 'don't care' fields, the degree to which rules are grouped, and the size of rule sets. For the purpose of the experiments, 'degree of grouping' means the percentage of rules within a rule set that are in a shared group. The resource usage figures include resource to support the I/O to RAM, which is a fixed overhead and is insignificant when compared with the overall resources required by a rule set. All the experiments are performed on a Celoxica RC1000-PP reconfigurable hardware platform that contains a Xilinx Virtex XCV1000 FPGA device.

5.1 Resource Usage

Figure 7 shows the resource usage for rule sets with different degrees of grouping, and the effects of 'don't care' fields. The resource usage of regular CAM remains unchanged as the degree of grouping varies.

When the degree of grouping is at 0% as shown on the left-hand-side of Figure 7 (a) and (b), there is no reduction in resource usage in the case of no 'don't care' fields, but there is around 45% reduction in the case with 'don't care' fields for both the Siamese Twins and the Propaganda structures. A rule set that does not contain any 'don't care' fields in all of its rules is unrealistic. In reality, most rule sets contain a certain amount of 'don't care' fields. This suggests that both Siamese Twins and Propaganda will achieve reduction in resource usage over a regular CAM, whenever a 'don't care' field exists in a rule set.

For the parallel versions, the resource usage of Siamese Twins and Propaganda are about the same as shown in the lower parts of Figure 7 (a) and (b), where their corresponding graphs almost overlap. For the pipelined version, Propaganda uses noticeably more resources than Siamese Twins. This is due to the additional pipeline registers. This suggests that both Siamese Twins and Propaganda are suitable for implementations involving parallel structures. However, if an implementation must involve pipelining and when resource usage is the main concern, Siamese Twins is the preferred choice.

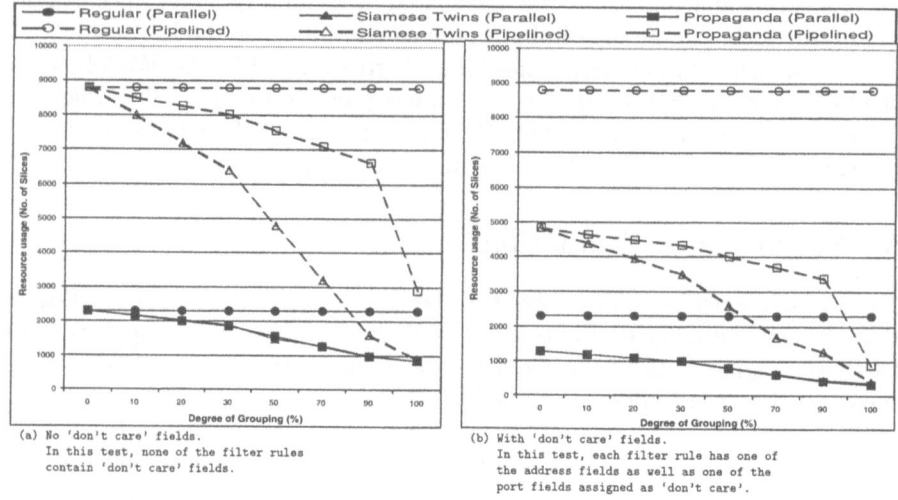

Fig. 7. Resource usage versus degree of grouping. Resource usage of both Siamese Twins and Propaganda reduce approximately linearly with the degree of grouping. For the parallel versions, both structures perform almost identically. For the pipelined versions, Siamese Twins is considerably better. Note that for this test, one of the address field is not shared in a group. Since the address field is the largest field in a filter rule, it gives the worst case resource usage for a single non-shared field.

When the degree of grouping is low (less than 10%), the pipelined versions consume 3.8 times more resources then their parallel counterparts. This shows the well-known trade-offs between speed and size. However, when the degree of grouping is high (larger than 70% in the case of no 'don't care' fields, and larger than 50% in the case with 'don't care' fields), the pipelined versions of Siamese Twins consume comparable or fewer resources than the parallel versions of the regular CAM. These two figures correspond to 138% (in the case of no 'don't care' fields) and 188% (in the case with 'don't care' fields) of the speed of the regular CAM. This suggests that, in situations when both size and speed should be optimised, a pipelined version of Siamese Twins can be better than a parallel version of the regular CAM.

5.2   Speed

Figure 8 shows the speed performance for rule sets with different degrees of grouping, and the effects of 'don't care' fields. The maximum operating frequency of regular CAM remains unchanged as the degree of grouping varies.

Results for the parallel versions are shown in the lower parts of Figure 8 (a) and (b). Both Siamese Twins and Propaganda performs approximately the same as the regular CAM. Results for the pipelined versions are shown in the upper parts of Figure 8 (a) and (b). While Propaganda performs comparable

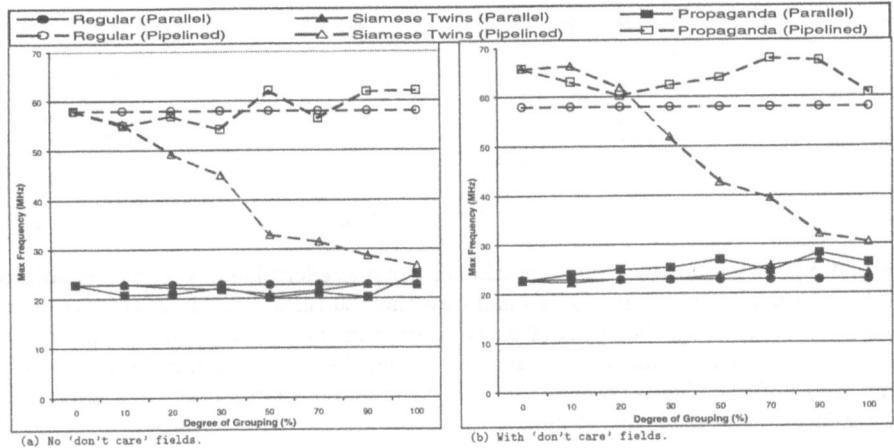

Fig. 8. Speed performance versus degree of grouping. Propaganda consistently achieves comparable performance to the regular CAM. Siamese Twins, while having approximately the same performance as regular CAM in the parallel version, suffers from performance degradation when the degree of grouping increases in the pipelined versions. Note that for this test, all the shared rules using the Siamese Twins structure are grouped into a single CAM location. This produces the highest propagation delay and so the lowest performance.

to or sometimes slightly better than the regular CAM, the Siamese Twins suffers from performance degradation when the degree of grouping increases. This reduction in performance is due to the increased routing and propagation delay of the enlarged OR-structure inside the Siamese Twins. When the degree of grouping is low (less than 10%), both structures are 2.5 times faster than their parallel counterparts. When the degree of grouping is at 100%, the performance of Siamese Twins is reduced by nearly 50% to have similar performance to its parallel counterpart. This suggests that both Siamese Twins and Propaganda are suitable for implementations involving parallel structures. However, if implementations involve pipelining and when speed is also a major concern, Propaganda can be a better choice.

Figure 9 shows that maximum operating frequency is determined not only by the degree of grouping, but also by the maximum group size. For the parallel versions, both Siamese Twins and Propaganda do not vary much with the degree of grouping. For the pipelined versions as shown in the top-left parts of Figure 9 (a) and (b), when the group size is small, the performance of Siamese Twins is comparable to the regular CAM even at 100% degree of grouping. When the group size is large (100 rules/location), its performance decreases by nearly 50%.

The effects of maximum group size suggest that there can be a trade-off between resource utilisation and the maximum operation frequency. In order to avoid performance degradation at a high degree of grouping, one can choose

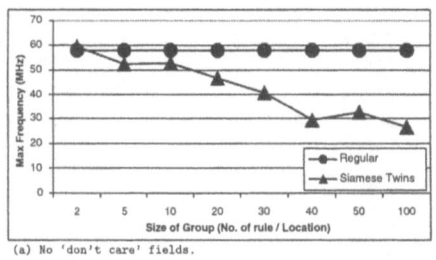 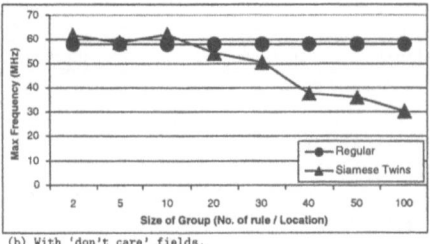

(a) No 'don't care' fields.   (b) With 'don't care' fields.

Fig. 9. Speed performance versus maximum group size for pipelined implementations. When the maximum group size is small (less than 20 rules/location in the case of no 'don't care' fields, and less than 30 rules/location in the case with 'don't care' fields) the performance of Siamese Twins and the regular CAM are comparable. In this test, the degree of grouping is always 100%, but the group of shared filter rules are broken down into a number of smaller groups.

to impose either a ceiling group size or a maximum degree of grouping for the pipelined versions of Siamese Twins. For example, groups with number of rules exceeding the ceiling group size can be broken down into several smaller groups. This method can maintain performance, but at the expense of using more hardware resources to implement the additional groups.

5.3   Results Summary

The analysis results are discussed in Section 5.1 and Section 5.2. The maximum reduction in hardware usage and maximum group size before performance degradation are shown respectively in Table 1 and Table 2. For the purpose of the experiments, performance degradation is defined as no more than 10% reduction in speed, when compared to corresponding regular CAMs.

Table 1. Maximum reduction in hardware usage before performance degradation.

		Reduction in hardware usage	Degree of grouping
Siamese Twins	Parallel	84%	100%
	Pipelined	60% with 'don't care'	30%
		18% without 'don't care'	20%
Propaganda	Parallel	84%	100%
	Pipelined	90%	100%

Table 2. Maximum group size before performance degradation (Siamese Twins pipelined).

	Rule / Location	Reduction in hardware usage
With 'don't care'	30	90%
Without 'don't care'	10	85%

6   Conclusion

We have presented the Siamese Twins and the Propaganda irregular CAM structures. These two structures employ resource sharing to reduce hardware usage for packet-filtering firewalls. Experiments show that resource usage reduces approximately linearly to the degree of grouping of the filter rules in a rule set. These two irregular CAM structures offer various trade-offs between speed and size, under different situations involving parallel and pipelined implementations. Both structures are capable of reduction, up to 90%, of hardware resources of regular CAMs without losing performance.

Current and future work includes the use of bit-level sharing to achieve further reduction in hardware usage, and global and local optimisations of irregular CAM using the Siamese Twins and the Propaganda structures.

Acknowledgements

The support of UK Engineering and Physical Sciences Research Council (Grant number GR/R 31409, GR/R 55931 and GR/N 66599), Celoxica Limited and Xilinx, Inc. is gratefully acknowledged.

References

1. Celoxica Limited, *Handel-C v3.1 Language Reference Manual*, http://www.celoxica.com/.
2. N. Damianou, N. Dulay, E. Lupu and M Sloman, "The Ponder Policy Specification Language", in *Proc. Workshop on Policies for Distributed Systems and Networks*, LNCS 1995, Springer, 2001, pp. 18-39.
3. J. Ditmar, K. Torkelsson and A. Jantsch, "A Dynamically Reconfigurable FPGA-based Content Addressable Memory for Internet Protocol Characterization", *Field Programmable Logic and Applications*, LNCS 1896, Springer, 2000.
4. P.B. James-Roxby and D.J. Downs, "An Efficient Content-addressable Memory Implementation Using Dynamic Routing", in *Proc. IEEE Symp. on Field-Programmable Custom Computing Machines*, IEEE Computer Society Press, 2001.
5. T.K. Lee, S. Yusuf, W. Luk, M. Sloman, E. Lupu and N. Dulay, "Compiling Policy Descriptions into Reconfigurable Firewall Processors", in *Proc. IEEE Symp. on Field-Programmable Custom Computing Machines*, IEEE Computer Society Press, 2003.
6. J.T. McHenry and P.W. Dowd, "An FPGA-Based Coprocessor for ATM Firewalls" in *Proc. IEEE Symp. on Field-Programmable Custom Computing Machines*, IEEE Computer Society Press, 1997.
7. R. Sinnappan and S. Hazelhurst, "A Reconfigurable Approach to Packet Filtering", *Field Programmable Logic and Applications*, LNCS 2147, Springer, 2001.
8. Xilinx Inc., *Designing Flexible, Fast CAMs with Virtex Family FPGAs*, 1999, http://www.xilinx.com/.

Compiling
for the Molen Programming Paradigm

Elena Moscu Panainte[1], Koen Bertels[1], and Stamatis Vassiliadis[1]

Computer Engineering Lab
Electrical Engineering Department, TU Delft, The Netherlands
{E.Panainte,K.Bertels,S.Vassiliadis}@et.tudelft.nl

Abstract. In this paper we present compiler extensions for the Molen programming paradigm, which is a sequential consistency paradigm for programming custom computing machines (CCM). The compiler supports instruction set extensions and register file extensions. Based on pragma annotations in the application code, it identifies the code fragments implemented on the reconfigurable hardware and automatically maps the application on the target reconfigurable architecture. We also define and implement a mechanism that allows multiple operations to be executed in parallel on the reconfigurable hardware. In a case study, the Molen processor has been evaluated. We considered two popular multimedia benchmarks: mpeg2enc and ijpeg and some well-known time-consuming operations implemented in the reconfigurable hardware. The total number of executed instructions has been reduced with 72% for mpeg2enc and 35% for ijpeg encoder, compared to their pure software implementations on a general purpose processor (GPP).

1 Introduction and Related Work

In the last decade, several approaches have been proposed for coupling an FPGA to a GPP. For a classification of these approaches the interested reader is referred to [1]. There are four shortcomings of current approaches, namely:

1. **Opcode space explosion**: a common approach (e.g. [2], [3], [4]) is to introduce a new instruction for each portion of application mapped into the FPGA. The consequence is the limitation of the number of operations implemented into the FPGA, due to the limitation of the opcode space. More specifically stated, for a specific application domain intended to be implemented in the FPGA, the designer and compiler are restricted by the unused opcode space.
2. **Limitation of the number of parameters**: In a number of approaches, the operations mapped on an FPGA can only have a small number of input and output parameters ([5], [6]). For example, in the architecture presented in [5], due to the encoding limits, the fragments mapped into the FPGA have at most 4 inputs and 2 outputs; also, in Chimaera [6], the maximum number of input registers is 9 and it has one output register.

P.Y.K. Cheung et al. (Eds.): FPL 2003, LNCS 2778, pp. 900–910, 2003.

3. No support for **parallel execution** on the FPGA of sequential operations: an important and powerful feature of FPGA's can be the parallel execution of sequential operations when they have no data dependency. Many architectures [1] do not take into account this issue and their mechanism for FPGA integration cannot be extended to support parallelism.

4. No **modularity**: each approach has a specific definition and implementation bounded for a specific reconfigurable technology and design. Consequently, the applications cannot be (easily) ported to a new reconfigurable platform. Further there are no mechanisms allowing reconfigurable implementation to be developed separately and ported transparently. That is a reconfigurable implementation developed by a designer A can not be included without substantial effort by the compiler developed for an FPGA implementation provided by a designer B.

A general approach is required that eliminates these shortcomings. In this paper, a programming paradigm for reconfigurable architectures [7], called the Molen Programming Paradigm and a compiler are described that offer alternatives and a solution to the above presented limitations.

The paper is organized as follows: in the next section, we discuss related research and present the Molen programming paradigm. We then describe a particular implementation, called the Molen processor that uses microcoded emulation for controlling the reconfigurable hardware. Consequently, we present the two main elements of the paper, namely the Exchange Register mechanism and the compiler extension for the Molen processor. We finally discuss an experiment comparing the Molen reconfigurable processor with the equivalent non-reconfigurable processor, using two well-known multimedia benchmarks: mpeg2 and ijpeg.

2   The Programming Paradigm

The Molen programming paradigm[7] is a sequential consistency paradigm for programming CCMs possibly including a general purpose computational engine(s). The paradigm allows for parallel and concurrent hardware execution and it is intended (currently) for single program execution. It requires only a one time architectural extension of few instructions to provide a large user reconfigurable operation space. The added instructions include:

- Two instructions[1] for controlling the reconfigurable hardware, namely:
  - SET < *address* >: at a particular location the hardware configuration logic is defined
  - EXECUTE < *address* >: for controlling the executions of the operations on the reconfigurable hardware

---

[1] Actually, five if partial reconfiguration, pre-loading of reconfiguration and executing microcode are also explicitly assumed [7].

- Two move instructions for passing values of to and from the GPP register file and the reconfigurable hardware.

Code fragments constituted of contiguous statements (as they are represented in high-level programming languages) can be isolated as generally implementable functions (that is code with multiple identifiable input/output values). The parameters stored in registers are passed to special reconfigurable hardware registers denoted as Exchange Registers(XRs). The Exchange Register mechanism will be described later in the paper. In order to maintain the correct program semantics, the code is annotated and CCM description files provide the compiler with implementation specific information such as the addresses where the SET and EXECUTE code are to be stored, the number of exchange registers, etc. It should be noted that this programming paradigm allows modularity, meaning that if the interfaces to the compiler are respected and if the instruction set extension (as described above) is supported, then:

- custom computing hardware provided by multiple vendors can be incorporated by the compiler for the execution of the same application.
- the application can be ported to multiple platforms with mere recompilation.

Finally, it is noted that every user is provided with at least $2^{(n-op)}$ directly addressable functions, where n represents the instruction length and 'op' the opcode length. The number of functions can be easily augmented to an arbitrary number by reserving opcode for indirect opcode accessing. From the previous discussion, it is obvious that the programming paradigm and the architectural extensions resolve the aforementioned problems as follows:

- There is only a one time architectural extension of few new instructions to include an arbitrary number of configuration.
- The programming paradigm allows for an arbitrary (only hardware real estate design restricted) number of I/O parameter values to be passed to/from the reconfigurable hardware. It is only restricted by the implemented hardware as any given technology can (and will) allow only a limited hardware.
- Parallelism is allowed as long as the sequential memory consistency model can be guaranteed.
- Assuming that the interfaces are observed, modularity is guaranteed because the paradigm allows freedom of operation implementation.

Parallelism and Concurrency: As depicted in Figure 1, the split-join programming paradigm suggests that the SET instruction does not block the GPP because it can be executed independently from any other instruction. Moreover, a block of consecutive resource conflict free SET instructions (e.g. set op1, set op2 in our example) can be executed in parallel. However, the SET-instruction (set op3) following a GPP-instruction can only be executed after the GPP-instruction is finished. As far as the EXECUTE-instruction is concerned, we distinguish between two distinct cases, one that adds a new instruction and one that does not:

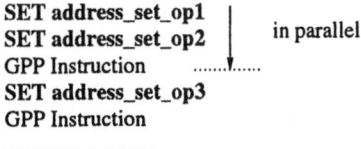

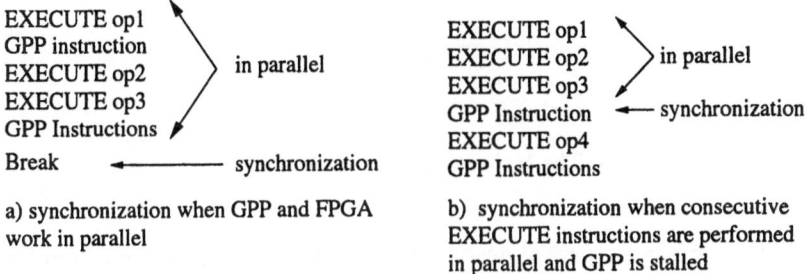

Fig. 1. SET instructions performed concurrently with GPP instructions

a) synchronization when GPP and FPGA work in parallel

b) synchronization when consecutive EXECUTE instructions are performed in parallel and GPP is stalled

Fig. 2. Models of synchronization

1. If it is found that there is a substantial performance to be gained by parallel execution between GPP and FPGA, then the GPP and EXECUTE-instructions can be issued and executed in parallel. The sequence of instructions performed in parallel is initiated by an EXECUTE instruction. The end of the parallel execution requires an additional instruction (BREAK in the example) indicating where the parallel execution stops (see Figure2 (a)). A similar approach can be followed for the SET instructions.

2. If such performance is not to be expected (which will most likely be the case for reconfigured "complex" code and GPP code with numerous data dependencies), then a block of EXECUTE-instructions can be executed in parallel on the FPGA while the GPP is stalled. An example is presented in Figure 2(b) where the block of EXECUTE instructions which can be processed in parallel contains the first three consecutive EXECUTE instructions and it is delimited by a GPP instruction.

   We note that parallelism is guaranteed by the compiler, that checks whether there are data dependencies and whether the parallel execution is supported by the reconfigurable unit. Moreover, if the compiler detects that a block of SET/EXECUTE instructions cannot be performed in parallel, it separates them by introducing appropriate instructions. In the remaining of the paper, we assume that the separating instruction for SET/EXECUTE is a GPP instruction.

The Molen Reconfigurable Processor: The Molen $\rho\mu$-coded processor has been designed having in mind the programming paradigm previously presented. The Molen machine organization is depicted in Figure 3.

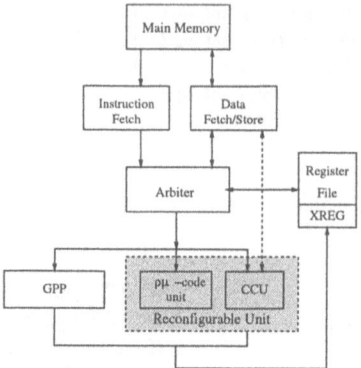

Fig. 3. The Molen machine organization

The arbiter performs a partial decoding of the instructions fetched from the main memory and issues them to the corresponding execution unit. The parameters for the FPGA reside in the Exchange Registers. In the Molen approach, an extended microcode - named reconfigurable microcode - is used for the emulation of both SET and EXECUTE instructions. The microcode is generated when the hardware implementation for a specific operation is defined and it cannot be further modified.

3   Compiler Extensions

In this section we present in detail the mechanism and compiler extensions required to implement the Molen programming paradigm.

The Exchange Registers: The Exchange Registers are used for passing operation parameters to the reconfigurable hardware and returning the computed values after the operation execution. In order to avoid dependencies between the RU and GPP, the XRs receive their data directly from the GPP registers. Therefore, move instructions have to be provided for this communication.

During the EXECUTION phase, the defined microcode is responsible for taking the parameters of its associated operation from XRs and returning the result(s). A single EXECUTE does not pose any specific challenge because the whole set of exchange registers is available. However, when executing multiple EXECUTE instructions in parallel, the following conventions are introduced:

- All parameters of an operation are allocated by the compiler in consecutive XRs and they form a block of XRs.
- The (micro)code of each EXECUTE instruction has a fixed XR, which is assigned when the microcode is developed. The compiler places in this XR a link to the block of XRs where all parameters are stored. This link is the number of the first XR in the block.

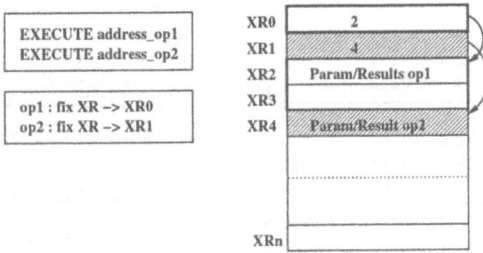

Fig. 4. Exchange Registers allocation by the compiler

Based on these conventions, the parameters for all operations can be efficiently allocated by the compiler and the (micro)code for each EXECUTE instruction is able to determine the associated block of parameters. An example is presented in Figure 4, where two operations, namely *op1* and *op2,* are executed in parallel. Their fix XRs (XR0 and XR1) are communicated to the compiler in a FPGA description file. As indicated by the number stored in XR0, the compiler allocates for operation *op1* two consecutive XRs for passing parameters and returning results, namely XR2 and XR3. The operation *op2* requires only one XR for parameters and results, which in the example is XR4, as indicated by the content of XR1.

Compiler Extensions: The compiler system relies on the Stanford SUIF2[8] (Stanford University Intermediate Format) Compiler Infrastructure for the front-end, while the back-end is built over the framework offered by the Harvard Machine SUIF[9]. The last component has been designed with retargetability in mind. It provides a set of back-ends for GPPs, powerful optimizations, transformations and analysis passes. These are essential features for a compiler targeting a CCM. We have currently implemented the following extensions for the x86 processor:

- Code identification: for the identification of the code mapped on the reconfigurable hardware, we added a special pass in the SUIF front-end. This identification is based on code annotation with special pragma directives (similar to [2]). In this pass, all the calls of the recognized functions are marked for further modification.
- Instruction Set extension: the Instruction Set has been extended with SET/ EXECUTE instructions at both MIR (Medium Intermediate Representation) level and LIR (Low Intermediate Representation) level.
- Register file extension: the Register File Set has been extended with the XRs. The register allocation algorithm allocates the XRs in a distinct pass applied before the GPR allocation; it is introduced in Machine SUIF, at LIR level. The conventions introduced for the XRs are implemented in this pass.
- Code generation: code generation for the reconfigurable hardware (as previously presented) is performed when translating SUIF to Machine SUIF IR, and affects the function calls marked in the front-end.

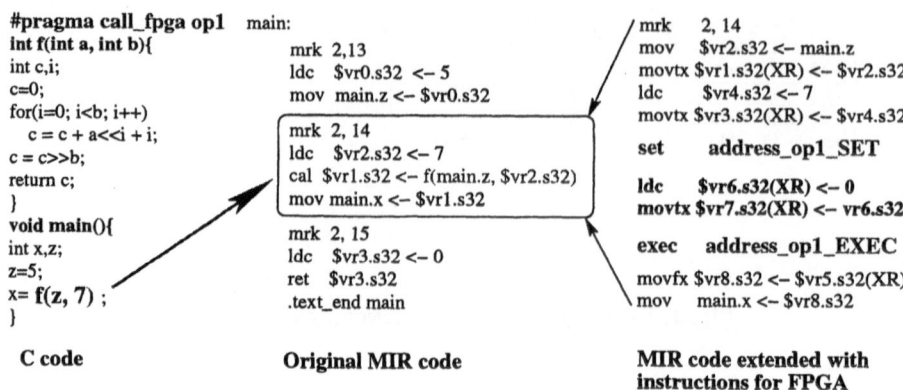

C code · Original MIR code · MIR code extended with instructions for FPGA

Fig. 5. Code Generation at MIR level

An example of the code generated by the extended compiler for the Molen programming paradigm is presented in Figure 5. In the first part, the C program is given. The function implemented in reconfigurable hardware is annotated with a pragma directive named *call_fpga*. It has incorporated the operation name, *op1* as specified in the description file. In the central part of the picture,the code generated by the original compiler for the C program is depicted. The pragma annotation is ignored and a normal function call is included. The last part of the picture presents the code generated by the compiler extended for the Molen programming paradigm; the function call is replaced with the appropriate instructions for sending parameters to the reconfigurable hardware in XRs, hardware reconfiguration, preparing the fix XR for the microcode of the EXECUTE instruction, execution of the operation and the transfer of the result back to the GPP. The presented code is at MIR level and the register allocation pass has not been applied.

The compiler extracts from a description file the information about the target architecture such as microcode address of SET and EXECUTE instructions for each operation implemented in the reconfigurable hardware, the number of XRs, the fix XR associated with each operation, etc.

4   A Case Study

In order to evaluate the performance improvements provided by the Molen processor, we used two well-known multimedia benchmarks, namely *mpeg2enc* and *ijpeg* for which we perform a pure software analysis. We made the following assumptions:

- the parts of the applications which can be implemented in the reconfigurable hardware are isolated in functions. This constitutes the base model for the comparison between the GPP and the Molen processor;
- the input data are:

- for mpeg2enc: the frames included in the benchmark
- for ijpeg: specmun, 1024 * 688

The parts of the applications that are candidates for the reconfigurable hardware implementation are the well-known time-consuming multimedia operations[7]: SAD (sum of absolute-difference), DCT (2 dimensional discrete cosine transform), IDCT (inverse DCT) and VLC (variable length coding). In order to study the performance improvements, we use the *Halt* library[10] available in Machine SUIF and which we modified to suit our purpose. This library is an instrumentation package that allows the compiler to change the code of the program being compiled in order to collect information about the program own behavior (at run-time).

For the above considered applications, the following is measured for their pure software implementation on the GPP (x86):

- The exact types and numbers of instructions - generated by the compiler-which are executed in the whole application and in each chosen function for hardware implementation plus their exact number of calls
- The number of cycles for the whole application and for each function chosen for hardware implementation

Based on these data, the following information can be computed for the Molen reconfigurable processor:

1. The code reduction as a result of implementation of parts of the application in reconfigurable hardware
2. An approximation of the maximum performance improvement of processor cycles for the whole application and for a particular implementation of one operation

However, because we lack a real implementation of the Molen processor, we cannot yet provide the second set of data for a particular implementation. We therefore restrict ourselves to indicating what functions are most likely to yield the highest performance improvement.

We introduced an additional pass in order to instrument the basic blocks of a program with the number and type of the included instructions. We also developed two sets of run-time analysis routines. The first set of routines is used to collect the type and number of instructions executed in the whole application and each specific function; it uses the instrumentation pass previously mentioned in this section. The second set of run-time analysis routines provides the number of cycles spent in the whole program or in a specific function. The measurements for the processor clock cycles have been performed on a Pentium II at 300MHz and we used the Pentium benchmarking instruction RDTSC - Read Time Stamp Counter - which returns the number of processor clock cycles since the CPU was reset. In this manner, the finest granularity is achieved (the code instrumentation does not affect the results).

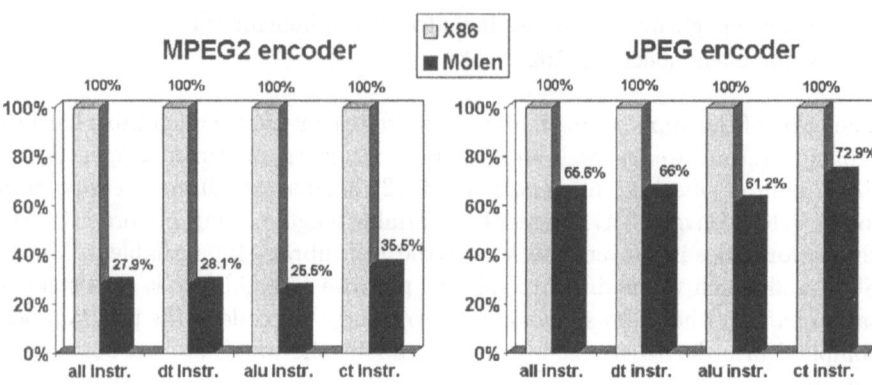

Fig. 6. mpeg2enc and ijpeg encoder instruction results

For example, we compute the total number of instructions executed in the Molen approach, when the above described functions have been implemented in hardware as follows:

$$n_{all(MOLEN)} = n_{all(GPP)} - \sum_{i=1}^{N} n_{f_i(GPP)} + \sum_{i=1}^{N} n_i(N_{call_i(MOLEN)} - N_{call_i(GPP)})$$

where n_{all} is the number of all instructions executed by the application, N is the number of functions implemented on the reconfigurable hardware, n_{f_i} represents the total number of instructions executed in the function f_i for all its calls, n_i is the number of calls of function f_i and N_{call} is the fixed number of instructions used for passing parameters, function call .

The measured data for the GPP alone and the computed data for the Molen processor are compared for mpeg2enc and ijpeg in Figure 6. The most important categories of instructions have been considered, namely data transfer (dt) instructions, arithmetic and logical (alu) instructions and control transfer(ct) instructions. From these pictures, a substantial reduction of the number of instructions is achieved by the Molen reconfigurable processor compared to the GPP: 72.1% for mpeg2enc and 34.4 % for ijpeg encoder. Also it is obvious that in both cases the alu instructions are the most reduced category of instructions, while the ct instructions are the least reduced instructions. This conclusion is confirmed by the inspection of the function code since it contains a large number of arithmetical computation and only a small number of branches. In table 1, the cycle measurements are reported. From these results, we can identify those functions that potentially give the highest performance improvement, given an efficient hardware implementation. The numbers suggest that the SAD function is the most promising candidate for hardware implementation, while the rest of the functions can provide at best a moderate performance improvement.

Table 1. mpeg2enc (left) and ijpeg encoder (right) cycle result

Fct	Cycles	% Total
SAD	149.947.461	55.2 %
DCT	42.529.647	15.7 %
VLC	3.946.954	1.4 %
IDCT	3.693.986	1.36 %
mpeg2enc Application	271.616.655	100 %

Fct	Cycles	%Total
DCT	40.206.773	12.5 %
VLC	36.571.622	10.5 %
ijpeg enc Application	341.316.466	100 %

5   Conclusions

In this paper, we presented the Molen Set-Execute paradigm that addresses a number of previously unresolved issues such as parameter passing and parallel execution of operations into the reconfigurable hardware. The paradigm involves the instruction set extension and requires on behalf of the FPGA developers only the address where the configuration(SET) and execution(EXECUTE) code is stored. A particular architectural implementation was presented, where the microcoded emulation of the SET and EXECUTE instructions are included.

The compiler extensions allow to generate code where the functions mapped on the reconfigurable hardware are automatically (rather then manually) substituted by the appropriate SET-EXECUTE instructions. It has been shown through experimentation that the compiler can be used as an important tool to support the design process focusing on the identification of good candidates for the reconfigurable hardware implementation. The presented results show a substantial reduction of the executed number of instructions and potential reduction of processor cycles for two multimedia benchmarks for their execution on the Molen reconfigurable processor compared to their pure software implementation on the GPP.

References

1. M. Sima, S. Vassiliadis, S.Cotofana, J. van Eijndhoven, and K. Vissers, "Field-Programmable Custom Computing Machines – A Taxonomy," in *12th International Conference on Field Programmable Logic and Applications (FPL)*, Montpellier, France, Sep 2002, pp. 79–88.
2. M. Gokhale and J. Stone, "Napa C: Compiling for a Hybrid RISC/FPGA Architecture," in *Proc. IEEE Symp. on Field-Programmable Custom Computing Machines*, Napa, California, April 1998, pp. 126–137.
3. S. Hauck, T. W. Fry, M. M. Hosler, and J. P. Kao, "The Chimaera Reconfigurable Functional Unit," in *Proc. IEEE Symp. on Field-Programmable Custom Computing Machines*, Napa, California, 1997, pp. 87–96.
4. A. L. Rosa, L. Lavagno, and C. Passerone, "Hardware/Software Design Space Exploration for a Reconfigurable Processor," in *Proc. of the DATE 2003*, 2003, pp. 570–575.

5. F. Campi, R. Canegallo, and R. Guerrieri, "IP-Reusable 32-Bit VLIW Risc Core," in *Proc. of the 27th European Solid-State Circuits Conference*, Villah, Austria, Sep 2001, pp. 456–459.
6. Z. Ye, N. Shenoy, and P. Banerjee, "A C Compiler for a Processor with a Reconfigurable Functional Unit," in *ACM/SIGDA Symposium on FPGAs*, Montery, California, USA, 2000, pp. 95–100.
7. S. Vassiliadis, S. Wong, and S. Cotofana, "The MOLEN $\rho\mu$-Coded Processor," in *11th International Conference on Field Programmable Logic and Applications (FPL)*, Springer-Verlag LNCS, vol. 2147, Belfast, UK, Aug 2001, pp. 275–285.
8. http://suif.stanford.edu/suif/suif2.
9. http://www.eecs.hardward.edu/hube/research/machsuif.html.
10. M.Mercaldi, M. D. Smith, and G. Holloway, "The Halt Library," in *The Machine-SUIF Documentation Set*, Hardvard University, 2002.

Laura:
Leiden Architecture Research and Exploration Tool

Claudiu Zissulescu, Todor Stefanov, Bart Kienhuis, and Ed Deprettere

Leiden Embbeded Research Center,
Leiden Institute of Advanced Computer Science (LIACS),
Leiden University, The Netherlands
{claus,stefanov,kienhuis,edd}@liacs.nl

Abstract. At Leiden Embedded Research Center (LERC), we are building a tool chain called *Compaan/Laura* that allows us to map fast and efficiently applications written in Matlab onto reconfigurable platforms. In this chain, first the Matlab code is converted automatically to executable Kahn Process Network (KPN) specification. Then a tool called *Laura* accepts this specification and transforms the specification into design implementations described as synthesizable VHDL. In this paper, we present our methodology implemented in the *Laura* tool, to automatically convert KPNs to synthesizable VHDL code targeted for mapping onto FPGA-based platforms. With the help of *Laura*, a designer is able to either fast prototype signal processing and multimedia applications directly in hardware or to extract very fast valuable low-level quantitative implementation data such as performance in terms of clock cycles, time delays and silicon area.

1 Introduction

The potential of achieving high-performance implementations onto FPGA-based systems (platforms) has been demonstrated by the FPGA research community for applications in the domain of signal processing, multimedia, and imaging. These performance improvements depend very much on the expertise of the hardware designer, who has to possess an accurate knowledge of the underlying FPGA platform and the application. Moreover, the mapping of applications onto this type of platforms is in most cases done manually, which leads to a slow, difficult, and error prone design process. Therefore, we have developed a methodology that allows fast and efficient mapping of a class of multimedia and signal processing applications onto FPGA-based platforms. Part of this methodology is captured in the *Laura* tool that we present in this paper. Central to our methodology is the use of the Kahn Process Network (KPN) [3] model of computation to specify applications. The Laura tool accepts applications written in this KPN model and produces synthesizable VHDL code that implements the application for a specific FPGA platform.

Our methodology uses the KPN model of computation as it is a convenient model to specify imaging applications like Stereo Vision, multimedia applications like MJPEG, and classical signal processing applications like Digital Beam-forming. The model reveals the inherent parallelism of an application that is exploited when mapping the application onto FPGA platforms that are inherently fine-grained parallel platforms.

P.Y.K. Cheung et al. (Eds.): FPL 2003, LNCS 2778, pp. 911–920, 2003.

The KPN specification represents an application in terms of distributed control and distributed memory, which in our case is derived from a sequential code written in Matlab using a tool called *Compaan*. The distributed control and distributed memory are key to obtain efficient implementations on FPGAs for stream oriented applications. This is in great contrast to the original Matlab code that is using a single thread of control and shared memory. Other work describing the mapping of Matlab code (or C for that matter) onto FPGA uses other computational models like CDFG [2] or CSP [5]. These models are well suited for control dominated applications, but less for stream oriented applications.

We present our methodology to map an application written in Matlab onto an FPGA platform in Section 2. In Section 3, we look in more detail at the Laura tool that we have developed. In Section 4, we explain in more detail, using a running example, how Laura constructs an architecture in VHDL. In Section 5, we present experiments that have been obtained by using Laura for three applications. We conclude this paper in Section 6.

2   Integrating *Laura* in an FPGA-Based Design Flow

The *Laura* tool takes as an input a KPN specification of a given application and generates synthesizable VHDL code that targets a specific FPGA platform. In general, specifying an application as a KPN is a difficult task. Therefore, we use our compiler called *Compaan* [4] that fully automates the transformation of Matlab code into Kahn Process Networks (KPNs). The applications Compaan can handle, have to be specified as parameterized static nested loop programs, which is a subset of the Matlab language. We have designed the Laura tool to operate as a back-end of the Compaan compiler, realizing a fully automated design flow that maps sequential algorithms written in Matlab onto reconfigurable platforms. This design flow is shown in Figure 1.

In the first part of the design flow, an application specification is given in Matlab. This is because Compaan only accepts Matlab code. Nevertheless, the design flow is equally applicable to C code or Java code, as their model of computation is equal to the imperative model of computation of Matlab. The Compaan compiler itself is composed of a number of tools. One tool in Compaan performs an aggressive array-dataflow analysis by exploring all data-dependencies in the original program. The result of this tool is a data structure representing the dependence graph of the program. Another tool in Compaan converts this data structure into a KPN specification.

In the second part of the design flow, Laura transforms a KPN specification together with predefined IP cores into synthesizable VHDL code. The IP cores are needed as they implement the functionality of the functions used in the original Matlab program. They are provided to Laura by the *IP cores* box in Figure 1.

In the third part of the design flow, the generated VHDL code is processed by *Commercial Tools* to obtain quantitative results. These results can be interpreted by designers, leading to new design decisions. These decisions are reflected by writing a new Matlab program that exposes, for example, more or less parallelism. For that purpose, we have developed a tool called *MatTransform* that manipulates the Matlab input specification in a Source-to-Source fashion to generate more instances of the application, in which each instance exposes a different level of concurrency without altering the algorithm's behavior [8]. The concurrency is altered by performing high-level transformations like loop unrolling (unfolding), retiming (skewing), and code merging. By rewriting Matlab code, we can explore different mappings of a Matlab algorithm in an efficient way. When an obtained algorithm instance meets the requirements of the designer, the corresponding VHDL output is synthesized by a commercial tool and mapped onto an FPGA platform.

3   The Laura Tool

The KPN model of computation [3] assumes concurrent autonomous processes that communicate in a point to point fashion over unbounded FIFO channels, using a *blocking-read* synchronization primitive. Each process in the network is specified as a sequential program that executes an internal function. At each execution (also referred to as an iteration) this function reads/writes data from/to different FIFO channels. Because of the unboundedness of the FIFO channels, the KPN cannot be translated directly into a VHDL representation and mapped onto a hardware platform. Instead, a *blocking-write* primitive is needed next to the *blocking-read*. Also, the FIFO channel sizes now need to be fixed such that no deadlock occurs. Using the method presented in [6], we find a bound on the size of the FIFOs such that the network will not deadlock.

To convert a KPN specification into hardware, we have implemented in Laura a strategy that divides the conversion process into two parts: a platform independent part and a platform dependent part. In the platform independent part, we define an abstract model of the architecture on which we map a KPN application. The model of architecture defines the key components of the architecture and their attributes. It also defines the semantic model, i.e., how the various components interact with each other. Hence, the architecture also implements autonomous processes, that communicate over channels using blocking read and blocking write semantics.

The abstract architecture model is captured in Laura in terms of a class-hierarchy. This class hierarchy describes a network of virtual processors. Each of them is composed of four units: a *Read unit*, a *Write unit*, an *Execute unit* and a *Controller unit*. The first three units are synchronized by the Controller unit of the processor. Each FIFO channel in the KPN specification is represented by a *Hardware Channel* unit.

In the platform dependent part we start to add information to the abstract architecture model that is specific for the target platform. At this stage, we include IP cores in the Execute units that implement the functions of the original application. Also, we set attributes of the components like bit-width and size of the Hardware Channels.

When an architecture model is established for a given KPN specification, we convert the architecture model into VHDL code using a Visitor Design Structure. For each component in the abstract architecture, we have a small piece of VHDL code that expresses how to represent that component on the target architecture. The visitor structure gives Laura a lot of flexibility. If needed, the output can easily be convert to other formats like Verilog or SystemC.

The steps that make up the Laura tool are shown in Figure 2. In the first step, the *KPN-ToArchitecture* method converts the given KPN specification into an equivalent network of virtual processors (*Network of Virtual Processors*). This is a platform independent

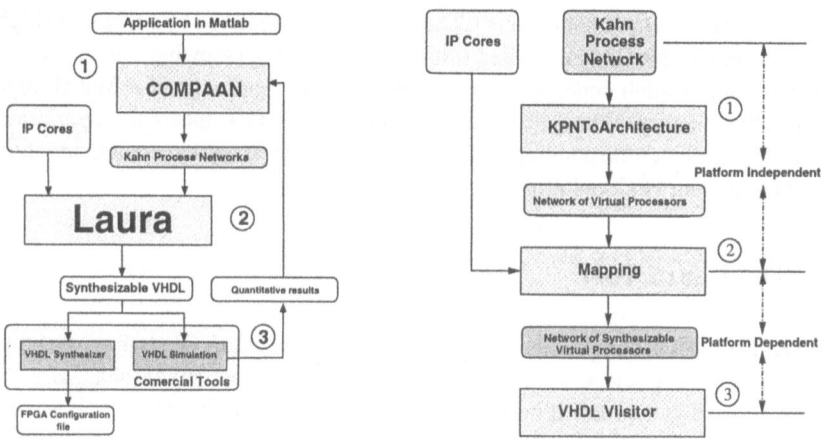

Fig. 1. The Compaan/Laura tool chain   **Fig. 2.** Steps in Laura

step as no information on the target platform is taken into account. In the second step, platform specific information is mapped onto the abstract architecture model leading to a network of Synthesizable Processors (*Network of Synthesizable Processors*). In the third step, the architecture model is visited by a VHDL visitor to generate the VHDL code.

4   Laura in Action

To make clear how a Matlab program is converted into a VHDL code, we explain the steps done in Laura using the very simple Matlab program given in Figure 4. This program consists of three loops. In the first loop, variable a(j) is initialized using function Init, which represents a *Source*. In the second loop, the function Compute performs an operation on a(j-1), introducing a self-loop. Finally, the last loop takes the result of a(6) using function Pass, representing a *Sink*. The Matlab program is given to the Compaan compiler that converts it into a KPN representation consisting of three different processes. A graphical representation of this KPN is given in the top-part of Figure 3. One process (P1) is the Source, one process implements the Compute function (P2), and one process is the Sink (P3). The picture clearly shows the self-loop of function Compute. As said before, each process contains a sequential program. In Figure 5, the sequential program for process P2 is given in C++ using the YAPI [1] format.

The sequential program produced by Compaan always follows a particular sequence of events. These events are highlighted by the three different boxes in Figure 5. The first box, contains the code that reads data from input ports. The actual computation takes place in the second box (i.e., performing the function Compute from the Matlab program of Figure 4). In the third box, we show the code that writes out data produced by the computation. The three boxes are enclosed by a for-loop, indicating that the sequence of events needs to be repeated for a given number of times. As a consequence, this

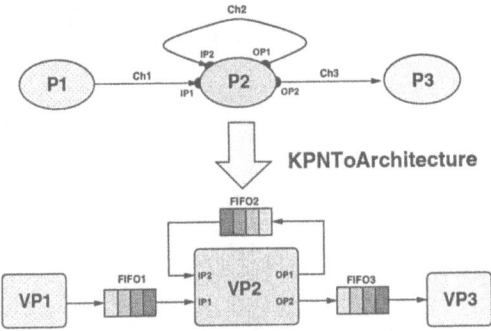

Fig. 3. An example of *KPNToArchitecture* step in Laura

process operates in a stream based fashion, an operation model which is very applicable to multi-media and digital signal processing applications.

4.1   KPNToArchitecture

The KPN shown in the upper part of Figure 3 is mapped by the KPNToArchitecture step in Laura onto an abstract architecture model. This model is composed of *Virtual Processors* and bounded hardware communication channels. The lower part of the Figure 3 represents the network of virtual processors that has the same topology as the input KPN. This is because Laura currently performs a *one-to-one* mapping. The three processes *P1*, *P2*, and *P3* are mapped onto the virtual processors *VP1*, *VP2*, and *VP3*, respectively. The KPN unbounded FIFO channels *Ch1*, *Ch2*, and *Ch3* are mapped onto the bounded hardware FIFOs *FIFO1*, *FIFO2*, and *FIFO3*, respectively.

Every virtual processor is composed of four units: a *Read* unit, a *Write* unit, an *Execute* unit, and a *Controller unit*, as shown in Figure 6. The Execute unit is the computational part of a virtual processor. It has *Input Arguments* that provide to the unit the necessary data for execution and *Output Arguments* that are the result of the computation process. In our model, the Execute unit fires when all the input arguments have data and always produces data to all the output arguments. The Read unit is responsible for assigning all the input arguments of the Execute unit with valid data. Since there are more input ports than arguments, the Read unit has to select from which port to read data. This information is stored in the *Control Table* of the Read unit. The *Input Port* is the input interface that connects the virtual processor with a communication channel. The *Output Port* is the output interface that connects the virtual processor with a communication channel. The Write unit is responsible for distributing the results of the Execute unit to the relevant processors in the network. A write operation can be executed only when all the output arguments of the execute unit are available for the write unit. A *Control Table* is used to select the proper Output Port according to the current iteration of the virtual processor.

The virtual processor's *Controller* synchronizes all the processor's units and keeps track of how many times the processor has already fired. The Read unit and the Write unit can block the next firing when a blocking-read or a blocking-write situation occurs,

```
1 void P2 ::main() {
2 for (int i = 2 ; i <= 6 ; i += 1 ) {
```

```
3      if (i-2 == 0) {                        READ
4          //reads a token from a channel
5          in_0 = read(IP1);
6      }
7      if (i-3 >= 0) {
8          //reads a token from a channel
9          in_0 = read(IP2);
10     }
```

```
11     out_0 = Compute(in_0) ;        EXECUTE
```

```
12     if (-i+5 >= 0) {                      WRITE
13         //writes a token to a channel
14         write( OP1, out_0);
15     }
16     if (i-6 == 0) {
17         //writes a token to a channel
18         write( OP2, out_0);
19     }
```

```
20 } // for i
21 }
```

```
for j = 1 : 1 : 1,
    [ a(j) ] = Init;
end
for j = 2 : 1: 6,
    [a(j)] = Compute (a(j-1));
end
for j =6 : 1 : 6,
    [] = Pass(a(j));
end
```

Fig. 4. A very simple Matlab Program   **Fig. 5.** Process P2

thereby stalling the complete processor. A blocking-read situation occurs when data is not available at a given input port. A blocking-write situation occurs when data cannot be written to a particular output port.

Let us consider the P2 process as it is specified by the sequential code given in Figure 5. This code is analyzed in the KPNToArchitecture step to instantiate the corresponding components of the virtual processor. The Read unit is generated based on the information contained between lines 3 and 10. Two input ports, **IP1** and **IP2**, are required to read the input argument **in_0** of the Execute unit. Because a 2-to-1 relationship exists between the input ports and the input argument, a Control Table is needed to select the proper input port for reading the input argument at a particular iteration of the processor. For the example, the Control Table $c = [1, 0, 0, 0, 0]$ is derived based on the number of firings (line 2) and the **if** statements from lines 3 and 7. The Write unit is instantiated according to the lines 12 to 19. It requires two output ports **OP1** and **OP2** to write the output argument **out_0** of the Execute unit. Again a Control Table is derived based on the number of firings (line 2) and the **if** statements from lines 12 and 16. The Control table is equal to $d = [0, 0, 0, 0, 1]$. The Control unit of the processor is instantiated as a counter that iterates i from 2 to 6. For the Execute unit an interface is defined, based on the information contained in line 11. This interface is used again in the Mapping step (Figure 2) when an IP core is connected to the Execute unit. The complete virtual processor that corresponds to process P2, is shown in Figure 7.

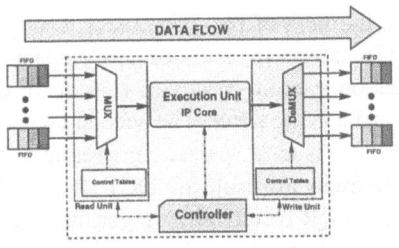 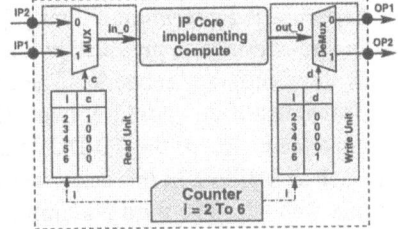

Fig. 6. The Virtual Processor model   **Fig. 7.** VP2 in the example shown in Fig. 3

4.2 Mapping

The Mapping step is used to include additional information to the abstract architecture model. This is information about the IP cores used by the virtual processors and the bit-width of data. At this step, the width and size of a hardware channel is provided. Furthermore, the notion of a clock event is taken into consideration.

We use IP cores in designing new hardware applications to reduce the design time. This means that we add in the Mapping step the functionality of the Execute unit in terms of an IP core. In order to select the appropriate IP core, the Mapping step searches through a library of predefined cores until it matches the required functionality. The found IP core is subsequently associated to the Execute unit of the virtual processor. For the IP cores which are pipelined, additional information needs to be provided to the Control unit to accommodate the control for the pipelining.

The final result of the mapping step is an annotated architectural model called *Network of Syntesizable Virtual Processors* (NSVP) that is targeted to a particular FPGA platform.

4.3 Visitor

The last step of Laura generates the correct VHDL code for the NSVP structure. First, the communication network is generated, followed by the various processors, and finally a test bench. Within the Visitor there is a well-defined relationship between the components of the abstract architecture model and its representation in VHDL. This means that we have a VHDL template for each component. For example, there is a template for the various units in a processor as well as for the processor itself. The relationship is often one-to-one, but for example in the case of the hardware communications channels, a one-to-many relationship exists. A hardware communication channel operates as a data buffer that can be realized using flip-flops, a look-up table, or internal BRAM memory. This gives the Visitor a lot of flexibility to derive alternative VHDL code taking advantage of specific elements of the target platform.

5   The Experiments

Our experimental results are obtained by evaluating the synthesizable VHDL code generated by Compaan/Laura for three computational intensive algorithms. The first one is the *Sequence Alignment* algorithm [7] from the field of bio-informatics. Using the unfolding transformation provided by the MatTransform [8] tool box of Compaan, we generate three different networks. The application specific processor uses an IP core called *Match* that is composed of two adders and a comparator. The second algorithm is the implementation of the 2D-DCT function that is used in data compression algorithms such as MJPEG. In this case, we used the freely available 2D-DCT IP core from the Xilinx web site. The third one implements the QR factorization algorithm used in signal processing applications. It has two IP cores, *Vectorize* and *Rotate*, provided by QinetiQ, Ltd [10]. Table 1 shows the complexity of the input KPNs given by the number of processors and communication channels that has to be handled by the Laura tool. The complexity of the IP cores used to implement the application specific processor is given by the number of hardware multipliers and the pipeline depth used to implement the core.

Table 1. The Process Network complexity

Experiment	*No. of Processors*	*No. of Channels*	*Pipeline Stages*	*Multipliers*
Sequence Alignment	7	13	0	0
Seq. Alignment Unfold 2x2	10	40	0	0
Seq. Alignment Unfold 3x3	15	83	0	0
2D-DCT	4	4	92	6
QR(Rotate, Vectorize)	5	18	55, 42	8, 8

For each benchmark algorithm, a description of the algorithm in Matlab was written and passed through Compaan and Laura. We verified the hardware in two ways. The first way is by simulating the generated hardware using a VHDL simulator and comparing the results to the output of the algorithm executed in the Matlab interpreter. The second way is by implementing the generated hardware onto our reconfigurable platform and comparing the results to the Matlab output. The VHDL simulator provided the total number of cycles needed to execute a given algorithm, as shown in the **Cycles** column of Table 2. We use the XST synthesizer and the Xilinx Foundation 5.1i tool to synthesize, place, and route the output of Laura. The clock delay and the total amount of slices needed to implement the networks onto a Virtex II-6000 are also provided in Table 2.

Table 2. Experimental Results

Experiment	*Cycles*	*Clock delay (ns)*	*Used Slices*	*Used Area Virtex II-6000*
Sequence Alignment	865	16.030	1321	3%
Seq. Alignment Unfold 2x2	466	15.751	3127	9%
Seq. Alignment Unfold 3x3	293	18.511	5874	17%
2D-DCT	364	19.733	1610	4 %
QR(N=7,T=21)	19181	24.390	11270	33 %

To study the overhead introduced by our methodology in terms of cycle delays and area (i.e., used slices), we conducted a second experiment. In this experiment, we compare a single IP core with the same core embedded in a network. For a single IP core we determine its clock speed and area and compare this to the speed and area taken by the same IP core used in an application network. This gives an indication about the overhead introduced by our methodology. Table 3 shows the delays and the area used by the IP cores, the influence of communication on clock delay (**Delay Overhead**), and the used area (**Area Overhead**). We notice that for fine-grained core implementations the area needed to communicate data in a distributed way is dominant. For example, in case of Sequence Alignment, 20 times more area is needed then a stand-alone version of the *Match* IP core. The communication takes more that 2 times longer in terms of clock-delay than the stand-alone version, due to the routing of the hardware channels on the FPGA. The network of embedded coarse-grained cores, i.e., 2D-DCT, Vectorize and Rotate, introduce considerable less clock-delay than the network of embedded fine-grained cores, i.e., Match. The area overhead depends mainly on the network complexity in terms of channels used. See the difference in number of channels between 2D-DCT and QR in Table 1.

Table 3. Trade off between Computation and Communication

Experiment	Working Processor	Clock Delay	Slices	Delay Overhead	Area Overhead
Sequence Alignment	Match	6.156	66	2×	20×
Seq. Alig. Unfold 2x2	4×Match	6.156	264	2×	11.8×
Seq. Alig. Unfold 3x3	9×Match	6.156	594	3×	10×
2D-DCT	2D-DCT	13.656	1365	1.4×	1.17×
QR	Vectorize, Rotate	15.862	3442	1.5×	3.27×

6 Conclusions and Limitations

In this paper, we have presented the Laura tool that implements our methodology to map KPNs generated by the Compaan tool onto a reconfigurable platform such as FP-GAs. Although the tool generates only VHDL code, it can be reconfigured to generate other kinds of output, such as Verilog or SystemC. A number of experiments have been conducted for applications in the field of bio-informatics, image processing, and signal processing. The experiments show that we are able to derive fully automatically a hardware implementation from Matlab code. Because Laura implements Kahn Process Networks into hardware, it is well suited for stream oriented applications. Laura is not suited to map control dominated applications. To study the impact of the KPN model on the hardware realization, we investigated the trade off between a stand-alone IP core and an integrated IP core. We found that for more coarse-grained IP cores, the presented methodology gives the best results.

A number of limitation can still be found in Laura. The first issue is that Laura can handle only FIFO communication between processors. High-level code transformations, such as unfolding and skewing, can introduce out-of-order communication between processors [9]. In such case a FIFO can no longer be used in the communication between processes. Future work includes extending the communication components to include this out-of-order communication. The second issue is that Laura generates hardware implementations for non-parameterized KPN models, while Compaan is capable of deriving parameterized descriptions. Future work will focus on generating parameterized hardware networks. The third issue is that communication channels are not always used at their full capacity. We would like to collapse some of these channels onto one channel to share its hardware to reduce communication requirements.

Acknowledgments

We would like to acknowledge Alexandru Turjan of the LERC group, Leiden University, for his very valuable input and sharing his insights on the Laura work and his substantial effort to integrate the Laura work with Compaan. Also, we would like to thank to Steven Derrien for his insides toward the processor synchronization issues.

References

1. E. de Kock, G. Essink, W. Smits, P. van der Wolf, J.-Y. Brunel, W. Kruijtzer, P. Lieverse, and K. Vissers. YAPI: Application modeling for signal processing systems. In *Proc. 37th Design Automation Conference (DAC'2000)*, pages 402–405, Los Angeles, CA, June 5-9 2000.
2. M. Haldar, A. Nayak, A. Choudhary, and P. Banerjee. A system for synthesizing optimized fpga hardware from matlab. In *Proc. Int. Conf. on Computer Aided Design*, San Jose, CA, Nov. 2001.
3. G. Kahn. The semantics of a simple language for parallel programming. In *Proc. of the IFIP Congress 74*. North-Holland Publishing Co., 1974.
4. B. Kienhuis, E. Rypkema, and E. Deprettere. Compaan: Deriving process networks from matlab for embedded signal processing architectures. In *Proceedings of the 8th International Workshop on Hardware/Software Codesign (CODES)*, San Diego, USA, May 2000.
5. I. Page. Constructing hardware-software systems from a single description. In *Journal of VLSI Signal Processing, 12(1):87–107*, 1996.
6. T. Parks. *Bounded Scheduling of Process Networks*. PhD thesis, University of California at Berkeley, 1995.
7. T. Smith and M. Waterman. Identification of common molecular subsequences. *Journal of Molecular Biology, 147:195-197*, 1981.
8. T. Stefanov, B. Kienhuis, and E. Deprettere. Algorithmic transformation techniques for efficient exploration of alternative application instances. In *10th Int. Symposium on Hardware/Software Codesign (CODES'02), pp. 7-12, Estes Park, Colorado, USA*, May 6-8, 2002.
9. A. Turjan, B. Kienhuis, and E. Deprettere. Realizations of the extended linearization model in the compaan tool chain. In *proceedings of the 2nd Samos workshop*, Samos, Greece, Aug. 2002.
10. R. Walke, R. Smith, and G. Lightbody. 20Gflops QR processor on a Xilinx Virtex-E FPGA. In *Proc. SPIE Advanced Signal Processing Algorithms, Architectures, and Implementations X*, pages 300–310, 2000.

Communication Costs Driven Design Space Exploration for Reconfigurable Architectures

Lilian Bossuet, Guy Gogniat, and Jean-Luc Philippe

LESTER Lab
University of South Brittany,
Lorient, France
{lilian.bossuet, guy.gogniat, jean-luc.philippe}@univ-ubs.fr

Abstract. In this paper we propose a design space exploration method targeting reconfigurable architectures that takes place at the algorithmic level and aims to rapidly highlight architectures that present good performance vs. flexibility tradeoffs. The exploration flow is based on a functional model to describe the architectures that the designer wants to compare. The paper mainly focuses on the projection step of our flow and presents an allocation heuristic that is based on communication costs reduction.

1 Introduction

The new telecommunication and multimedia applications need to have a reduction of actual systems on chip (SoC) power consumption and an increase of SoC flexibility. Although the evolutions of integration and design are more and more important, they are not sufficient to face these challenges. It is necessary to provide novel approaches that work at the system level to design more efficiently, and to target new technologies.

Reconfigurable architectures are becoming more attractive in terms of capacity, performances, low-power consumption and flexibility (through the possibilities of run-time reconfiguration and multi-granularity resources) [1][2]. They correspond to an efficient solution to the SoC challenge and will be unavoidable in a near future. But the design space of reconfigurable architectures is very large, because these architectures can be extremely heterogeneous in term of processing, memory and routing resources. Hence, it is very complex to find the best reconfigurable architecture for a panel of applications where each application can be dynamically configured on the architecture.

In order to help the designer it is necessary to develop tools that compare several architectures for different applications. We propose in this paper an original method of design space exploration for reconfigurable architectures that works at the algorithmic level.

The paper is organized as follows. Section 2 describes related work dealing with design space exploration methodologies. Section 3 describes major issues in design space exploration at the algorithmic level and section 4 presents our approach. Section 5 details the projection step. Section 6 gives some results for different reconfigurable architectures. Finally, section 7 concludes the paper and exposes future direction.

P.Y.K. Cheung et al. (Eds.): FPL 2003, LNCS 2778, pp. 921–933, 2003.

2 Related Work

Many research teams are focusing on reconfigurable architecture [1]. Some are working on design space exploration methodology in order to find the best architecture for a panel of applications.

Two ways are possible to explore reconfigurable architectures. The first one is to synthesize all the applications for the different target architectures, and to compare the overall performance results. In that case the results are very accurate, but it is necessary to have a specific synthesis tool for each architecture (which is not always available in the case of architecture exploration) or to use generic synthesis tools [3][4]. However, synthesis steps use very complex algorithms, which conducts to a limited and slow exploration. Furthermore it is necessary to have a very good knowledge of the target architectures when using generic synthesis tools since it is necessary to provide them a model of the target architectures. Hence, this method is not really adapted for a large and rapid architecture exploration and is more dedicated to do some architecture refinement steps.

The second way is to perform estimations. In that case it is necessary to consider a generic architecture model to describe the different architectures to target. The objective is to make relative performance estimations (speed, power consumption and area) in order to compare very quickly different architectures. Although the estimations do not give necessarily real and accurate performance results, it is enough to compare architectures since the important point in that case is that estimations are faithful and an absolute error is not the major concern.

Both exploration methods require having an architecture model. It is possible to consider a physical model. Then it is necessary to know precisely the physical parameters of the architecture (technology, routing type and size, routing switch resources, clusters size, etc). Versatile Place and Route (VPR) tool, developed at the Toronto University, is a very interesting approach that works on a physical model [3]. VPR is a synthesis tool that works at the logic level and is oriented for island style fine-grained architectures (as FPGA). It is not suitable for coarse-grained architectures. It is also possible to model architecture with a functional model. Each element of the architecture is described by the functions it can execute. The functional model enables to describe a large panel of architectures and the description are technological independent. This model is used in the generic place and route tool for fine-grained reconfigurable architectures called Madeo-Bet [4] that works at the logic level.

VPR and Madeot-Bet are not the only tools that use an architecture model, but there are very representative. Both are FPGAs oriented, but other approaches target reconfigurable architectures. In [5] the design space exploration flow targets mesh architecture called KressArray - a fast reconfigurable ALU. The exploration tool, Xplorer works at the algorithmic level and aims to assist the designer in finding a suitable architecture for a given set of applications. This tool is architecture-dependent, but the use of fuzzy logic to analyze the results of the exploration is a very attractive approach.

[6] presents the design space exploration for the Raw Microprocessor as an example of a tiled architecture. The Raw Microprocessor is reminiscent of coarse-grained FPGA and comprises a replicated set of tiles coupled together by a set of

compiler orchestrated, pipelined, switches. Each tile contains a RISC-like processing core and SRAM memory for instructions and data.

3 Design Space Exploration Flow Principles

Design space exploration can be performed at different levels of abstraction in order to reduce progressively the number of solutions. More the abstraction level is refined more accurate results can be obtained since a lower number of solutions need to be considered.

At the algorithmic level the objective is to rapidly identify target architectures that present a high performance and versatility potential. To reach such a goal, design space exploration methods must promote the flexibility, the rapidity and the fidelity. Encouraging (i) the flexibility means that performance and versatility can be estimated for a wide variety of reconfigurable architectures, (ii) the rapidity enables to estimate performances without the time-consuming computation of programs such as Place & Route algorithms and (iii) the fidelity points out that the relative comparisons between two alternative architectures must be close to the relative errors that would be obtained after the synthesis steps even if there may be significant absolute errors in the performance estimation at the algorithmic level.

Another major concern is to promote the interactivity with the designer at all the abstraction levels in order to take benefit from his experience. Thus, the refinement process at a given abstraction level can be performed through several runs of the exploration method in order to converge progressively to an efficient mapping between the application and the architecture. Between several runs, the designer can improve the architecture model according to the previous results of the exploration. Once, the designer has selected some efficient architectures he can refine his results by decreasing the abstraction level and thus using more accurate architecture model and exploration tools.

In this paper we proposed a design space exploration method targeting reconfigurable architectures that addresses the previous highlighted principles. Our method is based on an estimation approach and takes place at the very first steps of the design flow since it works at the algorithmic level. A functional model is used to describe the target architectures since such model as proven its efficiency to characterize a wide variety of reconfigurable architectures [7].

4 Proposition of a Design Space Exploration Flow

The design space exploration flow that we propose is depicted on the left of the figure 1. The specification is provided in a high level language (C language) and is first translated into an intermediate representation - the HCDFG model. This model is a Hierarchical Control and Data Flow Graph allowing efficient algorithm characterization and exploration of complex applications including control flow and multi-dimensional data.

The first part of the flow (figure 1) is the **System Estimation** step [8][9], during which the application is characterized and scheduled. The results computed are defined as **Costs Profiles** i.e., scheduling for all the resources used by the application and for different time constraints. The available processing and memory resources to perform the scheduling are defined in a file called **User Abstract Rules**.

Relative Estimation, the second step of the flow (figure 1), aims to estimate the application performances on several reconfigurable architectures. Relative Estimation gives designer information to improve progressively the architecture definition with several runs in the design exploration flow.

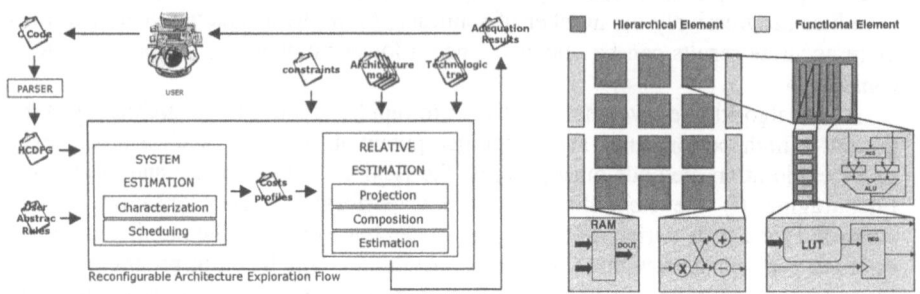

Fig. 1. The design exploration flow and an example of reconfigurable architecture that can be specified with the HF model.

4.1 Specification

The application is specified with the C language. Once it has been functionally validated, it is translated into an intermediate model, which is a Hierarchical Control and Data Flow Graph (HCDFG) [9][10]. For sake of simplicity we can say that an application is modeled with several DFG connected through control structures, hierarchy and dependence relations. An important characteristic of the HCDFG model is that processing and data are explicitly represented with nodes in the graph. This decomposition enables to emphasize the hierarchy and the potential parallelism of the application, which is an essential characteristic to perform an efficient algorithmic and architecture exploration.

4.2 System Estimation

System Estimation [8] consists in two steps; **Characterization** and **Scheduling**. During the Characterization step, the application orientation is analyzed through three axes - processing, control and memory. Specific metrics are used to find this orientation [9]. Once the application orientation has been exhibited the scheduling of the application is performed accordingly. For example, if the application is processing oriented, the processing resources are first scheduled and the memory resources are scheduled in a second step (the opposite if the application is memory oriented). The main objective of the System Estimation is to show up the intrinsic processing and memory parallelisms of the application and to give some guidance on how to build the architecture (pipeline, parallelism, memory hierarchy etc.). Further details on the system estimation step are beyond the scope of this paper. Interested reader can refer to [9].

4.3 Relative Estimation

Relative Estimation is linked with the System Estimation through the cost profiles. The cost profiles describe the scheduling results (for all the processing and memory operations) for different time constraints. These values characterize the application to be implemented.

This step is composed of three tasks: **projection, composition** and **estimation** that are performed sequentially. In order to evaluate an application on different architectures, it is necessary to specify the target reconfigurable architectures. For this, a functional model is used since such model is suitable for rapid relative comparison and is efficient to describe architectures at a high abstraction level. This model called **HF model** [7], enables to describe functionally the elements of the architecture and to represent different architectural styles and different architecture elements. Although, the routing resources are not explicitly described, they are taken into account in the estimation flow with connection costs in the HF model. There are two types of elements in the model: the hierarchical elements and the functional elements. The hierarchical elements are used to describe the architecture hierarchy. A hierarchical element can be composed of functional elements and other hierarchical elements. The functional elements are used to describe the architecture resources. They can be logical, input/output, memory and processing resources. An extension of the HF model exists to model tile-based architectures, where the communications between the tiles must be explicitly modeled [11]. Figure 1 (on the right) shows an example of architecture with several levels of granularity that can be modeled with the HF model.

Relative Estimation begins with the **Projection** step that makes the link between the functional needs (processing and memory) of the application and the available resources of the architecture. If a difference of resource granularity exists between the functional needs (from the application) and the available resources (in the architecture), the necessary resources can be split as fine-grained resources using a library called **Technological Trees** (the same method is used with the PipeRench reconfigurable architecture [12]).

The **Composition** step takes into account the application scheduling obtained during the System Estimation step in order to refine previous results. Since it is necessary to add resources dedicated to realize the scheduling like multiplexer, register or states machine. These additional resources are taken into account in the last step of the estimation process.

The **Estimation** step computes the global application performances on the selected architectures. The estimations take into account the static costs of the model (interconnect costs between two hierarchical elements and the costs of the functional elements), and the dynamic costs of the application (critical path, operator communications, memory reads/writes). The results of this step are gathered into a file where the application is characterized for the target architectures. Composition and estimation steps are not the topic of this article.

5 Projection Algorithms

In this section we focus on the projection step which is particularly important in the flow since it has a strong impact on the quality of the final estimated performances. The goal of this step is to allocate the resources of the architecture that will support the application operations (processing and memory). The allocation algorithms aim to assign in a same hierarchical level of the architecture the resources that communicate the most in order to reduce the cost overhead due to communications. In our model architecture hierarchy corresponds to routing topology. Communication costs are smaller inside a hierarchical element than between two hierarchical elements. More hierarchical levels are crossed by a communication higher is the communication cost.

Why focussing on communication costs for allocation algorithms? Studies on power repartition in fine-grain reconfigurable architecture like FPGA show that the major contribution to power consumption is due to routing resources (wires and switch). It is always better to reduce communication paths [13][14] since implementations are more energy efficient when wires are short and number of switches is low. The clock frequency can also be increased when communication paths are short since critical path is reduced.

In our approach we enhance the spatial locality between resources that most communicate. Different cost functions have been defined, to estimate the communication costs, which are computed in three algorithms that give respectively a lower and an upper bound and a mean value for the total communication costs.

Since our approach works at the algorithmic level it does not target a specific synthesis tool and does not consider an accurate physical architecture model. Hence instead of giving designers a single communication cost value that may present a significant absolute error due to backend synthesis algorithms and architecture refinement we propose to give them some bounds and an average value. Such approach as shown figure 2 enables designers to select architectures at the algorithmic level with the guaranty that the final performance will belong to the estimated performance interval.

Such approach also enables to give designer metrics on allocation algorithm impact. On the example (figure 2) the architecture **C** has a narrow performance interval so allocation algorithms will have a small effect on the final performance and low complexity algorithms can be considered. The architecture **B** has a large performance interval so allocation heuristics will have a strong impact on final performance and it might be important to consider better allocation algorithms.

Results of the relative estimation step provide designers resources utilization rate and estimations of communication costs. Based on these results designers can perform a first architectures performance comparison and remove architectures that present a poor synergy with the application (e.g. unadapted granularity) or too important communication costs.

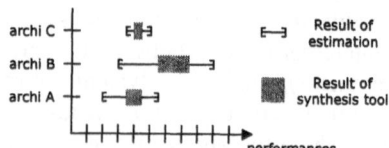

Fig. 2. Bound performance results.

5.1 Average Communications Graph (*ACG*)

In order to make this projection, the first step builds, from the HCDFG model, a new graph with only processing nodes, i.e. a graph without memory node and control structure. The edges between two nodes represent the communications. This graph is then reduced in order to take into account the scheduling result, the final graph is called Average Communications Graph (ACG). This graph exhibits how each type of processing resources communicates with the other types of processing resources. This graph is used during the projection step to enhance the spatial locality of communicating resources. In the ACG each edge represents the communications between two types of processing nodes.

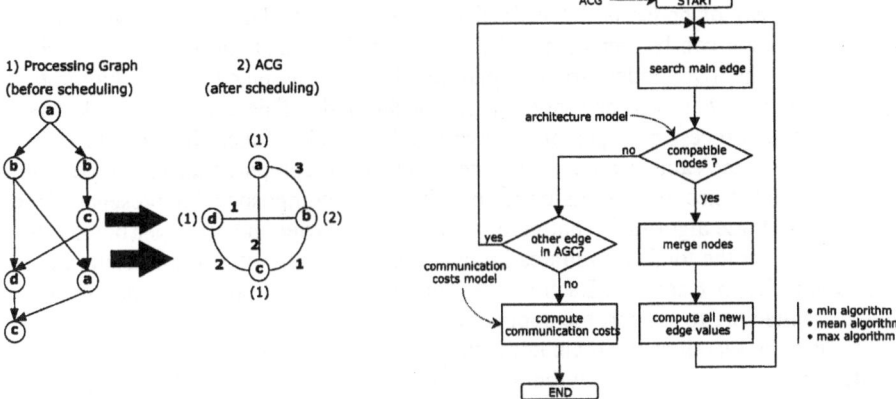

Fig. 3. Transformation of the processing graph (1) into ACG (2), and the generic projection algorithm

Figure 3 shows on a simple example how to transform a processing graph in an ACG. In the ACG, each node corresponds to a type of processing resources. In our example the type of the processing resources correspond to a letter (a, b, c or d). Several differences exist between the two graphs. The ACG is realized after the processing graph scheduling. There are fewer nodes in the ACG than in the processing graph, since the ACG has only one node for one type of processing resource. Its edges are not oriented, the communications are take into account in all directions. Several attributes are added in the ACG to better describe the communications inter-node.

The number in brackets beside a node is the number of operators that the scheduling has allocated (see section 4.3). The boldface number beside an edge is the total number of communications between two processing types. In order to know what pair of nodes communicates the most, it is necessary to compute the relative number of communications between two processing types. This value is obtained with the following expression:

$$RelativeComm_{Op1-Op2} = \frac{TotalComm_{Op1-Op2}}{NumberOp1 + NumberOp2}$$

Where $RelativeComm_{Op1-Op2}$ is the relative number of communications, $TotalComm_{Op1-Op2}$ is the total number of communications, $NumberOp1$ and $NumberOp2$ are the numbers of allocated operators of each type. It is very fast to build the ACG from the HCDFG of the application. The ACG is the input of the projection algorithms.

5.2 Generic Projection Algorithm

The projection step makes the link between the necessary (application) and the available (architecture) resources with the challenge that the most communicating resources must be assigned in a same hierarchical level of the architecture. Figure 3 shows the proposed algorithm. The algorithm begins with the ACG, and the first step searches in the graph the pair of nodes that communicates the most. This step searches the edge with the highest relative value.

When a pair of nodes is determined it is necessary to know if the two types of processing node can be implemented by functional elements in a same hierarchical element. If the two nodes are compatible, they are assigned to the hierarchical element, and they form a new node, a composite node. This composite node has as parameter the processing types of the two previous nodes. Since the ACG has a new node, it is not the same graph, so it is necessary to re-computed all this edge values and make all the necessary transformations due to the composite edge presence. To do this, we use three algorithms that give respectively a lower and an upper bound and a mean value for the total communication number. These algorithms are detailed in the paragraphs 5.3, 5.4 and 5.5. Each of them gives the total number of communications in the architecture to support the application.

If, after the search of the main edge, the pair of nodes is not compatible, the search re-starts with other nodes. If all the ACG edges have already been selected, and if any pair of nodes is compatible (it is not possible to implement it in functional elements in a same hierarchical element), then the projection algorithm stops.

To compute the communication costs a communication costs model that is based on the architecture hierarchy must be considered. A communication between two hierarchical elements does not represent the same cost than a communication in a single hierarchical element. More details are given in the section 5.6.

5.3 Min Algorithm (Lower Bound)

To illustrate the execution of the min algorithm, figure 4 shows with a very simple example the different steps of computation. The modifications of the architecture are presented on the right side of the figure. The architecture has two hierarchical levels (represented by two hierarchical elements H1 and H2). The hierarchical element H2 has three functional elements, one functional element can realize one multiplication and the two other functional elements can realize one addition or one subtraction (depends on the configuration). During the process the functional elements are progressively allocated (in grey). On the left side of the figure the modifications of the ACG during the process are detailed. At the beginning of the process, the ACG has no composite node. At each step one composite node is created by merging two nodes, and all the ACG edge values are re-compute to take into account the new composite node.

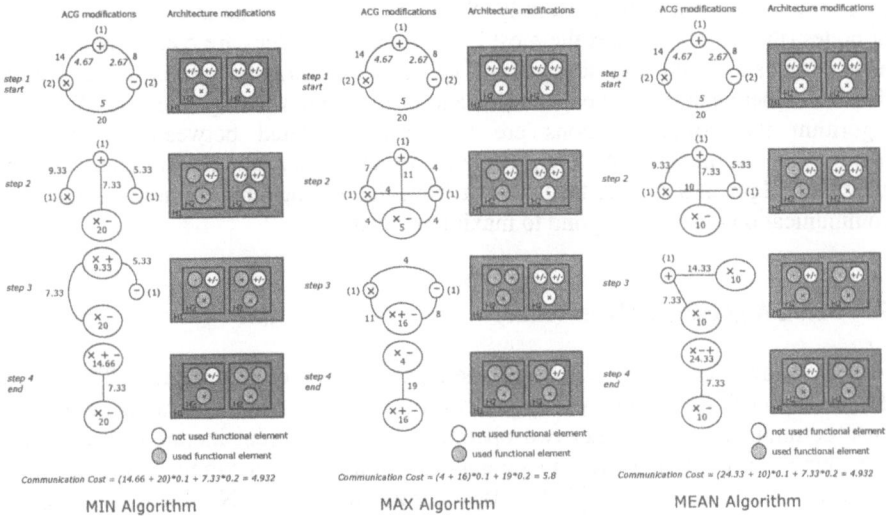

Fig. 4. The Min, Max and Mean algorithms.

The strategy of the min algorithm is to consider that if two operators of distinct processing types can be assigned in a same hierarchical element, they will perform all the communications between all the nodes of both processing types. It is the reason why in the start of the algorithm no edge is created between the new composite node (that describe a hierarchical realization) and one of the two nodes that communicate the most. For the example of the figure 4, the node multiplier and the node subtracter shape the pair of nodes that communicates the most (the relative value of the edge between this pair of node is the higher of the ACG). So in the second step of the algorithm, a composite node is created with two operations; one multiplication and one subtraction. The numbers (in brackets) of allocated operator multiplier and subtracter are decremented. The edge between the pair of nodes is deleted but this value (20) is transformed into internal communications of the new composite node.

However if another node has some communications (i.e. an edge) with one of the two previous nodes, then this node has an edge with the new composite node, and it preserves its other edges but with new values. It is the case of the node adder in the example of the figure 4.

The process stops when it is not possible to merge nodes. When the process ends, the communication cost is computed. In this example, the cost to communicate between two hierarchical elements H2 (in the H1 element) is 0.2, and the cost to communicate in a H2 element is 0.1. With this algorithm most communications are executed in composite nodes. As the communication costs in composite nodes are low (because there is not communication across several hierarchical levels) the total cost of communication will represent a lower bound.

5.4 Max Algorithm (Upper Bound)

As in the case of the min algorithm, figure 4 shows with the same example the different steps of computation. The strategy of the max algorithm is to consider that all operators communicate uniformly. When a composite node is created with a pair of nodes (that communicates the most in the ACG) two edges are created between the new composite node and the pair of nodes. The communications are uniformly distributed between the two edges value and inside the new composite node. With this algorithm the communications are uniformly executed between the different hierarchical elements, which corresponds to an upper bound. In that case the allocation algorithm do not take benefit from the architecture hierarchy, hence, the communication costs correspond to maximum costs.

5.5 Mean Algorithm (Mean Value)

Figure 4 shows for the same example the different steps of computation. The idea of the mean algorithm is to consider that two operators in a same hierarchical element must communicate more than two operators in two different hierarchical elements. The number of communication in a composite node, is the maximum communication value between two nodes. That is the reason why the MIN function is used to determine the internal communications in the new composite node.

We can see on the figure 4 that contrary to the min and max algorithms, with the mean algorithms, the two first created composite nodes are with one multiplier and one subtracter, that is most appropriated to this application. But as this example is very simple, when the communication costs are compute, the result is the same with the mean that with the min algorithm. But with a bigger application it is not the case as we can see in the section 6.

5.6 Communication Costs

On figure 3, the final step of the projection algorithm uses a communication costs model to compute the total communication cost. This model describes the costs to perform a data transfer in a hierarchical element and between several hierarchical elements. These costs depend on the routing topologies (mesh, segmented base, hierarchical, etc) and on the routing resources (channel of wires, bus, crossbar, etc). A survey of interconnects architectures for reconfigurable architecture is done in [15]. In our approach these costs are relative since we want to compare two architecture styles instead of giving an absolute performance value. However, if the designer has a good knowledge of the target architecture, precisely costs can be considered.

Once the communication costs are determined, it is necessary to compute the number of communication between each hierarchical element in function of the graph result (result of one of the three previous algorithms). The total communication cost is given by the following expression:

$$CommCost = \sum_i^n \left(\sum_j^{m_i} C_j \right)_i \times k_i$$

Where there are n hierarchical elements in the target architecture. The total number of communication in the element i is the sum of communication C_j inside each of the m_i hierarchical element i. The cost for one communication in the hierarchical element i is k_i. In this expression the fisrt sum with the i index allows to scan all the architecture hierarchical elements. The second index j allows to sum all the communication inside each hierarchical element i.

6 Application and Results

The one dimension Lee-DCT, a typical application of video compression [16], has been chosen to exhibit the exploration process and its ability to compare several architectures. Figure 5 presents the processing graph and the ACG for this application. The processing scheduling during the system estimation step has given four multipliers, four subtracters and three adders to perform this application. Four architectures have been targeted during the projection step, with different hierarchical levels, and cluster sizes (number of functional elements for a hierarchical element). Figure 5 depicts it.

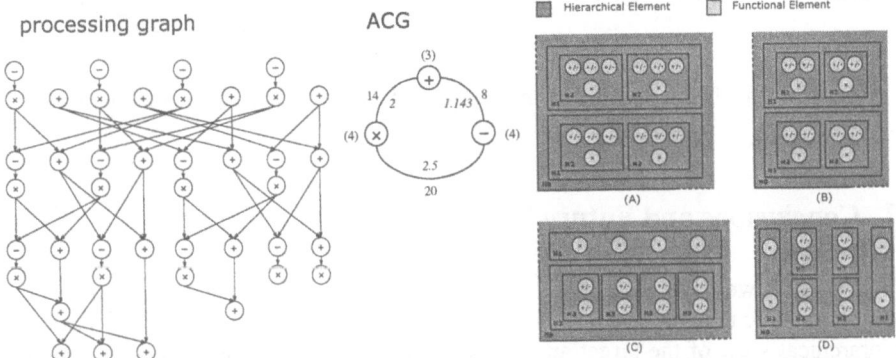

Fig. 5. Processing graph and ACG for the 1-D Lee DCT, and four examples of target architectures.

The results of the projection step is the total number of communications between resources weighted by the costs to communicate between hierarchical elements (communication costs model). For the considered architectures the communication costs model is the following (for the example of architecture A): Cost of communication internal H2 = 0.1, cost of communication H2 to H2 = 0.2, cost of communication H2 to H1 = 0.3

Results are given in figure 6. They highlight several interesting elements. Architectures **A** and **B** are better for this application than architectures **C** and **D**, since in the application the base structure is a MAC so it is necessary to have in a same hierarchical element the possibility to assign one multiplier, one adder and one subtracter.

We can notice that the results are close between the MIN and the MAX algorithm for the architecture **C** and **D**. Since these architectures have a limited exploration potential for this application, so the allocation will not have a strong impact on the final performances.

The last row of the table gives the utilization rate of functional elements in the architecture. Although the **A** architecture is better to reduce communication costs, it has not a high functional element utilization rate. The **A** architecture is larger than the other architectures, so its static power consumption is higher than the other architectures. With the utilization rate information, we can see that **B** architecture is a good tradeoff between communication cost reduction vs. functional element utilization rate.

In order to show the ability of our approach to make architecture exploration, we have computed the mean communication cost for architectures **A** and **B** with several numbers of hierarchical elements H2 in the hierarchical H1. We can see, on figure 6, that the larger is the cluster H1, the better is the communication cost. Indeed if H1 has many H2, local communications are promoted. This experiment show how is easy to explore some architecture characteristics.

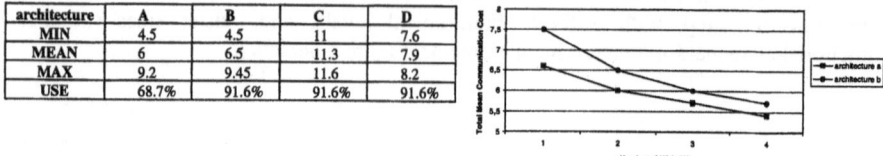

Fig. 6. Table of projection algorithms results, and curves of exploration of cluster size.

7 Conclusions and Future Work

In this paper we have presented a rapid allocation algorithm that takes into account the communications between application's operations. This algorithm uses the hierarchical view of the target architecture that is modeled with a functional model. Communication costs are taken into account with a communication cost model that depends on the routing structure in the target architecture. This algorithm is part of a design exploration flow that works at the algorithmic level and enables to compare quickly several reconfigurable architectures.

This approach which has been integrated in the codesign environment *Design Trotter* [8], enables to explore a large design space at an early stage of the design cycle and to characterize each solution in terms of area versus delay. Future work will complete the design exploration flow development and test it on a tile-based architecture [11].

References

[1] R. Hartenstein. A Decade of Reconfigurable Computing : a Visionary Retrospective. *In DATE'01, Munich, Germany, 13-16 March, 2001.*
[2] J. M. Rabaey. Reconfigurable Processing: The Solution to Low-Power Programmable DSP. *In IEEE, ICASSP'97, Munich, Germany, April 1997.*
[3] V. Betz, J. Rose, A. Marquart. *Architecture and CAD for Deep Submicron FPGAs.* Kluwer Academic Publishers, 1999.
[4] L. Lagadec, D. Lavenier, E. Fabiani, B. Pottier. Placing, Routing and Editing Virtual FPGAs. *In FPL'01, August 2001.*

[5]   U. Nageldinger. Coarse-Grained Reconfigurable Architecture Design Space exploration. *Ph.D. Thesis, University of Kaiserlautern, Germany, June 2001.*

[6]   C. A. Moritz, D. Yeung, A. Agarwal. Exploring Optimal Cost-Performance designs for Raw Microprocessors. *In FCCM'98, Napa, CA, USA, April 1998.*

[7]   L. Bossuet, G. Gogniat, J.P. Diguet, J.L. Philippe. A Modeling Method for Reconfigurable Architecture. *In IWSOC'02, Banff, Canada, July, 2002.*

[8]   Y. Le Moullec, J.P. Diguet, J.L. Philippe. Design Trotter: a Multimedia Embedded Systems Design Space Exploration Tool. *In IEEE MMSP'02, Virgin Island, USA, 9-11 December, 2002.*

[9]   Y. Le Moullec, N.Ben Amor, J.P. Diguet, M. Abid, J.L. Philippe. Multi-Granularity Metrics for the Era of Strongly Personalized SOCs. *In DATE 03, Munich, Germany, 3-7 March, 2003.*

[10] J. P. Diguet, G. Gogniat, P. Danielo , M. Auguin, J.L Philippe. The SPF model. *In FDL'00, Tübingen, Germany, September, 2000.*

[11] L. Bossuet, W. Burleson, G. Gogniat, V. Anand, A. Laffely, J.L. Philippe. Targeting Tiled Architectures in Design Exploration. *RAW'03, Nice, France, April, 2003.*

[12] M. Budiu, S. C. Goldstein. Fast Compilation for PipeRench Reconfigurable fabrics. *In ACM/SIGDA FPGA'99, Monterey, CA, USA, February, 1999.*

[13] L. Shang, A. Kaviani, K. Bahala. Dynamic Power in VirtexTM-II FPGA Family. *In ACM/SIGDA FPGA'02, Monterey, Californie, USA, February, 2002.*

[14] A. Garcia, W. Burleson, J.L. Danger. Power Modeling in Field Programmable Gate Arrays (FPGA). *In FPL'99, Glasgow, Scotland, September 1999.*

[15] H. Zhang, M. Wan, V. George, J. Rabaey. Interconnect Architecture Exploration for Low-Energy Reconfigurable Single-Chip DSPs. *IEEE Computer Society Workshop on VLSI, April 1999.*

[16] V. Bhaskaran, K. Konstantinides. *Image and Video Compression Standards : Algorithms and Architectures.* Kluwer Academic Publishers, 1995.

From Algorithm Graph Specification to Automatic Synthesis of FPGA Circuit: a Seamless Flow of Graphs Transformations

Linda Kaouane[1], Mohamed Akil[1], Yves Sorel[2], and Thierry Grandpierre[1]

[1] Groupe ESIEE-Laboratoire A2SI, BP 99 – 93162 Noisy-le-Grand, France
{kaouanel,akilm,grandpit}@esiee.fr
[2] INRIA Rocquencourt-OSTRE, BP 105 – 78153 Le Chesnay Cedex, France
yves.sorel@inria.fr

Abstract. The control, signal and image processing applications are complex in terms of algorithms, hardware architectures and real-time/embedded constraints. System level CAD softwares are then useful to help the designer for prototyping and optimizing such applications. These tools are oftently based on design flow methodologies. This paper presents a seamless design flow which transforms a data dependence graph specifying the application into an implementation graph containing both data and control paths. The proposed approach follows a set of rules based on the RTL model and on mechanisms of synchronized data transfers in order to transform automatically the initial algorithmic graph into the implementation graph. This transformation flow is part of the extension of our AAA (Algorithm-Architecture Adequation) rapid prototyping methodology to support the optimized implementation of real-time applications on reconfigurable circuits. It has been implemented in SynDEx[1], a system level CAD software tool that supports AAA.

1 Introduction

The increasing complexity of signal, image and control processing algorithms in embedded applications requires high computational power to meet real-time constraints. This power can be achieved by high performance mixed hardware architectures built from different types of programmed components (RISC or CISC processors, DSP,..) to perform high level tasks and/or specific components (dedicated boards, ASIC, FPGA,...) used to perform efficiently low level tasks such as signal and image processing and devices control. Implementing these complex algorithms on such distributed and heterogenous architectures while verifying the severe real-time constraints is generally a difficult and complex task. This explain the need for dedicated high level design environnement based on efficient system-level design methodology to help the real-time application designer to solve the specification, validation and synthesis problems [1].

---

[1] http://www-rocq.inria.fr/syndex

P.Y.K. Cheung et al. (Eds.): FPL 2003, LNCS 2778, pp. 934–943, 2003.

In order to cope with these increasing needs, we have developped the AAA rapid prototyping methodology [2] wich helps the real-time application designer to obtain rapidly an efficient implementation of his application algorithm on his heterogenous multiprocessor architecture and to generate automatically the corresponding distributed executive. This methodology is based on an unified model of factorized graphs [3], as well to modelize the applicative algorithm and the multicomponent architecture, than to deduce the possible implementations in terms of graphs transformations.

Based on this model, we have extended the AAA methodology to support the implementation of real-time applications on reconfigurable circuits. This extension uses a single factorized graph model, from the algorithm specification down to the architecture implementation, through optimizations expressed in terms of defactorization transformations applied to the algorithmic graph [4]. This optimization aims to satisfy the real-time constraints while minimizing the required hardware resources. In prospect, this extension is expected to allow the AAA methodology to be used for optimized hardware/software codesign.

In this paper, we focus on the rules used to synthesize both the data and the control paths of the circuit corresponding to an algorithm specified as a factorized data dependence graph. It is known that control path synthesis is more difficult to carry out than data path synthesis. We show here that it is possible to synthesize the control path in a secure and systematic way by using a technique of data transfers synchronization based on the RTL model. This approach allows us to carry out an automatic generator of synthesizable VHDL in a simple way. The remainder of the paper is organized as follows: in the next section, we briefly present the transformation flow used by our extented methodology to automate the hardware implementation process of an application algorithm on reconfigurable circuits. In section 3, we present the factorized data dependence graph model used to specify the application algorithm. As critical portions of control, signal and image processing algorithms often consist of regular computations generally expressed as nested loop, we will use a motivating example of matrix-vector product to illustrate the proposed transformation design flow. Section 4 gives rules to automate the synthesis of data and control paths extracted from the algorithm specification while the principles of optimization by defactorization are shown in section 5. We also show in section 6, the results of the implementation of the matrix-vector product algorithm onto a *Xilinx* FPGA following these rules. Finally, we conclude and discuss future work in section 7.

2   AAA Methodology for Circuits

Given an algorithm graph $G_{al} = (O, D)$ specifying the application, we transform it to an implementation graph $G_{im} = (O", D^*)$ following a set of graphs transformations as described in Fig.1. This seamless transformation flow is composed of the generation of the data-path graph $G_{dp} = (O"_1, D"_1)$ and the control-path graph $G_{cp} = (O"_2, D"_2)$ $(G_{im} = G_{dp} \cup G_{cp})$. Data-path transformations are quite simple, but control-path transformations are not trivial and require

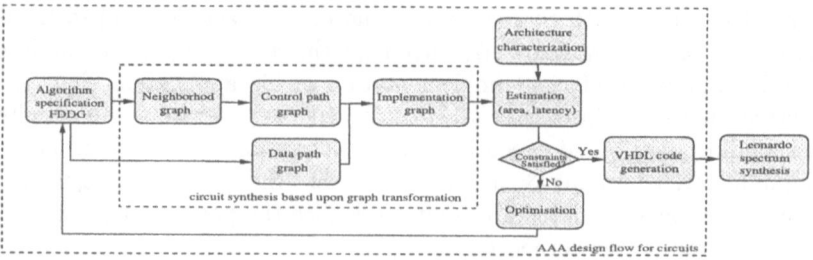

Fig. 1. The AAA methodology for circuits

to build first a neighborhod graph $G_{ng} = (O', D')$. Finally the implementation graph containing both data and control graphs is then charaterized in order to estimate time and area performance. If the deduced implementation does not meet the user specified constraints, we apply a defactorization process in order to reduce the latency by increasing the hardware resources. Since there is a large number of possible defactorized implementations with different characteristics (FPGA area required, latency,..) among which we need to select the most efficient one, we need to use heuristics guided by their cost function. Finally, the VHDL code corresponding to the optimized FPGA implementation is generated.

3   Algorithm Model

The algorithm specification is the starting point of the process of hardware implementation of an algorithm application onto an architecture. According to the AAA methodology, the algorithm model is an extention of the directed data dependence graph (direct acyclic hypergraph DAG), where each node models an operation (more or less complex, e.g. an addition or a filter), and each oriented hyperedge models a data, produced as output of a node, and used as input of an other node or several other nodes (data diffusion). The extended model provides specification of loops through factorization nodes (fork, join, iterate, diffuse), leading to an algorithm model called Factorized Data Dependence Graph (FDDG) [3]. This algorithm graph may be specified directly by the user or it may be generated from high level specification languages such as the synchronous languages (Esterel, Signal,...), which perform formal verifications in terms of events ordering in order to prevent dead-locks [5].

3.1   Factorized Data Dependence Graphs Model

As described in [3], an algorithm specification contains regular parts (repetitive subgraph) and non-regular parts. In fact, these spatial repetitions of operation patterns (identical operations that operate on different data) are usually reduced by a factorization process to reduce the size of the specification and to highlight its regular parts. Graph factorization consists in replacing a repeated pattern,

- slow-downstream: "slow" side of a consumer FF;
- fast-upstream: "fast" side of a producer FF;
- fast-downstream: "fast" side of a consumer FF;
- slow-upstream : "slow" side of a producer FF.

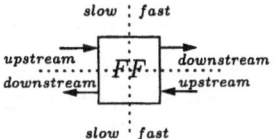

Fig. 2. Node of neighborhood graph for a frontier FF

i.e. a subgraph, by only one instance of the pattern, and in marking each edge crossing the pattern frontier with a special "factorization" node, and the factorization frontier itself by a dashed line crossing these nodes. The type of the factorization node depends on the way the data are managed when crossing a factorization frontier, it may be: a Fork node 'F' (array partition in as many subarrays as repetitions of the pattern), a Join node 'J' (array composition from results of each repetition of the pattern), a Diffusion node 'D' (diffusion of a data to all repetitions of the pattern) or an Iterate node 'I' (data dependence between iterations of the pattern).

Note that from the algorithm specification point of view, the factorization reduces only the size of the specification, without any modification of its semantics. However, from the implementation point of view, the factorization describes also in intention all the possible implementations, from the entirely parallel one to the entirely sequential one, with all the intermediate cases mixing both sequential and parallel. Obviously, each of these implementation will have different characteristics in terms of area and response time.

3.2   Neighborhood Graph

Every factorization frontier may be a consumer (located downstream) or/and a producer (located upstream) relatively to another frontier according to the data dependences relating them. Two frontiers are neighbors if there is at least one relation of direct dependence that does not cross a third frontier.

Based on these neighborhood relations between the factorization frontiers, we build a neighborhood graph denoted $G_{ng} = (O', D')$. The nodes $o'_{F_i} \in O'$ of such graph represent the factorization frontiers and the oriented edges $d'_i = (o'_{F_i}, o'_{F_j}) \in D'$ represent the data flow between factorization frontiers. The edge orientation describes the consumption/production relation: an edge starts at a producer (o'_{F_i}) and ends at a consumer (o'_{F_j}).

In the case of a sequential implementation, every factorization frontier, called FF, separates two regions, the first one called "fast", being repeated relatively to the second one, called "slow". These slow and fast sides of a frontier are due to the difference of data rate on each side of the factorization frontier. Every node of the neighborhood graph is then subdivided in four parts (see Fig.2).

This neighborhood graph, deduced automatically from the algorithm graph (FDDG), is used during the implementation in order to establish the control relationships between frontiers.

3.3   Example: Specification of MVP (Matrix-Vector Product)

We now use a Matrix-Vector Product example (MVP) to illustrate the algorithm model of specification and its use for the building of the neighborhood graph. So the MVP of one matrix $A \in R^m \times R^n$ by a vector $B \in R^n$ gives a vector $C \in R^m$, and can be written in a factorized form as follows: $C = \left[\sum_{j=1}^{n} a_{ij} b_j \right]_{i=1}^{m}$

This equation allows us to obtain the graph corresponding to the algorithm specification of the factorized MVP (Fig. 3). The interface with the physical environment is delimited by input (F_A^∞ et F_B^∞) and by output (J_C^∞). It corresponds to the factorization frontier of the infinitely repeated pattern of the graph (FF_1) since we deal with reactive applications that interact infinitely with the physical environement. The square brackets $[\;]_{i=1}^{m}$ correspond to a second frontier (FF_2), delimited by factorization nodes of a finitely repeated pattern. This frontier selects the m lines of the matrix A (F_{21}), diffuses the vector B (D_{21}) and collects the result vector C (J_{21}). The functor $\sum_{j=1}^{n}$ corresponds to a third frontier (FF_3), also delimited by factorization nodes of a second finitely repeated pattern. This frontier selects the a_{ij} elements of the ith line of the matrix A (F_{31}) and the elements b_j of the vector B (F_{32}) and it supplies the result of the sum of products between a_{ij} et b_j for every line of matrix A (I_{31}). The "slow" and "fast" sides of each frontier are labeled "s" and "f", respectively.

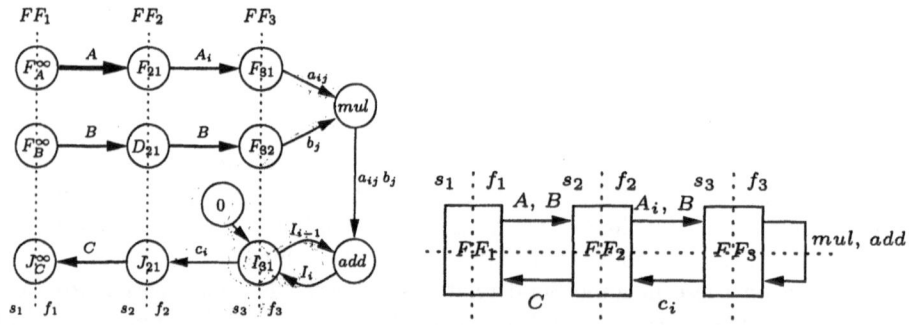

Fig. 3. Factorized data dependence graph of MVP

Fig. 4. Neighborhood graph of MVP: relations between frontiers

The neighborhood graph between factorization frontiers, obtained from the factorized data dependence graph specifying the MVP algorithm, is shown in Fig.4. Because the factorization frontier FF_1 is infinite, it does not have neighbor on its "slow" side which corresponds to the physical environment. FF_1 is, at the same time, a producer (edges A and B) and a consumer (edge C) compared to FF_2. FF_2 is also a producer (edges A_i and B) and a consumer (edge C_i) compared to FF_3. FF_3 is a producer and a consumer, compared to itself through the arithmetic operations mul and add.

4   Circuit Synthesis from a Neighborhood Graph

High-level circuit synthesis transforms a high-level behavioral specification into a register-transfert-level implementation (RTL) [6]. The resulting RTL design containing both the data path and control path can then be synthesized using logic synthesis tools that map components such as adders, multipiers,... to gates, perform optimization and finally generate the netlist of the final design. The automation of this synthesis process reduces significantly the development cycle of the circuit, and allows the exploration of different hardware implementations, seeking for an efficient compromise between area and response time of the circuit. Afterward we will present the rules used to automatically generate the data path (denoted by G_{dp}) and the control path (denoted by G_{cp}) of the circuit, from the factorized data dependence graph G_{al} and the neighborhood graph G_{ng}.

4.1   Data Path Synthesis

The hardware implementation of operations consists in providing for every node ($o_i \in O$) of the algorithm graph $G_{al} = (O, D)$ a matching operator ($o"_{op_i} \in O"_1$) (by instanciating the corresponding component of a VHDL library). The matching operator is a logic function in the case of an operation node, or it is composed of a multiplexer and/or registers in the case of a factorization node (F is implemented by a multiplexer, J by a demultiplexer with a memory array, I by a register with initialization value,..). The hardware implementation of the data dependences between operations consists in providing, for each edge ($d_i \in D$) of the G_{al}, a matching connection ($d"_i \in D"_1$) between the corresponding operators. The resulting graph G_{dp} of operators ($O"_1$) and their interconnections ($D"_1$) composes the data path of the circuit.

4.2   Control Path Synthesis

The control path corresponds to the logic functions that must be added to the data path, in order to control the multiplexers and the transitions of the registers composing the factorization operators. It is then obtained by data transfer synchronization between registers. However, two conditions must be satisfied to allow a register to change state: the new upstream data to the register must be stable, and all downstream consumers of the register must have finished the use of previous data. If moreover upstream data of a circuit comes from various producers with different propagation time, it is necessary to have a synchronized circuit. This synchronization is possible through the use of a request/acknowledge communication protocol. Consequently, the synchronization of the circuit implementing the algorithm is reduced to the synchronization of the request/acknowledge signals of the set of factorization operators.

These operators are gathered in factorization frontier and their data consumption and production are done in a synchronous way at the level of the frontier. We propose then a control system where each factorization frontier will have its own control unit. This delocalized control approach allows the CAD tools used for the synthesis to place the control units closer to the operators to control rather then a centralized control approach.

Control Unit: As mentioned above, each factorization frontier has upstream and downstream relations on both sides, "slow" and "fast". The relations between upstream/downstream and request/acknowledge signals on both sides of a frontier are implemented by the "control unit" of the factorization frontier (Fig.5). This control unit contains a counter C with d states (corresponding to the d factorized repetitions) and an additional logic function in order to generate, in the one hand the communication protocol between frontiers (the slow and fast, request and acknowledge signals at the upstream and downstream sides), and in the other hand the counter value cnt and the enable signal (en), that control the frontier operators. The counter value (cnt) controls the frontier operators: F, J and I. The enable signal (en) determines the clock cycles where the registers of the frontier operators (J, I, F^∞"sensor", J^∞"actuator") will change state. Note that, the signal ($init$) resets the counter while the signal (end) indicates that the counter is in its last state ($d-1$).

All the other signals are the request (r) and acknowledge (a) signals generated by the frontier(s) located upstream or diffused to the frontier(s) located downstream. They are separated in two groups: those which relate to the frontier(s) located on the "slow" side and those which relate to the frontier(s) located on the "fast" side, corresponding to the four parts of the control unit: slow-upstream (su), slow-downstream (sd), fast-upstream (fu) and fast-downstream (fd).

Thus, the control path, modeled by the graph $G_{cp} = (O"_2, D"_2)$, is mainly composed of the set of control units associated to the corresponding factorization frontiers. These control units are then inter-connected in an automatic way based on relationships between the factorization frontiers deduced from the neighborhood graph. In this control path graph, the nodes $o"_{UC_i} \in O''_2$ correspond to the control units and the edges $d^{rd-ru}_{UC_i,UC_j} \in D''_2$ correspond to the request signals transmitted between the control units. The acknowledge signals are transmitted, in the opposite direction of the associated request signals, between the same control units. When several signals occur at the same input of a control unit, the conjunction is performed by a logical gate AND.

5 Implementation Optimization: Principles

As previouly mentioned, the optimized implementation of a factorized algorithm graph onto an application specific integrated circuit or a FPGA, is formalized in terms of graph transformations, i.e defactorization. When we defactorize a graph we expect to reduce the latency by increasing the number of hardware resources. Thus, the implementation space, which must be explored in order to find the best solution, is composed of all the possible defactorizations of a factorized algorithm graph. For instance, for a given algorithm graph with n frontiers, we have at least 2^n defactorized implementations. Moreover, each frontier can be partially defactorized: a factorization frontier of r repetitions can be decomposed in f factorization frontiers of r/f repetitions. Consequently, for a given algorithm graph, there is a large, but finite, number of possible implementations which are more or less defactorized, and among which we need to select the most efficient one, i.e which satisfies real-time constraints (upper bound on latency), and which uses as less as possible the hardware resources, logic gates for ASIC and number of Configurable Logic Blocks CLB for FPGA. This optimization problem is known to be NP-hard, and its size is usually huge for realistic applications. This is why we use heuristics guided by their cost function, in order to compare the performances of different defactorizations of the specification. These heuristics, using tricks related to practice, allows us to explore only a small subset of all the possible defactorizations into the implementation space. These cost functions take into account the characteristics of an implementation: hardware resources required (number of gates or CLBs) and latency [3].

6   Example: Synthesis of MVP Implementation on FPGA's Circuit

The Fig.6 represents the hardware implementation of the factorized MVP corresponding to the algorithm specification given in Fig.3. The data path is composed of the factorization frontier operators ($F_{i,j}$, $D_{i,j}$, $J_{i,j}$ and $I_{i,j}$) and of the combinatorial operators mul and add delimited by a dotted box. The control path is composed of the control units UC_1, UC_2 and UC_3, and of the control signals r (request), a (acknowldge), cnt and en. The interconnections between the request and acknowledge signals, is based on the relationships between the factorization frontiers, namely the neighborhood graph (Fig.4) built from the algorithm graph.

In Fig.7 we present the hardware implementation of a defactorized solution corresponding to the partial defactorization of the frontier FF_2 by a factor of 2. The $FF2$ frontier has been replaced by two frontiers FF_{2a}, FF_{2b} being repeated $m/2$ times (m : factor of repetitions of FF_2). The factorization frontier FF_3 remain inchanged but it has been duplicated (FF_{3a}, FF_{3b}) due to the partial defactorization of FF_2. The data path is then composed of the factorization frontier operators, the combinatorial operators delimited by dotted boxes and of the operators X_1 (array-decomposition operation), M_1 (array-composition operation). The control path is composed of the control units UC_1, UC_{2a}, UC_{2b}, UC_{3a} and UC_{3b}, The synchronisation of frontiers FF_{2a}, FF_{2b} is assured by the AND gate at the upstream request and the downstream acknowledge of UC_1.

Tab.8 shows the synthesis result of the generated VHDL code of hardware implementation of MVP (6 × 6 matrix and 6 elements vector, coded on 3 bits) onto a *Xilinx* FPGA XL4000XL-3 4005xIPC84, using the CAD tool *Leonardo Spectrum 2003*, developed by *Exemplar Logic Inc.*. The implementation results are presented in function of, the area (hardware ressources: number of CLBs), the number of clock cycles required by the algorithm execution, the maximum frequency of operators in *MHz*, and finally the latency in *ns*.

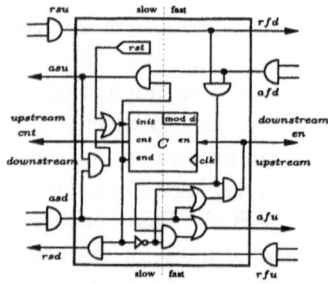

Fig. 5. Control unit

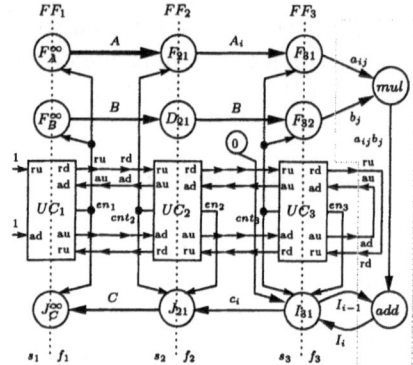

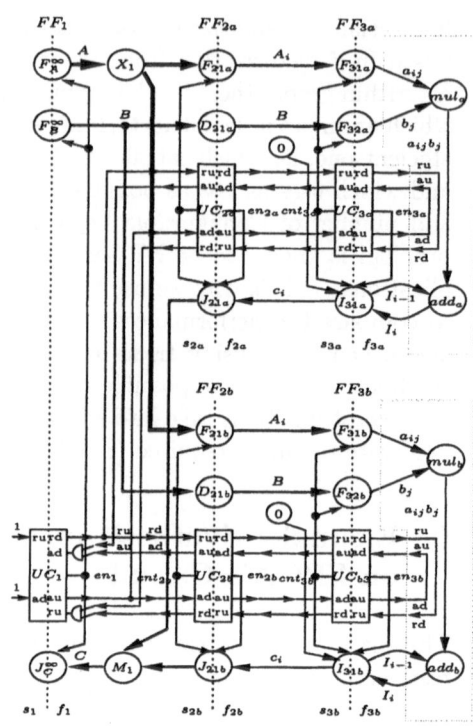

Fig. 6. Implementation graph of MVP

Fig. 7. A defactorized implementation graph of MVP

7 Conclusion and Future Work

We have shown that from an algorithm specification based on a factorized data dependence graph model it is possible to generate automatically a hardware implementation onto an FPGA circuit, employing a set of rules for the data path and the control path synthesis. The delocalized control approach presented in this paper allows the CAD tools used for the synthesis to place the control units closer to the operators to control rather then a centralized control approach. We validated this seamless graph based design flow on several examples of low-level image processing applications that includes interesting cases of data factorization like: mean filtering [4], edge detector operators (sobel, deriche [7], ...).

This work is part of the extension of the AAA methodology implemented in the software tool SynDEx to support implementation onto reconfigurable circuits. Basically, AAA/SynDEx for multiprocessors, allows to generate automatically the dead-lock free executive for the optimized implementation of the given algorithm onto the specified multiprocessor architecture [8].

The principles described in this paper allowed us to carry out an automatic generator of structural synthesizable VHDL for mono-FPGA (one FPGA) architectures, that has been added to SynDEx [9]. The generated VHDL code which

Implementation	Area (CLB)	Nb. cycl.	Freq. (MHz)	Lat. (ns)
Factorized Spec.	76	36	12,4	2916
Part.defac. by FF_2	99	18	13,5	1332
Fully. defac. by FF_2	168	6	14,3	420
Part. defac. by FF_3	92	30	10,8	2790
Fully. defac. by FF_3	79	6	9,0	660
Fully. defactorized	234	1	11,4	87

These results represent some possibles implementations explored by the optimization heuristic by partial defactorization (as described in[4]) of the initial factorized implementation. Note that these defactorized solutions allow to reduce the latency of the implementation, but they increase the number of required hardware resources (CLB).

Fig. 8. Optimization results for the implementation of MVP onto FPGA

corresponds to the optimized FPGA implementation obtained by successive defactorizations of the factorized algorithm graph, is then used by a CAD tool in order to generate the netlist needed for the FPGA configuration.

Currently we are working on the control involved by the conditioning in the algorithm specification, in addition to the control involved by repetition of operation. After that we plan to extend the proposed methodology to the case of multi-FPGAs architectures. To support such architectures, the optimization heuristic will adress both defactorization and partitioning issues.

References

1. S. Edwards, L. Lavagno, E.A. Lee, A. Sangiovanni-Vincentelli. *Design of embedded systems: formal models, validation, and synthesis.* Proc. of IEEE, v.85, n.3, March 1997.
2. T. Grandpierre, C. Lavarenne, Y. Sorel. *Optimized rapid prototyping for real-time embedded heterogeneous multiprocessors.* CODES'99 7th Intl. Workshop on Hardware/Software Co-Design, Rome, May 1999.
3. L.Kaouane, M. Akil, Y. Sorel, T. Grandpierre. *A methodology to implement real-time applications on reconfigurable circuits.* ERSA'03, Intl. Conference on Engineering of Reconfigurable Systems and Algorithms, Las vegas, USA, June 2003.
4. A. F. Dias, C. Lavarenne, M. Akil, Y. Sorel. *Optimized implementation of real-time image processing algorithms on field programmable gate arrays.* Proc. of the 4th Intl. Conference on Signal Processing, Beijing, Oct. 1998.
5. N. Halbwachs. *Synchronous programing of reactive systems.* Kluwer Academic Publishers, Dordrecht Boston, 1993.
6. D. Gajski, F. Vahid. *Specification and design of embedded hardware-software systems.* IEEE Design & Test of Computers, Spring 1995, p. 53-67.
7. L.Kaouane, M. Akil, Y. Sorel, T. Grandpierre. *An automated design flow for optimized implementation of real-time image processing applications on FPGA.* Eurocon'03, Intl. Conference on computer as a tool, Ljubljana, Slovenia, sept. 2003.
8. T. Grandpierre, Y. Sorel. *From algorithm and architecture specifications to automatic generation of distributed real-time executives: a seamless flow of graphs transformations.* MEMOCODE03, Intl. Conference on Formal Methods and Models for Codesign, Mont Saint-Michel, France, June 2003.
9. R. Vodisek, M. Akil, S.Gailhard, A.Zemva. *Automatic Generation of VHDL code for SynDEx v6 software.* Electro technical and Computer Science conference, Portoroz, Slovenia, september 2001.

Adaptive Real-Time Systems and the FPAA

Stuart Colsell and Reuben Edwards

Department of Communication Systems, Lancaster University,
Lancaster, Lancashire, U.K., LA1 4YR
s.colsell@lancaster.ac.uk, r.edwards@lancaster.ac.uk

Abstract. Adaptive real-time systems are often typically considered by design engineers to be a solely digital preserve. However, in recent years analogue technologies such as the Field programmable Analogue Array (FPAA) have been developed to support real-time reconfiguration. These technologies can offer an analogue platform for a vast number of applications requiring adaptive processing and form the subject of this paper. Although in terms of their technology and software/algorithmic support, FPAAs are still in the relative infancy compared with digital techniques, we aim to show their advantages and how present devices can be exploited in real-time applications.

1 Introduction

The application of adaptable devices has migrated from the prototyping stages of design, into the final commercial architectures of real-time systems. This is particularly so, within the field of communications where flexible systems such as Software Defined Radio (SDR), require the ability to support of multiple-standards and protocols. Modern design techniques and the demand for fast time-to-market products have driven the development of reconfigurable devices. These devices can, and have been used in very diverse products, both in the development stage and within the finished article. Until recently these flexible devices have materialized predominately within the digital domain. But now analogue circuitry has achieved the adaptability required, by using FPAAs. This paper discusses the challenges faced by both FPAA device designers and end users.

2 Real-Time Requirements of FPAA-Based Systems

There are a number of real-time applications, which would benefit greatly from the flexibility of FPAA devices. The key application for this research is the use of these devices within real-time communication systems. Such systems require that; any reconfiguration of its current processing architecture should be carried out with minimum disruption to the data flow. This is very challenging as the buffering of data is not always an option. In order to meet the system performance criteria the reconfiguration of the devices needs to be as fast as possible with minimal degradation to the implementation. Often the algorithms used to ensure a high

P.Y.K. Cheung et al. (Eds.): FPL 2003, LNCS 2778, pp. 944–947, 2003.

performance implementation can hinder the speed in which the device can be configured. This presents an argument for more application specific architectures in order to minimize the analysis of place and route options, thus moving toward tuning a circuit to suit the new requirements rather than having an architecture that can be completely changed. Another alternative would be to use parallel circuits within a single device using redundant resources. This method would ensure that the replacement circuit could be placed and routed carefully before removing the previous circuit from the data path.

Current and Chu have presented a new maintenance approach for increasing the reliability of analogue ICs by the direct self-calibration and built-in self-test of off line replacement sub-circuits for an analogue function block, and the commutation, switching, of replacement sub-circuits into the signal path without disruption of the analogue signal path or any loss of analogue functionality. The commutation scheme connects the replacement sub-circuit in parallel with the active sub-circuit to be replaced before it is removed. The new commutation control signal switching edge is carefully designed to minimize "clock" feed-through noise. The strategy for on-chip analogue function maintenance is to create analogue functions with adjustable figures of merit, test and adjust (calibrate) redundant functional blocks while they are off-line, and replace on-line active functional blocks with calibrated redundant blocks when desired. Steps required in the use of the strategy are to select an analogue function that is to be maintained on-chip, define the figures of merit that indicate acceptable performance, coalesce figure-of-merit-adjustment circuitry with the analogue function, create the necessary on-chip excitation and measurement circuitry, create circuitry for delivery off-chip of the analogue test results of digital indications of the test results [1].

The use of parallel devices to support seamless reconfiguration may seam like an expensive approach in terms of power consumption, PCB real estate, and cost. But in fact, when using a filter implementation to compare analogue and digital reconfigurable solutions [2], using two analogue devices consumes less power, PCB real estate, and costs less than using one FPGA. Often with smaller FPGAs only one filter can be implemented on a device. So we have a trade-off between the digitals implementation accuracy and software support and the analogue's power efficiency, smaller size, reduced cost and lack of latency. The question is, can the analogue devices provide acceptable accuracy to out weigh the digital advantage? This will be explored in the following section.

3   Circuit Performance Repeatability

It is important to consider the performance repeatability of circuits based within reconfigurable technologies. When a designer complies a circuit with a performance criterion, he or she requires that it be maintained each time the circuit is downloaded or transferred from one device to another. This is particularly important when considering multipath systems because they rely on the real and complex paths being very closely matched. This is particularly important when considering matched filters within I and Q paths of a communication receiver.

Matched filter imperfections can have a detrimental effect on the recovery of the wanted signal. Imperfections materialize in the form of Amplitude and phase error.

Amplitude error is the main cause of signal mirroring errors in the passband and it is caused solely by resistor mismatch. Phase errors however occur at the edges of the passband and are caused by both resistor and capacitor mismatch.

If FPAA technology is to be used in complex multipath systems, it is important to investigate how well it can support the filtering requirements. For the purpose of this experiment a 9 KHz channel filter design was chosen to demonstrate the ability of an FPAA to produce matched filters suitable for an Intermediate Frequency stage of a DRM or HF receiver. The filter design was implemented twice within a single switched capacitor based Anadigm AN10E40 FPAA.

4   Results

The matched filters within the FPAA were tested using both a frequency sweep from a spectrum analyser. In this section we will show the result of the measurements. A frequency sweep was made first to obtain the response of one of the filters and compare it to the noise floor. The plot from this test can be seen in figure 1. It is clear from this plot that there is some non-linearity in the response of the filter; this however, can be corrected with more accurate impedance matching. The plot also shows that the filter has a dynamic range of approximately 25dB in the passband. This is due to gain compression and the high noise floor. This high level of noise can be partially attributed to the chips clock signal and internal switching. This is partly because the clock input pin is positioned next to the analogue output pins of the device. The clock problem and the internal noise have been addressed in the new Anadigm Vortex family of devices.

The frequency response of both filters was measured and can be seen in figure 2. The plot shows a very slight amplitude mismatch when studied closely. This however can be removed easily using an Active Gain Control algorithm within the DSP section of a communications receiver. Using an oscilloscope, an attempt was made to observe any phase mismatch between the two filters but it was too slight to view and record.

5   Conclusions

Currently FPAA technologies can be used for a wide range of applications including instrumentation, control systems, analogue baseband and Low/Medium frequency transceiver front-end circuitry. The most recent field programmable analogue devices have shown great improvements in their ability to support more practical applications. This is most apparent when considering the 400% increase in bandwidth over the last six years. However it would be unrealistic to expect such a high rate of improvement to continue unless a routing scheme is developed, which introduces a minimal amount of performance degrading effects. We can see from the results of the circuit performance repeatability experiment that when using switched capacitor based FPAAs the matching of circuit performance is quite good and can be deemed acceptable for communication systems such as DRM and HF.

FPAA technologies can provide adaptive processing without compromising battery life and portability. It is clear that further development is required especially to the software support/design tools and speed of configuration. Comparisons based on circuit complexity, chip size, power consumption and latency show that reconfigurable analogue can be considered a very efficient alternative solution to digital devices such as the FPGA. This is particularly apparent when considering operations such as filtering [2] within power critical mobile applications.

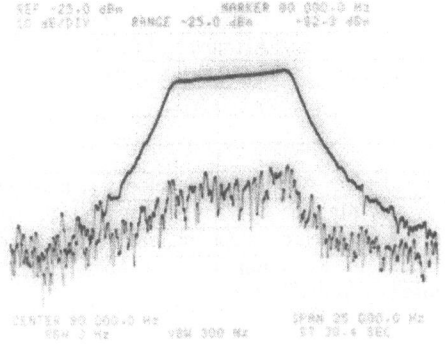

Fig. 1. Filter frequency response & noise floor.

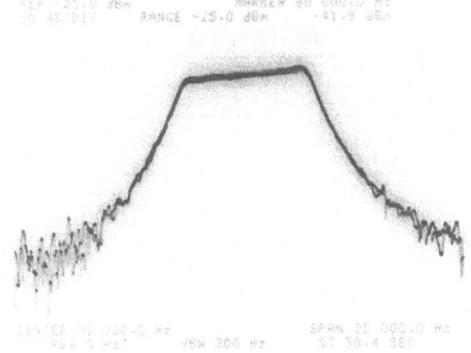

Fig. 2. Overlaid frequency resonses of I & Q filters.

References

1. Current, K. W. & Chu, W. 2001, Demonstration of an Analogue IC function maintenance strategy, including direct calibration, built-in self-test, and commutation of redundant functional blocks, Analogue integrated circuits and signal processing, No.26, 129-140.
2. Colsell, S. A. & Edwards, R. 2002, FPAD Versus FPAA For Future Mobile Communications, *Proc.* 9th International workshop on systems, signals and image processing (IWSSIP), Manchester, UK, pp. 181-185.

Challenges and Successes
in Space Based Reconfigurable Computing

Mark E. Dunham, Michael P. Caffrey, and Paul S. Graham

Los Alamos National Laboratory
Los Alamos, NM 87545
mdunham@lanl.gov, mpc@lanl.gov, grahamp@lanl.gov

Abstract. For 6 years Los Alamos has developed space-compatible versions of Reconfigurable Computing (RCC). Several such designs are now operational. We describe the key research steps required to make commercial silicon processes amenable to Radiation Tolerant operations, and the limits on algorithms imposed by reliability concerns.

1 RadHard-by-System-Design for an FPGA

Over the last 6 years a new means of using near-commercial silicon processes for Radiation Tolerant applications in space has emerged[1], which we call Radiation Hardness by System Design (RHSD). This method allows the identical mask sets used for commercial products to be used in creating Radiation Tolerant FPGA versions on epitaxial or SOI substrates, thereby removing the enormous technology lags now seen in space processing ASICs, while offering useful total dose levels (TLD) and latch-up immunity (SEL). Single event upsets due to sporadic radiation must still be tolerated however; which is the subject of this FPGA research work.

Unlike hardware approaches to mitigating upsets, our techniques do not attempt to make FPGA circuits 100 % tolerant of SEUs, but instead allow them to operate through SEU without requiring system resets. While a number of techniques are used to achieve high SEU reliability, our primary means of mitigation depends on being able to read back the state of the FPGA continuously, which is a Xilinx feature. Heavy-ion testing has shown that the Xilinx XQR300 average saturation cross-section per bit is 8e-8 cm^2 and that the cross-section measured for the Single Event Functional Interrupts (SEFIs) is 1e-5 cm^2 total per device[2]. The set of all upsets includes categories with different consequences and different cross-sections. Most categories are common with many digital devices. But the Virtex has some unique sensitivities such as the configuration bitstream, half-latches, and the configuration management controller. Results from several Los Alamos tests of the Virtex XQV300 FPGA are shown in the overleaf table. Lookup table (LUT) and configuration bits are both observable in the configuration readback bitstream, which the device can provide through the SelectMAP configuration interface while the design is in operation. This makes the vast majority of static upsets directly observable while the system is in service. Primary bit stream and LUT errors are corrected via hardware monitoring techniques described in Section 3. Recent Los Alamos work has shown that "half-

P.Y.K. Cheung et al. (Eds.): FPL 2003, LNCS 2778, pp. 948–951, 2003.

Table 1. Xilinx XQVR1000 SEU Test Results, partitioned by source in the FPGA

Resource	Contribution in Bits	Fraction of Total
Configuration	5,603,456	91%
LUT Bits	393,216	6.4%
Block Select RAM	131,072	2.1%
User Flip-Flops	26,112	0.4%
Single Event Functional	?	<.0021%
Transients	?	?
Half Latches	?	?

latches," are also susceptible to SEUs, and are a significant contributor to residual SEU not addressed by hardware monitoring. When upset, the output values of these circuits remains inverted until the device is fully reprogrammed, and this inversion is not directly observable by reading back the configuration bitstream. To address this problem we created a tool, RadDRC,[3] which parses the XDL representation of a design to locate half-latch issues and generates an FPGA Editor script to automate their removal .

2 RHSD Architectures for Signal Processing

In all FPGA instantiations, many Virtex resources are left unused, so not every upset results in incorrect processing. The Virtex SelectMAP interface allows the device's configuration data to be read back while the device is in use, the primary feature we exploit to detect upsets. In addition, the Virtex can be partially configured, allowing rewrite of only the corrupted segment. These are key features of our Radiation Tolerant space processor designs, allowing mitigation of SEU without significant interruption in dataflow through the processor.

Our reconfigurable processor module shown in Figure 1 uses three Xilinx Virtex XQVR1000 FPGAs as the data processors. The FPGAs each have identical pin-outs so they can share configuration files. The module has two high bandwidth TTL busses for data input and output, and a ring bus for inter-processor data flow. The busses run at 50 MHz to deliver 200 Mbytes/sec per bus. Signaling is compatible with the VME FPDP specification, with 32 data lines, strobe and data valid. Each module also has a resident Actel RT54SX32S device that acts as a microprocessor interface and board controller. The Actel provides watchdog monitoring for the three Xilinx FPGAs as well as a configuration interface. It also manages the Xilinx bitstream SEU detection by continuously reading each Xilinx configuration, available for reconfiguration). Each FPGA configuration is read every 180 msec calculating a cyclic-redundancy check (CRC) for each frame (the finest granularity while the device is in operation, with no interruption in service, resulting in better than .999 unperturbed dataflow for typical low earth and geostationary orbits.

A disadvantage of SRAM-based FPGA technology for space is the significant power consumption and corresponding concerns for thermally induced mechanical failure. One key consideration is that the power consumption is a function of the processing

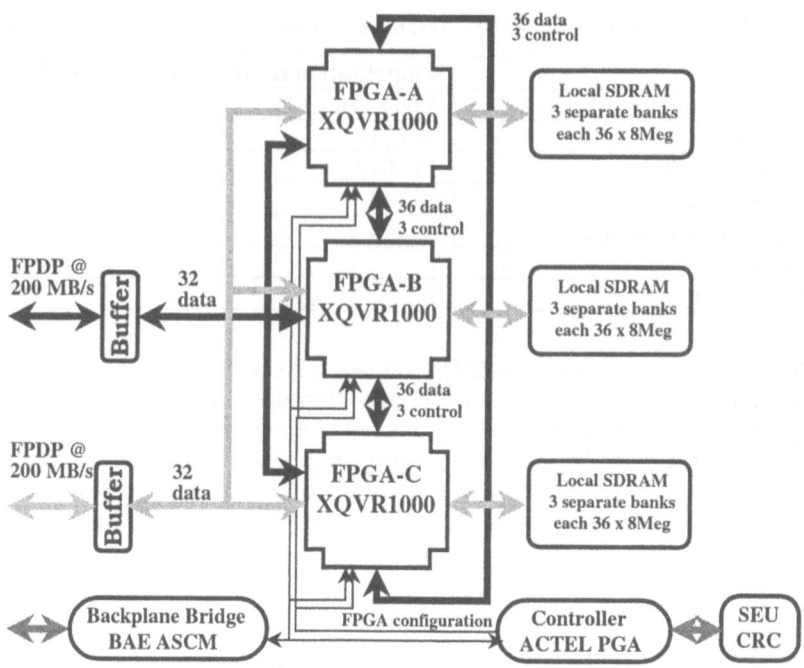

Fig 1. CIBOLA Space Based Reconfigurable Computer Module

being performed. Different configurations will have widely varying power requirements. We are limiting peak power consumption to ~6W / FPGA, which is still significant in light of a 560 grid array package that is 42.5 mm^2, cavity down. See references for detail on how we have chosen to manage assembly thermal issues[4].

3 Gabor Software Radio as an Application Example

The overlapped FFT process, more formally known as a Gabor Transform when near-perfect reconstruction is demanded of the processed data, is a highly useful practical example[5]. Here we have done extensive work, driven by our interest in software radio. Modern Gabor processing is derived from previous work in Trans-Multiplexer filter banks, but uses frame theory to derive more rigorous design methods for the reconstruction window operator. Two benefits accrue from such a rigorous formalism: 1) Ability to reconstruct analyzed data to arbitrarily fine precision using modern window designs, and 2) ability to relax the required block oversampling from 2:1 to 4:3 or even 5:4, with the same savings in processing rates. For input resolutions of up to 16 bits, we find that rate 4:3 gives errors of less than –100 dB RMS, which is more than adequate for most practical applications.

To the basic Forward Gabor Transform, we add two key steps in order to achieve a compressed output datastream, suitable for further processing. A magnitude and phase representation of the complex data is much more natural than rectangular format for most applications. CORDIC operators are efficient means of implementing polar conversion in an FPGA. Then, log compressing the magnitude only arrives in a natural domain for multiplicatively modulated signals. This sequence of operations, and the effect on convolution of two signals is shown below diagrammatically.

$$S_1(t) \otimes S_2(t) \xrightarrow{\;FFT\;} S_1(f) S_2(f) \xrightarrow{\;Polar\;} |S_1||S_2|(\angle_1 + \angle_2)$$

&

$$|S_1||S_2|(\angle_1 + \angle_2) \xrightarrow{\;Log\;} (Log|S_1| + Log|S_2|)(\angle_1 + \angle_2)$$

A case study for the current XQVR1000 architecture shows a 4096 point FFT with 1/3 overlap, windowing, and log magnitude/phase post-processing has been designed over 2 FPGA, operating at a 100 MHz data input rate with ~12 watts dissipation. For this design, it was necessary to use internal, dual-ported Block SelectRAM and internal LUT RAM to provide the required rates. However the cores run at 100 MHz, beyond the required 67 MHz rate, leaving idle time to run test vectors. This demonstrates a second means of bitstream SEU test, allowing the use of fast LUT RAM in a radiation environment. The work also shows how valuable new hybrid ASIC/FPGA designs will be in reducing die power dissipation.

4   Conclusions and Directions

We have described an exciting new capability for space processing, that provides many useful features needed for current and future commercial payloads. This capability is now moving into use in various government and commercially sponsored projects. A major attribute of RCC is that new generations of FPGA are constantly being released, but each new release stretches the boundaries of RHSD design. Therefore, space RCC is far from mature, and many topics for further research and development remain. We gratefully acknowledge the contributions of a large team in performing this work, and the sponsorship of the US Department of Energy.

References

1. Graham, P. et. al. Reconfigurable Computing in Space: From Current Technology to Reconfigurable Systems on Chip. *Proc. IEEE Aerospace Conference, Big Sky, MT.* (2003).
2. Fuller, E. et. al.: Radiation Testing Update, SEU Mitigation, & Availability Analysis of the Virtex FPGA for Space Reconfigurable Computing. *Proc. 3rd Conf. On Military & Aerospace Programmable Logic*. Columbia, MD (2000)
3. Graham, P.: RadDRC software LACC-02-39. Available via a license through Los Alamos National Laboratory, Industrial Business Development Office. Los Alamos, NM (2003)
4. Caffrey, M.: A Space Based Reconfigurable Radio. In *Proc. Engineering of Reconfigurable Systems and Algorithms*. Las Vegas, NV (2002)
5. Qian, S., Chen, D.: **Joint Time-Frequency Analysis**. Prentice-Hall PTR, Upper Saddle River, NJ (1996) 45-74

Adaptive Processor:
A Dynamically Reconfiguration Technology for Stream Processing

Shigeyuki Takano

Graduate School of Computer Science and Engineering
University of Aizu
Aizu-Wakamatsu, Fukushima-ken 965-8580, Japan
d8021103@u-aizu.ac.jp

Abstract. In order to improve the performance of reconfigurable computing, the number of reconfigurable units is increased with advance of semiconductor technology. The array of reconfigurable units can be configured to application-specific pipelined processing datapath. Then configuration overhead will be critical overhead of total execution time for dynamic reconfiguration based system. In this paper, models of efficient configuration methodology and application-specific pipelined stream processing are proposed. Adaptive processor architecture is also proposed, and discussed in summary.

1   Introduction

Performance is improved by three factors of array processing, clock cycle time, and overhead. Increasing the number of reconfigurable units provides higher performance of the array processing. The array construct an application-specific pipelined datapath that makes clock cycle time be short and hides the global communication delay. In addition, set of data may be composed to a stream that flows the configured datapath. Generally, an overhead decreases the effects of array processing and application-specific pipelining. Regarding reconfigurable computing, the reconfiguration time is critical overhead of total execution time with enlarging the size of array. However, the overhead reduction is not yet discussed well.

Section 2 proposes computation model, how to construct a stream processing and its coarser data granularity datapath efficiently. Section 3 proposes a processor architecture that focuses onto compiler workload reductions, eliminating place and route, and eliminating communication and interaction between kernel on host and kernel on reconfigurable unit. Finally, the model and processor architecture are summarized.

2   Computation Model

Concept of Object

In order to sequence configuration for coarser data granularity datapath, concept of an *object* is proposed. The *logical* object consists of set of information, result data and status.

P.Y.K. Cheung et al. (Eds.): FPL 2003, LNCS 2778, pp. 952–955, 2003.

An object ID (a tag) is assigned to each logical object to identify. By the addressing of object ID, the logical object is loaded from main memory into *physical* object(s) which is prefabricated hardware. Then configuration data called as static configuration data addressed by object ID is also loaded from main memory into the physical object. A dynamic configuration data is also loaded into the physical object from instruction at the same time. Physical object consists of storages for the logical object, object ID, static and dynamic configuration data, router, and reconfigurable fabric. Let's call pair of logical object and its configuration data as object. Any hardware unit in the processor is an object. The set of objects constructs an object space in the processor.

Instruction Scheduling and Configuration Cache

Application program is partitioned into two parts, set of logical object and instruction stream. Object specifies an operation. The detail of operation is specified by the static and dynamic configuration data. Instruction consists of processing object's ID field, two referenced object's ID fields for binomial model, and dynamic configuration data. There is no opcode, and instruction decoder and its pipeline stage are not necessary. In order to schedule instruction, general resource systems proposed by [1] is applied to it. Resource is the objects. The systems perform request, acquirement, and release of objects. Output data dependency is equivalent to resource conflict in the systems. At the resource conflict, adaptive processor pipeline waits and stalls for release of acquired object. The adaptive processor pipeline is as follows.

- Pipeline Stage 1: Request
  Instruction requests the objects. If all requested objects are in the object space and processing object is not acquired, then goto Pipeline Stage 2. If requested object is not in the object space, then the object is stored into physical object. If there is no physical object for the storing, then not acquired object is replaced for the store.
- Pipeline Stage 2: Acquirement
  During the acquirement phase, object processes operation and sends result data and local control signal to chained following object(s). When release token is fired, then goto Pipeline Stage 3.
- Pipeline Stage 3: Release
  The firing of release token sends result data, local control signal, and release token to chained following object(s). Object does not process configured operation.

In order to explain stream processing and its configuration methodology, execution of dot product is used as an example. Figure 1 (a) shows the dot product program. The necessary objects makes instruction stream as shown in Figure 1 (b). Name of object is shrinked as "Obj" in the figure. Object ID is named by a number. As shown in Figure 1 (b), the object ID is assigned to each necessary object. Then the ID addresses the logical object and its static configuration data. Instruction stream constructs coarser data grain datapath, a data flow graph based on node of object. True data dependency between instructions is used for chaining objects to construct the datapath. Based on the instruction scheduling, instruction stream is sequenced and it constructs the datapath in the object space as shown in Figure 1 (c), (d), and (e). Then streaming of vector elements is overlapped with the following sequence including the configuration. Black circle on paths from object 5 to two sequencers shows release token. Scalar data always has the release token. Flow of the release token and its firing technique makes in-order release for objects. Although in this example object keeps result, there are cases of when the vector store is used for generating vector result.

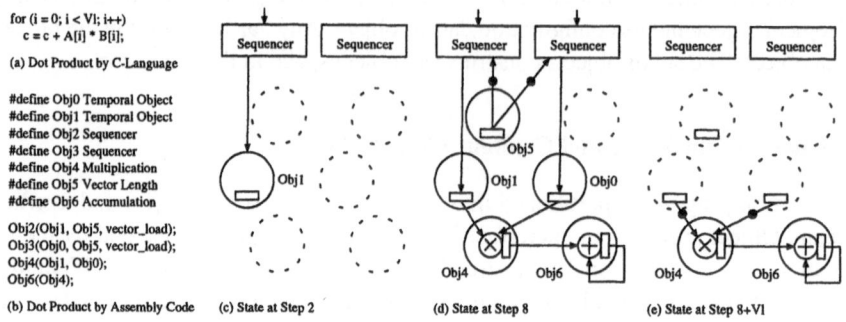

```
for (i = 0; i < Vl; i++)
    c = c + A[i] * B[i];
```
(a) Dot Product by C-Language

```
#define Obj0 Temporal Object
#define Obj1 Temporal Object
#define Obj2 Sequencer
#define Obj3 Sequencer
#define Obj4 Multiplication
#define Obj5 Vector Length
#define Obj6 Accumulation

Obj2(Obj1, Obj5, vector_load);
Obj3(Obj0, Obj5, vector_load);
Obj4(Obj1, Obj0);
Obj6(Obj4);
```
(b) Dot Product by Assembly Code   (c) State at Step 2   (d) State at Step 8   (e) State at Step 8+Vl

Fig. 1. Dot Product Execution

Logical object information and its configuration data are in object space after the processing. When the object is requested by other following instruction, the object is in the object space. This is a cache hit, called as configuration cache technology [2]. Placement location and replacement are decided by the cache technology. The replacement decides unused object for placement of requested and cache miss object. Not acquired objects are idle, however, they play as role of the configuration cache.

3   Adaptive Processor Architecture

Figure 2 (a) explains basic organization without instruction sequence datapath to simplify the figure. It consists of instruction register (IR), working set register file (WRF), cache miss handler (CH), and physical objects. The IR is similar to conventional processor. WRF holds acquired instructions. Acquiring an instruction is to store it into the WRF. Search for requests, cache hit signals, and acquirement signal from a selected register of WRF perform an instruction scheduling, the pipeline transition, and does chaining objects based on true data dependency between intructions. After a register of WRF receives release token from a fired object, the register is free, the instruction (objects) is released. CH performs replacement and placement of requested cache miss object. The adaptive processor architecture is tightest coupling model, and there is no host processor. There is also no register file, thus, no spill and fill instructions, and no register allocation. Figure 2 (b) explains a router simplified basic organization. A bus is only used to chain objects for one true dependency, no arbitration exist. Physical objects, set of buses, planes, and working set registers are replicated to scale the processor.

The configuration cache architecture extends the replacement algorithm studied by [3]. The contents of list are a set of information which physical object has.

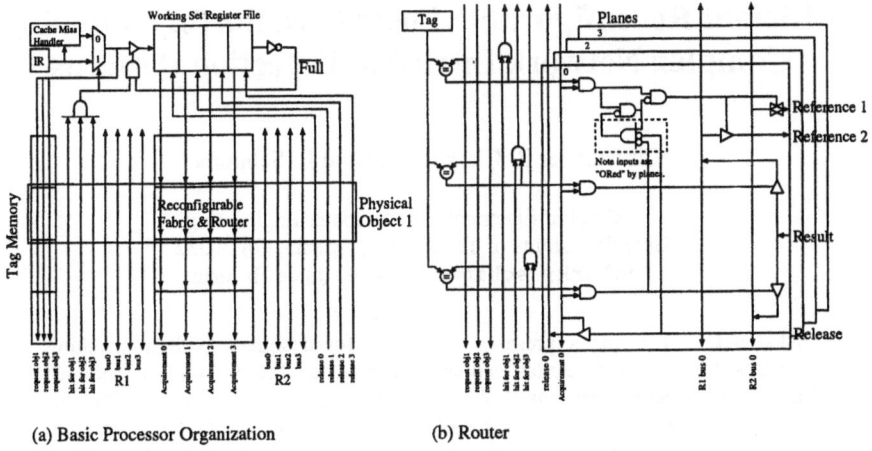

(a) Basic Processor Organization   (b) Router

Fig. 2. Adaptive Processor Basic Organization

4 Summary

In this paper, models of efficient configuration methodology and application-specific pipelined stream processing were proposed. Concept of object, release token technique, and configuration cache, are introduced and used respectively for the sequence. Instruction is scheduled by general resource systems. Configuration cache efficiently uses silicon resource for physical objects. The adaptive processor architecture is also proposed, and it is tightest coupling model that does not need to partition application to kernel on host and kernel on reconfigurable fabric, and not need to analyze the interaction. It does not need to place and route for the objects. The instruction stream (excepts for a part of dynamic configuration data) and set of logical object are compatible information for any adaptive processor architectures. We can select an interconnection network for chaining and reconfigurable fabric architectures on demand. Compared to other models, speedup for first architecture comes from overhead reduction, overlapping configuration and processing, reducing reconfigurations, temporal variables, memory accesses, eliminating register file and its workload, and eliminating instruction decoder and its pipeline stage. Therefore, total compiler workload is less than in other models. Future work is to develop simulator and to design a vector cache.

References

1. Richard C. Holt, Some Deadlock Properties of Computer Systems, Computing Surveys, Vol. 4, No. 3, pp. 179-196, September 1972
2. Micheal J. Wirthin and Brad L. Hutchings, DISC: The dynamic instruction set computer, Field Programmable Gate Arrays (FPGAs) for Fast Board Development and Reconfigurable Computing, Proc. SPIE 2607, pp. 92-103, 1995
3. R. L. Mattson, J. Gecsei, D. R. Slutz, and I. L. Trainger, Evaluation techniques for storage hierarchies, IBM Systems Journal, Volume 9, Number 2, pp. 78-117, 1970

Efficient Reconfigurable Logic Circuits for Matching Complex Network Intrusion Detection Patterns

Christopher R. Clark[†] and David E. Schimmel

School of Electrical and Computer Engineering,
Georgia Institute of Technology, Atlanta, GA 30332
{cclark, schimmel}@ece.gatech.edu

Abstract. This paper presents techniques for designing pattern matching circuits for complex regular expressions, such as those found in network intrusion detection patterns. We have developed a pattern-matching co-processor that supports all the pattern matching functions of the Snort rule language [3]. In order to achieve maximum pattern capacity and throughput, the design focuses on minimizing circuit area while maintaining high clock speed. Using our approach, we are able to store the entire current Snort rule database consisting of over 1,500 rules and 17,000 characters into a single one-million-gate FPGA while comparing all patterns against traffic at gigabit rates.

1 Introduction

Network intrusion detection systems (NIDS) have become critical network security components due to the rapidly growing rate of reported security incidents [1]. This trend, coupled with the fact that network traffic is increasing faster than computer performance [2], places an overwhelming burden on NIDS. Current NIDS are struggling to keep up with traffic rates above 100 Mb/s, while 1 Gb/s networks are common and 10 Gb/s networks are being deployed. It is clear that the gap between network traffic rates and NIDS analysis rates must be addressed. One of the most computationally-intensive tasks performed by rule-based NIDS, such as Snort [3], is pattern-matching on packet content [4]. Despite improved software pattern matching algorithms [4,5], pattern matching is still the limiting factor in the analysis of high-speed traffic. The goal of our research is to eliminate this bottleneck by offloading all the pattern matching tasks to a reconfigurable FPGA co-processor.

The effectiveness of NIDS can always be improved by adding patterns, therefore simply dedicating more hardware resources will not solve the problem; efficiency must be improved as well. The task of matching a large number of patterns against small data sets (packets) is different from classical research, which uses few patterns and large data sets. A method of generating circuits for regular expressions using nondeterministic finite automata (NFA) was presented in [6]. In [7], JHDL [8] was used to develop an NFA circuit generator that translated Snort rules into a pattern matching circuit. We use a similar approach, but our generator supports additional pattern matching functions and produces more efficient circuits. In addition, our design outputs a bit-vector indicating the Snort rules that match the input.

[†] This work was supported in part by NSF Grant 9876573 and by a grant from Intel Corporation.

2   NFA Circuit Optimization and Extension

Logic and Routing Resource Optimization. A circuit that implements an NFA for pattern matching consists of a pipeline of character match units. On a LUT-based FPGA (eg. Xilinx and Altera), one way to implement the match function using two logic elements is shown in general in [6] and specifically for Xilinx parts in [7]. We will now describe an optimization that allows a character match unit to fit into a single logic element.

The key observation leading to the reduction in area of a character match unit is that a full 8-bit comparison does not need to be performed by each unit. In fact, with a pattern set containing several thousand characters it is likely that there will be hundreds of identical 8-bit comparisons performed. These redundant comparisons waste valuable logic and routing resources and can be eliminated. Each character match unit only needs to know whether or not the input character matches the unit's programmed target character. This can be achieved with 8-to-256 decoder. Rather than broadcasting the 8-bit character to every unit, this approach sends the 1-bit output of the decoder corresponding to the value of the unit's target character. By sharing the character comparison results, it is possible to fit a character match unit into one logic element. Compared to the highest-capacity design found in previous works [7], our approach more than doubles the maximum pattern capacity of a given reconfigurable logic device. Fig. 1 illustrates the differences between the designs.

In addition, our approach uses routing resources more efficiently. While the distributed character approach requires $8*n$ connections from the input character to the character match units, the shared decoder approach only requires n connections, where n is the number of character match units. Since n increases linearly as the number of patterns is increased and the character match distribution is in the critical path, our approach scales much better in terms of capacity *and* operating frequency.

Case Insensitivity. If directly translating a regular expression to an NFA, a case-insensitive comparison performs separate comparisons for each case. However, there is a more efficient method that accomplishes the comparison with a single character match unit by using the fact that the ASCII codes for each case differ by only one binary digit. Using the distributed comparator, this can be implemented by having two 1-values in the LUT for the high-order bits of the character. Using the shared decoder, the OR of the match signals for each case is taken as the match input.

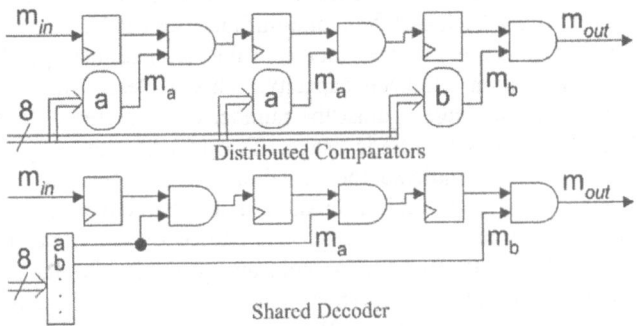

Fig. 1. Comparison of design approaches for the pattern "aab"

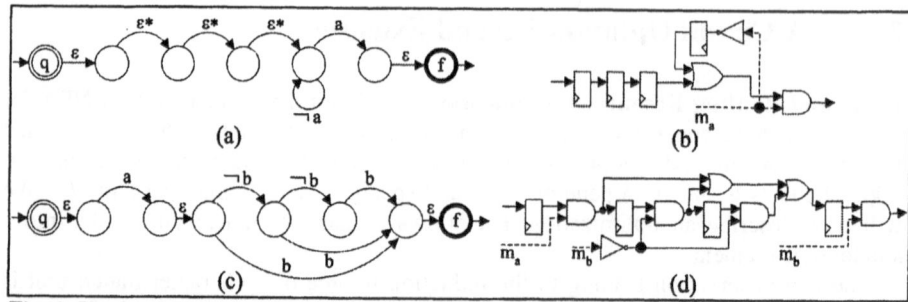

Fig. 2. Bounded-length wildcards. NFA (**a**) and circuit (**b**) for $(*_{,3})(a)$. NFA (**c**) and circuit (**d**) for $(a)(*_{,2})(b)$.

Bounded-Length Wildcards. Earlier work has shown how to implement wildcards for zero or more characters [6] and zero or one characters [7] using NFAs. Here we show how to implement wildcards whose length is specified by either a lower bound or an upper bound. In regular expressions, we use $(*_{,n})$ to denote any character sequence with a minimum length of n characters and $(*_{,n})$ for a sequence with a maximum length of n. In NFA diagrams we use $*$ to represent a character that equivalent to , but whose transition cannot be eliminated in the conversion to logic, and we use $(\neg c)$ to label a transition for any character other than the character c. Examples of NFAs using this notation and corresponding circuits are shown in Fig. 2

3   Performance

We implemented and tested five designs. The first was based on the distributed comparator approach described in [7]. The second extended the first by adding a prefix tree to eliminate circuitry for patterns with common prefixes. The third and fourth designs were based on our shared character decoder approach, with the latter using the prefix tree optimization. The fifth added support for all Snort content rule options. Each design was tested with sets of rules from Snort 2.0. The full set of default rules contained 17,537 characters. JHDL was used for synthesis and Xilinx Foundation was used for mapping, placement, and routing. The same configuration parameters were used for every case, and included a target clock rate of 100 MHz (the maximum supported by our FPGA platform). The designs were verified to produce correct output on a Xilinx Virtex-1000, a one-million-gate FPGA.

The speed and area results are presented in Table 1. The distinguishing property between the designs is their character capacity. This is clearly illustrated in Fig. 3, which plots logic element usage against the number of characters. All designs have approximately linear increases in logic usage. The important message is the relative slopes of the lines, which indicate the scalability of the designs. A prefix tree improves efficiency somewhat, but it is apparent that a shared character decoder is the key to providing major gains in capacity. As shown by the plot, support for all pattern matching options adds only a small constant overhead. To our knowledge, the fifth design is the first published system that performs all the pattern matching operations of Snort against all the rules. By processing one 8-bit character each cycle at 100 MHz, its throughput is 800 Mb/s, which is sufficient for 1 Gb/s networks.

Table 1. Speed and Area Comparison of Different Design Approaches

Number of Characters	Comparators		Comparators Tree		Decoder		Decoder Tree	
	Area	Freq	Area	Freq	Area	Freq	Area	Freq
2,001	39%	100.8	30%	83.3	16%	100.6	17%	101.7
4,012	75%	100.9	52%	73.9	30%	101.4	25%	102.1
7,996	-	-	99%	69.5	52%	100.2	42%	101.1
17,537	-	-	-	-	99%	100.1	80%	100.1

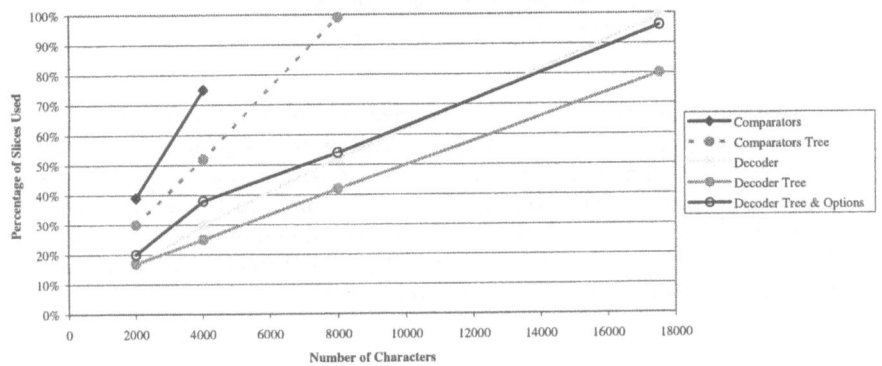

Fig. 3. FPGA slice usage for all designs

4   Conclusion

We have presented techniques that provide significant improvements in the efficiency and capacity of pattern matching circuits for reconfigurable logic. We have also extended the regular expression support of NFA-based designs. A co-processor for NIDS was implemented that is able to completely offload the task of pattern matching and compare the entire Snort rule set against packets at nearly 1 Gb/s. Our ongoing research indicates that speeds of 10 Gb/s and beyond are feasible with our approach.

References

1.   J. Allen, et. al., "State of the Practice of Intrusion Detection Technologies," Technical Report CMU/SEI-99-TR-028, 1999.
2.   L.G. Roberts, "Beyond Moore's Law: Internet Growth Trends," *IEEE Computer*, pp. 117-119, Jan 2000.
3.   Martin Roesch and Chris Green, "Snort User's Manual". http://www.snort.org.
4.   Mike Fisk and George Varghese, "Fast Content-Based Packet Handling for Intrusion Detection," Technical Report UCSD CS2001-0670, May 2001.
5.   C. Jason Coit, Stuart Staniford, and Joseph McAlerney, "Towards Faster String Matching for Intrusion Detection," *DARPA Information Survivability Conference*, June 2001.
6.   R. Sidhu and V.K. Prasanna, "Fast Regular Expression Matching using FPGAs," *Proceedings of IEEE FCCM 2001*, Apr. 2001.
7.   R. Franklin, D. Carver, and B.L. Hutchings, "Assisting Network Intrusion Detection with Reconfigurable Hardware," *Proceedings of IEEE FCCM 2002*, pp. 111-120, Apr. 2002.
8.   P. Bellows and B.L. Hutchings, "JHDL—An HDL for Reconfigurable Systems," *Proceedings of IEEE FCCM 1998*, pp. 175-184, Apr. 1998.

FPGAs for High Accuracy Clock Synchronization over Ethernet Networks

Roland Höller

Vienna University of Technology,
Institute of Computer Technology, ASIC Design,
Viktor-Kaplan Straße 2, 2700 Wiener Neustadt, Austria
roland.hoeller@tuwien.ac.at
http://agcad.ict.tuwien.ac.at

Abstract. This article describes the architecture and implementation of two systems on a programmable chip, which support high accuracy clock synchronization over Ethernet networks. The network interface node on one hand provides all necessary hardware support to be flexibly used in a broad range of applications. The switch add-on on the other hand accounts for the packet delay uncertainties of Ethernet switches and is crucial for high accuracy clock synchronization.

1 Introduction

The SynUTC (Synchronized Universal Time Coordinated) technology enables fault tolerant high accuracy distribution of GPS time and time synchronization of network nodes connected via standard Ethernet LANs [1–8]. By means of exchanging data packets in conjunction with hardware support at network nodes and network switches, an overall worst-case accuracy in the range of some 100 ns can be achieved, with negligible communication overhead. Applications can use the high-accuracy global time provided by SynUTC for event timestamping and event generation, both at hardware and software level [2, 3].

Furthermore the IEEE standard 1588 has been approved recently, which specifies a dedicated protocol, the Precision Time Protocol (PTP), for Ethernet networks to synchronize clocks in networked measurement and control systems [10]. Both the IEEE 1588 standard and the SynUTC technology are based upon inserting highly accurate time information into dedicated data packets at the media independent interface (MII) between the physical layer transceiver and the network controller upon packet transmission and reception, respectively. Implementing a clock synchronization service according to the ideas outlined above requires support on every node, which can be provided in a single SoPC (System on a Programmable Chip).

To alleviate the deterioration caused by network switches due to the undefined amount of time a packet stays on the switch before it is forwarded, a mechanism that measures this variable duration on-the-fly is necessary. This measurement is performed by another FPGA, which enhances standard Ethernet Switch functionality with a delay measurement mechanism to achieve highest precision and accuracy also in switched networks [8].

P.Y.K. Cheung et al. (Eds.): FPL 2003, LNCS 2778, pp. 960–963, 2003.

2   The Network Node FPGA

The network nodes support clock synchronization by making use of the network node FPGA. This FPGA provides the following functions [8]:

- A 96-bit adder-based local clock with a mechanism for linear continuous amortization.
- Two additional 64-bit adder-based clocks holding and automatically deteriorating the bounds on accuracy with respect to external reference time.
- An interface to a GPS timing receiver, made up of a 1-pps digital input and a RS232 serial interface.
- Application-level event generation and timestamping capabilities.
- A standardized interface for packet timestamping near the physical layer (IEEE 802.3 MII).
- An Ethernet MAC Controller.
- An interface to the CPU, which runs the synchronization algorithm and protocol stacks according to IEEE 1588.

For the first evaluation prototype the FPGA uses the PCI bus to communicate with the node's CPU, which run the protocol stacks and the synchronization algorithm. In this case the CPU is situated on a PC/104+ computer board.

However if a network node is intended to be used to connect to e.g. a sensor or an actuator in an industrial automation network, space and power requirements have to be taken into account. To be able to develop a system node that contains also the CPU on the chip, reconfigurable hardware is used as the centerpiece of the network interface card. It allows incremental realization of a corresponding system on chip without the need to develop a new printed circuit board by being able to skip the PCI interface and integrate a processor into the FPGA instead. This processor is an Altera NIOS 32-bit microcontroller core, which runs uCLinux as operating system (see Figure 1). In the first step the Ethernet MAC and the PCI interface are common of the shelf IP cores, that are connected via a 32-bit on-chip bus. The AHB bus (Advanced High Speed Bus) is used in this design.

3   The Switch Add-on FPGA

The switch add-on is used in conjunction with a common four port Ethernet switch. The switch add-on board provides eight 10/100-Mbit Ethernet physical layer interfaces of which four are connected to the Ethernet switch and four to the network nodes (see Figure 2). Whenever a dedicated clock synchronization packet passes through one of the four ports, which are connected to the network nodes, a 96-bit timestamp will be inserted on-the-fly into the packet. The packet's frame check sequence will be updated accordingly.

When after having been processed in the Ethernet switch, the clock synchronization packet is again traversing the switch add-on, now entering from one of the four ports connected to the Ethernet switch, the value of the timestamp in

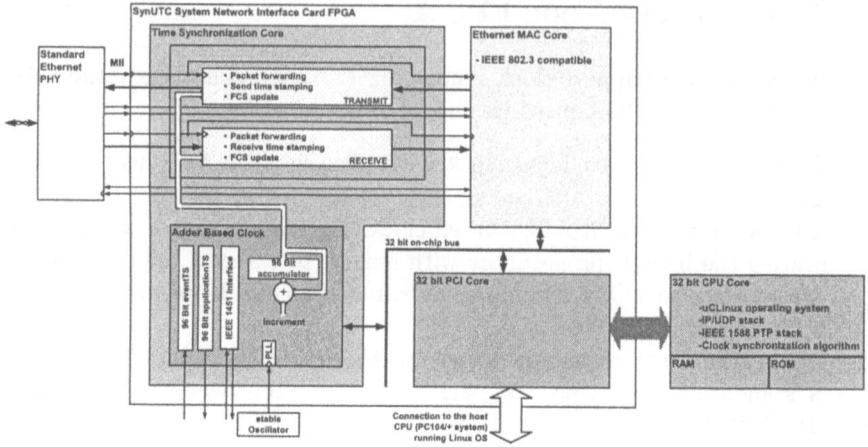

Fig. 1. Overview of the network interface card FPGA

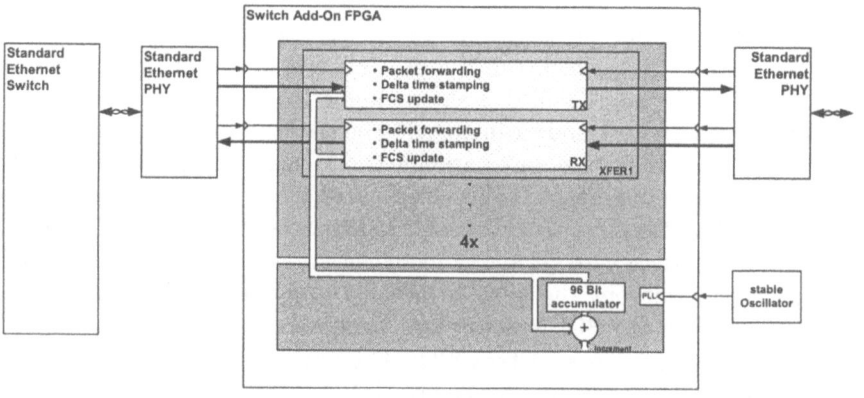

Fig. 2. Overview of the switch add-on FPGA

the packet will be subtracted from the current local time. This allows accurate measurement of the amount of time the packet needed to be forwarded by the switch. This delta time is now stored into the packet as it is sent to the destionation node. The clock synchronization algorithm is now able to take this variable delay into account [7, 8].

4   Conclusion and Future Work

This aritcle presented two FPGAs, that support the synchronization of clocks over Ethernet networks. The basic architecture of the designs has been presented and implementation results have been shown. The system containing the two integrated circuits will be presented at the first workshop on IEEE-1588 Standard for a Precision Clock Synchronization Protocol for Networked Measurement and Control Systems. Using reconfigurable hardware allows incremental realization of the system and future more complete protocol support (e.g. IEEE 1588 boundary clocks).

References

1. M. Horauer. *Hardware Support for Clock Synchronization in Distributed Systems.* Supplement of the 2001 International Conference on Dependable Systems and Networks, pp. A.10-A.13, Göteborg, July 2001.
2. M. Horauer, U. Schmid, K. Schossmaier. *NTI: A Network Time Interface M-Module for High-Accuracy Clock Synchronization.* Proceedings of the 6th International Workshop on Parallel and Distributed Real-Time Systems (WPDRTS), Orlando Florida USA, April 1998.
3. U. Schmid. *Orthogonal Accuracy Clock Synchronization.* Chicago Journal of Theoretical Computer Science 2000(3), 2000, pp. 3-77. IEEE Computer, 1990, Vol. 23 (10), pp. 33-42.
4. U. Schmid, M. Horauer, N. Kerö. *How to Distribute GPS-Time over COTS-based LANs.* 31st Annual Precise Time and Time Interval (PTTI) Systems and Applications Meeting, Dana Point - California, December 7-9 1999.
5. Ulrich Schmid, Klaus Schossmaier. *Interval-based Clock Synchronization.* Journal of Real-Time Systems 12(2), March 1997, pp. 173-228.
6. M. Horauer, R. Höller. *Integration of highly accurate Clock Synchronization into Ethernet-based Distributed Systems.* SSGRR 2002w, 2002, ISBN 88-85280-62-5.
7. R. Hoeller, M. Horauer, G. Griedling, N. Keroe, U. Schmid, K. Schossmaier. *SynUTC - High Precision Time Synchronization over Ethernet Networks.* Proceedings of the 8th Workshop on Electronics for LHC Experiments, 9-13 September 2002, Colmar, France, pages 428-432, ISSN 0007-8328, ISBN 02-9083-202-9.
8. M. Horauer, K. Schossmaier, U. Schmid, R. Höller, N. Kerö. *PSynUTC- Evaluation of a High Precision Time Synchronization Prototype System for Ethernet LANs.* Precise Time and Time Interval (PTTI), December 2002, Washington, USA.
9. IEEE Standard 802.3, 2000 Edition. *Carrier sense multiple access with collision detection (CSMA/CD) access method and physical layer specifications.* March 2000, New York, USA.
10. IEEE Standard 1588-2002. *Standard for a Precision Clock Synchronization Protocol for Networked Measurement and Control Systems.* November 2002, New York, USA.

Project of IPv6 Router
with FPGA Hardware Accelerator

Jiří Novotný[1], Otto Fučík[2], and David Antoš[3]

[1] Institute of Computer Science,
Masaryk University Brno,
Botanická 68a, Brno 602 00, Czech Republic
novotny@ics.muni.cz
[2] Faculty of Information Technology,
Brno University of Technology,
Božetěchova 2, Brno 612 66, Czech Republic
fucik@fit.vutbr.cz
[3] Faculty of Informatics,
Masaryk University Brno,
Botanická 68a, Brno 602 00, Czech Republic
xantos@fi.muni.cz

Abstract. This paper deals with a hardware accelerator as a part of the *Liberouter* project which is focused on design and implementation of a PC based IPv6 router. Major part of the Liberouter project is the development of a hardware accelerator – the PCI board called COMBO6 and its FPGA design which allows processing most of the network traffic in hardware.

Keywords. IPv6, router, Liberouter, Virtex II, FPGA.

1 Introduction

The paper describes a hardware accelerator based on FPGA that is designed to speed-up packet processing in a PC-based IPv6/IPv4 router. It is a part of the *Liberouter* project[1], which aims at developing a high-performance router with entirely open design.

Both the reliability and functionality of PC routers has proven to be fully comparable with commercial middle-class products [4]. Two main limitations keep from wider use of PC based routers. First, the PC routers have more difficult configuration than commercial routers. Second, the PC based routers have reached their theoretical throughput due to system resources saturation. Even the fast PCI bus (64 bit, 66 MHz) is not powerful enough.

The goal of the Liberouter project [5] is to develop an IPv6 PC based router, which solves both limitations discussed above. The configuration issues will be worked out by means of a uniform configuration environment based on XML. Performance will be improved using of a hardware accelerator that processes most of the network traffic in hardware.

---

[1] This research is supported by the FP5 project "6NET" (IST-2001-32603) and CESNET project "IPv6 implementation in the CESNET2 network" (02/2003).

P.Y.K. Cheung et al. (Eds.): FPL 2003, LNCS 2778, pp. 964–967, 2003.

2   Router Architecture

The hardware accelerator is a PCI card containing network interfaces, FPGA, memories (SSRAMs and DRAM), and necessary logic. The card has several expansion connectors dedicated for interface cards and future extensions [6]. All physical interfaces are mounted on an expansion daughter board which allows to use many different interface standards, either metallic or optical.

Packet switching and filtering itself will be performed by the accelerator. Software can do the rest, providing operations like routing paths calculations, router configuration, and statistics computing. Such operations are not time critical; and PC's resources are suitable for them. Communication through the PCI bus will be limited to board configuration, routing tables and firewall rules initializing and changing, statistics collecting as well as exceptions handling.

Moreover, if the accelerator behaves as usual network adapters from the point of view of the Unix system, we may use ordinary system mechanisms for routing and packet filtering tables maintenance. This way we obtain configurability of a software router and speed of a hardware one. The router configuration is available by means of usual software like ifconfig and routing daemons. Development of the PC based router with an accelerator requires the design methodology known as *hardware/software codesign*. We have begun with the PC based router fully implemented in software, moving more and more operations to hardware. We start from input and output stages and continue with more complicated blocks.

3   Packet Processing in Hardware

Packet processing in FPGA is done by a chain of dedicated processors – we call them *nanoprocessors* due the simplicity of their instruction sets. Each nano-processor has its own specialized instruction set designed for the particular purpose. The programs of nanoprocessors are stored in both FPGA's on-chip memories as well as in the external SSRAM memories.

Complexity of nanoprocessors lies "between a Finite State Machine (FSM) and RISC processors." The advantage of the this approach is the possibility to change functionality at runtime, as opposed to the case of FSMs. There is no need to rewrite the source code (e.g., in VHDL), synthesize it, place and route the design, and download the configuration data into FPGA. This differs from partial reconfiguration where all development steps must be done. On the opposite, partial reconfiguration can lead to smaller and more efficient design.

Let us now briefly describe the packet lifecycle in the proposed router. Complete information can be found in [1]. The packet processing is pipelined, the packet flows through the FPGA and memories. An incoming packet is received by the Input Packet Buffer and passed to the Header Field Extractor. The HFE pushes the body of the packet (including original headers) into the dynamic memory. Meanwhile, it parses the packet's headers and creates a *Unified-header* and a structure reflecting the actual arrangement of headers in the packet. The Unified-header is a fixed structure containing relevant information from packet

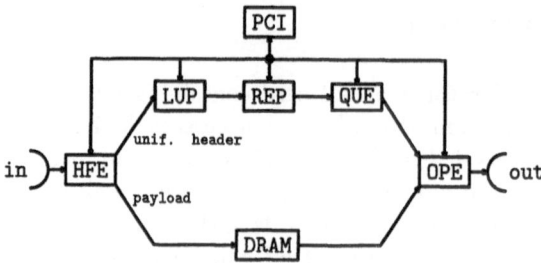

Fig. 1. Lifecycle of a packet traversing the router

headers, allowing to abstract the subsequent lookup operations from the actual header structure.

The Lookup Processor (LUP) uses CAM and SSRAM. LUP processes the Unified-headers by performing the lookup program and puts the result to Output Packet Editor. Using CAM is the fastest possibility when the application is small so that the lookup table fits into the memory. Unfortunately this is not the IPv6 case where we must match more that 440 bits of packet headers to decide what to do with a packet when it provides firewalling services.

Due to reasons discussed above we have developed a novel approach for the lookup machine. The word to be matched is stored in a sequence of registers. Instructions interpreted by the lookup machine are just conditional jumps. The lookup method is based on a (level compressed) search tree structure [1].

The Packet Replicator and Block of Output Queues (RQU) replicates the packet identification as well as the pointers to the editing programs into the dedicated queues. RQU computes the number of replicas of the packet that should be sent out. The Output Packet Editor (OPE) block modifies headers of packets. The OPE can also send a packet to the operating system, in the case when the packet destination is either the host computer itself, or the packet cannot be processed by hardware. This concept allows adding new features, step by step, during the PC based router development and use. The price paid for such approach include slower processing of software processed packets.

4   Software Support

Software drivers are developed first for NetBSD (FreeBSD) and ported to Linux. Driver operations include the FPGA chip configuration, accessing all memories mounted on the card as well as inside FPGA, and other hardware/software interface operations. The other part of router software should also be able to hide the card presence in the PC – the card should perform the same routing and filtering functionality as the operating system itself. The task is to develop a daemon which monitors the changes in both routing and filtering tables and, upon their update, it generates a nanoprogram for the LUP processor. To be able to make both routing and filtering by one searching operation, we have developed a concept of *routing/firewalling table*. The routing/firewalling table contains filtering rules applied apriori to the routing table rows, it can be represented as a tree structure and converted to a LUP nanoprogram [1].

PC based routers are running under various operating systems with various configuration files. It causes problems to network administrators. To overcome this we are working on a unified configuration environment [3]. We have selected the XML language to store the router configuration. The user interface will be implementing Command Line Interface, web interface, and SNMP. From XML, the router configuration will be generated depending on the current operating system. This approach simplifies the use of the router and the user interface can be compatible with these seen in industry standard routers.

5   Conclusion

The PC based router design is a complex and long time task. It requires cooperation of experts from many areas. Currently, the entire team has more than 35 people from organizations in the Czech Republic including Masaryk University, Brno Technology University, Czech Technical University in Prague, and Camea ltd., as well as several consultants from other countries. The whole Liberouter project is organised by CESNET.

Currently, the accelerator prototype has already been developed, and produced. Processing blocks described are simulated and tested in hardware. The IPv6 PC based router is primarily dedicated but not limited for the use in academic networks, with focus on research in the internet protocols.

The project is fully open and all materials including source codes are available in the Internet [5]. Due to its flexible architecture, the hardware accelerator with low level software utilities can serve as a general purpose hardware platform for computation acceleration as well. There are many research project teams interested in using the proposed system. Possible application include programmable OEO (optic-electric-optic) switch, network monitoring tools, evolvable hardware research, reconfigurable computing, digital signal processing, and encryption.

References

1. David Antoš, Jan Kořenek, Kateřina Minaříková, and Vojtěch Řehák. Packet header matching in Combo6 IPv6 router. Technical Report 1/2003, CESNET, 2003.
2. Jiří Barnat, Tomáš Brázdil, Pavel Krčál, Vojtěch Řehák, and David Šafránek. Model Checking in IPv6 Hardware Router Design. Technical Report 8/2002, CESNET, 2002.
3. Petr Holub. XML Router Configuration Specifications and Architecture Document. Technical Report 7/2002, CESNET, 2002.
4. Ladislav Lhotka. Software tools for router performance testing. Technical Report 10/2001, CESNET, October 2001.
5. Liberouter. Liberouter Project WWW Page. http://www.liberouter.org, 2003.
6. Jiří Novotný, Otto Fučík, and Radomír Kokotek. Schematics and PCB of COMBO6 card. Technical Report 14/2002, CESNET, 2002.

A TCP/IP Based Multi-device Programming Circuit

David V. Schuehler, Harvey Ku, and John Lockwood

Applied Research Laboratory, Washington University
One Brookings Drive, Campus Box 1045
St. Louis, MO 63130-4899 USA
{dvs1,hku,lockwood}@arl.wustl.edu
http://www.arl.wustl.edu/projects/fpx

Abstract. This paper describes a lightweight Field Programmable Gate Array (FPGA) circuit design that supports the simultaneous programming of multiple devices at different locations throughout the Internet. This task is accomplished by a single TCP/IP socket connection. Packets are routed through a series of devices to be programmed. At each location, a hardware circuit extracts reconfiguration information from the TCP/IP byte stream and programs other devices at that location. A novel feature of the Multi-Device Programmer is that it does not use a microprocessor or even a soft-core processor. All of the TCP/IP protocol processing and packet forwarding operations are handled directly in FPGA logic and state machines. This system is robust against lost and reordered packets, and has been successfully demonstrated in the laboratory.

1   Introduction

As large numbers of reconfigurable hardware devices are deployed throughout the Internet, it has been observed that point-to-point configuration mechanisms cannot program multiple devices quickly. The concept of programming a chain of devices located throughout the Internet can be seen in Figure 1. By programming multiple devices using a single TCP/IP data flow, the task of reprogramming large numbers of devices at different locations becomes manageable.

The Multi-Device Programmer is a circuit design which provides a mechanism to extract device programming information from a TCP/IP connection. This information is then utilized to program a FPGA. Because the Multi-Device Programmer circuit does not inhibit the flow of data between the source programmer and the connection end point, multiple devices can be chained together and placed at multiple locations throughout the Internet. Identical copies of the device configuration data is extracted at each location. The development of the Multi-Device FPGA Programmer leverages existing research performed at the Washington University Applied Research Laboratory. More specifically, the Washington University Gigabit Switch (WUGS) and the Field-Programmable Port Extender (FPX) [2] plug-in card are used as the hardware platform on

P.Y.K. Cheung et al. (Eds.): FPL 2003, LNCS 2778, pp. 968–971, 2003.

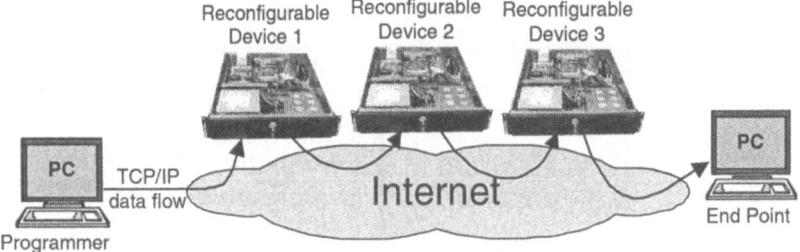

Fig. 1. Multi-Device Programmer

which the Multi-Device FPGA Programmer was developed and tested. The TCP-Splitter [3] and the Layered Protocol Wrapper [1] circuit components provide Internet Protocol packet processing functions. These circuit components provide a logic framework within which the Multi-Device Programmer resides.

2 Design

The Multi-Device Programmer is a circuit implemented in FPGA logic which extracts programming information from the network and programs a targeted FPGA or FPGAs. Specifically, it passes configuration and control data to other reconfigurable hardware devices in a system using information extracted from a TCP/IP byte stream.

In order to program a chain of devices, two software programs are run at the network endpoints. The source programmer application reads a formatted bitfile from disk and sends that data through a standard TCP/IP socket connection. A second program acts as a data sink and terminates the TCP/IP connection. The various components can be seen in Figure 1.

The source programmer can be executed on any workstation connected to a network. A TCP/IP socket connection is established to the connection endpoint. By design, this TCP/IP network connection follows a path through the network which contains the device(s) to be reprogrammed. This can be accomplished by either intelligently choosing the machines for the programmer application and the connection endpoint, or by a network administrator configuring static routes to ensure that the TCP/IP connection between the source programmer and the connection endpoint routes through the device(s) to be updated.

A FPGA on a FPX card is programmed by sending special control cells to the card. Information in these control cells are processed by logic on the card and used to program the target FPGA. The Multi-Device programmer circuit, containing the Protocol Wrappers, the TCP-Splitter, and the Multi-Device Programmer, was initially developed to execute on a FPX device. This circuit extracts device programming information from the TCP data stream and programs a second FPX module. By dropping selected packets, the TCP data

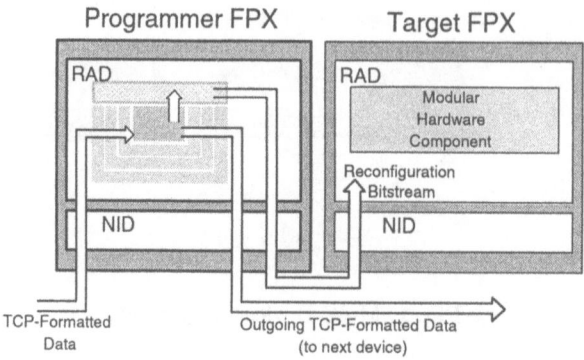

Fig. 2. Target Device Layout

stream can be reconstructed without requiring large reassembly buffers. This ensures that the system is robust against missing or reordered packets.

The layout of the target device, which consists of two components, can be seen in Figure 2. The Programmer FPX contains the Multi-Device Programmer which receives byte streams from the TCP-Splitter and extracts the appropriate device programming information. This FPGA reconfiguration data is then transmitted to one or more Target FPX devices. Upon receiving this reconfiguration data, the Target FPX device is reprogrammed.

The process to program a single device is identical to the process for programming 10, 100, 1000 or any number of devices. The programming application is executed on a node connected to the network and establishes a TCP/IP socket connection to the connection endpoint. This TCP/IP network connection is routed through each of the devices to be programmed. Once the connection is established, the programming application reads configuration information from a disk file and sends the configuration information down the network connection. For FPGAs that support partial reconfiguration, it is also possible to incorporate the Programmer circuit within a region of the target FPGA.

3   Results

The Multi-Device Programmer circuit has been successfully simulated, synthesized and tested in various configurations. The circuit has a post place and route operational frequency of 74 MHz on a Xilinx Virtex XVC2000E-6. All of the circuit components comprising the Multi-Device Programmer, including the TCP-Splitter and IP protocol wrappers, consume 22% of the BLOCKRAMs and 28% of the SLICEs on the target FPGA.

The TCP-Splitter and Multi-Device Programmer are designed to provide a peak theoretical throughput of 2.4 Gigabits/second using a 32-bit interface clocked at 75 MHz. The VirtexE FPGA on the FPX, however, has a reconfiguration throughput limited by the 8-bit interface of the SelectMAP interface. This limits the maximum possible reconfiguration to 400 Mbits/second.

The Multi-Device Programmer has successfully reprogrammed three target FPX devices simultaneously in as little as 1.102 seconds. The test configuration consists of a source programmer running on a Windows XP platform, a sink application running on a Linux machine, and a WUGS with three target FPX devices in a stacked configuration. Average programming times for this configuration are on the order of 1.3 seconds. A 2.2MByte configuration file containing command, control, and routing information was used in these tests. This file included a 1.2MByte bitfile. The configuration file was processed at and average bit rate of 16Mbits per second.

4 Conclusion

The Multi-Device Programmer provides an efficient mechanism for delivering FPGA reconfiguration information to a large number of devices over the Internet. This design feature provides a flexible programming solution which can integrate with other FPGA reconfigurations techniques. Partial FPGA reprogramming techniques such as incremental FPGA programming and FPGA plugin modules could easily be integrated with the Multi-Device Programmer.

The two main contributions of this research are (1) the successful demonstration of a FPGA application containing TCP/IP protocol processing in hardware (not a microprocessor or soft core), and (2) the demonstration of an application that can reliably program multiple devices utilizing a single TCP/IP connection. The design of the Multi-Device Programmer could easily be modified to send reprogramming information to any network attached device.

Acknowledgments

The authors of this paper would like to thank Florain Braun and James Moscola for the Layered Protocol Wrappers. We would also like to thank Todd Sproul for his work on NCHARGE and David Lim for his work on the NID.

References

1. F. Braun, J. Lockwood, and M. Waldvogel. Reconfigurable Router Modules Using Network Protocol Wrappers. In *Proceedings of Field-Programmable Logic and Applications*, pages 254–263, Belfast, Northern Ireland, Aug. 2001.
2. J. W. Lockwood, N. Naufel, J. S. Turner, and D. E. Taylor. Reprogrammable Network Packet Processing on the Field Programmable Port Extender (FPX). In *ACM International Symposium on Field Programmable Gate Arrays (FPGA'2001)*, pages 87–93, Monterey, CA, USA, Feb. 2001.
3. D. V. Schuehler and J. Lockwood. TCP-Splitter: A TCP/IP Flow Monitor in Reconfigurable Hardware. In *Proceedings of Symposium on High Performance Interconnects (HotI'02)*, pages 127–131, Stanford, CA, USA, Aug. 2002.

Design Flow for Efficient FPGA Reconfiguration

Richard H. Turner and Roger F. Woods

Programmable Systems Laboratory
Queen's University Belfast, Ashby Building, Belfast, BT9 5AH, N. Ireland
rht@salsabelfast.co.uk, r.woods@qub.ac.uk

Abstract. In Run Time Reconfiguration (RTR) systems, the amount of reconfiguration is considerable when compared to the circuit changes implemented. This is because reconfiguration is not considered as part of the design flow. This paper presents a method for reconfigurable circuit design by modeling the underlying FPGA reconfigurable circuitry and taking it into consideration in the system design. This is demonstrated for an image processing example on the Xilinx Virtex FPGA.

1 Introduction

Typically, the amount of configuration data needed bears little relation to the changes being made. This has been exacerbated in the most recent FPGA technologies where reconfiguration comprises fixed size frames which represent the smallest unit of configuration. In many cases, a large part of the configuration data is not required and there are no methods for increasing the density of the useful information in these blocks. Previous studies [1] have indicated that routing accounts for a large portion of this data when compared to that required by the logic.

This paper argues that this high level of reconfiguration results because the designer has no control of the circuit implementation. Previous work has shown that mapping reconfiguration using the reconfiguration mux (RC_MUX) to either the LUTs [1,2] or routing muxes [3] can produce large reductions in reconfiguration times. The main challenge is to develop an approach that allows reconfiguration to be considered as part of the design flow thereby allowing RTR control to be gained. The paper presents a unified approach and demonstrates it with an image processing example.

2 Addition of Redundancy and Control of Reconfiguration

To capture the concept of reconfiguration and treat it as part of the design flow, the well-established method by Shirazi et al. [4] for viewing reconfiguration using a multiplexer, termed RC_MUX, is used. In his original paper, it was envisaged the RC_MUX is used to represent reconfiguration. However, **Fig. 1** highlights a number of multiplexers that can be used to directly implement this mux and achieve a reduction in reconfiguration time. Courtney [3] demonstrated use of the routing multiplexer to reduce reconfiguration and Turner [1] embedded the RC_MUX within the LUT.

P.Y.K. Cheung et al. (Eds.): FPL 2003, LNCS 2778, pp. 972–975, 2003.

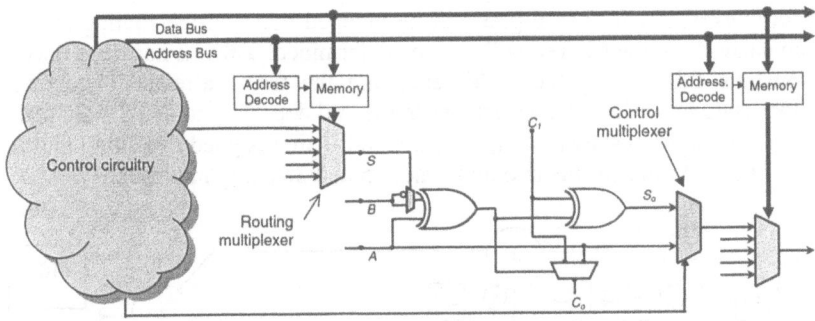

Fig. 1. Alternative view of the FPGA for RTR

A design that shares image filter functions, as illustrated in **Table 1**, originally developed by Shirazi et al. [4] is used to demonstrate the approach. A different implementation to Shirazi et al. [4] is shown in **Fig. 2** which has an extra addition that is not required in the adder branch of circuit **Fig. 2(b)**. This reduces the complexity and therefore size of the resulting circuit when the parts are combined. As the plan is to share the hardware, the circuit architecture has been developed as given by **Fig. 3**. Each of the multiplexers, labeled A, B, C, D and E, represent reconfiguration. Typically this would not be implemented but in our approach, these are efficiently mapped to the FPGA with the aim of minimizing the number of frames to be changed.

Table 1. Window functions for image processing

1	2	1		-1	0	1		1	2	1
2	4	2	,	-2	0	2	,	0	0	0
1	2	1		-1	0	1		-1	-2	-1

(a) Gaussian filter (b) Vertical edge detector (c) Horizontal edge detector

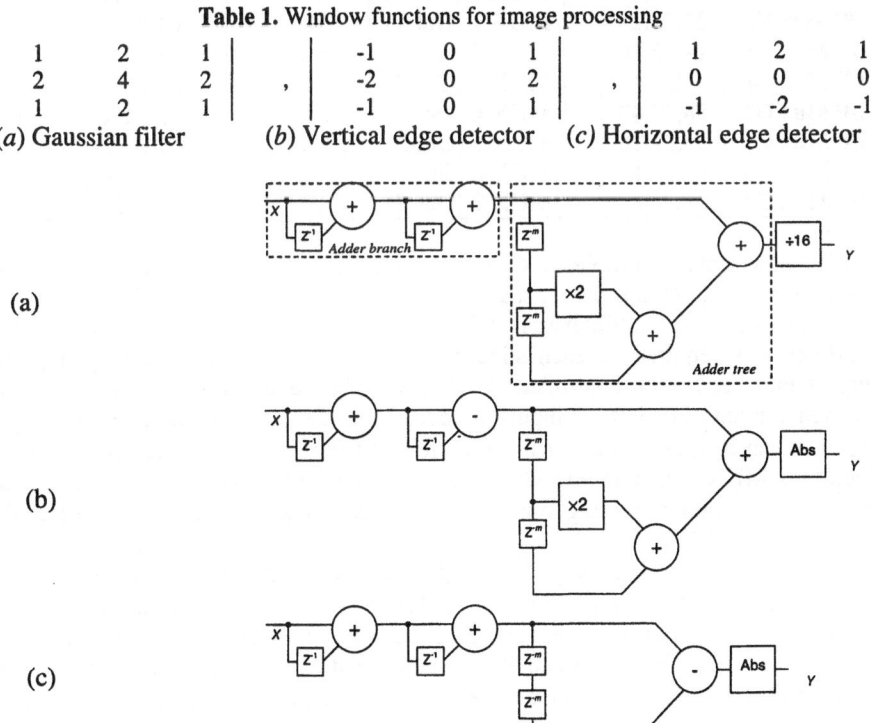

Fig. 2. (a) Gaussian filter, (b) Vertical edge detector and (c) Horizontal edge detector

Multiplexers *A* and *B* in **Fig. 3** will be mapped first and the routing multiplexers which share the same frames, will then be determined. **Table 2** has been drawn up to demonstrate this, giving the multiplexer and the frame affected (The table only includes state transitions that affect one frame and does not include the *Off* state). The proposed mapping is shown in **Fig. 4**. As *A* and *B* are assigned to routing multiplexers *Bx*, *By* of slice 0, this results in only frames 35 and 38 being modified.

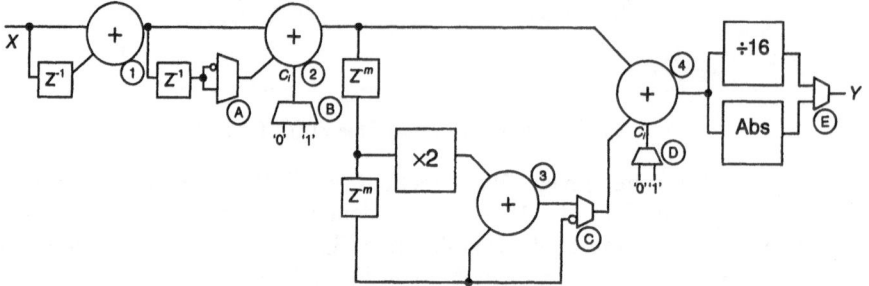

Fig. 3. Revised Combined circuit

Table 2. Frames which will be affected for possible routing multiplexer

Slice 1	Frame		Output	Frame			Frame
I/P Mux F1	38,39,40		MUX 0 (Out-0)	45		Slice 0 Bx	38
I/P Mux F2	26,27,29		MUX 1 (Out-1)	38		Inv MUX	
I/P Mux F3	14,15,16		MUX 2 (Out-2)	33,34		Slice 0 By	35
I/P Mux F4	2,3,5		MUX 3 (Out-3)	26		Inv MUX	
I/P Mux G1	38,39,40		MUX 4 (Out-4)	21		Slice 1 Bx	11
I/P Mux G2	26,27,28		MUX 5 (Out-5)	13, 14		Inv MUX	
I/P Mux G3	13,14,15		MUX 6 (Out-6)	9		Slice 1 By	13
I/P Mux G4	2,3, 4		MUX 7 (Out-7)	2		Inv MUX	

This now leaves multiplexer *C*. From **Table 2**, it can be seen that the routing multiplexers, *F1*, *G1* and *Out-1*, share the same frames as *Bx*, *By* of slice 0. For this case, output routing multiplexer *1* (MUX 1 (Out-1)) was chosen. A second output multiplexer is required as each slice performs two bits of the adder, giving two outputs that need to be controlled. Multiplexer *C* is required to switch between an inverted version of one of the inputs to adder 3 and the output of adder 3. Looking at **Fig. 4**, it can be seen that a signal can be routed from *BY* through the inverter to *YB*. Therefore, MUX 1 (Out-1) is used to select between the FU outputs *Y* and *YB* (of slice 1). The other output routing multiplexer, MUX 1 (Out-2) has to be changeable between *X* and *XB* (of slice 1).

The choice of using output routing multiplexer 2 can be seen from the Reconfiguration State Graph (RSG) in **Fig 5** which gives the necessary state transitions to change multiplexer settings. It can be seen to move between states *X* and *XB* (of slice 1) that frames 33 and 35 will be needed, where frame 38 is also used for the inverters. In **Fig. 6**, it can be seen that to change the state of output routing multiplexer 1 from *Y* to *YB* (of slice 1), requires frames 38, 42 and 47 to be changed where frame 35 is also used for one of the inverters.

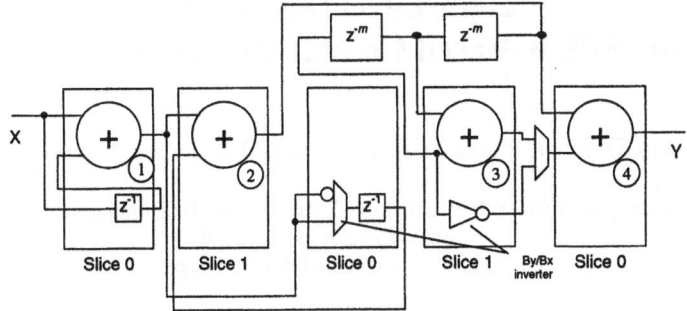

Fig. 4. Proposed mapping of the adders and multiplexers

The end result is that the RC_MUX has been mapped into routing multiplexers, taking account of the frames and minimising the number of frames that require to be changed. This can be extended to the input routing multiplexer. Alternatively, the RC_MUX can be treated as another logic component and mapped to the LUTs as shown in [2] therefore allowing the user to determine whether to implement the RC_MUX in logic or using reconfiguration prior to the place and route stage.

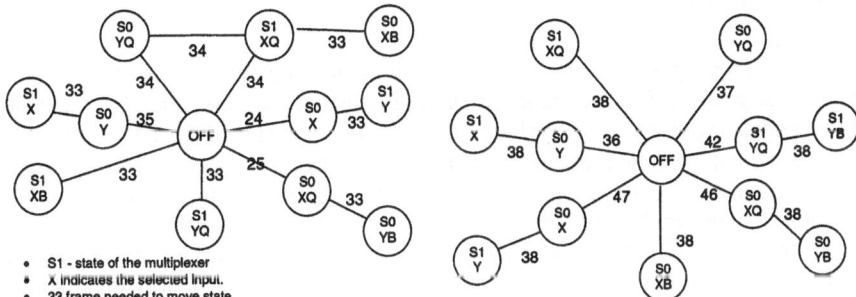

Fig. 5. RSG of o/p routing multiplexer 2   **Fig. 6.** RSG of o/p routing multiplexer 1

3 Conclusions

It has been shown that the process of mapping reconfiguration into the FPGA can be combined with the logic synthesis flow allowing optimisation, transformation, placement and routing to be done together and therefore minimizing reconfiguration.

References

[1] R. H. Turner, "Functionally Diverse Programmable Logic Implementations of Digital Signal Processing Algorithms", PhD Diss., Queen's University Belfast, 2002.

[2] C. N. Ang, R. H. Turner, T. Courtney and R. Woods, "Virtex FPGA Implementation of a Polyphase Filter for Sample Rate Conversion", 34th Asilomar Conf. on Signals, Syst. and Comp., USA, Oct. 2000, IEEE Comp. Soc., pp. 365-369.

[3] T. Courtney, R. H. Turner, and R. Woods, "An Investigation of Reconfigurable Multipliers for use in Adaptive Signal Processing", IEEE Symp. on FCCM, Napa, USA, May 2000, pp. 341-343.

[4] N. Shirazi, W. Luk, P. Cheung, "Automating Production of Run-Time Reconfigurable Designs", Proc. IEEE Symp. on FCCM, USA, 1998, pp. 147-156.

High-Level Design Tools
for FPGA-Based Combinatorial Accelerators

Valery Sklyarov, Iouliia Skliarova, Pedro Almeida, and Manuel Almeida

University of Aveiro, Department of Electronics and Telecommunications, IEETA,
3810-193 Aveiro, Portugal
{skl, iouliia}@det.ua.pt

Abstract. Analysis of different combinatorial search algorithms has shown that they have a set of distinctive features in common. The paper suggests a number of reusable blocks that support these features and provide high-level design of combinatorial accelerators.

1 Introduction

There are many practical problems that can be formulated over such mathematical models as graphs, discrete matrices, sets, Boolean equations, etc. The majority of these models are mutually interchangeable, i.e. any of them can be selected for similar purposes. Many practical problems can be solved by applying various combinatorial search algorithms over a chosen mathematical model. Such algorithms have two distinctive features. Firstly, they require a huge number of different variants to be considered. Secondly, these variants are frequently ordered and examined with the aid of a decision tree that provides an efficient way for handling intermediate solutions. Examples of combinatorial problems that can be solved with the aid of the algorithms mentioned above are Boolean satisfiability; covering of sets and Boolean matrices; graph coloring, mapping and partitioning, etc.

This paper suggests a number of reusable blocks that can be employed for constructing application-specific hardware accelerators (co-processors) that permit combinatorial problems formulated over Boolean and ternary matrices to be solved.

2 Basic Combinatorial Search Algorithm

The considered combinatorial search algorithm is based on such primary operations that involve the application of so-called reduction and selection rules [1]. The *reduction rules* enable an initial matrix and intermediate matrices (that are constructed on each iteration through the algorithm) to be simplified. The *selection rules* allow the problem to be decomposed into several sub-problems that are examined sequentially to determine that either a solution will be found, or that no solution exists. The use of these rules for a particular example of the covering problem was considered in [2].

P.Y.K. Cheung et al. (Eds.): FPL 2003, LNCS 2778, pp. 976–979, 2003.

Search algorithms (such as [2]) for solving different combinatorial problems have similar characteristics. Their distinctive feature is the execution of problem-specific operations, traversing a decision tree starting from its root by involving such procedures as forward search and backtracking (if the forward search fails). Any branch point can be considered to be extracting a sub-tree with a local root. Thus a recursive algorithm, such as [2] can be used very efficiently. Analysis of the basic operations for general search algorithms has shown that it is reasonable to construct a set of functional blocks that enable the design process for combinatorial accelerators (co-processors) to be simplified. These blocks permit the design process to be realized at a high level of abstraction without losing sight of the details of a particular problem, and without increasing the hardware resources required or reducing performance.

The search algorithm contains branch points that can specify two or more alternative branches. The proposed computational model permits a graph to be traversed using the following strategy. All alternative branches of the search tree that might lead to the solution have to be examined. If we can prove that a selected branch does not permit a solution to be found, control is returned to the nearest branch point. All data that are needed to restore the state of any branch point on a path leading to the current step are kept in a stack. In order to store/restore data at branch points, push/pop operations are executed. Operations over vectors (rows and columns of matrices) are executed in a special functional unit that takes into account the uniqueness of combinatorial computations. Analysis of different combinatorial search algorithms has shown that the following four types of application-specific blocks are required: storage for a matrix, stack memory, a functional unit for operations over vectors, and a control circuit that supports recursion.

3 Specification of Reusable Functional Blocks

Matrix storage is the block that affects the timing characteristics of the algorithms most significantly. The proposed architecture includes memory based on RAM, mask registers that permit some rows and columns to be excluded to select a sub-matrix, and the circuits that generate RAM addresses. The latter allow the rows/columns that are not required at the current step to be skipped.

A parameterizable stack memory (that can be used for any type of intermediate data that are needed for backtracking) was constructed as a library component in VHDL and Handel-C. Elementary computations over vectors have been described using bit-manipulation operations and a reprogrammable FSM (examples can be found in [3]).

A control unit is modeled by a hierarchical FSM (HFSM) [4]. An example of a HFSM considered in [4] for executing sorting algorithms demonstrates all the steps that are required for HFSM synthesis. A number of ISE 5.2 projects for FPGAs that implement recursive hierarchical algorithms are available in the tutorial section of [5]. The same method was used for describing combinatorial search algorithms.

All the blocks considered have been designed in Handel-C and in VHDL so that their dimensions can be parameterized and they can be reused for different kinds of combinatorial accelerators (co-processors).

4 Implementation of Combinatorial Accelerators

Three combinatorial accelerators that will find a minimal row cover of a Boolean matrix using the exact algorithm [1] have been constructed based on the Handel-C and VHDL. A combinatorial accelerator that solves the Boolean satisfiability problem has been designed on the basis of VHDL. They are intended to demonstrate how the proposed library can be used. The designed macros permit storage to be allocated either in an external RAM or in FPGA embedded memory blocks (block RAM). The first accelerator was implemented on the prototyping board RC100 of Celoxica [6]. It contains 2 banks of onboard static RAM (256K*36 bit each), FPGA XC2S200 and some other components. A number of auxiliary blocks allowing data input and output have also been designed. They enable us to communicate with the keyboard and mouse, to display matrices, intermediate and final results on a VGA monitor screen, to receive/send data from/to PC, etc. The available drivers of Celoxica [6] have been utilized. However these are just 10-15% of the originally designed circuits, which in addition allow the display and scrolling of matrices, highlighting (or selecting by color) of elements considered at any intermediate step, visualizing the contents of stacks, checking any currently executing operation, and many others.

For debugging purposes the first bank of the onboard RAM was used as a memory for the VGA monitor (this bank can be used as additional storage for matrices in the "release version"). The second bank stores the matrices. The basic operation of the covering algorithm [1] is counting the number of ones in the matrix rows. This operation is only performed at the beginning of the algorithm execution. The results for all the rows are kept in FPGA block RAM and they are corrected at each newly executed step. An auxiliary stack stores these data for each branch point. General-purpose registers permit the result to be accumulated, include a counter that enables us to count the number of ones, and some temporary registers.

The Handel-C project was translated to an EDIF file in the Celoxica DK1 design suite and the bitstream for FPGA was generated from the EDIF file with the aid of Xilinx ISE 5.2.

The second accelerator was implemented on the basis of the ADM-XPL PCI board [7] containing FPGA XC2VP7 of Virtex-II Pro family. Access to the FPGA from the PC is provided through AlphaData API functions [7] that are described in the C language. Communication from the FPGA to the PC is supported by a driver described in Handel-C. Many auxiliary C++ programs that provide support for experiments and establish a user friendly graphical interface have also been implemented. The FPGA clock frequency was set to 60 MHz. An initial matrix is loaded from the PC and the accelerator solves the problem and returns the results to the PC. Finally the results are displayed on the screen. The execution time for randomly generated matrices with dimensions up to 100x100 does not exceed 2 minutes.

The third accelerator was constructed on the basis of VHDL. Two VHDL-based circuits for solving the covering and satisfiability problems were tested in Xilinx FPGAs XC4010XL and XCV812E. The reusable blocks considered above were also implemented as VHDL library modules. They have been tested in FPGA XC2S300E (the prototyping board TE-XC2Se from Trenz Electronic [8]). Many examples of such blocks are available in [5].

The results of experiments have shown that the proposed technique makes it possible to shorten the design of combinatorial accelerators (co-processors). The proposed building blocks take into account specific features of combinatorial search algorithms and they have been optimized for the problems considered. This allows block-based high-level design to be provided, i.e. to concentrate the efforts of the designer on the algorithms being considered, and avoid (or at least minimize) the details of hardware implementation. This permits consideration of either the design flow on the basis of the system-level specification language or the widely used hardware description language. The technique was validated in a number of projects that were implemented for three types of FPGAs and prototyping boards [6-8]. A set of additional blocks that were described in Handel-C and VHDL [5] provide very effective visualization and debugging tools and they can be inserted into a "debug version" and removed from the "release version". They are also very useful for experimental purposes.

5 Conclusion

It has been shown that many problems formulated over Boolean and ternary matrices can be solved with the aid of search algorithms that have a number of distinctive features. Such algorithms are recursive and they execute periodically operations for simplifying an initial matrix and making a decision for future steps. The paper suggests four primary building blocks for the high-level design of combinatorial accelerators, which provide storage and unique operations on matrices, support stacks, and implement recursive control algorithms. These blocks were described in Handel-C and in VHDL. To validate the method, three combinatorial accelerators were designed, implemented in FPGAs on the basis of stand alone and PCI boards, and tested. This work was supported by the grants FCT-PRAXIS XXI/BD/21353/99 and POSI/43140/CHS/2001.

References

1. Zakrevski, A.D.: Logical Synthesis of Cascade Networks. Moscow: Science (1981)
2. Sklyarov V., Skliarova I.: Architecture of Reconfigurable Processor for Implementing Search Algorithms over Discrete Matrices. In: Proceedings of ERSA'2003 (Engineering of Reconfigurable Systems and Algorithms) (Las Vegas, USA, 2003)
3. Sklyarov V.: Reconfigurable models of finite state machines and their implementation in FPGAs. Journal of Systems Architecture, 47 (2002) 1043-1064
4. Sklyarov, V.: Hierarchical Finite-State Machines and Their Use for Digital Control. IEEE Transactions on VLSI Systems, vol. 7, n. 2 (1999) 222-228
5. http://webct.ua.pt, "2 semester", the discipline "Computação Reconfigurável", public domain is indicated by the letter "i" enclosed in a circle. Login and password for access to the protected section can also be provided (via e-mails: skl@ieeta.pt, iouliia@det.ua.pt)
6. HandelC, DK1, RC100. [Online]. Available: http://www.celoxica.com
7. Alpha Data. [Online]. Available: http://www.alpha-data.com
8. Spartan-IIE Development Platform. [Online]. Available: http://www.trenz-electronic.de

Using System Generator to Design a Reconfigurable Video Encryption System

Daniel Denning[1], Neil Harold[2], Malachy Devlin[2], James Irvine[3]

[1] Institute of System Level Integration, Alba Campus, Livingston, EH54 7EG, UK
daniel.denning@sli-institute.ac.uk
[2] Nallatech Ltd, Boolean House, One Napier Park, Cumbernauld, Glasgow, G68 0BH, UK
{n.harold, m.devlin}@nallatech.com
[3] EEE Department, University of Strathclyde, 204 George St., Glasgow, G1 1XW, UK
j.m.irvine@strath.ac.uk

Abstract. In this paper, we discuss the use of System Generator to design a re-configurable video encryption system. It includes the design of the AES (Advanced Encryption System) and Enigma encryption cores. As a result of using this design flow, we are able to efficiently implement our system and algorithms with a significant improvement on traditional design times, without compromise for performance.

1 Introduction

System Generator [1] from Xilinx is an extension of Simulink and provides a block-based high-level schematic tool that generates VHDL. The nature of System Generator makes it ideal for rapid development of data-path algorithms. One of the encryption algorithms used in this system is based on the Enigma machine, which is equivalent to an encrypter/decrypter typewriter whereby pressing a letter of the alphabet reveals a different letter of the alphabet. A much more recent product cipher is AES, which is based on the Rijndael algorithm developed by V.Rijmen and J.Daemen [2]. AES was chosen by NIST (National Institute of Standards and Technology) to replace the highly popular but less-efficient DES (Data Encryption Standard).

Xilinx's Xtreme DSP kit [3], developed in conjunction with Nallatech, contains a BenOne motherboard and a BenAdda module, which are part of the scalable DIME-II™ family.

In this paper we investigate the implementation issues of designing a reconfigurable video encryption system using System Generator. We map our system to a Virtex-II FPGA on a BenAdda module, housed on a BenOne motherboard. The system is fully reconfigurable in that we essentially have two identical designs but with different encryption cores, a modified Enigma algorithm and the AES algorithm. Video is transmitted over a wireless link and fed back into the same FPGA to be decrypted.

P.Y.K. Cheung et al. (Eds.): FPL 2003, LNCS 2778, pp. 980–983, 2003.

2 Modelling with System Generator

System Generator consists of a Simulink library called the Xilinx blockset that maps the Xilinx block elements defined in Simulink into architectures and entities, signals, ports, and attributes. It also produces command files for FPGA synthesis, HDL simulation and implementation tools. The tool keeps the Simulink hierarchy when converted into VHDL.

When designing in System Generator it is possible to access key features in the FPGA such as the high-speed multipliers, and it is also possible to incorporate user-defined VHDL blocks in to the model. For verification and testing the tool can automatically generate testbenches, where by the Simulink input stimuli to the input block can be recorded for the VHDL simulation. The outputs can then be compared with the recorded results from the Simulink simulation in the VHDL simulation.

3 Encryption Cores

The AES algorithm is a symmetrical block cipher. We have chosen the algorithm to use the 128 bit key length. This results in 10 rounds of encryption within the algorithm, with each round, except the last round, having 4 transformations as shown in Fig. 1. For this implementation we have chosen to design the traditional looped feedback for each incremental round without any pipelining. Each round has its own subkey, these are generated when needed and then stored locally in Block-RAM for the decryption core.

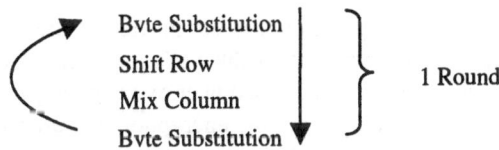

Fig. 1. AES transformations within one round.

Our modified Enigma algorithm is an 8-bit algorithm accepting values between 0 – 255 and uses 3 permanent rotors. For more information on the Enigma see [4].

4 System Implementation

A PAL video stream is captured by an ADC channel on the BenAdda, fed through the FPGA, encrypted, and presented at the DAC. This is then fed back through the second ADC to the FPGA to be decrypted. A high-level System Generator model can be seen in Fig. 2 on the following page. For reconfigurability the system is in two parts. The first part is an encryption system using the Enigma algorithm and the second is the same system except that the Enigma is substituted for the AES algorithm.

The AES algorithm uses the standard looped round design. This takes up the smallest amount of space on the FPGA but affects the throughput by the feedback latency. The S-boxes with in the subByte transformation are implemented using single port ROM blocks and can be seen as a one-dimensional array. Multiplication with in the mixColumns transformation is a combination of shift, XOR, and multiplexor blocks. The shiftRows transformation uses only slice and concatenate blocks.

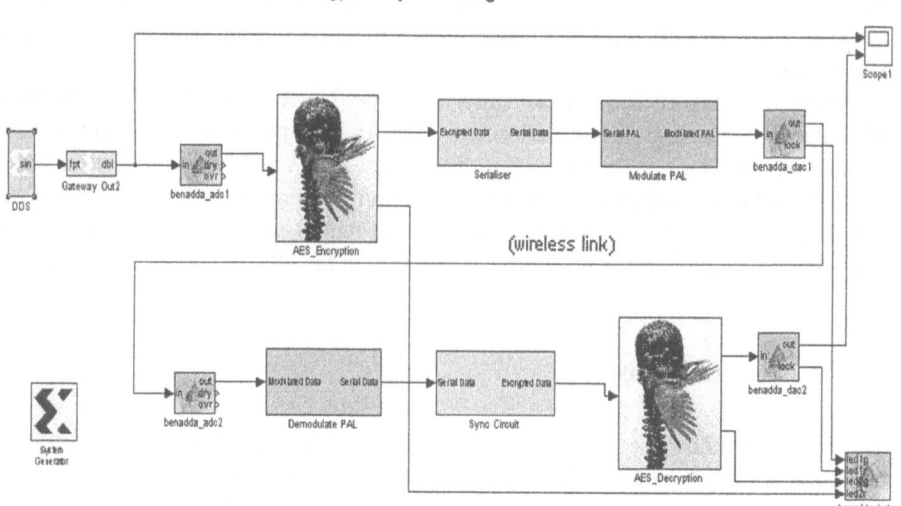

Fig. 2. High Level model of Video System using AES

The decryption core is the inverse of the encryption core [6]. The expansion of the key to produce the subkeys is preformed locally when needed and then stored in block-ROM to be accessed later by the decryption core. A buffering and de-buffering circuit is added before the encryption and after the decryption to buffer up and down 8-bits to 128-bits and vice-versa.

The data having been encrypted, start and stop bits are added prior to transmission. The data is then serialised and modulated at 100MHz (via DAC1) onto a laser diode module that transmits to a pin diode receiver connected to ADC2 on the BenAdda. The modulated data received on ADC2 is captured and passed to a synchronisation circuit, which identifies the boundary of each parallel word and removes the start and stop bits before applying the parallel data to the decryption block.

Following decryption, the reformed 8-bit PAL signal is fed out through DAC2 on the BenAdda to a suitable display. The entire system is bandwidth limited by the ADCs, which are clocked at 100MHz.

5 Results

Each system, encryption and decryption, fits onto one XC2V3000. The Enigma system takes up 4588 slices (32%) of the device while the AES system takes up 1719 slices (12%) of the device. The performance of both systems is limited by the wireless laser link and the ADCs. Without these limitations the performance would go as high as the performance of the encryption cores. The Enigma encryption core has a throughput of 1.25 Gbit/sec and is fully pipelined. The design of the encryption and decryption cores took one engineer just over one week. The AES encryption core has a throughput of 1.3Gbit/sec and requires 466 (3%) slices of the device. However, due to the use of block-RAMs for s-boxes in the subByte transformation, the core needs 37 out of the 96 block-RAMs available. The design of each encryption core took one engineer just over two weeks. These same statistics can be applied to both of the decryption cores.

6   Conclusions

As can be seen from the results, especially the implementation of the AES core, it is possible with System Generator to design high-speed cores for FPGAs. The standard AES encryption core has also been fully pipelined. The System Generator model took over 2 hours to loop-unroll. A synthesisable core has been produced with a throughput of 16.4Gbit/sec, but has been targeted at a XC2V6000. Although the design times cannot be compared with an equivalent VHDL design time, it is the authors' view that these design times have increased impressively.

The System Generator tool has proved to be very intuitive for datapath and algorithm design. It does produce many VHDL files, for example the AES encryptor/decryptor core has over 2000 VHDL files in its hierarchy yet only takes up 3% of the slices in a XC2V3000. In terms of slices for whole 128-bit processing and on-chip key expansion, this core is one of the most compact AES encryptor/decryptor implementations published. The average simulation time of the system took around 10-15 minutes. When the AES fully pipelined encryption model was simulated the tool took over 4 hours to simulate 15 clock cycles.

Acknowledgements

On behalf of Nallatech Limited the authors would like to thank the Ministry of Defence (UK), and the Engineering and Physical Sciences Research Council. And lastly Dave Shand, Derek Stark, and Eric Lord for their help and support.

References

1.   Xilinx Inc., System Generator Reference Guide,
     www.xilinx.com/ipcenter/dsp/ref_guide.pdf
2.   AES, Federal Information, Processing Standards Publication 197, Nov 26, 2001.
3.   Xilinx Inc. Xtreme DSP Development Kit, ww.xilinx.com/ipcenter/dsp/development_kit.htm.
4.   Bletchley Park, Enigma, www.bletchleypark.org.uk.

MATLAB/Simulink Based Methodology for Rapid-FPGA-Prototyping

Miroslav Líčko[1,2], Jan Schier[1,2], Milan Tichý[1,2], and Markus Kühl[3]

[1] Institute of Information Theory and Automation, Department of Signal Processing,
Pod vodárenskou věží 4, Prague 8, Czech Republic
{licko, schier, tichy}@utia.cas.cz
[2] Center for Applied Cybernetics,
Department of Control Systems, Faculty of Electrical Engineering,
Czech Technical University,
Karlovo nám. 13, Prague 2, Czech Republic
[3] FZI Forschungszentrum Informatik, Dept. of Electronic Systems and Microsystems,
Haid-und-Neu-Strasse 10-14, Karlsruhe, Germany
kuehl@fzi.de

Abstract. The paper is focused on rapid prototyping for FPGA using the high-level environment of MATLAB/Simulink. An approach using combination of the Xilinx System Generator (XSG) and Handel-C is reviewed. A design flow to minimize HDL coding is considered.

1 Introduction

For development of embedded applications, the methods of hardware-software co-design are often needed: on one hand, the algorithm design must be solved, on the other hand, proper balance between the flexibility (software implementation) and performance (hardware implementation) must be considered.

At the level of rapid algorithm design, MATLAB is often used for block specifications and for their inner analysis. This environment is characterized by its high-level scripting possibilities, strong support for matrix operations, object oriented approach, extensive graphing possibilities and rich set of application-specific toolboxes.

The application prototyping is typically performed using either some suitable high-level programming language, e.g. C or C++, or the MATLAB scripting language. On contrary, the hardware part of the target implementation has traditionally been designed in a Hardware Description Language (often, VHDL or Verilog). Hence, it was necessary to manually recode the specification. That is itself an error-prone process, leading to two versions of the same code (in different programming languages), which are difficult to keep up-to-date and corresponding to each other. There has been a number of attempts to solve this problem, either by designing languages that can be used for both hardware and software specification (the best-known example is probably the SystemC language) or by automating the HDL code generation by parsing the high-level specifications. In this paper we focus on the second option, namely on using Simulink for the

P.Y.K. Cheung et al. (Eds.): FPL 2003, LNCS 2778, pp. 984–987, 2003.

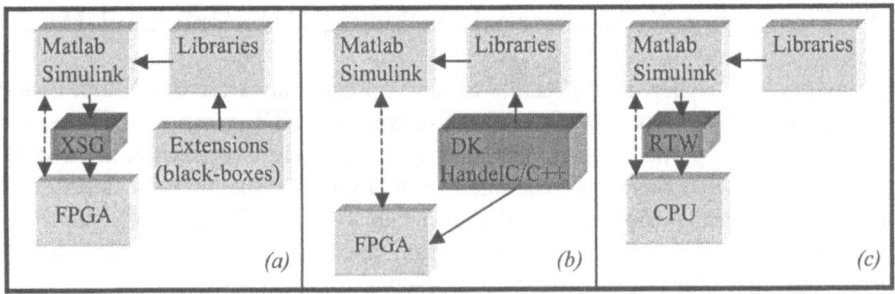

Fig. 1. Some possible high-level design flows for FPGA/CPU designs

initial specification of the system (Simulink is a MATLAB extension for visual programming). Using this tool, the system to be designed can be represented in a form of a block diagrams. In the paper, we revise some possible hardware design methodologies using Simulink for the system description and consider possible extensions. These extensions – aimed to provide support for a hardware-software co-design – are depicted in the diagrams given in Figure 1.

a) *Xilinx System Generator for DSP* (XSG) is an add-on tool for Simulink. The Core generator produces relationally placed macro components optimized for Xilinx FPGAs. With XSG, it is possible to generate VHDL source code out of Simulink block diagram. It allows to encapsulate user IPs too, using either VHDL or Verilog wrapper [3].

b) *Handel-C* is a HDL from Celoxica, based on the C language. It can be combined with C/C++ routines (representing parts of the design running in software). The compiler of the Handel-C language allows to generate directly the HDL description or to synthetise the EDIF code out of the C-like description. This design flow addresses HW/SW co-design problems as well [4].

c) *Real-Time Workshop* (RTW) represents generic methodology for parsing the Simulink block diagrams to high-level languages (primarily to the C-language). Parsers for different CPUs are provided. By adopting the methodology for Handel-C, System-C, or other appropriate language, it is possible to target the FPGA using the Simulink specification.

There are other tools that we mention only briefly:

– The *TargetLink* from dSpace (concept similar to XSG – Figure 1a) allows to use the Simulink block diagram to generate and optimize a C-code.

– The *Processor Expert* tool is a Delphi-like IDE for design of HW components. The concept used by ProcessorExpert is to encapsulate API into interactive IDE with graphical entities.

– *Forge* is a compiler from Xilinx for the FPGA designs written in Java language. It can be used with any Java-based IDE; it is possible to integrate it with the Matlab environment, which is based on the MathWorks proprietary JVM.

2   Designing XSG Extensions Using HW-Oriented HLL

As mentioned earlier, it is possible to develop extensions for the Xilinx System
Generator (XSG) using the XSG black-boxes and HDL wrappers. To general-
ize and speed-up the prototyping and design process, we have combined this
methodology with the tools available in the DK-environment from Celoxica (let
us recall that it is a toolset around the Celoxica Handel-C language): the toolset
was used to generate the VHDL code from the Handel-C description, and this
code was then plugged into the System Generator.

The extended design flow is shown in Figure 2 – the right part of this figure
is an extension of the flow presented in Figure 1 (a). We plan to investigate
the possibilities of RPM support in terms of partially reconfigurable design as
well [5].

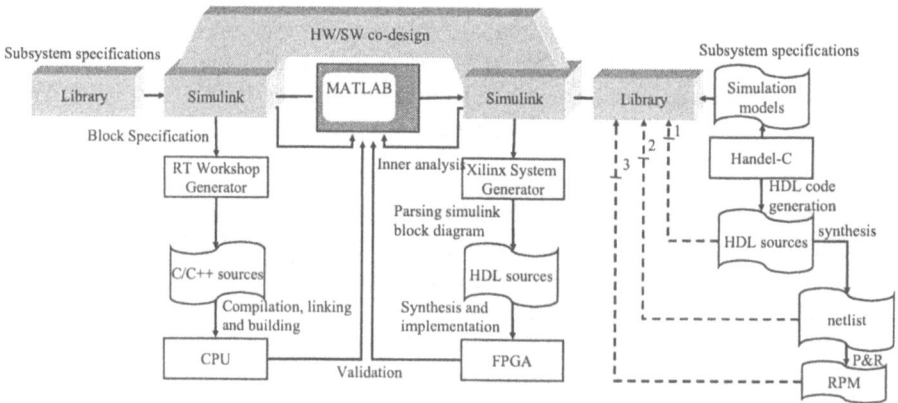

Fig. 2. Rapid-Prototyping design flow for the XSG Black-Boxes

The approach was tested on an example of a High-Speed Logarithmic Arithmetic (HSLA) unit, which has been developed in our group [1]. This unit represents an efficient way for implementation of floating-point arithmetic operations
in FPGA. The above specified design flow allowed us to prepare generic specifications using the MATLAB/Simulink IDE and to use them for XSG modeling
of applications requiring the floating-point data range precision.

3   HW/SW Co-design Options

The presented design flow could be used also for the HW/SW co-design. An
illustration example: in the complex signal processing applications, the HSLA-
cores can be used as a floating-point coprocessor. The load of the signal processor
can be reduced by allocating parts of the running code into the FPGA circuit.

Using the combination of RTW and XSG, we can go farther: the XSG represents the system-level based implementation on one or more FPGAs, while the RTW is used to co-design the control program for the processor (see left part of the Figure 2). It will be possible to extend this approach so that we can use these high-level tools for the System on Chip (SoC) designs. Finally, we are investigating the possibilities to simulate and analyze the Run Time Reconfiguration schedule (RTR, see [5]). Seamless implementation on the target is the most important factor we are taking into account.

4   Conclusions

A design methodology for rapid prototyping of HW/SW-implemented DSP systems has been presented. The approach presented in the paper uses a combination of the MATLAB/Simulink environment with the Xilinx System Generator for DSP and the Handel-C.

The methodology has been used for preparing a bit-exact and cycle-exact Simulink library of the HSLA functions with corresponding IP cores for FPGA, in a form suitable for DSP engineers. We have started with a sequential implementation using MATLAB with MEX functions. Later, we prepared library for visual programming using Simulink. Using its combination with Handel-C we could better explore algorithmic structure (parallelism). The behavioral simulation in Simulink was used to speed-up our design process.

Acknowledgments

This work has been partially supported by the Ministry of Education of the Czech Republic under the Project LN00B096, by the IST programme under the RECONF2 project (IST-2001-34016) and by the DLR WTZ-project CZE01/19.

References

1. HSLA homepage, Dept. of Signal Processing, ÚTIA AV ČR, Prague, Czech Republic [available online: http://www.utia.cas.cz/ZS/home.php?ids=hsla]
2. Coleman, J.N., Chester, E., Softley, C.I. and Kadlec, J. 'Arithmetic of the European Logarithmic Microprocessor', IEEE Trans. Comput. Special Edition on Computer Arithmetic, Vol. 49, No. 7, pp. 702–715 and erratum vol. 49, no. 10, p. 1152 (July 2000)
3. Xilinx Inc., 'Xilinx System Generator for DSP' [available online: http://www.xilinx.com/xlnx/xil_prodcat_product.jsp?title=system_generator]
4. Celoxica Ltd., 'Handel-C' [available onoine: http://www.celoxica.com]
5. RECONF Project homepage: IST programme 2002-34016 [available online: www.reconf.org]

DIGIMOD: A Tool to Implement FPGA-Based Digital IF and Baseband Modems

J. Marín-Roig[1], V.Torres[1], M.J.Canet[1], A.Pérez[1], T. Sansaloni[1], F. Cardells[3], F.Angarita[1], F.Vicedo[4], V.Almenar[2], and J.Valls[1]

[1] Dpto. Ingeniería Electrónica, Universidad Politécnica de Valencia, Gandia, Spain
[2] Dpto. Comunicaciones, Universidad Politécnica de Valencia, Gandia, Spain
[3] Inkjet Commercial Division (ICD) R&D Lab, Hewlett-Packard, Barcelona, Spain
[4] Dpto. Física y Arquitectura Computadores, Universidad Miguel Hernández, Elche, Spain

Abstract. This paper presents a software tool to design intermediate frequency and baseband digital transceivers on FPGA. Main characteristic of this tool is that an *ad-hoc* interpolation or decimation filter chain composed by CIC, polyphase, pulse shaping, matched filters and a CORDIC-based or ROM-based mixer can be selected. The tool allows the software radio designer to develop downconverters and upconverters and, finally, automatically to generate the VHDL code to implement the system on Xilinx FPGAs.

1 Introduction

During the next years it is expected that new communication systems will provide higher mobility and wider bandwidth than present systems. For these reasons those firms that develop future communication systems will have to face up to next challenges: continuous evolving standards, fast deployment of new services, higher spectral efficiency, high capacity and mobility data transmissions, and a demand of a high reliability in those services offered.

In order to meet all theses objectives it will be necessary the use of new methods of design and development. In recent years a new technology called software radio (SWR) has come up. The main idea behind SWR is to move the digital part of any communication system towards the antenna. This means that most of the analogue components from transmitters and receivers have been substituted by digital signal processing under FPGA or DSP devices. Working in this way it is possible to change the system configuration without changing the hardware.

This paper presents a tool that speeds up and makes easy the implementation of wireless communications systems. With this tool one can design a digital transmitter/receiver in intermediate frequency or in base band. By means of its graphical interface the user can choose all the subsystems required, and once the transmitter/receiver is completed, this tool can evaluate the performance of those employed IP-cores. Finally, if the specifications are met, the tool generates the VHDL code for the whole system.

P.Y.K. Cheung et al. (Eds.): FPL 2003, LNCS 2778, pp. 988–991, 2003.

2   Description of DIGIMOD Tool

Graphical user interface of DIGIMOD makes easy and fast the design and implementation of a transceiver for digital communications on FPGA. The transmitter, also called upconverter, is composed of several stages: binary data source, symbol mapping, pulse shaping, interpolation filters and mixer. Meanwhile, the receiver (or downconverter) is composed by a chain of elements that performs the reverse operations: a mixer to bring the signal to base band, decimation filters to cancel the double frequency image and to reduce the sampling rate, a matched filter adapted to the pulse shape, a demapper and a symbol detector. So, the designer of a digital communication modem can use this tool:

- to select the type of modulation between BPSK, QPSK, and QAM, and the pulse shape parameters;
- to evaluate what kind of filters are needed in the interpolation or decimation stages, CIC [1] or polyphase filters can be used;
- to design the selected filters by using MATLAB;
- to carry out a floating point simulation where the system performance is evaluated through two kinds of results: the Error Vector Measurement (EVM) of the generated signal, and the bit error rate (BER);
- to evaluate the finite precision for each block by comparing the use of fixed point operators with the floating point design;
- to generate a VHDL code of the fixed point system, in which each block is an area optimized relatively placed macro (RPM)

2.1   Design Flow

There are several steps during the design process with DIGIMOD. The first step is the selection of those blocks needed in the design: source, modulation mapping, pulse shaping, interpolation filters, (and mixer).

Once the filter chain is specified, a simulation using floating point precision is performed to evaluate if all the specifications are fulfilled. The tool generates the frequency response of the filter chain, as well as of individual filters, and two quality measurements: BER and EVM. If the obtained results do not match with system specifications the user can aggregate or delete blocks, or adjust block parameters until simulation gives the correct results.

After tuning the floating point design, the user can begin with the fixed point evaluation in a similar way. Simulations are performed in each step in order to asses the user to choose the number of bits needed.

If finite precision simulation accomplishes the specifications of the application, the tool is ready to generate the VHDL code of the whole system for a target FPGA device. After DIGIMOD generates the VHDL code, it can be synthesized with Synplify and implemented with ISE Xilinx tool.

2.2   Implementation Technology

Pulse shaping can be designed using two implementation methods: polyphase filter or look-up table [2]. Look-up method makes use of the embedded block select RAM available in a FPGA device. This method allows a higher interpolation factor than polyphase filter at a lower cost and simplifies the interpolation filters that come later in the transmitter chain.

Polyphase filters can be used for pulse shaping, matched filtering and interpolation, they are implemented with bit-serial or digit-serial distributed arithmetic [3].

Mixers can be built using two methods. The first performs a CORDIC-based mixer [4] and the second a ROM-based one [5]. This last method uses compression techniques to reduce the size of the required memories

The VHDL code generated by DIGIMOD tool instantiates different cores that perform the required operations. All the cores have been described in VHDL by using relative placed attributes. Furthermore, they are area efficient with respect to same blocks generated with Xilinx Coregenerator system [6].

3   A Design Example

For a better understanding of DIGIMOD characteristics and performance, in this section we will present a design example of a digital IF QPSK modulator using DIGIMOD tool. In order to benchmark the performance of our tool we will compare it with the results given by Xilinx System Generator [6]. The QPSK modulator parameters are: 2 Mbps of bit rate, QPSK with a roll-off 0.3 root raised cosine pulse shape and a length of 8 symbols, an IF of 10.7 MHz, and a DAC with 8 bits and 40 MHz of sampling rate.

The interpolation chain used consists of: a polyphase structure for pulse shaping with an interpolation factor of 4, a halfband 6th order FIR interpolator, and a 3rd order CIC with an interpolator factor of 5.

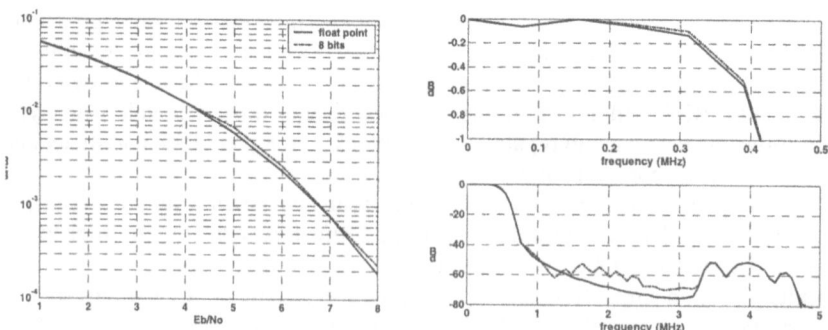

Fig. 1. Floating vs. fixed (dotted) point implementations: BER and signal spectrum

After some simulations fixed point parameters are: 8 bits for filter coefficient, both in pulse shaping and half band, and 8 bits for signal output for every block in the chain system. Figure 1 shows the comparison in BER between floating point design and 8 bits fixed point design (with a loss of less than 0.1 dB), and the signal spectrum at CIC filter output in both the passband (above) and the stopband.

Table 1 shows the results of three implementations of this design in a Xilinx VirtexE-8 device. The first one has been performed with Xilinx System generator, the second and third ones use DIGIMOD tool, where pulse shaping is performed with a polyphase filter and with the look-up table method. These results show that DIGIMOD implementation is more efficient than system generator one. It is achieved an area saving of 45% with the polyphase filter option. If the look-up table method is used, 2 Block Select RAMs and 27 slices are required. Finally, it can also be made out that DIGIMOD requires less area to implement each block and, furthermore, does not require extra resources to connect those blocks employed, as System generator does.

Table 1. Implementation results

Resource (slices)	System Generator	DIGIMOD	
		Polyphase PS	Look-up PS
Pulse shaping (PS)	201	118	27 (+ 2 BSRAM)
Half band	90	30	
CIC	67	67	67
DDS+MIXER	281	189	189
QPSK modulator	1137	627	350 slices + 2 BSRAM

4   Conclusions

This paper presents a software tool that allows software radio designers to develop digital transceivers: from simulation to VHDL code generation for Xilinx FPGAs. As a design example an IF QPSK modulator has been implemented. Obtained results are compared with those given by Xilinx System Generator, it is shown that our tool leads to an area efficient implementation.

References

1. E. Hogenauer, "An Economical class of Digital Filters for Decimation and Interpolation", IEEE Transactions on Acoustic, Speech and Signal Processing, vol ASSP-29, n°2, April 1981.
2. José Marin-Roig, Javier Valls, Vicenç Almenar, "LUT-based Up-converters for FPGA", Communication Systems, Networks and Digital Signal Processing Conference, July 2002
3. Stanley A. White, "Aplications of Distributed Arithmetic to Digital Signal Processing: A Tutorial Review", IEEE ASSP Magazine, July 1989
4. F. Cardells, J. Valls, "Optimisation of direct digital frequency synthesizers based on CORDIC", Electronic Letters, Vol. 37, no. 21, pp. 1278-1280, October 2001
5. F. Cardells, J. Valls, "Area-Optimized Implementation of Quadrature Digital Direct Synthesizer on LUT-based FPGAs", IEEE Trans. on Circuits and Systems II, Vol. 50, no. 3, pp. 135-138, March 2003
6. www.xilinx.com.

FPGA Implementation of a Maze Routing Accelerator

John A. Nestor

Department of Electrical and Computer Engineering
Lafayette College
Easton, Pennsylvania 18042 USA
nestorj@lafayette.edu

Abstract. This paper describes the implementation of the L3 maze routing accelerator in an FPGA. L3 supports fast single-layer and multi-layer routing, preferential routing, and rip-up-and-reroute. A 16 X 16 single-layer and 4 X 4 multi-layer router that can handle 2-16 layers have been implemented in a low-end Xilinx XC2S300E FPGA. Larger arrays are currently under construction.

1 Introduction

The classic Lee Algorithm for *maze routing* [1] remains popular in electronic design automation because it is guaranteed to find a shortest-path connection if one exists. The algorithm represents the routing surface as a rectangular grid. During the *expansion* phase, it searches outward from the *source* node of a desired connection in breadth-first fashion while labeling each encountered node to indicate the shortest path back to the source. When the *target* node is found, the *backtrace* phase selects a path by following these labels while marking the path as an obstacle. The *cleanup* phase clears the remaining labels before additional connections are routed.

Although popular, the Lee Algorithm is computationally expensive. For single layer routing, expansion is $O(d^2)$ for a connection of distance d, and cleanup is $O(N^2)$ for an N X N grid. Multilayer routing is even more costly. This has motivated several proposals for hardware accelerators. *Direct grid* accelerators (e.g., [2]) map each node into an array of simple processing elements (PEs). This reduces the expansion time to $O(d)$ and cleanup time to $O(1)$ but requires N^2 processing elements for a single-layer. *Virtual grid* accelerators (e.g. [3]) map more than one node onto each PE. This requires fewer PEs, but each PE must be significantly more complex to handle the multiple mappings. Other acceleration approaches include raster pipelines [4], specialized processors [5], and accelerators intended for routing FPGAs, which have a specialized routing structure [6].

This paper describes a return to the direct grid approach called L3. L3 improves over previous direct grid approaches by providing (1) efficient support for multiple layers; (2) a significant reduction in PE logic; and (3) support for quick initialization and removal of obstacles and connections during rip up and reroute. A preliminary version of L3 was described in [7]; this paper describes a refined version and its implementation in an FPGA.

P.Y.K. Cheung et al. (Eds.): FPL 2003, LNCS 2778, pp. 992–995, 2003.

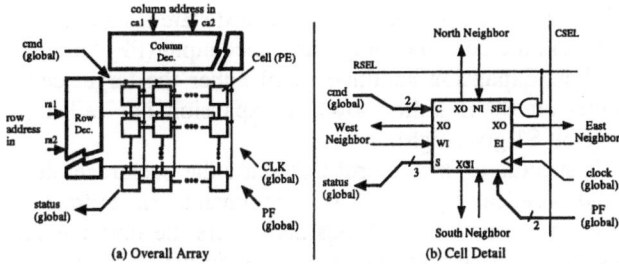

Fig. 1. L3S Organization. The attached control unit is not shown

2 The L3 Architecture

Figure 1 shows the general organization of L3S, the single-layer version of L3. Each *cell* (PE) is connected locally to neighboring cells in the grid using output XO. XO is true when a cell is labeled during expansion. It is connected to the WI, EI, NI, and SI inputs of the cell's west, east, north, and south neighbors, respectively. An attached control unit (not shown) communicates with the array of cells using a global command bus CMD, and a row and column decoder that allow the selection of either individual cells or rectangular regions of cells. Cells pass status information back to the control unit using and a global tristate/wired-OR STATUS bus. Global input PF supports preferential routing and will be discussed later.

Each cell is a finite state machine with 6 states: E (empty), BL (blocked), XE (expanded east), XW (expanded west), XN (expanded north) and XS (expanded south). Table 1 describes the function of each cell as it responds to one of four commands: CLEAR, SET, EXPAND, and TRACE.

Table 1. Cell state sequencing in response to CMD. X* indicates any expanded state

CMD	SEL	EW * PF1	WI * PF1	NI * PF0	SI * PF0	PS	NS	S1	S0
CLEAR	0	–	–	–	–	X*	E	1	1
	1	–	–	–	–	Any	E	1	1
SET	1	–	–	–	–	Any	XE	1	1
TRACE	1	–	–	–	–	Any	BL	D1	D0
EXPAND	–	1	–	–	–	E	XE	0	SEL'
	–	0	1	–	–	E	XW	0	SEL'
	–	0	0	1	–	E	XN	0	SEL'
	–	0	0	0	1	E	XS	0	SEL'
	–	–	–	–	–	X*	PS	1	SEL'
All other conditions							PS	1	1

To perform maze routing, the control unit first selects all cells and broadcasts a CLEAR command to set all cells to the E state. It then selects the source node and uses the SET command to set the source cell to the XE state. It next applies the EXPAND command while selecting the location of the target cell. During each successive clock cycle, a cell will enter an expanded state if one of its neighbors is in an expanded state and the corresponding preference input (PF1 for east/west, PF0 for north/south) is high. Expansion continues until either the target cell is reached (at which point status bit S0 is pulled low) or expansion has failed (in which case "watchdog" status bit S1 goes high).

If expansion is successful, the control unit starts the backtrace phase by selecting the target cell and broadcasting the TRACE command. In response, the target cell asserts its state code on the STATUS bus and enters the obstacle state (BL). The control unit uses the direction information encoded in the state code to determine the address of the next cell in the path and repeats this process until the source node is reached and the entire path is marked as an obstacle. To route another connection, the CLEAR command is applied with no cells selected to remove expansion labels (but not obstacles) and the process repeats. The entire process takes d clock cycles for expansion, d clock cycles for backtrace, and 1 clock cycle for cleanup.

L3M is the multi-layer version of the L3 architecture. It uses the same array structure as L3S but time-multiplexes cell hardware over multiple layers. Figure 2 shows the organization of a single L3M cell, which uses a shift register to store the states of each layer. The L3M cell processes states from bottom to top on each successive clock cycle. The state sequencer is similar to that of the L3S cell except that it has two additional states XU and XD to support vertical expansion. Expansion information from the layer "above" the current layer is taken from the next shift register stage. Expansion information from the level "below" the current layer is stored in a flip-flop at the end of the preceding cycle. Both calculations are suppressed by the /TOP signal when the top layer is reached. The preferential routing input PFV allows vertical expansion to be suspended; when zero the sequencer recirculates the current layer so that horizontal expansion can continue on successive clock cycles. The L3M array will find a connection in $O(L*d)$ cycles for L layers.

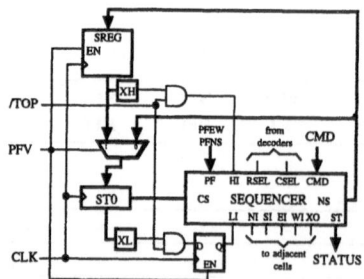

Fig. 2. The L3M Cell design. Layers are processed from bottom to top

3 Implementation Results

As a proof of concept, the modules of the L3S and L3M accelerators have been coded in Verilog HDL and synthesized using the Xilinx ISE 4.2i and Synopsys FPGA Express tools targeting a Xilinx XC2S300E FPGA [8] on a Memec Development Board [9]. The synthesized L3S cell requires 17 4-input lookup tables (LUTs), 3 D flip-flops (DFFs), and 3 tristate buffers (TBUFs). The L3M cell requires 32 LUTs (including 3 SRL16 shift registers), 3 DFFs, and 3TBUFs.

Table 2. L3S and L3M implementation costs and predicted performance

Array Size	Serial LUTs/FFs	Control LUTs/FFs	Array LUTs/Ffs	Total LUTs/FFs	Clock (ns)
4 X 4	173 / 141	167 / 15	260 / 48	600 / 204	38ns
8 X 8	173 / 141	171 / 19	1083 / 192	1427 / 352	40ns
16 X 16	173 / 141	370 / 23	4067 / 768	4610 / 932	51ns
4 X 4 X 4	173 / 141	238 / 23	425 / 50	836 / 214	41ns

Table 2 shows the results of synthesizing the complete L3 router including cell array, decoders, control unit, and a serial interface to communicate with a host computer. All designs were tested and work properly, although the 16 X 16 single-layer design failed to meet the timing constraint.

The LUT requirements of the L3S and L3M cells can be used to predict the size of router that can be accommodated by a larger FPGA. For example, the Virtex-II Pro XC2VP125 device [8] contains 111,232 LUTs and could accommodate an 82 X 82 array of single-layer L3S cells or a 58 X 58 array of L3M cells.

5   Conclusion

This paper has described a new architecture for a direct-grid hardware routing accelerator and its implementation using FPGAs. The L3 architecture supports multiple layers, preferential routing, and iterative rip-up-and reroute. A working 16 X 16 single-layer router and 4 X 4 multi-layer have been demonstrated in a low-end FPGA. Future work will include implementation of larger routers, further design refinement to improve clock cycle time, and detailed performance comparisons of the hardware router to software implementations.

References

1. Lee, C. Y. "An Algorithm for Path Connections and its Applications," *IRE Transactions on Electronic Computers* vol. EC-10, no. 2, pp. 346-365, 1961
2. Breuer, M., and Shamsa, K. "A Hardware Router," *Journal of Digital Systems*, vol. IV, no. 4, pp. 393-408, 1981
3. Ventkateswaran, R., and Mazumder, P., "Coprocessor Design for Multilayer Surface-Mounted PCB Routing,", *IEEE Trans. VLSI Systems*, vol. 1, no. 1, 1993
4. Rutenbar. R. and Mudge, T., "A Class of Cellular Architectures to Support Physical Design Automation", *IEEE Trans. CAD*, Vol. CAD-4, No. 4. October 1984
5. Won. Y, Sahni, S., and El-Ziq, Y., "A Hardware Acclerator for Maze Routing", *Proceedings Design Automation Conference*, June 1987
6. Huang, R, Wawwrzynek, J., and DeHon, A., "Stochastic, Spatial Routing for Hypergraphs, Trees, and Meshes", *Proc. International Symposium on FPGAs*, February 2003
7. Nestor, J. "A New Look at Hardware Maze Routing", *Proceedings Great Lakes Symposium on VLSI*, March 2002
8. Xilinx, Inc., *Xilinx Databook*, 2003. Available online at http://www.xilinx.com
9. Memec, Inc., *Spartan-IIE Development Board User's Guide*, 2002

Model Checking Reconfigurable Processor Configurations for Safety Properties*

John Cochran, Deepak Kapur, and Darko Stefanović

University of New Mexico, Albuquerque NM 87131, USA,
cochran@cs.unm.edu

Abstract. Reconfigurable processors pose unique problems for program safety because of their use of computational approaches that are difficult to integrate into traditional program analyses. The combination of proof-carrying code for verification of standard processor machine code and model-checking for array configurations is explored. This approach is shown to be useful in verifying safety properties including the synchronization of memory accesses by the reconfigurable array and memory access bounds checking.

1 Introduction

Reconfigurable computing is a rapidly evolving technology that has great potential for improved processing efficiency for important computational tasks. This improvement, however, comes at the expense of increased risk of problems from faulty or malicious programming. We are exploring model checking combined with proof-carrying code as a method of ensuring safety of reconfigurable processor programs. We show that significant safety properties can be *efficiently* and *automatically* verified by model checking.

2 Approach

The novel approach explored here uses model checking to verify safety properties of the reconfigurable array, and proof-carrying code (PCC) [Nec98] to verify safety properties of the standard machine code. In addition, proof-carrying code provides a context for model checking the reconfigurable array by providing preconditions, postconditions, memory partitions between standard processor executions and array executions, and local safety properties. This is achieved by extending PCC's context mechanism for function calls to deal with indeterminacy from reconfigurable array executions.

The extensions include new instructions for accessing array registers, loading configurations, and starting array execution. The semantics and symbolic execution of old instructions are updated to take into account reconfigurable array execution, including an ϵ-calculus based semantics for the symbolic evaluator to model inaccessible and indeterminate values during array execution. The extensions are documented in [Coc02].

The reconfigurable processor which we use as our example is the Garp processor [Hau00]. This processor has not been physically implemented but it has been thoroughly specified and documented, which is critical for proving safety properties.

---

* Partially supported by the NSF Grants nos. CCR-9996150 and ITR-CCR-0113611.

3   Model Checking

Model checking is used to verify safety properties of the reconfigurable array. Model checking is a formal method that automatically examines finite models of concurrent systems to verify properties of the models. In this work we use RTCTL [Cam96], a branching-time propositional temporal logic, as a property specification language. RTCTL uses bounded temporal quantifiers to implement bounded model checking.

We use the NuSMV system [CR98] for the verification of array configurations. NuSMV uses the language SMV to describe models. This language is similar to hardware description languages. NuSMV supports RTCTL as a specification language.

We build the model by translating bit-level encodings of configurations to the SMV language. Any information from the surrounding machine code that is needed to verify safety must be used in the safety properties for the configuration as detailed in [Coc02].

4   Properties to Be Model Checked

The generic properties that will be checked for *all* configurations include:
- At most one memory access is initialized per clock cycle
- At most one memory access is scheduled to use the bus for each cycle
- There is a memory item ready to read when a row reads one
- There is a row initiating a memory write when a row transfers to memory

The generic properties all deal with synchronizing memory accesses so that they are defined by the semantics of Garp's reconfigurable array. If any of these properties is false, then there can be undefined behavior from the array. From a list of which control blocks can initiate memory accesses, it is possible to deduce these specifications.

The context dependent safety properties for a *particular* configuration include:
- All memory accesses respect the memory partition
- All memory accesses respect the memory access safety properties
- The postcondition is true after array execution

The context dependent properties all rely on information from the safety policy and the symbolic evaluation of the program where the array configuration is executed.

Both types of properties can also rely on the value of the initial count for the array execution. This count gives the number of cycles that the array executes if it does not halt itself first. The actual or maximum count can be used as a bound for temporal quantifiers so that the behavior of the model after the array would halt is not checked.

5   Performance of NuSMV on Translated Input Files

The performance of the model checker, NuSMV 2.0 running on an AMD Athlon at 1900 MHz with 1024 MB of memory under Debian Gnu/Linux 2.2, is the main factor to be explored for performance. Four example configurations were checked for six properties, the first is a generic property and the rest are context dependent properties:
- Memory control safety (MC)
- Memory alignment (MA)
- Lower bounds for memory reads (LBR)

- Upper bounds for memory reads (UBR)
- Lower bounds for memory writes (LBW)
- Upper bounds for memory writes (UBW)

The four configurations include three that perform the same function, but have different control paths and preconditions. The application is to read 200 word-sized pixels from an array, lighten each color component, and write the results back to the array.

The first configuration (IM1) has the precondition that the register in the fifth row of the reconfigurable array is loaded with a value equal to the value loaded in the first row of the reconfigurable array minus 14. This is because the first row reads the pixel array, the fifth row writes the result back, each of them is incremented by 2 on each cycle, and it takes seven cycles for the computation. The second configuration (IM2) has the precondition that the values loaded into the first and fifth row registers are equal. This is because the fifth row does not start incrementing until it is signaled on cycle 7 by the control path. The third configuration (IM3) does not have any corresponding precondition because it passes addresses from the first row to the fifth alongside the computation so that the value in the fifth row is always correct.

The fourth configuration (HASH) is a simple hash table. It is included to check the effect of *computed addresses* on the model checking. It reads values from an array, computes a 10-bit offset by repeated shifts and exclusive ors, and writes the value to an address plus the offset.

Each of the configurations has preconditions to ensure that the control path is correctly initialized, and that the access locations fit into memory without wraparound. All of the configurations have the trivial postcondition *true* because they do not leave any values in the array registers for later use.

The results are presented in Table 1 for properties that did not require an execution count related bound, and Table 2 for execution count bounded properties. The memory write properties for IM1, IM3, and HASH do not appear in Table 2 because they did not finish model checking in under four days. The reasons for the poor performance on the write boundary properties (LBW, UBW) are varied. For IM1, the precondition requires a very long bit-level specification and therefor a large symbolic representation that makes even trivial specifications practically uncheckable. IM3 and HASH both have a write address that is dependent on several rows of registers and cycles. This seems to be more than the model checker can handle efficiently as there is less possibility for cone-of-influence reduction of the symbolic representation. Even for IM2 it took a significantly longer time to check the write boundary properties as they depend on more of the array than the rest of the properties. Further information is available in [CKS03].

Table 1. Results of Model Checking Example Configurations for Unbounded Properties. Model checking times are given in seconds.

Property	Configuration			
	IM1	IM2	IM3	HASH
MC	6.710	6.140	6.810	17.400
MA	6.740	13.690	14.140	33.670

Table 2. Results of Model Checking Example Configurations for Bounded Properties. Model checking times are given in seconds.

Configuration / Property	Time Step Bound						
	400	1000	1024	4000	4096	16000	16384
IM1 LBR	15.23	16.84	16.92	24.10	24.42	54.00	54.76
IM1 UBR	16.97	20.66	20.26	38.78	36.95	106.72	98.70
IM2 LBR	15.13	16.55	16.65	23.20	23.12	49.38	50.04
IM2 UBR	16.59	19.98	19.99	36.57	36.57	102.92	94.32
IM2 LBW	64.00	76.11	70.52	141.71	88.37	161.54	414.43
IM2 UBW	69.64	87.61	76.30	109.13	197.60	240.14	239.10
IM3 LBR	16.12	18.63	18.91	30.70	31.22	80.50	81.77
IM3 UBR	18.00	23.73	23.22	48.70	47.40	147.10	142.53
HASH LBR	27.27	29.30	29.20	38.84	39.29	78.56	79.93
HASH UBR	28.42	32.99	32.65	58.93	57.52	168.92	169.41

6   Conclusions and Further Work

The main result of this work is that model checking reconfigurable processor configurations is a viable verification method for important safety properties, in particular the memory control properties. Although some memory access boundary properties have been found to be too complex for efficient checking, there may be methods to mitigate this in many cases. As the examples show, configurations having equivalent computations with different control strategies can have very different behavior when model checked. This could be taken into account in a compiler designed to produce efficiently checkable configurations.

Integrating model checking into a synoptic system of program verification in order to solve these problems is being explored. In particular, replacing proof-carrying code based on first-order logic with proof-carrying code based on temporal logic [BL02] is expected to provide a greater range of safety properties that can be checked. This may allow model checking of easy-to-check properties, which are then integrated into the proof of safety attached to the program as a whole.

References

[BL02]   A. Bernard and P. Lee. Temporal logic for proof-carrying code. Technical Report CMU-CS-02-130, School of Computer Science, Carniege Mellon University, Pittsburgh, PA, 2002.

[Cam96] S.V. Campos. *A Quantitative Approach to the Formal Verification of Real-Time Systems*. PhD thesis, Carnegie Mellon University, 1996.

[CKS03] J. Cochran, D. Kapur, and D. Stefanović. Model checking reconfigurable processor configurations for safety properties. Technical Report TR-CS-2003-18, Computer Science Department, University of New Mexico, 2003.

[Coc02] J. Cochran. Towards provably safe reconfigurable processor code: A model checking and proof-carrying code approach. Master's thesis, University of New Mexico, 2002. Available as Technical Report TR-CS-2002-36.

[CR98]   A. Cimatti and M. Roveri. *NuSMV 1.1 User Manual*. ITC-IRST and CMU, 1998.

[Hau00] J.R. Hauser. *Augmenting a Microprocessor with Reconfigurable Hardware*. PhD thesis, University of California, Berkeley, 2000.

[Nec98] G.C. Necula. *Compiling with Proofs*. PhD thesis, Carnegie Mellon University, 1998.

A Statistical Analysis Tool
for FPLD Architectures

Renqiu Huang[1], Tommy Cheung[2], and Ted Kok[2]

[1] University of Cincinnati, Cincinnati, OH 45220,USA
`huangr@ececs.uc.edu`
[2] Hong Kong University of Science and Technology,
Clear Water Bay, Kowloon, HKSAR, China
{`eetommy,eekok`}`@ust.hk`

Abstract. This paper investigates an analysis tool for the routing resources in the FPLD architecture design. The developed tool can assess the performance of a given architecture specified by the physical configuration of logic blocks and the switch boxes topology. Two problems are mainly considered in this paper: given an architecture, the terminal distribution of each switch box is first determined via probabilistic assumptions, then the sizes of required universal switch boxes are evaluated for routing successfully. The estimations are validated by comparing them with the results obtained in the previous published experimental study on FPGA benchmark circuits. Moreover, our result confirms that the universal switch block is a good candidate for FPLD design.

1   Introduction

As design cycle is shortening and design complexity is increasing, system designers respond to these pressures by moving to the more cost-effective block-based designs. Field-Programmable-Logic Devices (FPLDs) and intellectual property (IP) techniques, without designing circuit from scratch, help designers to build the required complexity in a short time-to-market. Being a relatively new technology, FPLDs are constantly undergoing upheaval changes in their architectures to cope with the increasing component density and versatility as demanded by new IP functionality. This paper aims to develop an architectural analysis tool that provides the FPLD architects with the ability to perform trade-offs in designing a new FPLD architecture. The developed tool does not assume a particular architecture, which could be hierarchical and heterogeneous. However, it assumes the switch blocks to be universal [3]. Through a statistical analysis, information will be provided to the architect, identifying the possible deficiencies in the architecture such as routability reaching a low threshold value. This prompts the architect to improve the configuration of design.

2   Model and Analysis

A generic FPLD architecture is illustrated in Fig. 1. It consists of some number of different sizes of nonoverlapping polygons. There are two kinds of polygons.

P.Y.K. Cheung et al. (Eds.): FPL 2003, LNCS 2778, pp. 1000–1003, 2003.

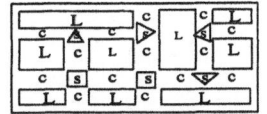

Fig. 1: A FPLD architecture (L for Logic
block, S for Switch box, and C for Channel)

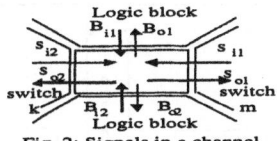
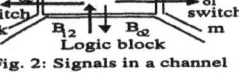

Fig. 2: Signals in a channel

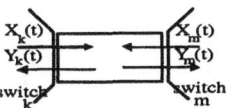

Fig.3: Modified signal flow
model for a channel

The one labeled with "L" is a configurable logic box, which may include look-up tables, flip-flops, and so on. The other labeled polygon with "S" is a switch box, which defines the connections between logic boxes. The white space between two sides of switch boxes is referred to as a routing channel. As for the switch boxes, they are assumed to be universal [2, 3]in this architecture. It has been proved in [3] that the universal switch blocks can accommodate significantly more routing instances than Xilinx XC4000-type switch blocks. Tsu, in his report [8], also concluded that the universal switch blocks require smaller silicon areas. The architectural model presented is generic, possibly hierarchical and heterogeneous.

The typical signals, shown in Fig. 2, may be related to Logic Blocks (B_{i1}, B_{o1}, B_{i2}, B_{o2}) or switches (S_{i1}, S_{o1}, S_{i2}, S_{o2}), where the subscript denotes the signal flow directions with "i" for going into the channel, and "o" for getting out of the channel. As a result: $B_{i1} + B_{i2} + S_{i1} + S_{i2} = S_{o1} + S_{o2} + B_{o1} + B_{o2}$, Since our objective is to analyze the routing requirement of a universal switch box, therefore, an equal model as shown in Fig. 3 is considered which only consider signals generated or terminated at the channel under consideration. To analyses the routing resource requirement in a given architecture, there are two basic assumptions: the signals are randomly generated at the side of a logic block according to the Poisson distribution and each signal connection is assumed to have a random path [4-7]. Let $X_k(t)$ denotes the number of non-terminating signals measured at location t in the channel switch k which enter the channel from k. Furthermore, let $Y_k(t)$ denotes the number of signals measured at location t of the channel from k that enter the switch box k. Assumed the signal entering the logic blocks from the channel has a random arrival. Therefore, it can be modeled by Poisson distribution with an arrival rate λ. Further assuming that the probability of a signal to make a connection after traveling a distance t follows the exponential distribution with an average length α. Same as [5], the total number of connections made at location t of the channel as measured from k is given by $Z_k(t)$, where $Z_k(t) = X_k(t) + X_m(t) + Y_k(t) + Y_m(t)$. Noted that $Z_k(t)$ is also Poisson with parameter $\lambda_{Z_k}(t) = E(Z_k(t))$. When $t = 0$, $Z_k(0)$actually denotes the number of signal emerging from the k-th side of the switch box. If the switch box has N sides, the total number of signal emerged from the switch box to neighboring channels are given by $\Gamma = \sum_{k=1}^{N} Z_k(0)$. Since $Z_k(0)$ is Poisson with parameter $\lambda_{Z_k(0)}$, Γ is also Poisson with parameter $\lambda_\Gamma = \sum_{k=1}^{N} \lambda_{Z_k(0)}$. Depending on the connection topology of the switch box, each of the signal in Γ can be connected by one of the K different types of connection. Let $A_i, i = 1, \ldots, K$, denotes the random variable for the numbers of i-th type switch used in the switch box. Therefore, the probability of successful connection for all the signals in Γ is given by the multinomial distribution

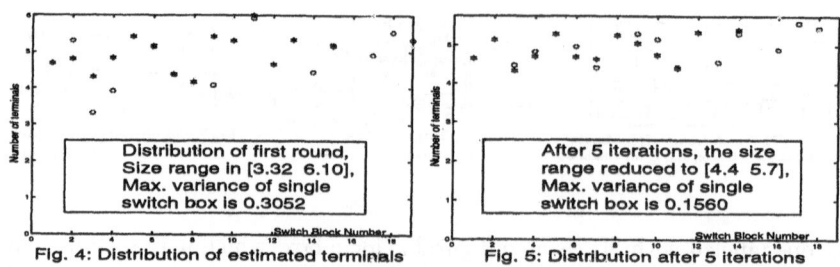

Fig. 4: Distribution of estimated terminals   Fig. 5: Distribution after 5 iterations

of the joint probability of A_i with a given required number, γ, of signals to be connected.

$$P\left(A_1 = n_1, \ldots, A_K = n_K \middle| \Gamma = \left(\gamma = \sum_{i=1}^{K} n_i\right)\right) = \frac{\left(\sum_{i=1}^{K} n_i\right)!}{n_1! n_2! \ldots n_K!} P_1^{n_1} \ldots P_K^{n_K} \quad (1)$$

where P_i are the probability of connecting a signal using the i-th switch, and, $\sum_{i=1}^{K} P_i = 1$. The above conditional probability can be simplified by observing that condition $\Gamma = \gamma$ is independent of the A_i process. Therefore, by sampling the Poisson process

$$P\left(A_1 = n_1, A_2 = n_2, \ldots, A_K = n_K\right) = \prod_{i=1}^{K} \frac{(P_i \lambda_\Gamma)^{n_i} e^{-P_i \lambda_\Gamma}}{n_i!} \quad (2)$$

As a result, the above equations can be used to estimate the routability of FPLDs with universal switch blocks.

3   Experimental Results and Conclusions

The first experiment demonstrates the application of the developed analysis tools for optimizing FPLD architecture to minimize the number of switches requirement. The switch requirement can be minimized by equalizing the number of signals emerging from each side of the switch box. This is important for universal switch design. The optimization begins by choosing the switch box that has the largest difference between the number of signals emerged from two sides of the switch box. The positions of the neighboring logic blocks are then swapped. After swapping the logic blocks, the numbers and types of connections at each box are computed again, followed by another round of logic block swaps. Several FPLD architectures that follow Fig. 1 are developed by randomly choosing the positions, sizes, and the number of logic blocks. The total numbers of switches required are reduced, observations are shown in Fig. 4 and Fig. 5 respectively. Further, the number of switch blocks is also reduced (18 verse 19). The parameters λ and α used in our simulation are predicted as in [4].

The second experiment estimates the switch resource requirement (listed in Table 1.) for a Xilinx XC4000 alike architecture but with the Xilinx clique switch replaced by the universal switch box. Noted that the number, N, is chosen, such

Table 1. Switch box size for successful routing of a given circuit and logic blocks

Circuit	logic blocks	Altor [10] Sroute [9]	VPR	Splace [9] Sroute	VPR	Channel Width Max.	Min.	Switch sizes Max	Min
9symml	70	7	6	7	5	8	4	10	8
apex7	77	9	9	6	4	8	4	10	8
example2	120	11	10	7	5	9	5	10	8
Alu2	143	9	8	8	6	9	5	10	8
teml	54	8	7	5	4	7	3	10	8

that the total numbers of logic blocks, $N \times N$, is large enough to implement the logic of the given circuits. Comparing the channel width requirement obtained after placement and routing (first and second row of 3-5th column) in Table 1 with the estimated minimum number of tracks in each channel, it is observed that the results are comparable except for the case of VPR [1] routing and placement. This is because VPR used simulated annealing in the placement and routing algorithm, thus archiving a global minimum in the switch size. Another observation is that all the requirements of switch size are same, there are two reasons: first, the algorithm only consider different switch box configuration, i.e. switch box with different number of switches of each type; second, the rounded values hide the tendency of increase. On the other hand, the relative steadiness of switch size indicates that the universal switch block is a good candidate topology for FPLD design. Listed in Table 1 is the bounds for the switch box size that can achieve higher than 99% of successful routing as estimated by eq.(2).

References

1. V. Betz, J. Rose, and A. Marquardt, *Architecture and CAD for deep-submicron FPGAs.* Kluwer Academic publishers, 1999.
2. Y. Chang, D. F. Wong, and C. K. Wong, *Universal Switch Modules for FPGA Design.* ACM Trans. on Design Automation of Electronic Systems, Jan. 1996.
3. M. Shyu, G. Wu, Y. D. Chang, and Y. W. Chang, *Generic universal switch blocks.* IEEE Trans. Comput., pp.348-359 Apr. 2000.
4. S. D. Brown, J. Rose, and Z. G. Vrabesic, *A stochastic model to predict the routability of field programmable gate arrays.* IEEE Trans. CAD, Dec. 1993.
5. A. A. El. Gamal, *Two-dimensional stochastic model for interconnections in master slice integrated circuits.* IEEE Trans. CAS-I, no. 2, pp. 127-138, 1981.
6. A. A. El. Gamal, *A stochastic model for interconnections in custom integrated circuits.* IEEE Trans. CAS-I, pp.888-894, Sept. 1981.
7. S. Sastry and A. C. Parker, *Stochastic models for wirability analysis of gate arrays.* IEEE Trans. Computer-Aided Design, pp.52-65, Jan. 1986.
8. W. Tsu, *A comparison of universal and Xilinx switches.* CS294-7 project report, Univ. of California-Berkeley, Spring 1997.
9. S. Wilton, *Architectures and Algorithms for Field-Programmable Gate Arrays with Embedded Memories.* Ph.D. Dissertation, University of Toronto, 1997.
10. J. Rose, W. Snelgrove and Z. Vranesic, *Altor: an automatic standard cell layout program.* Proc. of Canadian Conf. on VLSI, 1985, pp. 169-173.

FPGA-Implementation
of Signal Processing Algorithms
for Video Based Industrial Safety Applications

Jörg Velten and Anton Kummert

Communication Theory,
University of Wuppertal,
42119 Wuppertal, Germany
{velten,kummert}@uni-wuppertal.de

Abstract. Conventional protective devices as light curtains allow safe but often inconvenient flow of work. Unfortunately uncomfortable safety devices are often bypassed or simply switched off. Consequently, the design of a video based protective device, which avoids inconvenient processing steps is of special interest. The present paper describes favorable combinations of FPGA-hardware and algorithms, which allow safeguarding of work places if several constraints are met. The methods were originally developed for surveillance of press brakes, but it is easily adaptable to different types of machines or work places.

1 Introduction

Development of video based industrial safety devices is still in its beginnings. It is discussed intensively by responsible authorities for the aim of safeguarding complex situations and enable convenient work flows. Problems arise from the combination of demands for high recognition reliability and low device prices, i.e. ability to implement detection algorithms in low cost hardware. Especially a high-speed implementation of video data processing at convenient frame rates represents a challenging barrier. The present paper describes the implementation of suitable combinations of algorithms and hardware for a video based protective device at press brakes in FPGA architecture. A camera takes images from a bird's eye view perspective and delivers the usual R- G- and B- signals with 8-bit resolution each. Fig. 1 shows at the lefthand-side the intensity information of an example image. Two independent algorithms, which follow different principles for ensuring an inviolated "region of danger" are described below. The first algorithm follows the principle of detecting any object intruding a buffer zone between worker and machine, while the second one verifies hands at risk-less positions, as described in [2].

Selection of algorithms has been performed with respect to usual FPGA-elements, mainly 4:1 LUTs with subsequent latch, in this paper referred to as Logic Element or short LE. Consequently, realizations are not restricted to a specific FPGA-family.

P.Y.K. Cheung et al. (Eds.): FPL 2003, LNCS 2778, pp. 1004–1007, 2003.

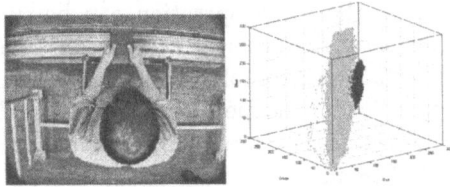

Fig. 1. Example image of a press brake (grayscale) at the left-hand side and the representation of the colour-image in RGB color space at the right-hand side

2   Surveillance of a Danger Zone

Surveillance of a danger zone can be established by comparison of camera images with a previously acquired image of the empty workplace. For reduction of illumination influences, differentiation between edge-images instead of intensity images is performed. The Sobel operator (cf. [4]) uses a 3 x 3 pixel filter mask containing the values ± 2 , ± 1 and 0, and is consequently suitable for an FPGA implementation. Detected differences are subsequently compared to predefined image masks that represent warning and danger areas. Further independence from illumination conditions is obtained by additional binary morphological operations, namely dilation of the edge image and erosion of the difference image, each performed by a 3 x 3 pixel structuring element (cf. [3]). The whole processing cycle is shown in Fig. 2.

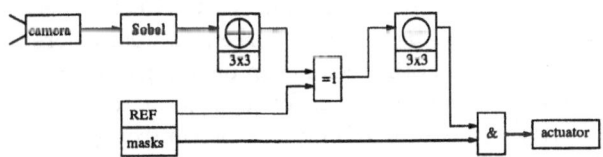

Fig. 2. The processing cycle of "surveillance of a danger zone"

The Sobel operator filter mask can be transferred directly to an FPGA implementation by using a usual 2D-FIR filter for serial image data, where delays in horizontal direction are realized by time delays T_1, equal to the reciprocal of the sampling frequency, and delays in vertical direction are realized by $T_2 = NT_1$, where N denotes the number of pixels in a row. Size of actual FPGAs allow a direct implementation of the filter. Multiplications are realized by shift operations, which are realized by fixed by wiring. The delays T_1 and T_2 are either implemented by using LE-latches, or external shift registers. Several FPGAs allow the conversion of one LE into a 16-bit shift register, which leads to the possibility of combining several LEs to achieve a long delay T_2. By exploiting these abilities, the number of necessary delays for implementation of the 3x3 pixel filter mask and 8 bit grayscale image data of size 284x288 pixel, amounts to 423 LEs. Furthermore, the Sobel operation can be simplified by replacing the square root expression of the gradient calculation by an absolute sum and the search for a maximum by a threshold operation. The full implementation of the simplified Sobel operator for the mentioned input signal needs only 532 LEs (synthesis by LeonardoSpectrum). Morphological operations also can be implemented in a similar 2D-FIR filter structure (cf. [5]) to act in the same way as the Sobel operator, except that signal types are binary, which drastically reduces the necessary hardware effort. Only 100 LEs are used to implement both morphological operations. Comparison with filter masks is done by a bit-wise AND concatenation of the output signal with suitably synchronized read out mask information. The masks as well as the reference image have to be stored in an external memory. The length of each mask word is determined by the necessary mask configuration, i.e. how many different areas have to be distinguished. Timing simulations of this implementation can be performed at frequencies of up to 86 MHz (ModelSim). The number of latency clock pulses is 1184, which leads for 50 Hz cameras and the above mentioned image size to a total recognition time of only 20,22 ms.

3   Detection of Endangered Extremities

The most endangered extremities at press brakes are hands and fingers, which can be recognized by skin color detection. For assessing minimal hardware effort in the present paper, application conditions are limited to a non-varying lighting spectrum and favorable background colors. Fig. 1 shows a color image (in grayscale), as well as its representation in RGB color-space. Furthermore, significance can be enhanced by coercing workers to wear colored gloves. Rough distinction between skin and other colors can be applied in the presented example in RGB space by using a linear plane for separation, which can be established by a very simple ANN consisting of only one perceptron (cf. [1]). This leads to detection of skin color in a non-optimal, but "hardware efficient" way. Fig. 3 shows at the top left hand side the result of a skin color detection performed in the above presented way. Several misclassified pixels have to be eliminated by considering regional information using a binary morphological erosion operation.

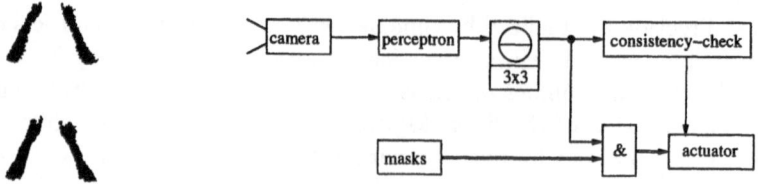

Fig. 3. Respective result images before (top) and after the erosion (botom) are shown at the left-hand side. The block diagram at the right illustrates the processing cycle of the hand detection

The respective result image is shown at the lower left-hand side. The latter is compared to image masks that represent danger and warning zones as proposed in section 2. Further processing is necessary to consider detectability of hands and foresighted danger recognition. Evaluation of these properties however, doesn't need to take place at video frame rate, since respective attributes don't change very fast. For this reason, detectability of hands can be measured evaluating size and number of detected skin regions. The whole processing flow is shown in Fig. 3, at the right-hand side.

The implementation of the perceptron requires at least a resolution of 11 bits for the input weights. Synthesis of a VHDL coded perceptron with three 8-bit input signals leads to the requirement of 444 LEs.

The morphological operation is realised in a filter structure and requires 50 additional logic elements. The comparison with appropriate filter masks is done by bit-wise AND concatenation of the output signal with suitably synchronized read out mask information. Like in the implementation of the first algorithms, masks have to be stored in an external memory, using 8-bit words for each mask pixel. Detection and mask-comparison functionality can be implemented using 564 logic elements. The timing simulation allows frequencies of up to 100 MHz. The number of latency clock pulses is 397, which for 50 Hz cameras and the above mentioned image size leads to a total recognition time of 20,08 ms

4   Conclusion

The development of video based safety devices based on the proposed algorithms leads to extended possibilities for safeguarding of dangerous workflow. Improvements in reliability are necessary for industrial application. A first possible improvement in this respect, forms the combination of both implementations. Although development of video based protective devices is still in its beginnings, the proposed combinations of algorithms and hardware represent promising results for further development. The latter includes adaptive learning of skin color in the FPGA and evaluation of motion tracking algorithms for the purpose of preemptive danger avoidance.

References

1. S. Haykin. *Neural Networks*. Prentice Hall, New Jersey, USA, 1999.
2. W. Lehner, A. Kummert, and J. Velten. Device and method for monitoring a detection range on an operating tool. Application for US patent, 2001.
3. W. K. Pratt. *Digital Image Processing*. Wiley-Interscience, New York, USA, second edition, 1991.
4. I. Sobel. *Camera Models and Machine Perception*. PhD thesis, Stanford University, Stanford, USA, 1970.
5. J. Velten and A. Kummert. FPGA-Implementation of high speed n-D binary morphological operations. In *Proceedingts of the IEEE International Conference on Acoustics, Speech and Signal Processing, (ISCAS'03)*, Bangkok, Thailand, 2003.

Configurable Hardware Architecture
for Real-Time Window-Based Image Processing

Cesar Torres-Huitzil and Miguel Arias-Estrada

Computer Science Department, INAOE, Apdo. Postal 51 & 216, México
ctorres@ccc.inaoep.mx, ariasm@inaoep.mx

Abstract. In this work, a configurable hardware architecture for window-based image operations for real-time applications is presented. The architecture is based on an array of elemental processors under a systolic and pipeline approach to achieve a high rate of processing. A configurable window processor has been developed to cover a broad class of image processing algorithms and operators. The system is modeled in a Hardware Description Language and has been prototyped on an FPGA device. Some implementation and performance results are presented and discussed.

1 Introduction

Window-based image processing requires a large amount of low level processing on massive amounts of data in order to analyze the image contents and recover useful information [1]. The processing is characterized by a large amount of data processed at small neighborhoods by relatively simple operations [2]. In window-based operations the input is an image, and the output, is some symbolic quantity denoting features or location of objects in the image.

The window-based image processing complexity is expressed in terms of the elementary arithmetic operations required to process an image. A window-based operator has a computational complexity of $O(w^2MN)$ for an MxN image with a wxw mask. The complexity is further increased since most vision systems are intended for real time operation [3]. Over 50 million operations are required to process an image of 640×480 pixels, and a computational power of tents of giga-operations per second is required to achieve real-time performance. The time requirements and the inherent limitations on the computational power of conventional processors have motivated the development of dedicated coprocessors for image processing. The nature of window-based operations is inherently local and suitable for data parallel processing. Combining data overlapping and parallel processing, it is possible to achieve high performance in window-based image processing.

The rest of the paper is organized as follows. Section 2 presents the hardware architecture developed for window-based image processing. Section 3 presents some implementation and performance results. Finally, section 4 presents some conclusions and further work derived from this research.

P.Y.K. Cheung et al. (Eds.): FPL 2003, LNCS 2778, pp. 1008–1011, 2003.

2   Configurable Hardware Architecture

A simplified block diagram of the proposed hardware architecture is shown in figure 1. The architecture reads data from the input memory where input image pixels and mask window pixels are stored, denoted as P and W, respectively, in figure 1. The pixels are read in a column-based scan and transmitted to the array of processors to compute, in parallel, several window operations [4]. The processed data is captured by the global data collector and stored in the output memory bank.

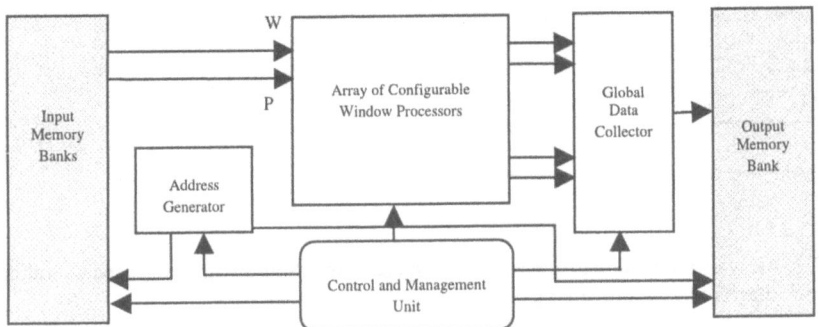

Fig. 1. Proposed hardware architecture for window-based image processing

The architecture contains four separate functional units: *control and management unit, address generator, global data collector*, and an *array of configurable window processors*. The *control unit* manages the data flow and synchronizes the different operations performed in the architecture. The *address generator* module generates the addresses for accessing image data. The input image memory and the window memory are read in a column-based scan, but a row-based scan is possible. Also, the address generator generates the addresses for the output memory to store the results produced by the architecture. The *global data collector* module collects the processed image data over the entire array and the result is sent to the output image memory.

The *array of processors* is the computation core of the architecture. The processors are arranged under a 2-D systolic approach. A closer view of the array of processors is shown in figure 2. The modules labeled by CWP and D denotes a processing element and a delay line, respectively. A linear systolic array of CWPs is interconnected to another through a delay line D for transmission of the mask window coefficients [5][6]. The delay line size is dependent on the number of rows processed in parallel. For an array of wxw CWPs the delay line is composed of (w-1) registers.

Figure 2 also shows a block diagram of the CWP designed to cover most window-based operations. It is called Configurable Window Processor (CWP). An accumulator, a register and an ALU-like processor (AP) integrate a CWP. Each CWP in the array computes a window operation progressively. At each cycle, one CWP receives a different value of the window mask and all the CWPs receive the same pixel value from the input image. After a latency period, the CWPs deliver their results progressively. Once a result is produced and captured by the data collector, the CWP is ready to start a new computation. The architecture works progressively in the same manner until the end of the data in the input image.

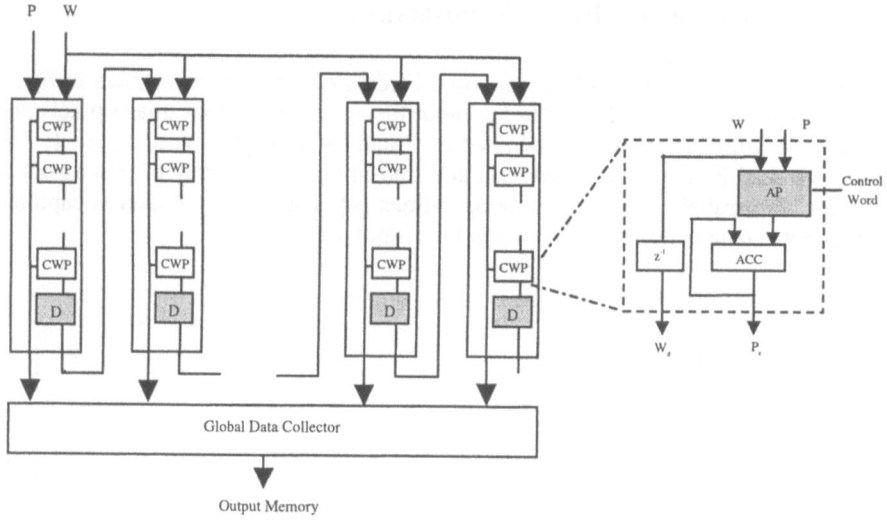

Fig. 2. Array of processing elements for fast speed computing of window-based operations and a block diagram of the processing element

3   Implementation and Results

The architecture was modeled in VHDL and synthesized for a VirtexE FPGA device using an array of 7×7 CWPs. The architecture was prototyped on an RC1000PP board from AlphaData [7]. At a frequency of 40 MHz, the processing time for a 512x512 image is 12.6 milliseconds, so the architecture processes around 80 images per second. The frequency is limited by the time delay introduced by the AP module. Table 1 shows a summary of the hardware resources and performance characteristics. Table 2 compares he architecture to other systems reported in [8][9]. The architecture achieves similar performances with less hardware resources.

Table 1. Summary of hardware resources utilization and performance characteristics of the architecture when mapped to a VirterE XCV2000E-6 BGA560 device

Complete Architecture (7×7 CWPs)	2604 slices
Single CWP	54 slices
FPGA percentage	13 %
Maximum frequency	40 Mhz

4   Conclusions and Further Work

In this work a configurable hardware architecture for window-based algorithms has been presented. The regular structure of the architecture allows the array of CWPs to be expanded and configured to adapt to different mask sizes and performance requirements. The use of a large number of CWPs in the architecture to increase the number of rows processed in parallel, can reduce the processing time of an image to hundred of microseconds. The architecture has been tested on algorithms such as convolution, correlation, 2-D filtering, feature detection, template matching, and gray-level morphology. The architecture functionality can be extended to applications such as motion estimation, and stereo disparity. As a further work the integration of the architecture on a smart camera will be addressed and aspects of dynamic reconfiguration, at the processing element level, will be explored for a more efficient use of hardware resources

Table 2. Comparison performance of the proposed architectures with other systems using image convolution on a 512×512 gray-level image

System	Architecture	Window size	Timing
PDSP16488, 40 MHz	ASIC	8x8	6.56 ms
LSI Logic's L64240, 20 MHz	ASIC	8×8	13.11 ms
DECchip 21064, 200MHz	Multiprocessor	5×5	220 ms
MAP1000, 200 MHz	VLIW	7×7	7.9 ms
Proposed architecture, 40 MHz	FPGA-based	7×7	12.6 ms

References

1   N. Ranganathan, *VLSI & Parallel Computing for Pattern Recognition & Artificial Intelligence*, Series in Machine Perception and Artificial Intelligence, Volume 18, World Scientific Publishing, 1995.

2   R. Jain, R. Kasturi and B. S. Shunck, *Machine Vision*, McGraw-Hill, International Edition, 1995.

3   Phillip A. Laplante and Alexander D. Stoyenko, *Real-time Imaging: Theory, Techniques, and Applications*, IEEE Press, 1996.

4   Miguel Arias Estrada, and César Torres Huitzil, *"Real-time Field Programmable Gate Array Architecture for Computer Vision"*, Journal of Electronic Imaging, Volume 10, number 1, January 2001, pp. 289-296

5   H. T. Kung, *"Why systolic architectures?"*, IEEE Computer, pp. 37-46, January 1982.

6   Peter Pirsh, and Hans-Joaching Stolberg, *"VLSI Implementations of Image and Video Multimedia Processing Systems"*, Transactions on Circuits and Systems for Video Technology, Vol. 8, No. 7, November 1998, pp. 878-891

7   Charles Sweeney, and Bill Blyth, *"RC1000-PP Hardware Reference Manual"*, Celoxica Limited, Version 1.22, 2000

8   M. Y. Siyal, M. Fathy, *"A Programmable Image Processor for Real-time Image Processing Applications"*, Microprocessors and Microsystems 23, 1999, pp. 35-41

9   N. Ratha and A. Jain, *"FPGA-based computing in Computer Vision"*, Fourth IEEE International Workshop CAMP, p. 128-137, IEEE Computer Society, 1997.

An FPGA-Based Image Connected Component Labeller

K. Benkrid, S. Sukhsawas, D. Crookes, and A. Benkrid

School of Computer Science, The Queen's University of Belfast, Belfast, BT7 1NN, UK
(k.benkrid, s.sukhsawas, d.crookes, a.benkrid)@qub.ac.uk

Abstract. This paper describes an FPGA implementation of a Connected Component Labelling algorithm (CCL), developed at Queen's University Belfast. The algorithm iteratively scans the input image, performing a non-zero maximum neighbourhood operation. It has been coded in Handel C language and targeted Celoxica RC1000-PP PCI board. The whole design was fully implemented and tested on real hardware in less than 24 man-hour. It uses a Virtex-E FPGA and two banks of off-chip memory. For 1024x1024 input images, the whole circuit consumes 583 FPGA slices and 5 Block RAMs and can run at 72 MHz, leading to a 68 pass/sec performance. The FPGA implementation outperforms, easily, an equivalent software implementation running on a 1.6 GHz Pentium-IV PC. A 10-fold speed up has been realised in many instances.

1 Introduction

Connected Component Labelling is an important task in intermediate image processing with a large number of applications [1][2]. The problem is to assign a unique label to each connected component in the image while ensuring a different label for each distinct object as illustrated in Figure 1.

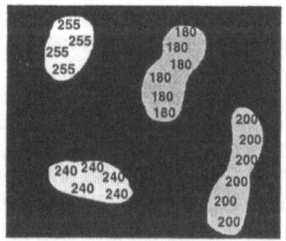

Fig. 1. A labelled Image example

To date, many algorithms have been developed to handle this problem [3][4]. This paper describes an FPGA implementation of an architecture based on a serial iterative algorithm for CCL developed at Queen's University Belfast [5].

In the following section, we describe the algorithm used. We then describe the hardware architecture, after which we give details of the FPGA implementation and its performance. Finally conclusions are drawn.

P.Y.K. Cheung et al. (Eds.): FPL 2003, LNCS 2778, pp. 1012–1015, 2003.

2   The Proposed CCL Algorithm

Given an arbitrary input grey-scale image, the CCL algorithm which we use is an iterative one and consists of the following steps:

Step 1:   Threshold the input image to obtain a binary image. This will make all pixels in the objects equal to 1 and all other pixels equal to 0.

Step 2:   The thresholded binary image is initially labelled. The initial labelling technique we adopt is: firstly to give the first non-zero pixel the highest label, and decrease the label value for subsequent non-zero pixels. Secondly, and in order to reduce the number of bits per pixel, we give adjacent non-zero pixels, within the same column (assuming a vertical scan), the same label as we know they are connected.

Step 3:   Apply an iterative 'non-zero maximum' neighbourhood operation on the image, using the window given in Figure 2. During this operation, each result pixel is stored back in the source image. A complete forward pass is followed by an inverse pass in which the image is scanned in reverse order.

Step 4:   Repeat Step 3 until there is no change in the image.

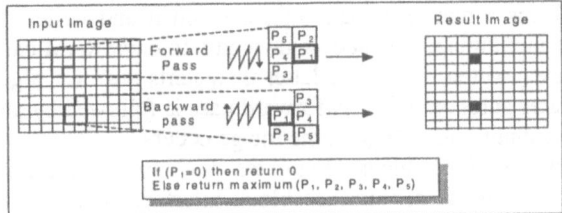

Fig. 2. CCL neighbourhood operation

3   The Hardware Architecture

The hardware architecture for the algorithm proposed above is illustrated in Figure 4. This architecture is a serial one, where pixels are fed to the FPGA one at a time. During the first pass, the input image is first thresholded, to produce the binary image, then initially labelled. In subsequent passes, these operations are bypassed. In order to eliminate the propagation of the pixel label from the bottom of an image column to the top of the following one, a column counter is provided to inhibit this propagation. To cater for the forward/backward multi-pass scheme, the image is read in the reverse order from which it has been stored during a backward pass.

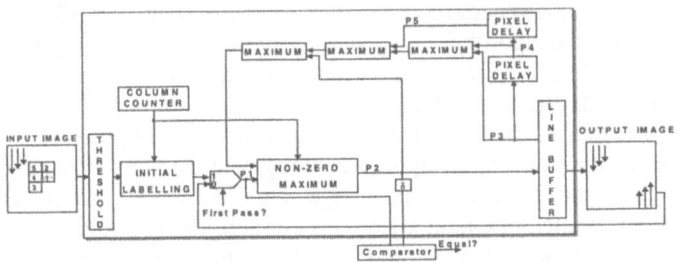

Fig. 4. Architecture of the Labelling Unit

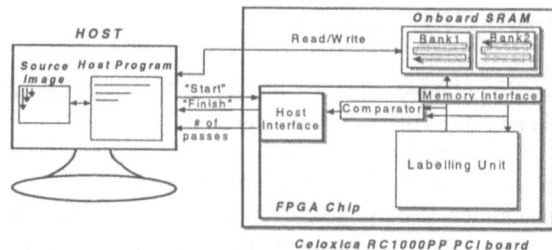

Fig. 5. Hardware organisation of the CCL implementation on Celoxica RC1000-PP PCI board

Detecting if a pass has resulted is done on the fly, to avoid a separate pass for this purpose. A flag is maintained during processing, and is set to 1 if and when any result pixel differs from its original value. To test for termination of the whole algorithm, this flag must be 0 at the end of a pass (either forward or backward).

The organisation of the proposed architecture on Celoxica's RC1000-PP PCI board is given in Figure 5. The architecture makes use of the onboard XCV2000E-6 Virtex chip and two 4MB SRAM banks. The FPGA reads the input image alternatively from one bank, performs a non-zero maximum pass on it and writes the result onto the other bank. The input image is first downloaded into one SRAM bank on the board by the host software (written in C). The latter then starts the CCL algorithm. After that, the FPGA operates autonomously. A forward pass is followed by an inverse one. The process is repeated until no change in the image occurs.

The memory interface block generates the proper off-chip RAM address sequences, whereas the Host interface block interfaces between the Host program (written in C) and the FPGA hardware through a number of registers.

4 Real Hardware Implementation

The CCL algorithm presented above has been fully coded in Handel C language and implemented and fully tested using Celoxica DK1 suite and RC1000-PP PCI board in less than 24 man-hour. The circuit description is scaleable and parameterisable in terms of the: the image size, the input image pixel word length, the processing word length and the threshold level (for binarisation).

The whole algorithm needed ~250 lines of Handel C code. The **par** instruction for expressing parallel flows, as well as the **if** statement for expressing conditions and selecting alternative flows based on these conditions, have been used extensively. The circuit was simulated conveniently at the Handel C source level. Hardware, in the form of EDIF netlist, is generated automatically from Handel C by Celoxica tools.

We have tested the algorithm using many combinations. Table 1 gives a sample of the results for different image sizes using 8-bit/pixel input images, a user-tuneable threshold level and a processing word length of 32 bits.

Figure 6 gives an example of an input image and the corresponding labelled image after the application of the CCL algorithm, along with the number of passes required. It takes ~1.57 sec to label this image on FPGA. An equivalent software implementation (written in C), on a 1.6 GHz Pentium-IV PC, takes ~18.17 sec. Thus, a 10-fold speed up is possible with the FPGA-based implementation.

Table 1. Implementation results of CCL algorithm on the RC1000-PP PCI board

Image size	Slices (out of 19200)	BRAMs (out of 160)	Speed (MHz)	Number of passes/sec
256x256	526	1	78	1190
512x512	553	3	73	278
1024x1024	583	5	72	68

 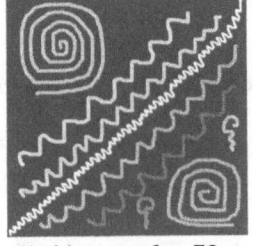

Original 1024x1024 image   Labelled image after 78 passes

Fig. 6. Sample result of the implementation of our CCL algorithm

5 Conclusion

In this paper, we have presented an FPGA implementation of an architecture based on a serial iterative algorithm for Connected Component Labelling (CCL). The circuit has been coded in Handel C language and fully tested on Celoxica Virtex-E based RC1000-PP PCI board in less than 24 man-hour. Handel-C proved extremely convenient for this kind of control-intensive algorithms.

The paper presented implementation results for different image sizes. For instance, the implementation of the proposed algorithm for 1024x1024 input images consumes 583 slices and 5 Block RAMs on the FPGA and can run at 72 MHz. This results in a maximum number of passes equal to 68 passes/sec. The FPGA implementation, easily, outperformed a 1.6 GHz Pentium-IV PC implementation. A 10-fold speed up has been realised in many instances.

References

[1] D. L. Milgram, "Region extraction using convergent evidence", *Computer Graphics and Image Processing*, vol. 5, no. 2, pp. 561-572, 1988.
[2] L. C. Sanz and D. Petkovic, "Machine vision algorithms for automated inspection of thin-film disk heads", *IEEE Transactions on Pattern Analysis and Machine Intelligence*, Vol. 10, no. 6, pp. 830-848, 1988.
[3] H. M. Alnuweiri, and V. K. Prasanna, "Parallel architectures and algorithms for image component labeling", *IEEE Transactions on Pattern Analysis and Machine Intelligence*, Vol. 14, No. 10, pp. 1014-1034, Oct. 1992.
[4] R. Klette, P. Zemperoni, "Image Processing Operators", John Wiley & Sons, New York, 1995.
[5] D. Crookes, K. Benkrid, "An FPGA Implementation of Image Component Labelling", SPIE, Vol. 3844, USA, 1999.
[6] Celoxica Handel-C and RC1000-PP PCI board Production Information, Celoxica Ltd. http://www.celoxica.com

FPGA Implementation
of Adaptive Non-linear Predictors for Video Compression

Rafael Gadea-Girones[1], Agustín Ramirez-Agundis[2],
Joaquín Cerdá-Boluda[1], and Ricardo Colom-Palero[1]

[1] Department of Electronic Engineering, Universidad Politécnica de Valencia, Camino de Vera
s/n,46020, Spain
rgadea@eln.upv.es, rcolom@eln.upv.es
[2] Department of Electronic Engineering, Instituto Tecnológico de Celaya, Av. Tccnológico s/n,
38010, México
aagundis@itc.mx

Abstract. The paper describes the implementation of a systolic array for a non-linear predictor for image compression. We can implement very large interconnection layers by using large Xilinx and Altera devices with embedded memories and multipliers alongside the projection used in the systolic architecture. These physical and architectural features create a reusable, flexible, and fast method of designing a complete ANN (Artificial Neural Networks) on FPGAs. Our predictor, a MLP (Multilayer Perceptron) with the topology 12-10-1 and with training on the fly, works, both in recall and learning modes, with a throughput of 50 MHz, reaching the necessary speed for real-time training in video applications.

Introduction

In recent years, it has been shown that neural networks can provide solutions to many problems in the area of image compression. Software simulations are useful for investigating the capabilities of ANN models and creating new algorithms; but hardware implementations remain essential for taking full advantage of the inherent parallelism of ANN.

Traditionally, ANNs have been implemented directly on special-purpose digital and analogue hardware. More recently, ANNs have been implemented with re-configurable FPGAs. Although do not achieve the power, clock rate, or gate density, of custom chips; they are much faster than software simulations [1]. Until now, a principal restriction to this approach has been the limited logic density of FPGAs.

This paper offers advances in two basic respects to previously reported neural implementations on FPGAs. The first is the use of an aspect of backpropagation and stems from the fact that forward and backward passes of different training patterns can be processed in parallel [2] .The second point we contribute is to produce a completed ANN with on-chip training, and good throughput for the recall phase – on a single FPGA. This is necessary, for example, in industrial machine vision, and for the training phase, with continual online training (COT) [3]. In our research, we need this properties in order to design a adaptive non linear predictor for video compressionNon-linear Predictor Application

P.Y.K. Cheung et al. (Eds.): FPL 2003, LNCS 2778, pp. 1016–1019, 2003.
© Springer-Verlag Berlin Heidelberg 2003

Non-linear Predictor Application

In this section we present results obtained by experimenting with the adaptive non linear prediction approach proposed by Marusic and Deng [4] together with the encoder method proposed by Howard and Vitter [5] in order to obtain a complete loss-less compression system (Fig. 1).

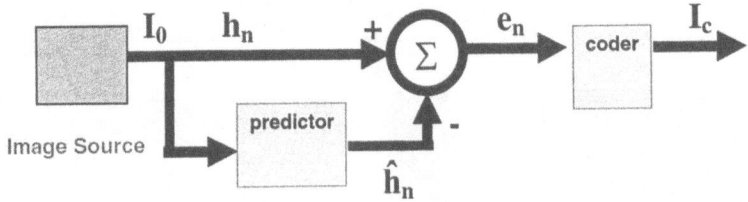

Fig. 1. Structure of the Predictor System

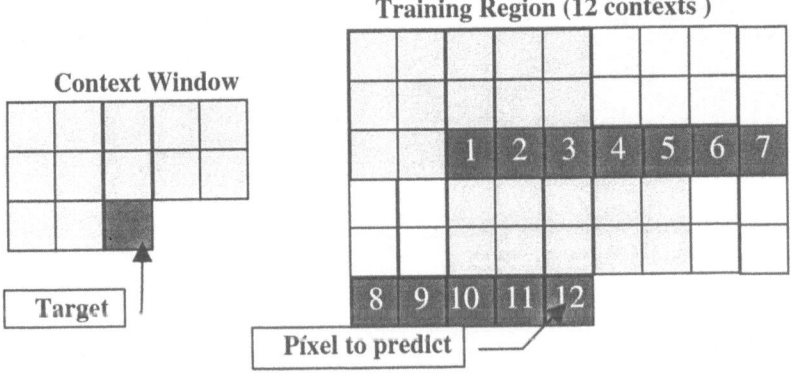

Fig. 2. Area for adaptive training.

The prediction stage is an adaptive predictor that uses a dynamic training region for each pixel to be estimated. The training region has twelve context windows, each one with twelve pixels (Fig. 2).

The image is scanned, pixel by pixel, from left to right and from top to bottom. On each pixel, the training region is represented by twelve training vector; these are used to train a 12-10-1 two layer perceptron. Once the network has been trained, the pixel is predicted and the current weight matrix is used as the initial one for the next pixel (Fig.2).

We can evaluate a predictor method by its ability to reduce the correlation between pixels (entropy) as well as the PSNR for the predicted image ($\bullet_n$). We can observe a summary of these parameters for different topologies of the MLP of the Adaptive Non-linear Predictor in Table 1. We obtain, with the on line version of the Backpropagation algorithm, better results than with batch line version and we show that the number of hidden layers (20 or 10) is not relevant for the prediction performance. These two ideas are very important for our hardware implementation, both in throughput and in area.

Table 1. Results of prediction with differents topologies with Lena256x256 Image

Hidden Neurons	10	20	10	10	20
Context (inputs)	12	12	4	12	12
PSNR	27.8575	27.9511	27.4993	28.5662	28.5537
Entropy of residue	4.9858	4.9763	5.0409	4.855	4.8648
Entropy of original	7.5888	7.5888	7.5888	7.5888	7.5888
Version Learning	Batch line	Batch line	Batch line	On line	On line

 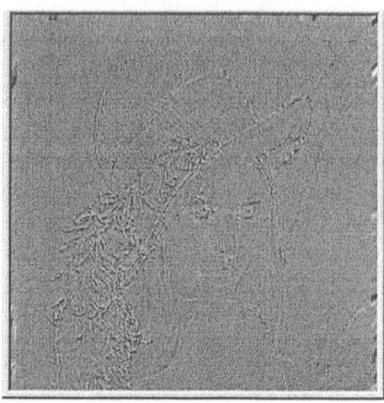

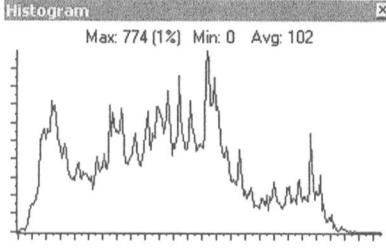 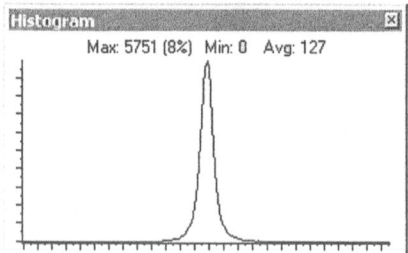

Fig. 3. Original (h_n) and residue (e_n) images, and its histograms.

Table 2. Design summary for MLP(12-10-1) .

X2V3000BF957-6 (VIRTEX2)			
Number of Slices:	3,601 out of	14,336	25%
Number of Flip-Flops:	2,114 out of	28,672	18%
Number of Mult18x18s:	75 out of	96	78%
Number of RAMB16s:	48 out of	96	50 %

The results of the implementation of the MLP(12-10-1) are shown in Table 2 and 3. This implementation works, both in recall and learning modes, with a throughput of 50 MHz (only possible with our pipeline version), reaching the necessary speed for real-time training in video applications and enabling more typical applications (wavelet transform and run-length coder) to be added to the image compression.

Table 3. Design summary for MLP(12-10-1) .

EP1S60B856C7 (STRATIX)			
Number of Logic Cells: 7,362 out of		57,120	12%
Number of Flip-Flops: 2,381 out of		59,193	4%
Number of DSP blocks: 133	out of	144	92%
Number of m512s: 47	out of	574	8%

Conclusions

Until now, mask ASICs offered the preferred method for obtaining large, fast, and complete ANN for designers who implement neural networks with full and semi-custom ASICs. Now, we can exploit all the embedded resources of the new programmable ASICs (FPGA) and the enormous quantities of logic cells to obtain complete applications of neural networks, with real-time training on the fly, and with topologies (size) that were impossible to achieve just two years ago.

Of course other certain drawbacks must be overcome in addition to size when we face up to the implementation of a real application. Decrease the throughput of the training is very important for the MLP in the transmit part of the prediction system and we attain it with the pipeline version of the algorithm; but also we must resolve the reception part, and for this, the latency of the training must be improved. Therefore we have to increase the parallelism of our implementation in the future.

References

[1] S. Hauck, "The Roles of FPGAs in Reprogrammable Systems" *Proceedings of the IEEE*, 86(4), April 1998, pp. 615-638.
[2] R. Gadea, A. Mocholí, "Forward-backward parallelism in on-line backpropagation", *Lecture Notes in Computer Science* , Vol. 1607, June 1999, pp. 157-165.
[3] B. Burton, R.G. Harley, G. Diana, and J.R. Rodgerson, "Reducing the computational demands of continually online-trained artificial neural networks for system identification and control of fast processes" *IEEE Transaction on Industry Applications*, Vol 34. no.3, May/June 1998, pp. 589-596
[4] Marusic, S., Deng, G. "A Neural Network based Adaptive Non-Linear Lossless Predictive Coding Technique". *Signal Processing and Its Applications, 1999. ISSPA '99.*
[5] Howard, P.G., Vitter, J.S, "Fast Progressive Lossless Image Compression". *Image and Video Compression Conf. Symp. On Electronic Imaging*, SPIE-2186, San Jose, C. A., 1994

Reconfigurable Systems in Education

Valery Sklyarov and Iouliia Skliarova

University of Aveiro, Department of Electronics and Telecommunications, IEETA,
3810-193 Aveiro, Portugal
{skl, iouliia}@det.ua.pt

Abstract. This paper describes methods and tools that have been used for teaching disciplines dedicated to the design of reconfigurable digital systems. It demonstrates students' projects, disseminates experience in the integration of different disciplines, and gives examples of stimulating student activity. A set of animated tutorials for students that are available on WebCT with a number of practical projects that cover a variety of topics in FPGA-based design can be seen as the most valuable contribution to the area considered.

1 Introduction

Today, FPLDs are considered to be an alternative to ASICs, and they have already been very efficiently used in a large number of practical applications, such as co-processors for general-purpose computers, problem-oriented digital systems, embedded controllers, and so on. Since the scope of potential applications is growing rapidly, a large number of well-prepared engineers are needed in the relevant areas. Thus, new trends must be reflected in the respective pedagogical activity of universities concerned with these subjects. Note that the domain of reconfigurable systems design is very dynamic and many-sided. Periodic upgrading of the pedagogical plans is essential in order to mirror recent advances in FPLD technology, design methods and CAD tools. Consequently, it is very important to provide an exchange of experience in pedagogical activity. This paper describes a methodology that has been used for more than 5 years for teaching reconfigurable systems at the Department of Electronics and Telecommunications of Aveiro University.

2 Basic Directions in Teaching Reconfigurable Systems

The major topics that have been considered as a base for the disciplines within the scope of reconfigurable systems design include design tools, prototyping boards with FPLDs and methods that provide an understanding of how FPLD-based circuits communicate with the external world, i.e. with peripheral electronic devices.

Design tools have been selected so that it is possible to learn both system-level specification languages and traditional HDL design flows. Two system-level specification languages, SystemC [1] and Handel-C [2], have been taught to students. SystemC is a library that permits hardware components to be modeled using a

P.Y.K. Cheung et al. (Eds.): FPL 2003, LNCS 2778, pp. 1020–1023, 2003.

standard C/C++ compiler and it was considered just at a description level. Handel-C permits digital circuits to be described in a C-based style and synthesized using Celoxica tools [2]. The latter together with Xilinx ISE 5.2 software [3] were used for the design, modeling, and implementation of digital systems based on FPGAs.

A number of *prototyping boards* have been employed for testing circuits in hardware. Recently used boards contain FPGAs from two Xilinx series; the Spartan-II/Spartan-IIE (the boards TE-XC2Se [4] and RC100 [2]) and the Virtex-EM/Virtex-IIPro (the PCI boards ADM-XRC and ADM-XPL [5]).

For the majority of practical applications, FPGA-based circuits have to interact with external devices. Three groups of *interfaces* have been studied. They are widely-used standard protocols, such as parallel, RS232, USB, etc; PCI; and those that provide interactions between FPGAs and external microchips, such as static memory, LCD controllers, microprocessors, etc.

Three kinds of applications have been proposed to students, combinatorial accelerators, hardware/software co-simulation, and processors with customized sets of instructions. During practical classes, students have to implement individual blocks for the systems mentioned above. Complete systems have to be constructed within semester projects.

The classes given to the students have a number of distinct features. The lectures are well-covered by a set of animated tutorials and examples of FPGA-targeted projects. All the required supplementary materials are available on the WebCT.

3 Tutorials and WebCT

In order to maximize the effectiveness of the classes, the students should have access to all the required materials. The materials can be divided into the following basic groups: manuals about FPGAs and the corresponding computer-aided design systems (ISE 5.2, ModelSim and DK1 in our case); supplementary documents (manuals on peripheral microchips, specification of interfaces, etc.); descriptions of auxiliary equipment that is required, such as logic analyzers; methods and tools that increase the productivity of education through facilities such as the extensive use of animated tutorials and providing materials for distance learning. For example, the section *Tutorials* in the public domain of [6] contains 10 examples. Each of them includes an animated tutorial in PowerPoint and examples of ISE 5.2 projects for VHDL-based design flow. They address the following topics:

- A sequence of steps for beginners to design, implement, and test in an FPGA, a trivial circuit based on a very simple VHDL code;
- The interaction of an FPGA with components such as LEDs, push buttons and DIP switchers through a CPLD;
- Design and functionality of a simple arithmetical circuit, which displays the results on a LCD (2 lines with 16 characters each). This tutorial explains how to use the ISE 5.2 schematic editor, the Core Generator, Xilinx libraries, hierarchical design, combining different components in the same project, interaction with an LCD controller, etc.;
- Synthesis of finite state machines (FSM) and the use of Xilinx StateCAD;

- Synthesizable VHDL in alphabetical order. It allows any letter (for instance, *G* - *Generic*) to be chosen to learn the corresponding topic, to run a relevant project in ISE 5.2, to load the generated bitstream into the FPGA, and to test the circuit in hardware;
- Simulation of VHDL descriptions in ModelSim;
- Interacting with a touch panel [7] through an RS232 interface;
- Parameterizable (generic) VHDL code for a reprogrammable FSM and examples of the FSM interacting with a datapath;
- FPGAs interacting with two LCDs (4 lines 20 characters in each and 2 lines 16 characters in each). The examples explain a number of control sequences for scrolling, editing, etc.;
- A very detailed and relatively complex example that explains how to use recursive hierarchical control. It shows a C++ program that illustrates algorithms, describes two recursive sub-algorithms, demonstrates how to construct a recursive hierarchical FSM that builds a special binary tree from a sequence of arbitrary integers and sorts integers on the basis of this binary tree. The results (i.e. the sorted data) are displayed on an LCD.

All the tutorials make use of different animation effects available in PowerPoint (Windows XP). This enables many processes to be demonstrated in a step by step manner, such as all the events appearing in each clock cycle; how VHDL code activates these events and reacts; how various bits in interface lines are changed, etc.

Similar tutorials have been prepared for explaining the functionality of FPGA DLLs, demonstrating an interaction with a mouse, a keyboard, a VGA monitor, and for a number of Handel-C topics. They are available in English and in Portuguese.

4 Stimulation of Student Activity and Integration with Other Disciplines

Two types of evaluation have been proposed to the students. The first is a traditional examination. The examination can be replaced by the second type of evaluation through an individual project that is suggested in the middle of a semester. According to the requirements, students have to design, implement, and test a digital system based on commercially available FPGAs. Potential projects are discussed with the students. This allows a task to be chosen from the area that is of the most interest to a particular student. The results of the projects have to be demonstrated in a working FPGA-based device and presented in a written report before the end of the examination period. The best projects are recommended for publications in the magazine "Electrónica e Telecomunicações" that is issued by the Department. All these publications can be accessed through the WebCT [6].

The methodology provides a very important opportunity, especially for final year students. It permits integration between different disciplines to be established. The group of disciplines considered in this paper suggests methods and tools; the other groups of disciplines offer applications. This approach provides additional motivation to the students because reconfigurable systems can be linked with practical work in other disciplines that particular students are interested in.

Our experience has shown that there are some auxiliary methods that stimulate the work of students. First of all, the result of the work should be visible and touchable especially at the beginning. That is why it is reasonable to use stand-alone boards that are cheap and provide a number of interactions such as communication between an FPGA and a mouse, a keyboard, a VGA monitor, LCD panels, etc. Only after some period of time hidden PCI-based prototyping boards can be used. As a rule they are much more expensive and do not permit the results to be appreciated visually. The work is organized through a set of API functions and it looks like programming. On the other hand, PCI-based boards contain much more powerful FPGAs and they are recommended for experienced students, especially those in Ph.D. and M.Sc. scholarships.

For example, in 2002/2003 a combinatorial processor that implements an exact algorithm for solving the covering problem has been proposed as a project for final year students. Initially, a stand-alone RC100 prototyping board with an FPGA from the Spartan-II family was used. All the individual components of the project were described in Handel-C and carefully tested using available peripheral devices and drivers supplied by Celoxica. Finally the entire circuit was implemented. After that the same circuit was constructed using the ADM-XPL PCI board [5] containing FPGA XC2VP7 of Virtex-II Pro family. This circuit allows much more complicated problems to be resolved [8]. The previous experience gained with a relatively cheap stand-alone board provides a basis for achieving results rapidly, and very similar methods and tools can be used for the most advanced FPGAs available on the market.

4 Conclusion

This paper disseminates experience in teaching reconfigurable systems, summarizes the pedagogical methods that have been adopted, the organization of classes, the basic directions of student's projects, and many other aspects. This work was supported by the grants FCT-PRAXIS XXI/BD/21353/99 and POSI/43140/CHS/2001.

References

1. SystemC. [Online]. Available: http://www.systemc.org/
2. Handel-C, DK1, RC100. [Online]. Available: http://www.celoxica.com/
3. ISE 5.2, Xilinx series FPGA. [Online]. Available: http://www.xilinx.com/
4. Spartan-IIE Development Platform. [Online]. Available: www.trenz-electronic.de
5. Alpha Data. [Online]. Available: http://www.alpha-data.com
6. http://webct.ua.pt, "2 semester", the discipline "Computação Reconfigurável", public domain is indicated by the letter "i" enclosed in a circle. Login and password for access to the protected section can also be provided (via e-mails: skl@ieeta.pt, iouliia@det.ua.pt)
7. EA KIT240-7, EA DIP204-4. Electronic Assembly. Available: http://www.lcd-module.de
8. Sklyarov, V., Skliarova, I. Almeida, P., Almeida, M.: High-Level Design Tools for FPGA-based Combinatorial Accelerators. Proceedings of FPL'2003 (Lisbon, 2003)

Data Dependent Circuit Design: A Case Study

Shoji Yamamoto, Shuichi Ichikawa, and Hiroshi Yamamoto

Department of Knowledge-based Information Engineering
Toyohashi University of Technology
1-1 Hibarigaoka, Tempaku, Toyohashi, Aichi 441-8580, JAPAN
{shoji, ichikawa, i8hyama}@ich.tutkie.tut.ac.jp
http://ich.tutkie.tut.ac.jp/en/

Abstract. Data dependent circuits are logic circuits specialized to specific input data. They are smaller and faster than the original circuits, although they are not reusable and require circuit generation for each input instance. This study examines data dependent designs for subgraph isomorphism problems, and shows that a simple algorithm is faster than an elaborate algorithm. An algorithm that requires many hardware resources consumes an accordingly longer circuit generation time, which outweighs the performance advantage in execution.

1 Data Dependent Hardware

If any input of a logic circuit turns out to be constant, the circuit can be reduced. For example, if any inputs of an AND gate turn out to be zero, the output becomes zero (*constant propagation*). This reduction can be applied recursively, consequently reducing the logic scale of the circuit. The derived circuit would operate at a higher frequency than the original, because the logic depth and wiring delay would also be reduced by this reduction.

Since the consequent circuit becomes dependent on the input data instance, such a circuit is called a *data dependent circuit* in the following discussion. The obvious drawback of a data dependent approach is that the derived circuit is *not reusable*. This naturally means that (1) the circuit must be generated for each input instance, and (2) reconfigurable devices such as FPGA must be used.

The total execution time T of a data dependent circuit is given by the sum of the circuit generation time T_{gen} and the execution time T_{exec}. T_{gen} consists of the time for HDL source code generation, logic synthesis, technology mapping, placement, routing, and FPGA configuration. T_{gen} depends on the logic scale, since a larger circuit usually requires accordingly larger generation time. T_{exec} is the product of cycle count and cycle time. Here, the cycle count depends on the algorithm, and the cycle time depends on the implementation. Fast algorithms can make T_{exec} smaller, but they often require more hardware resources and make T_{gen} larger. The total execution time T is thus not so obvious without empirical studies.

P.Y.K. Cheung et al. (Eds.): FPL 2003, LNCS 2778, pp. 1024–1027, 2003.

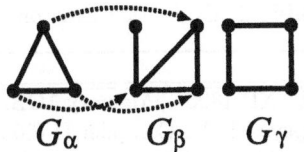

$$G_\alpha \qquad G_\beta \qquad G_\gamma$$

Fig. 1. Subgraph Isomorphism

2   A Case Study: Subgraph Isomorphism Problem

Hereafter, we examine a problem called a *subgraph isomorphism problem* as an example application. A subgraph isomorphism problem is a simple decision problem. Given two graphs G_α and G_β, it is determined whether G_α is isomorphic to any subgraph of G_β (Fig. 1). In Fig. 1, G_β has a subgraph that is isomorphic to G_α, while G_γ does not.

Ullmann [1] proposed a depth first search algorithm with a smart pruning procedure (*refinement procedure*) for subgraph isomorphism problems. He pointed out that his procedure can be implemented with parallel hardware, but Ichikawa et al. [2] later revealed that his circuit is too large to handle practical problems with the state-of-the-art FPGAs. Ichikawa, Udorn, and Konishi [3] proposed a new algorithm (Konishi's algorithm), which has a simpler pruning procedure than Ullmann's. Konishi's algorithm is generally slower than Ullmann's, but it can be implemented in a much smaller logic circuit than Ullmann's.

Ichikawa et al. [4] [5] previously suggested that data dependent implementations of Ullmann's circuit can be much smaller than the original circuit. The present study confirms this by showing the evaluation results with a Xilinx Virtex-II FPGA.

A data dependent Konishi circuit has not yet been investigated. This study also shows the implementation results of data dependent Konishi circuits, and compares them with data dependent Ullmann circuits. As the original Konishi circuit [3] was not suited for data dependent implementation, we designed a brand-new logic circuit with Konishi's algorithm in this study. In this design, the adjacency check circuits are implemented by parallel hardware. Although this design is an interesting example of a data dependent circuit, we do not have the space to detail it here.

3   Evaluation

This section describes the evaluation results for data dependent circuits. Each result in this section is the average of 100 pairs of G_α and G_β, which are randomly generated. Let p_α and p_β be the number of vertices of G_α and G_β, respectively. We only deal with the cases of $p_\alpha = p_\beta$ in this study. We implemented data dependent circuits for various graph sizes, and measured the execution time on a XC2V1000 FPGA. Our evaluation environment is summarized in Table 1.

Table 1. Evaluation Environment

Item	Note
Circuit	Athlon XP 1800+, Memory 1GB, Windows2000 SP3
Generation	Synopsys FPGA Compiler II (2001.08-FC3.7)
	Xilinx ISE 4.2i (Target device: Virtex-II XC2V1000)
FPGA platform	Insight MicroBlaze Development Kit (XC2V1000, 24MHz)
Software	Celeron 1.2GHz, Memory 512MB, Red Hat Linux 7.2
Implementation	Written in C, compiled with gcc-2.95.3

We examined the cases of $(ed_\alpha, ed_\beta) = (0.3, 0.6)$, where ed_α and ed_β indicate the edge density of G_α and G_β, respectively. Edge density ed is defined by the equation $ed = (2\,q)/(p\,(p-1))$, assuming that p is the number of vertices and q is the number of edges. In other words, ed is the ratio of the number of edges to that of the perfect graph K_p. It is clear that $0 \le ed \le 1$ holds.

In this study, we examine 4 designs. **Ko** and **Uo** designate the original Konishi circuit and the original Ullmann circuit, respectively. **Kd** and **Ud** designate the data dependent versions of a Konishi circuit and Ullmann circuit for the above-mentioned input graph set.

Figure 2 (left) displays the average logic scale of Ud and Kd, shown by the number of *slices* of Virtex-II FPGA. For the same number of vertices ($8 \le p_\alpha = p_\beta \le 16$), the logic scale of Uo is estimated to be 2.4–3.6 times larger than Ud. Meanwhile, Ko is 1.7–1.9 times larger than Kd.

Figure 2 (right) displays the average execution time on XC2V1000 FPGA with a 24 MHz system clock. For comparison, the software implementation of Ullmann's algorithm was also evaluated. The evaluation environment is summarized in Table 1. The execution time of software is denoted by S in Fig. 2 (right). For $p_\alpha = p_\beta = 16$, Ud and Kd is 63 and 32 times faster than S, respectively. This performance gain becomes larger in larger graphs.

Figure 3 (left) displays the circuit generation time. It is readily seen that the circuit generation time of Ud is far larger than that of Kd. This comes from the difference of logic scale. Figure 3 (right) shows the average total execution time

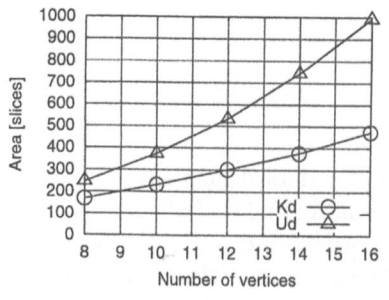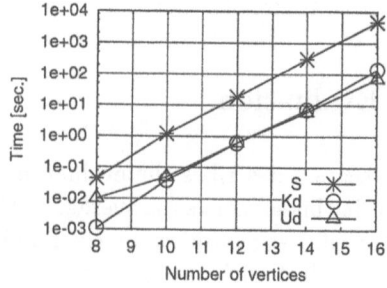

Fig. 2. Logic Scale (left) and Execution Time (right)

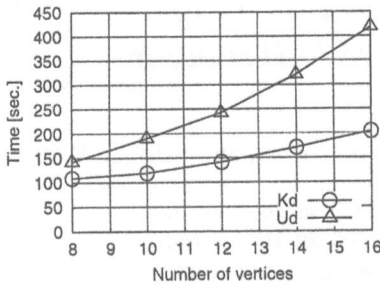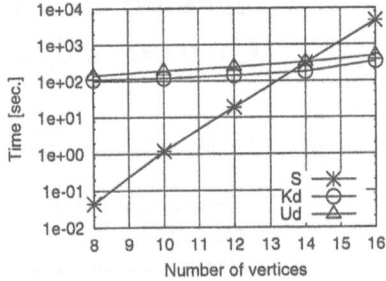

Fig. 3. Circuit Generation Time (left) and Total Execution Time (right)

of Ud and Kd, which is the sum of the circuit generation time and the execution time. The software execution time (S) is also shown for comparison.

Ud and Kd are faster than the software for $p_\alpha = p_\beta > 14$, even reckoning the circuit generation time. When $p_\alpha = p_\beta = 16$, Kd and Ud are 13.3 and 9.4 times faster than the software, respectively. As is readily seen, this performance advantage becomes larger when p_α and p_β are larger. It is also worth noting that Kd is faster than Ud after all, because the long circuit generation time of Ud outweighs its performance advantage over Kd.

Acknowledgment

This work was partially supported by a grant from the Okawa Foundation for Information and Telecommunications. The custom circuits in this study were designed with Synopsys CAD tools through the chip fabrication program of VDEC (the University of Tokyo).

References

1. Ullmann, J.R.: An algorithm for subgraph isomorphism. J. ACM **23(1)** (1976) 31–42
2. Ichikawa, S., Saito, H., Udorn, L., Konishi, K.: Evaluation of accelerator designs for subgraph isomorphism problem. In: Proc. FPL2000. LNCS1896, Springer (2000) 729–738
3. Ichikawa, S., Udorn, L., Konishi, K.: An FPGA-based implementation of subgraph isomorphism algorithm. IPSJ Trans. High Performance Computing Systems **41(SIG5)** (2000) 39–49 (in Japanese).
4. Ichikawa, S., Yamamoto, S.: Data dependent circuit for subgraph isomorphism problem. In: Proc. FPL2002. LNCS2438, Springer (2002) 1068–1071
5. Ichikawa, S., Yamamoto, S.: Data dependent circuit for subgraph isomorphism problem. IEICE Trans. Information and Systems **E86-D(5)** (2003) 796–802

Design of a Power Conscious, Customizable CDMA Receiver

Maurizio Martina, Andrea Molino, Mario Nicola, and Fabrizio Vacca

Dipartimento di Elettronica, Politecnico di Torino,
Corso Duca degli Abruzzi 24, 100129 Torino, Italy,
{maurizio.martina,andrea.molino}@polito.it,
{mario.nicola,fabrizio.vacca}@polito.it

Abstract. 2G wireless systems have gained a widespred diffusion. Due
to this fact, the transition to 3G ones can be critical. A possible solution
to the interoperability problem can came from the Software Defined Radio paradigm. In this paper a complete, reconfigurable CDMA receiver
implementation over a Xilinx XCV300E FPGA is described.

1 Introduction

Software Defined Radio(SDR) paradigm [1] is one of the most interesting topics
in wireless communications, since it can help the transition from 2G to 3G. SDR
main idea is to employ the reconfigurability as a key feature for implementation
of wireless terminals. Existing works [2] have demonstrated that actually a DSP
implementation isn't feasible, due to high processing demand. On the other hand,
FPGAs grant good performances and reconfigurability, but waste a great amount
of static power [3]. So FPGAs are an interesting platform for SDR: in this paper
a reconfigurable CDMA receiver architecture based on FPGAs is described, and
some strategies to dynamically manage the power consumption are advised.

2 CDMA Fundamentals

CDMA is a technique to access a shared channel, spreading trasmitted signals
with orthogonal codes [4]. Spreading can be performed multiplying the signal
$x[n]$ and the code $w[n]$. The $x[n]$ rate is the *bit–rate*, while $w[n]$ rate is the
chip–rate: their ratio is called *spreading factor* (G). The receiver can recover the
information from k–th user correlating over G chips the received bit stream r
and the despreading code w_k. CDMA poses many design challenges, including
the need of synchronization between received signal and despreading sequence.
Now an implementation of a BPSK-CDMA receiver is presented (figure 1(a)).

3 Receiver Architecture

3.1 Digital Down Conversion

The DDC is usually a critical block since input signal is at the ADC rate. To perform filtering operation at this rate, Cascaded Integrator Comb (CIC) filters [5]

P.Y.K. Cheung et al. (Eds.): FPL 2003, LNCS 2778, pp. 1028–1031, 2003.

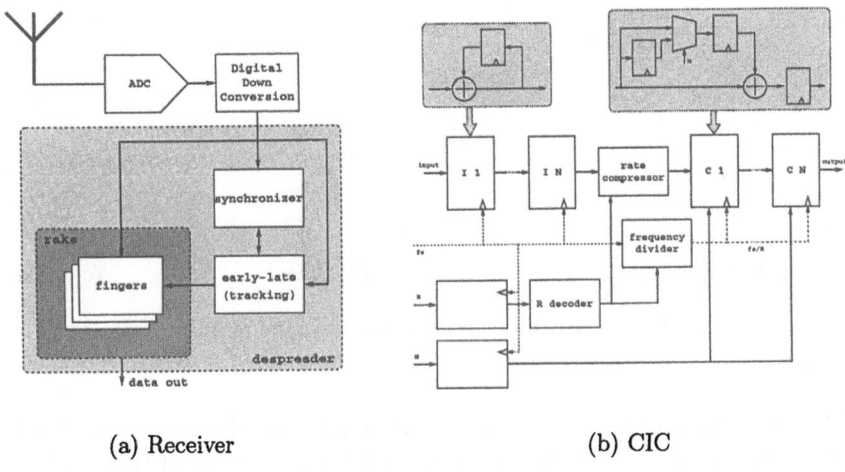

(a) Receiver   (b) CIC

Fig. 1. System architecture with detailed view of downconverter

have been used (figure 1(b)). A CIC filter is made of integrators and combs connected in cascade. Integrator can be implemented using an accumulator, while the comb stage consists in a digital programmable delay chain followed by a subtractor. In our architecture both downsampling ratio (R), and length of comb's delay chain (M) can be programmed. Some previous works concerning hardware implementations of CIC filters can be found (e.g. [6]), but these architectures are optimized to be implemented in ASIC.

3.2 Synchronizer

An exhaustive search over all possible synchronization schemes is used in synchronizer (figure 2(a)): it consists in evaluating energy of correlation, since it shows a peak when synchronization is reached [4]. Correlation is evaluated for both quadrature and in–phase branches in a proper number of chip cycles ($Ntest$) to achieve reliable results, then the contributes are squared and added together. After an appropriate number of measures (L), the final result is compared with a programmable threshold. In order to increase system's flexibility, both $Ntest$ and L can be programmed by the control unit. Code generator is implemented using reconfigurable Linear Feedback Shift Registers (LFSR) that can be enabled/disabled by a control unit: this approach is different from existing works suggesting use of clock tree structures [7].

3.3 Early–Late DLL

After synchronization is reached, an Early–Late DLL is needed to keep the receiver tracked. This block (figure 2(b)) is made of four main functional blocks: the correlation detector, the loop filter, the Numerically Controlled Oscillator

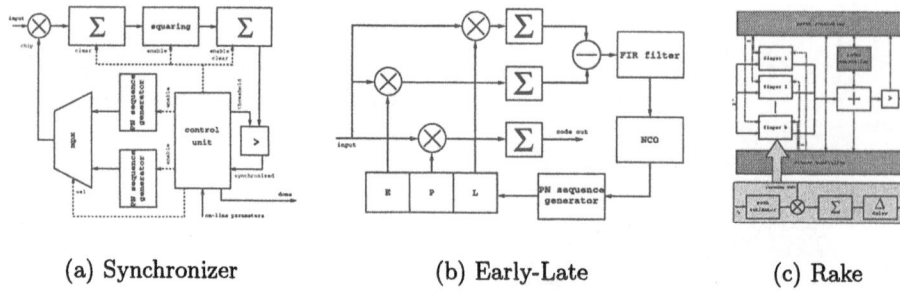

(a) Synchronizer   (b) Early-Late   (c) Rake

Fig. 2. Synchronization units and rake receiver

(NCO) and the Early-Punctual-Late code generator. DLL loop filter has been implemented as an FIR filter with a reconfigurable number of taps and a programmable number of computing resources. The filter output controls the NCO block, an accumulator with programmable increment, which is the block devoted to drive the sequence generator. The tracking loop is closed over two correlators, evaluating correlation of signal respectively with early and late replica of local code. The difference between these two values is the filter input signal.

3.4   Rake Receiver

The rake receiver is devoted to increase the received signal to noise ratio in a multipath environment. In particular it estimates both the attenuation and the delay parameters for a set of given paths: equations can be found in literature [4]. It's worth noting that estimation is obtained through an average over N_p chip: this parameter, crucial to satisfy both high performance and precision demand, is programmable in this architecture. In the block scheme (figure 2(c)) three main parts can be identified: the rake receiver core, composed by a programmable number of *fingers*, the adder, and the programmable comparator. In each finger phase delay and attenuation are estimated, then outputs from distinct fingers are added together, obtaining a value that can be used to recognize received bit. To reduce power consumption, fingers are turned off and on by a power controller, accordingly to the energy of received signal.

4   Results and Conclusions

This architecture supports both "on–line" and "off–line" reconfigurability: the former is the most valuable one and requires more resources, while the latter needs an FPGA reconfiguration to change parameters. In this work the "off–line" reconfigurability is used to pose some well known bounds, given a particular scenario (e.g. UMTS, CDMA2000, ...), while the "on–line" parameters enable the support for different operative profiles (table 1).

Table 1. Reconfigurable parameters description

Block	Parameter	Description	Type on-line	off-line
all blocks	#PIPE	Pipeline levels in data path unit		⋆
	DATA_W	Width of elaborating data		⋆
	G	Spreading/Despreading factor (B_{chip}/B_{bit})	⋆	
	CODE_LENGTH	Despreading code length	⋆	
Digital Down	R	Output data rate (rate downsapling factor)	⋆	
Converter	M	Differential Delay	⋆	
Synchronizer	THR	Threshold for synchronization decision	⋆	
	$Ntest,L$	Length of energy measure	⋆	
Early-Late	#TAP	Number of taps in loop FIR filter	⋆	
	N_MAC	Number of MAC units for the loop filter		⋆
	COEFF[]	Coefficient values of loop FIR filter	⋆	
	NCO_INC	Increment value in NCO block	⋆	
Rake Receiver	#FGR	Number of fingers		⋆
	#AFGR	Number of turnd-on (active) fingers	⋆	
	N_p	Number of chips used for the estimation	⋆	

Starting from this parametric description, the whole CDMA receiver architecture has been tested and validated using the CoCentric System Studio environment from Synopsys. The logical synthesis has been carried out with Synplify Pro v7.0 by Synplicity. The complete flow over the FPGA has been performed with Xilinx ISE tools. Whole receiver occupies roughly 88% of a Xilinx XCV300E, achieving a maximum 96 MHz clock frequency and an estimated power consumption of 410 mW plus 546 mW of static power dissipated by FPGA. These results have been obtained using a 4 stages DDC, 8 tap FIR for tracking, and 4 fingers rake. Data are in 16 bit, fixed point format.

The experimental results prove the feasibility of a 3G wireless receiver on a medium–sized FPGA. This can open the possibility of a systematic reconfigurable fabrics exploitation on wireless communications terminals.

References

1. Buracchini, E.: The software radio concept. IEEE Communications Magazine **38** (2000) 138–143
2. Dent, P.: W-CDMA reception with a DSP based software radio. International Conference on 3G Mobile Communication Technologies (2000) 311–315
3. Martina, M., Masera, G., Piccinini, G., Vacca, F., Zamboni, M.: Energy Evaluation on a Reconfigurable Multimedia-Oriented Wireless Sensor. International Conference on Field Programmable Logic and Application (2002)
4. Viterbi, A.: CDMA: principles of spread spectrum communications. Addison-Wesley Publishing Company (1995)
5. Hogenauer, E.: An economical class of digital filters for decimation and interpolation. IEEE Trans. Acoustic, Speech and Signal Processing **29** (1981) 155–162
6. Key-Yong, K., et al.: Efficient High Speed CIC Decimator Filter. In: IEEE Asic Conference. (1998) 251–254
7. Kitsos, P., et al.: A reconfigurable linear feedback shift register (LFSR) for the Bluetooth system. IEEE International Conference on Electronics, Circuits and Systems **2** (2001) 991–994

Power-Efficient Implementations
of Multimedia Applications on Reconfigurable Platforms[1]

K. Tatas, K. Siozios, D. Soudris, and A. Thanailakis

VLSI Design and Testing Center, Department of Electrical and Computer Engineering,
Democritus University of Thrace, 67100, Xanthi, Greece
{ktatas, ksiop, dsoudris, thanail}@ee.duth.gr

Abstract. The power-efficient implementation of motion estimation algorithms on a system comprised by an FPGA and an external memory is presented. Low power consumption is achieved by implementing an optimum on-chip memory hierarchy inside the FPGA, and moving the bulk of required memory transfers from the internal memory hierarchy instead of the external memory. Comparisons among implementations with and without this optimization, prove that great power efficiency is achieved while satisfying performance constraints.

1 Introduction

In this paper data memory hierarchy power exploration at the high levels of design and register transfer (RT) level implementation are combined in order to achieve the optimum implementation in terms of power consumption while meeting performance constraints. An FPGA has been used for the implementation of two common motion estimation algorithms with and without memory power optimization.

2 Target Architecture

The architecture we have considered is illustrated in Fig. 1. It consists of an FPGA which implements the required logic and an external (off-chip) memory, which stores the data of the application.

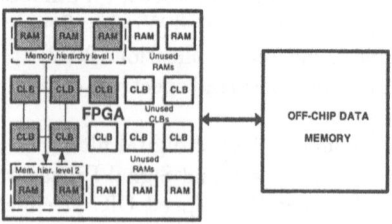

Fig. 1. Target architecture

---

[1] This work was partially supported by the project IST-34793-AMDREL which is funded by the E.C.

P.Y.K. Cheung et al. (Eds.): FPL 2003, LNCS 2778, pp. 1032–1035, 2003.

3 Target Application and Optimization Methodology

We have chosen two popular multimedia kernels for implementation, namely the full-search (FS) and hierarchical search (HS) motion estimation algorithms [1], [2], [3].

The optimization methodology is based on the fact that a large, off-chip memory consumes greater power per memory transfer, than a small, on-chip one, therefore the methodology attempts to move the greatest number of transfers from the off-chip memory to on-chip ones, by locating data that are used often.

It has been proven [4], [5] that a total of 21 possible data memory hierarchies for our applications exist. These candidate hierarchies are implied by the 21 possible data-reuse transformations, which are applied on the high-level description of the kernels. The power consumption of each memory hierarchy was estimated using Landman's memory model [6]. We have selected the one that exhibited the lowest power consumption. It implies two levels of on-chip memory hierarchy: A line of candidate blocks which is loaded from the external memory and a single candidate block is loaded from the line of candidate blocks.

4 Experimental Results

For illustration purposes, greyscale frames of 144×176 pixels were used, therefore an external memory of 2×25344×8 bit is necessary in order to contain two such grey-scale type of frames. The selected low-power transformation, was described in VHDL and implemented in various Xilinx devices. For the on-chip memory hierarchies plus the motion vector storing and the subsampled frames storing in the hierarchical search motion estimation kernel, an appropriate number of Xilinx BlockSelectRAMs were used in their 512×8 configuration [7].

Power consumption for the FPGA was estimated using Xpower [8]. The power consumption of the external memory was estimated using Landman's model [6].

Table 1 provides the total power consumption of all implementations in detail, at a frequency of 25 MHz. The last column presents the percentage gains of a specific optimized FPGA implementation in comparison to the corresponding non-optimized implementation on the same device.

Table 2 shows the device utilization for all implementations in number of slices and number of BlockRAMs used. The number of required resources (slices/BlockRAMs), the number of total available resources and the percentage of utilization is given.

Performance measurements can be seen in Table 3. Let us consider the perform-ance measures for slow-motion video (10 FPS), video conference (15 FPS), video file transfer (15 FPS) and digital video (30 FPS) according to [5]. The original FS algo-rithm can meet all applications except full-motion digital video in most devices.

The power-optimized FS scheme cannot meet the real-time constraint in any appli-cation due to the performance overhead. The original HS motion estimation, on the other hand, more than adequately satisfies the performance constraints in all devices.

The power-optimized version barely satisfies the performance required for full-motion video, but it is more than adequate for the remaining applications.

Table 1. Total power consumption comparison of alternative implementations

Application	Device	FPGA power (mW)	External memory power (mW)	Total power (mW)	Power gain (%)
FS (original)	xc2s15	160.88	703.76	864.64	
	xc2s200	275.81	703.76	979.57	
	xcv50	217.25	703.76	921.01	
	xcv1000	343.62	703.76	1047.38	
	xcv400e	586.49	703.76	1290.25	
	xc2v10000	593.02	703.76	1296.78	
FS (optimized)	xc2s15	120.38	211.93	332.31	61.56
	xc2s200	239.70	211.93	451.63	53.89
	xcv50	172.22	211.93	384.13	58.29
	xcv1000	295.2	211.93	507.13	51.58
	xcv400e	555.66	211.93	767.59	40.50
	xc2v10000	590.98	211.93	802.91	38.08
HS (original)	xcv400e	637.61	1093.88	1731.49	
	xc2v10000	592.85	1093.88	1686.73	
HS (opt.)	xc2v10000	1170.53	110.39	1280.92	24.05

Table 2. Area comparison of alternative implementations

Application	Device	Slice count (% utilization)	BlockRAM count (% utilization)
FS (original)	xc2s15	164/192 (85%)	2/4 (50%)
	xc2s200	164/2,352 (6%)	2/14 (14%)
	xcv50	164/768 (21%)	2/8 (25%)
	xcv1000	164/12288(1%)	2/32 (6%)
	xcv400e	164/4800 (3%)	2/40 (5%)
	xc2v10000	164/61440 (0.002%)	2/192 (0.01%)
FS (optimized)	xc2s15	166/192 (86%)	4/4 (100%)
	xc2s200	166/2352 (7%)	4/14 (21%)
	xcv50	166/768(22%)	4/8 (50%)
	xcv1000	166/12288 (1%)	4/32 (12%)
	xcv400e	166/4800 (3%)	4/40 (10%)
	xc2v10000	166/61440	4/192 (0.02%)
HS (original)	xcv400e	761/768 (99%)	36/40 (90%)
	xc2v10000	761/61440 (1%)	36/192 (18%)
HS (opt.)	xc2v10000	16817/61440 (27%)	161/192 (84%)

5 Conclusions

A power-efficient implementation of two popular multimedia applications based on exhaustive high-level data memory power exploration and RTL design on FPGAs was presented. Simulation results indicated significant power gains at the expense of performance due to larger hardware complexity.

Table 3. Performance comparison of alternative implementations

Application	Device	# cycles	Clock freq. (MHz)	FPS
FS(original)		5702400	25	4.38
	xc2s50		Max (87.030)	15.24
	xc2s200		Max(80.97)	14.19
	xcv50		Max(69.152)	12.11
	xcv1000		Max(91)	15.94
	xcv400e		Max(95)	16.64
	xc2v10000		Max(103)	18.0
FS (optimized)		9554688	25	2.61
	xc2s50		Max (54.44)	5.68
	xc2s200		Max(57.47)	6
	xcv50		Max(54.8)	5.72
	xcv1000		Max(51.16)	5.34
	xcv400e		Max(66.29)	6.92
	xc2v10000		Max(81)	8.45
HS (original)		320544	25	77.99
	xcv400e		Max(44.08)	137.53
	xc2v10000		Max(78.58)	245.13
HS (opt.)	xc2v10000	950994	Max(25.24)	26.44

References

1. V. Bhaskaran and K. Kostantinides: Image and Video Compression Standards. Kluwer Academic Publishers (1998)
2. K. M. Nam, J.-S. Kim, Rae-Hong Park, and Y. S. Shim: A fast hierarchical motion vector estimation algorithm using mean pyramid. IEEE Transactions on Circuits and Systems for Video Technology, vol. 5, no. 4 (1995) 344–351
3. Peter Kuhn: Algorithms, Complexity Analysis and VLSI, Architectures for MPEG-4 Motion Estimation. Kluwer Academic Publishers (1999)
4. D. Soudris, N. D. Zervas, A. Argyriou, M. Dasygenis, K. Tatas, C. Goutis, A. Thanailakis: Data-Reuse and Parallel Embedded Architectures for Low-Power, Real-Time Multimedia Applications. IEEE International Workshop on Power and Timing Modeling, Optimization and Simulation (PATMOS), Göttingen, Germany (2000) 243- 254
5. N. D. Zervas, K. Masselos, C.E. Goutis: Data-reuse exploration for low-power realization of multimedia applications on embedded cores. Proc. Of 9th Int. Workshop on Power and Timing Modeling, Optimization and Simulation (PATMOS'99) (1999) 71-80
6. P. Landman: Low-power architectural design methodologies. Doctoral Dissertation, U.C. Berkeley (1994)
7. http://direct.xilinx.com/bvdocs/publications/ds003.pdf
8. http://support.xilinx.com/support/sw_manuals/xilinx4/manuals.pdf

A VHDL Library
to Analyse Fault Tolerant Techniques*

P.M. Ortigosa, O. López, R. Estrada, I. García, and E.M. Garzón

Dpt. of Computer Architecture and Electronics, University of Almería,
Almería 04120,Spain. {pilar,inma,ester}@ace.ual.es

Abstract. This work presents an initiative to teach the basis of fault tolerance in digital systems design in undergraduate and graduate courses in electrical and computer engineering. The approach is based on a library of characteristic circuits related to fault tolerance techniques which has been implemented using a Hardware Description Language (VHDL). Due to the properties of the design tools associated to these languages, this approach allows with ease: (1) to implement faults tolerant digital systems; (2) to d etermine the behaviour of system when faults are presented; (3) to evaluate the additional resources and response time linked to any fault tolerance technique in the laboratory.

1 Introduction

Due to the great importance of faults tolerance techniques, current electrical and computer engineers have to know the fundamental topics related to this field. Therefore, Testing and Fault Tolerance belong to the set of areas associated with the body of knowledge of Computer Engineering defined by the Joint IEEE Computer Society and ACM Task Force on Computing Curricula [1].

Most fault tolerance techniques are based on the addition of not necessary resources for a common system function. These additional resources can be defined as a collection of characteristic circuits only used in this context. Our approach is based on the design of this specific library by a hardware description (VHDL in particular) to facilitate: (1) the design of systems to which fault tolerant techniques will be applied; (2) the insertion of redundant hardware systems; (3) the design, instantiation and use of generic systems; (4) the analysis of the area or logic cells needed to implement a determined system, and consequently, the analysis of the additional resources when fault tolerant techniques are used; (5) the analysis of the changes in the system response time when applying any of the fault tolerant technique; (6) the testing of the different fault tolerant techniques by using test bench included in the library and specifically designed to be used in the educational context; and (7) the development of the laboratory exercises due to the availability in a library of all basic components needed for the design of fault tolerant systems.

---

* This work was supported by the Ministry of Education of Spain (TIC2002-00228)

P.Y.K. Cheung et al. (Eds.): FPL 2003, LNCS 2778, pp. 1036–1039, 2003.

Table 1. Main hardware redundancy techniques and the circuits for implementing

Passive hardware redundancy

Technique	Circuits
N-Modular Redundancy (NMR)	Majority Voters
	Mid-value selectors
	Flux-summer

Active hardware redundancy

Technique	Circuits
Duplication with Comparison	Comparators
Standby Replacement or Sparing	Faults Detectors
	Reconfiguration Switches
Pair and a Spare	Comparators
	Faults Detectors
	Reconfiguration Switches

Hybrid hardware redundancy

Technique	Circuits
N-Modular Redundancy with Spares	Comparators
	Faults Detectors
	Reconfiguration Switches
	Majority Voters
Triple-duplex Redundancy	Reconfiguration Switches
	Flux-summer
Sift-Out Modular Redundancy	Comparator/Faults-Detectors
	Collector

In this work, we present our own initiative to support the teaching of faults tolerance techniques. This initiative is based on a VHDL library of circuits related to fault tolerance techniques, and a set of exercises to develop in the laboratory. Section 2 describes the circuits included in the library. An illustrative example of fault tolerant circuits implemented with the library is described. Finally, the main conclusions are described in Section 3.

2   Circuits to Implement Fault Tolerance

One common approach to reduce the probability of a system failure is based on including redundancy. Most references [2, 4, 3] distinguish four kinds of redundancy (1) hardware, (2) information, (3) software and (4) time. Therefore, it must be emphasised that the redundancy usually has a relevant impact on system qualities, such as performance, size, weight, power consumption, reliability and so on. Thus, a measure of this impact must be included in the evaluation of systems with redundancy. This work is focused on hardware and information redundancy, as well as, a framework which provides measures of additional resources and response times related to fault tolerance techniques. Moreover, the framework provides a collection of characteristic circuits to implement fault tolerance.

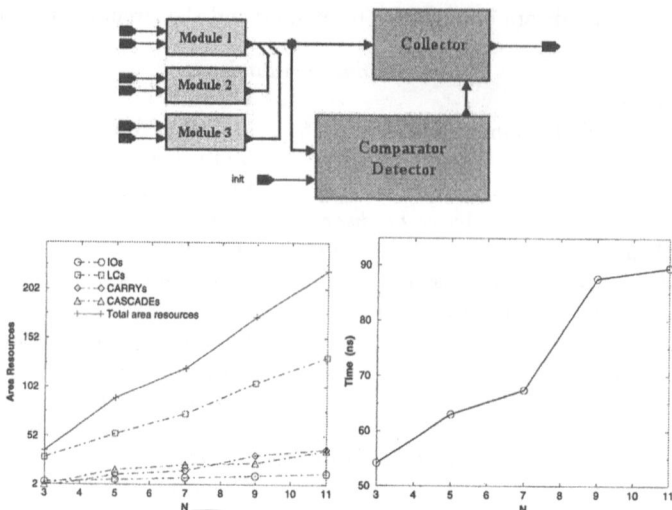

Fig. 1. Block diagram (top) and experimental results for with sift-out modular redundancy

Table 1 enumerates the main fault tolerance techniques based on *hardware redundancy* and the modules related to these techniques [2, 4, 3]. It must be emphasised that some characteristic circuits can be related to several techniques, such as voters and comparators. So, a library that includes these characteristic circuits can help students to develop exercises related to this kind of fault tolerance techniques.

An approach to decrease the data corruption possibility is related to the addition of redundant information for data. This kind of redundancy is defined in [2, 4, 3] as *information redundancy*. Wide diversity of error detecting and/or correcting codes has been developed for specific applications. The following codes are the most extended [2, 4, 3, 5]: parity codes, m-of-n codes, duplication codes, checksums, cyclic codes, arithmetic codes, Berger codes. These codes require length or code word greater than that of the original data. Thus, the systems must include additional resources to carry out the following stages (1) data encoding, (2) data processing, storage or transmission and (3) data decoding. Stages 1 and 2 are carried out by circuits (encoders and decoders) specified by the used code. Consequently, these specific modules to data encode and decode can be included in the proposed library to develop exercises related to these redundancy techniques.

Next, an example related to a hybrid hardware redundancy method which is called *sift-out modular redundancy* is described. This technique was proposed by de Sousa and Mathur [6], and it is described in classical references as [2]. Once the students know the theoretical model, they must design the fault tolerant system using both the components from fault tolerant library and the modules that are able to fault. These modules can either be instantiated from a basic digital circuit library or be designed by the students. As the specific components of the fault tolerant system (i.e. collector and comparator detector) have defined the number of possible inputs as a generic parameter, the students can easily design systems with different redundancy levels using the same specific components (modifying the generic parameter) and just replicating new input modules. When a system is designed, it can be simulated in order to analyse its behaviour and performance and, later on, it can be synthesised for a particular FPLD. With the synthesis tool, analysis of consumed area and required times can be extracted.

Illustrating this kind of comparisons, Figure 1 shows block diagram and area and time results for five couples of collector and comparator/detector modules whose number of input modules (N) range from 3 to 11. Specifically, the consumed area resources for the FLEX10K70 board are shown. These resources can be classified by the number of Input/Output Ports (IOs), number of Combinatorial Logic Blocks (LCs), Carry bits (CARRYs) and Cascade bits (CASCADEs). It can be seen that the area resources and the response time increase as the number of redundant modules does.

3   Conclusions

In this article we have described an educational approach to analyse fundamental fault tolerance topics. This initiative mainly consists of a library of characteristic circuits which are present in specific fault tolerant designs, and the guidelines to develop several kinds of exercises in the laboratory. The library circuits and the systems designed by the students have been implemented using a Hardware Description Language and a specific design CAD tool. We consider the approach introduced as the core of a specific framework which facilitates the comprehension and the analysis of the main issues in fault tolerant design. It could be included in undergraduate and graduate courses related to Digital Systems Design and Fault Tolerant Systems in Electrical and Computer Engineering.

References

1. The Joint Task Force on Computing Curricula IEEE Computer Society ACM "Computing Curricula 2001. Computer Engineering Main Report". October 25, (2002) URL: www.eng.auburn.edu/ece/CCCE/MainReport/MainReport.PDF
2. Barry W. Johnson. Design and Analysis of Fault-Tolerant Digital Systems. Addison-Wesley Publishing Company. (1989)
3. Pralag K. Lala. Self-Checking and Fault-Tolerant Digital Design. Morgan Kaufmann Publishers. (2001)
4. Dhiraj K. Pradhan. Fault-Tolerant Computer Systems Design. Prentice Hall. (1996)
5. Martin L. Shooman. Reliability of Computer Systems and Networks: Fault Tolerance, Analysis and Design. John Wiley & Sons. (2001)
6. P.T. de Sousa and F. P. Mathur. Sift-out modular redundancy IEEE Transactions on Computers C-27 7 (1978) 624–627

Hardware Design with a Scripting Language

Per Haglund, Oskar Mencer, Wayne Luk, and Benjamin Tai

Department of Computing, Imperial College, London SW7 2BZ, UK

Abstract. The Python Hardware Description Language (PyHDL) provides a scripting interface to object-oriented hardware design in C++. PyHDL uses the PamDC and PAM-Blox libraries to generate FPGA circuits. The main advantage of scripting languages is a reduction in development time for high-level designs. We propose a two-step approach: first, use scripting to explore effects of composition and parameterisation; second, convert the scripted designs into compiled components for performance. Our results show that, for small designs, our method offers 5 to 7 times improvement in turnaround time. For a large 10x10 matrix vector multiplier, our method offers respectively 365% and 19% improvement in turnaround time over purely scripting and purely compiled methods.

1   Introduction

Existing HDLs based on Java, C, or C++, such as JHDL [1], Handel-C [2], or PAM-Blox [6], are compiled languages for hardware design. The compilation stage of compiled HDLs delays the design process. Software programmers frequently use scripting languages for rapid application prototyping.

Previous work highlights how designers can benefit from scripting common hardware design tasks. For example, Luk [5] uses scripting to automate core testing, and Ho [4] links together stages in the tool-chain with scripts. However, we are not aware of methods that involve scripting a high-level structural description of circuits, such as proposed in this paper.

We propose a two-step methodology for capturing the advantages of scripting at the hardware design stage. First, designers use a scripting language to prototype designs using existing compiled components. Scripting removes the compilation stage and substantially accelerates the exploration process. Second, on completion of the exploration process, designers convert their scripted designs into new compiled components.

We implement our scripting methodology by extending the Python scripting language [7,8] with the features of the C++ HDL PamDC/PAM-Blox [6]. The contributions of our hardware scripting language PyHDL are:

- Extending a scripting language with existing hardware design libraries.
- Evaluating the impact of hardware scripting on design time.

The remainder of this paper is organised as follows. Section 2 briefly describes our implementation. Section 3 evaluates PyHDL through the presentation and analysis of experimental results from several circuit designs. Section 4 offers conclusions.

P.Y.K. Cheung et al. (Eds.): FPL 2003, LNCS 2778, pp. 1040–1043, 2003.
© Springer-Verlag Berlin Heidelberg 2003

2   Hardware Scripting

We develop a scripting HDL by *extending* a scripting language with the facilities of existing compiled hardware design libraries. The extension method requires a compiled HDL, a scripting language and an API to allow interaction between the scripting language and the HDL. We use the Python/C API to extend the scripting language Python with the C++ hardware design libraries of PamDC and PAM-Blox. PamDC provides primitives such as wires and registers. PAM-Blox is a collection of parameterisable, object-oriented hardware components built from PamDC primitives.

PyHDL consists of five components: mapping classes, interface, PAM-Blox library, design scripts, and output scripts. *Mapping classes* are a set of Python classes that replicate the class hierarchy of PAM-Blox. The mapping classes type-check and propagate inheritance before forwarding control to the interface. The *interface* manages the flow of control between Python and PAM-Blox. The interface associates instances of PyHDL mapping classes with PAM-Blox hardware objects. The *PAM-Blox library* contains the primitives and compiled hardware modules found in PamDC and PAM-Blox. *Design scripts* contain circuit designs written in Python. Python circuit designs typically contain both primitives and PAM-Blox components. An *Output script* defines the parameters for a particular design and drives the simulator or EDIF generator.

3   Evaluation of PyHDL

We use the following metrics to evaluate the effectiveness of our approach: (a) *turnaround time* is the time taken to generate new simulation output after a design change; (b) *hardware performance* results obtained from FPGA-vendor tools; (c) *lines of code* to indicate potential syntactical differences.

We implement three circuits in Python and C++: A greatest common denominator finder (GCD), a credit card validator (CARD), and a scalable matrix vector multiplier (MATMUL) using 16-bit combinational multipliers. The GCD and CARD circuits are small and allow us to evaluate the effectiveness of hardware scripting for testing and prototyping components. The MATMUL circuit is scalable and allows us to evaluate how the performance of hardware scripting changes with circuit size.

We present three implementations of the MATMUL circuit. Each implementation uses a serial shift-add multiplier. We design the multiplier using either primitives within Python or PAM-Blox components.

The first MATMUL implementation uses pure Python and illustrates how the performance of a scripting-only approach scales poorly with circuit size. Porting the Python multiplier to C++ produces a new compiled PAM-Blox multiplier. The second MATMUL uses the new PAM-Blox multiplier and thus illustrates the effectiveness of extending a scripting language with compiled components. The third MATMUL implementation is written entirely in C++ and shows the base case performance resulting from using only compiled C++ code.

Table 1. The implementation results of three circuits: Greatest common denominator (GCD), credit card validation (CARD), and a 10x10 matrix vector multiplier (MATMUL). The three implementations of MATMUL are: pure Python (multiplier made from primitives), Python using C++ components (PAM-Blox multiplier), and pure C++ (PAM-Blox multiplier). The hardware results are based on targeting Xilinx XCV1000E and XC2V6000 devices

	GCD		CARD		MATMUL (10x10)		
Circuit design language	Python	C++	Python	C++	Python		C++
Component design language	C++	C++	C++	C++	Python	C++	C++
Turnaround (seconds)	0.80	5.60	1.02	5.84	69.1	18.9	22.5
Slices	38	40	283	285	25170	25170	20702
LUTs	71	72	482	482	43570	43569	34051
FFs	21	22	25	26	3781	3781	4660
Clock speed (MHz)	71.5	69.1	4.3	4.1	18.2	18.3	29.8
Gate count	843	865	3469	3485	387981	387981	339252
Lines of code	86	84	128	135	169	104	109

We use GNU Cygwin on a 1.7 GHz Pentium 4 and Xilinx ISE 5.1 to produce our results. We target a Xilinx XCV1000E device for the GCD and CARD circuits, and a XC2V6000 device for the MATMUL circuit.

Table 1 presents our results for the GCD, CARD, and MATMUL circuits. For the two small circuits, Python reduces the turnaround time by over 80%. The results for the three implementations of the matrix vector multiplier show that scripting performance depends on both the design language and the component implementation. Pure C++ is faster than pure Python, but slower than Python combined with compiled components. Figure 1 shows how the turnaround time scales with the size of the matrix vector multiplier. Scripting-only becomes inefficient for vectors of length 4 or greater. However, using compiled components in a scripted circuit design is consistently faster than using either pure Python or pure C++.

The hardware results show that PyHDL produces hardware similar to PAM-Blox. The discrepancy is mostly due to differences in buffers added between the logic and technology mapping, which does not yet occur identically in all setups.

The experimental results confirm that the proposed approach, described in Section 2, can provide significant benefits. First, hardware scripting eliminates most of the compilation overhead, as seen in Table 1 and Figure 1. Second, converting a Python component into a compiled component preserves the performance advantage of hardware scripting for larger designs. The curve *Pure Python* in Figure 1 illustrates how the performance of a design approach based only on scripting deteriorates with increasing circuit size. The curve *Python with C++ components* in Figure 1 shows how scripting performance improves after converting the pure Python multiplier into a compiled PAM-Blox component.

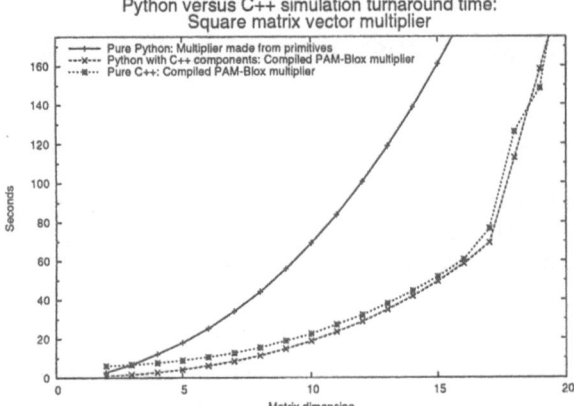

Fig. 1. *The turnaround times of three implementations of a scalable matrix vector multiplier using Python and C++. Each design uses a structurally identical multiplier implemented in either Python or C++.*

4 Conclusions

This paper shows how hardware scripting facilitates rapid prototyping and exploration of component characteristics. We evaluate our hardware scripting language PyHDL, an extension of the popular scripting language Python with the compiled PAM-Blox framework. We find that hardware scripting reduces the design time for small circuits, and that satisfactory circuit descriptions can be ported to C++ and compiled to reduce the design time for large circuits also.

References

1. P. Bellows and B. Hutchings, "JHDL - an HDL for reconfigurable systems," in *IEEE Symposium on FPGAs for Custom Computing Machines*, K. L. Pocek and J. Arnold, Eds.   Los Alamitos, CA: IEEE Computer Society Press, 1998, pp. 175–184.
2. Celoxica, "Celoxica homepage," http://www.celoxica.com/.
3. R. Goering, "Engineer creates open-source hdl in Ruby language," http://www.eetimes.com/story/OEG20020807S0019.
4. C. Ho, P. Leong, K. Lee, K. Tsoi, R. Ludewig, P. Zipf, A. Ortiz, and M. Glesner, "Fly - a modifiable hardware compiler," in *Field-Programmable Logic and Applications*. Springer Verlag, 2002, pp. 381–390, LNCS 2438.
5. W. Luk, D. Siganos, and T. Fowler, "Automating qualification of reconfigurable cores," in *Reconfigurable Systems*.   IEE Digest, 1999, pp. 4/1–4/6.
6. O. Mencer, M. Morf, and M. J. Flynn, "PAM-Blox: High performance FPGA design for adaptive computing," in *IEEE Symposium on FPGAs for Custom Computing Machines*.   IEEE Computer Society Press, 1998, pp. 167–174.
7. Python Software Foundation, "Python homepage," http://www.python.org.
8. G. van Rossum, "An introduction to python for UNIX/C programmers," *Proc. NLUUG - Dutch Unix User Group Conference*, 1993.

Testable Clock Routing Architecture for Field Programmable Gate Arrays

L. Kalyan Kumar, Amol J. Mupid, Aditya S. Ramani, and V. Kamakoti

Department of Computer Science and Engineering, Indian Institute of Technology
Madras, Chennai 600036, Tamilnadu, India.
{kalyan,amol,ramani}@peacock.iitm.ernet.in, kama@iitm.ernet.in

Abstract. This paper describes an efficient methodology for testing dedicated clock lines in Field Programmable Gate Arrays (FPGAs). A H-tree based clocking architecture is proposed along with a test scheme. The H-tree architecture provides optimal clock skew characteristics. The H-tree architecture consumes at least 25% less of the routing resources when compared to conventional clock routing schemes. A testing scheme, which utilizes the partial reconfiguration capabilities of FPGAs through selective re-programming of the Complex Logic Blocks, to detect and locate faults in the clock lines is proposed

1 Introduction and Previous Work

An FPGA, like any other semiconductor device is affected by faults occurring during the components lifetime. Though most faults occur at the time of fabrication, occasional operational faults can result after extended usage. Unlike manufacturing defects that can be avoided by using spare routing wires and programmable fuses, operational failures need to be addressed by generating a new programming configuration [3]. *Online checkers* are used in the design for detecting such operational/run-time errors [1]. FPGA testing is divided into testing of logic blocks, interconnects, embedded memory, I/O block and miscellaneous testing which includes clock and powerlines [1]. To the best of our knowledge, there is no result reported in the literature, which addresses the testing of clock lines. The objective of any clock design is to minimize clock skew, clock delay, clock area, power, and noise while maximizing clock reliability. A clock routing architecture for ASICs called H-tree is described in [2]. *This paper proposes a H-tree based clock routing structure for symmetric FPGA architectures and derives bounds to prove its efficiency. A BIST based testing algorithm is also presented to test the proposed H-tree architecture.*

2 H-Tree Clock Routing Architecture for FPGA

This architecture essentially consists of an H shaped routing structure that is repeated with successively halving edge lengths. The starting point of the clock routing architecture, called the *root node*, is located in the geometric center point

P.Y.K. Cheung et al. (Eds.): FPL 2003, LNCS 2778, pp. 1044–1047, 2003.

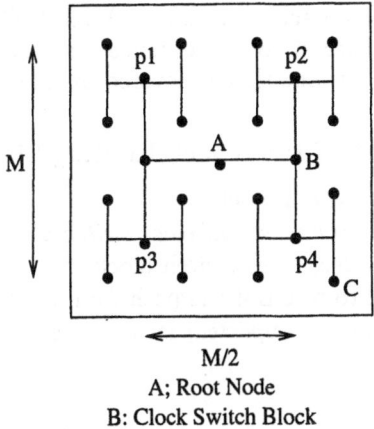

Fig. 1. 2-level H-Tree Clock Routing Structure

of the FPGA chip and the H-tree structure distributes it across to all CLBs. This distribution needs to be done till each CLB has an adjacent clock line from which it can tap a clock input. The external clock can be tapped either from the periphery of the FPGA chip (Flip I/O Technology) or from the geometric center point in the area of the FPGA chip (Area I/O technology) and connected to the root node. If clocks are predetermined, then, they may be embedded (*in-core* clock) inside the FPGA chip at the geometric center (root node). Additionally we assume that each CLB has an output that can be tapped to aid the testing procedure. The H-tree constructed has 4 end points p_1, p_2, p_3 and p_4 as shown in Figure 1. The algorithm recursively proceeds to construct H structures as in the figure with p_1, p_2, p_3 and p_4 as the center points and the lines of these structures, each covering $M/4$ CLBs. This is done till each of the CLBs receive a clock input.

The following bounds hold good for a $(M X M)$ CLB matrix.

1. The maximum Clock-to-leaf switch length of the clock line is $\leq M$.
2. Cumulative Length of Clock Routing Lines $= \frac{3}{2}(\frac{M}{2} - 1)M$

3   BIST for Testing H-Tree Structure

The BIST for testing the H-tree structure uses some of the CLBs available in the FPGA by configuring the Look-Up Tables (LUTs) in them in one of the following modes.

Normal Pass Mode: In this mode the LUT is loaded with a function, which passes one of its specified inputs to the output.

Inverse Pass Mode: In this mode the LUT is loaded to output the complement of one of its inputs.

The following definitions are used in the testing algorithms that follow.

Clock Switch Block Cover: A clock switch block C covers a CLB A if A is one of the 4 CLBs nearest to it and hence can take a clock input from it.
Leaf CLB: A leaf CLB is one that is covered by a leaf clock switch block

Algorithm 1 (Test All H-tree Clock Routing Lines)

1. Unmark all leaf clock switch blocks.
2. **while** (*there exists an untested leaf clock switch block*)
   - Select an unmarked leaf clock switch block L.
   - Apply Algorithm 2 to test L if the path from clock to L is fault free.
   - **if** (*a faulty clock line was reported*) locate clock segment using Algorithm 4.
   - Mark L.

Algorithm 2 (Testing of path P from Root node to any clock switch block L)

1. Select a CLB M covered by L.
2. Connect the output of M, by programming the interconnect structure, to an I/O port of the chip denoted by Tapped Output.
3. Let Clock be one of the inputs to CLB M.
4. Configure M in pass mode so as to pass clock to its output.
5. Apply the logical signal 1 at the Root node and observe the Tapped Output.
6. **if** (*Tapped Output =0*) Fault is in the Path P or Tapped Output.
7. Find region of fault using Algorithm 3 with input as (M,0).
8. **if** (*fault is in Tapped Output*)
   - Mark M as un-usable.
   - Select an unmarked/usable CLB M covered by L and go back to Step 2.
   - **if** (*no such CLB exists*)
     - report clock switch block to be un-testable and **return.**
   - **if** (*fault is in path P*) report that (P is faulty) and **return.**
9. Repeat steps 5 to 8 by applying the logical signal 0 at the Root node.
   (Parallel case replacing $0 \rightarrow 1$ and $1 \rightarrow 0$.)
10. **if** (*no faulty outputs have been obtained*), mark M as usable, report P is fault-free and **return.**

Algorithm 3 (Locating the fault region through a leaf CLB M)

1. **if** (*stuck value = 0*)
   - Let Clock be one of the inputs to CLB M.
   - Configure M in inverse pass mode so as to pass inverted clock to its output.
   - Apply logical signal 0 at Root node.
   - **if** (*Tapped Output = 1*) fault is in the path P.
       else fault is in the Tapped Output Line.
2. Repeat Step 1 checking for stuck value = 1
   (Parallel case replacing $0 \rightarrow 1$ and $1 \rightarrow 0$).

It is easy to see that the inverse pass mode detects the position of the fault as the output is dependant on whether the fault was encountered before or after inversion. The Algorithm 4 is similar to a binary search procedure to locate the faulty segment.

Algorithm 4 : Lookup(P) (Locate the faulty segment in a faulty path P)

1. Let $l_1 - l_2 - l_r$ be the segments making up P, where l_1 is the end-point closest to the root node.
2. Note that every l_i connects two clock switch blocks, c_{i-1} and c_i where c_0 is the Root node.
3. **if** ($r = 1$) report faulty segment is l_r.
   else
   - Let $c_{r/2}$ be the clock switch block corresponding to segment $l_{r/2}$ in the path P.
   - Identify a usable/untested CLB M covered by $c_{r/2}$ and perform the testing procedure in Algorithm 2, which will report either the path from - Root Node to CLB M is faulty or not.
   - **if** (*faulty clock path was reported by Algorithm 2*) $P = l_1 - l_2 - \cdots - l_{r/2}$
   - **if** (*fault-free clock path was reported by Algorithm 2*)
     $P = l_{r/2} - l_{r/2+1}, \cdots - l_r$.
   - **if** (*clock path was reported to be un-testable by Algorithm 2*) **exit**.
4. **LookUp**(P)

4   Conclusion

In this paper we presented a novel method to detect and locate faults in clock lines of FPGA based systems based on an efficient in-built dedicated routing architecture that we proposed. The number of clock switch blocks in any path from a clock Root node to a CLB, in a MXM CLB array, which contributes to the major portion of the propagation delay is upper bound by $\log_2 M$. We exploited the selective re-programming capabilities of FPGAs to arrive at an efficient fault detection and location procedure. The time required to locate the faulty clock segment, after identification of a faulty clock path is upper bound by $\log_2 \log_2 M$.

References

1. Subhasish Mitra, Philip P. Shirvani and McCluskey, J. E.: Fault Location in FPGA Based Reconfigurable Systems, BMDO/IST 1999.
2. Ullman, D. J.: Computational aspects of VLSI Design, Computer Science Press., pp84.
3. Vijay Lakamraju and Tessier, R.: Tolerating Operational Faults in Cluster-based FPGAs., FPGA2000, Monterey,CA, USA.

FPGA Implementation of Multi-layer Perceptrons for Speech Recognition

E.M. Ortigosa[1], P.M. Ortigosa[2], A. Cañas[1], E. Ros[1], R. Agís[1], and J. Ortega[1]

[1] Dept. of Computer Architecture and Technology,
ETS Ingeniería Informática. University of Granada, E-18071 Granada, Spain
{eva, acanas, eros, ragis, jortega}@atc.ugr.es
[2] Dept. of Computer Architecture and Electronics,
University of Almería, E-04120 Almería, Spain, ortigosa@ual.es

Abstract. In this work we present different hardware implementations of a multi-layer perceptron for speech recognition. The designs have been defined using two different abstraction levels: register transfer level (VHDL) and a higher algorithmic-like level (Handel-C). The implementations have been developed and tested into a reconfigurable hardware (FPGA) for embedded systems. A study of the two considered approaches costs (silicon area), speed and required computational resources is presented.

1 Introduction

An Artificial Neural Network (ANN) is an information processing paradigm that is inspired by the way biological nervous systems process information. An ANN is configured for a specific application, such as pattern recognition or data classification, through a learning process. The work presented in this paper addresses the study of the implementation viability and efficiency of ANNs into reconfigurable hardware (FPGA) for embedded systems, such as portable real-time Automatic Speech Recognition (ASR) systems for consumer applications. Let us focus on *a voice controlled phone dial system* (this can be of interest for drivers that should keep their attention in driving). This particularizes the Multi-Layer Perceptron (MLP) parameters to be implemented and the word set of interest (numbers from 0 to 9).

The multi-layer perceptron neural network model consists of a network of processing elements or nodes arranged in layers. The principle of the network is that when data from an input pattern is presented at the input layer the network nodes perform calculations in the successive layers until an output value is computed at each of the output nodes. The weights of the connections define the behavior of the network and are adjusted during training through a supervised training algorithm called back-propagation [1]. In the "forward pass" an input pattern vector is presented to the input layer. For successive layers, the input to each node is the sum of the scalar products of the incoming vector components with their respective weights (1),

$$sum_i = \sum_j w_{ij} out_j \, , \tag{1}$$

P.Y.K. Cheung et al. (Eds.): FPL 2003, LNCS 2778, pp. 1048–1052, 2003.

$$f(sum_i) = \frac{1}{1 + e^{-sum_i}} \qquad (2)$$

where w_{ij} is the weight connecting node j to node i and out_j is the output from node j. The output of a node i is $out_i = f(sum_i)$, which is then sent to all nodes in the following layer. The function f denotes the activation function of each node. A sigmoid activation function is frequently used, eq. (2).

2 Hardware Implementations

For our test bed application we need a MLP with 220 data inputs (10 vectors of 22 features extracted from speech analysis) and 10 output nodes in the output layer (corresponding to the 10 recognizable words). After testing different architectures, the best results (96.83% correct classification) have been obtained with 24 nodes in the hidden layer.

For the MLP implementation we have chosen fixed point computations with two's complement representation and different bit depths for the stored data (inputs, weights, activation function, outputs, etc). It is necessary to limit the range of different variables: inputs to the MLP (8 bits), output of the activation function (8 bits), weights (8 bits), and inputs to the activation function. Apparently we need 23 bits for the input of the activation function, but because we are using the sigmoid waveform, most of the values are repeated and only a small transition zone (≈1% of values) needs to be stored. After taking all these discretization simplifications the model achieves similar classification results. The results of the hardware system differ in less than 1% from the software full resolution results.

2.1 Register Transfer Level (VHDL)

The Register Transfer Level design of MLP has been defined using standard VHDL as hardware description language. As it can be seen in equation (1), the basic computations of a single neuron are the multiplication of the outputs from the connected neurons (synaptic signals) by their associated weights, and the summation of these multiplied terms. The Functional Unit (processing element) is composed by an 8-bit multiplier and a 24-bit accumulative adder.

The serial version of the MLP consists on a single functional unit that carries out all computations, for all neurons. The inputs of the functional unit are both the synaptic signals and their associated weights, which are stored in separate RAM modules. The output of the functional unit is connected to the activation function module. The activation function output is stored either in the hidden or in the final neuron RAMs, depending on the layer of the computed neuron.

The proposed parallel architecture describes a kind of node parallelism, in the sense that requires one functional unit per neuron when working at a determined layer. With this strategy, all neurons of a layer work in parallel and therefore produce their outputs simultaneously. This is not a fully parallel strategy because the outputs for different layers are obtained in a serial way. For our particular MLP where 24 neurons exist at the hidden layer and 10 ones at the output layer, 24 functional units are required. All of them will work in parallel when computing the outputs of the hidden layer and only 10 of them will work when the output layer is computed.

2.2 High Level Description (Handel-C)

The high level design of MLP has been defined using Handel-C [2] as a system-level specification language. Based on ANSI-C, Handel-C includes a simple set of extensions required for hardware development. The whole design processes have been defined with DK1 Design Suite tool from Celoxica [2]. Sequential and parallel designs have been finally compiled using the development environment *Xilinx Foundation 3.5i* [3].

In the sequential version, the MLP computes the synaptic signals for each neuron in the hidden layer by processing the inputs sequentially; and later on, the obtained outputs are similarly processed by the output neurons. In the parallel version, all neurons belonging to the same layer compute their results simultaneously, in parallel, except for accessing to the activation function that is done in a serial way.

3 Comparative Results

The systems have been designed using the development environments FPGA advantage and DK1.1 to extract the EDIF files. All designs have been finally placed and routed in a VirtexE 2000 FPGA, using the development environment *Xilinx Foundation 3.5i* [3]. Table 1 shows the implementation results obtained after synthesizing both sequential and parallel versions of the MLP defined using VHDL. These results are characterized by the following parameters: number of slices, number of EMB RAMs, minimum clock period, number of clock cycles required for each input vector (220 input components) evaluation, and total time consumed for each input vector evaluation. The computing time for each input vector (last column) is much shorter (20 times) in the parallel version. In this way we are taking advantage of the inherent parallelism of the ANN computation scheme.

Table 1. Implementation characteristics of the designs with VHDL

MLP design	# slices	% slices	# EMB RAMs	% EMB RAMs	Clock (ns)	# Cycles	Evaluation time (μs)
Serial	379	1.5	19	11	49.024	5630	276.005
Parallel	1614	8.5	26	16	53.142	258	13.710

Table 2 presents the results obtained after synthesizing, the sequential and parallel versions of the MLP defined using Handel-C. When defining the MLP we have to decide how to store the data in RAM. Different strategies are considered: (a) only distributed RAM for the whole designs have been utilized, (b) the weights associated to synaptic signals (large array) make use of EMB RAM modules, while the remaining data are stored in a distributed mode; and finally, (c) only EMB RAM modules are used. Results in Table 2 show that independently of the distribution memory option, the parallel version requires more area resources than its corresponding serial one. As happened in VHDL description, the computing time for each input vector is much shorter (about 20 times) in the parallel version. The choice of a determined option in the memory implementation will depend on the area and time constraints.

Table 2. Implementation characteristics of the designs with Handel-C. (a) Only distributed RAM. (b) Both EMB RAMs and distributed RAM. (c) Only EMB RAM

MLP design	# Slices	% slices	# EMB RAMs	% EMB RAMs	Clock (ns)	# cycles	Evaluation time (µs)
(a) Serial	2582	13	0	0	50.620	5588	282.864
Parallel	6321	32	0	0	58.162	282	16.402
(b) Serial	710	3	24	15	62.148	5588	347.283
Parallel	4411	22	24	15	59.774	282	16.856
(c) Serial	547	2	36	22	65.456	5588	365.768
Parallel	4270	22	36	22	64.838	282	18.284

When comparing Tables 1 and 2, where results for a RTL and a higher level description are shown respectively, it can be seen that the RTL implementation leads to a more optimized approach for the final system. However, one of the main advantages of the high level description is the design time of a system. So, for our MLP designs, it must be known that the design time for the serial case when using high level description has been about 10 times shorter than the RTL one. For the parallel case, the design time for the high level it has been relatively short (just to introduce the *"par"* directives) while for the RTL description a new architecture and control unit have been designed; this fact implies a larger difference in the designing time.

4 Conclusions

We have presented the FPGA implementation of MLP for speech recognition applications. Both sequential and parallel versions of the MLP have been described using two different abstraction levels: register transfer level (using VHDL) and a higher algorithmic-like level (using Handel-C). Results show that RTL implementation produces more optimized system; however, one of the main advantages of the high level description is the time consumed to design a system. For the speech recognition application we obtain a correct classification rate of 96.83% with a computation time around 14-16 microseconds per sample, which fulfills by far the time restrictions imposed by the application.

Acknowledgement

The work has been supported by CICYT TIC2002-00228 and SpikeFORCE (IST2001-35271).

References

1. Widrow, B., Lehr, M.: 30 years of adaptive neural networks: Perceptron, Madaline and Backpropagation. Proceedings of the IEEE, vol. 78, no. 9, pp.1415-1442 (1990)
2. Celoxica, http://www.celoxica.com/
3. Xilinx, http://www.xilinx.com/

FPGA Based High Density
Spiking Neural Network Array

Juan M. Xicotencatl and Miguel Arias-Estrada

Computer Science Department, INAOE, Apdo. Postal 51 & 216, Mexico
jperez@inaoep.mx, ariasm@inaoep.mx

Abstract. Pulsed neural networks can be applied to the design of dense arrays using minimum hardware resources in the interconnection among neurons. Using statistical saturation in pulse frequency coded neurons, a minimum size hardware neuron can be implemented. The proposed neuron is compact enough to be included in large arrays. The presented architecture has additional interesting characteristics like unrestricted topology and scalability. In this paper, the design and implementation of a high density spiking neural array is presented.

1   Introduction

ANNs are parallel processing structures that have a large number of processors and a large numbers of interconnections among them. In an ANN each processor is connected to its neighbors, so there are more interconnections than processors. The ANN computational power lies in the large number of interconnections. There is an associated synaptic strength or a weight with each connection. ANNs digital implementation focuses on the architecture design. A digital implementation has some important characteristics like flexibility, high accuracy and repeatability [2].

This paper presents an FPGA based high-density neural network array, which is a generic platform for applications that require over 1000 neurons in hardware. The rest of the paper is organized as follows: section 2, 3 and 4 discuss the architectural concepts proposed in our approach at the synapse, soma and interconnections respectively. Next section presents the architecture simulation and a discussion is given. Finally, some conclusions and future work are presented.

2   Synapse Design and Weight Storage

Practical ANNs dense implementations are possible only if the circuitry devoted to multiplication involved in the synapse is reduced. In this work, the instantaneous value of the neuron activity is represented as the instantaneous frequency of digital pulses (PFM, Pulse Frequency Modulation). Since the signal level is expressed by the frequency, a simple frequency converter replaces the multiplier in the synapse. Figure 1 shows the implemented synapse design.

P.Y.K. Cheung et al. (Eds.): FPL 2003, LNCS 2778, pp. 1053–1056, 2003.

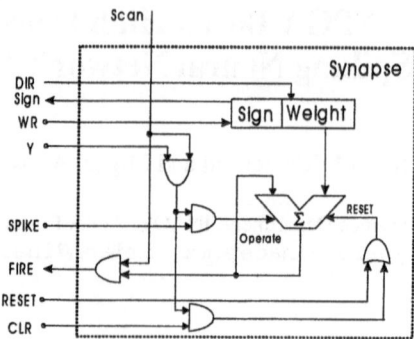

Fig. 1. The synapse module is implemented with a register and an accumulator.

The designed module loads a signed weight in serial form. The weight is stored in a serial input parallel output (SIPO) register to be accessible by the accumulator. The accumulator increases its value with each input pulse (SPIKE). If the accumulator reaches a maximum, it produces an overflow. Finally, the generated overflow is present in the FIRE terminal to be externally processed when the SCAN line is one.

3   Soma Design, Summation of Synaptic Contribution and Non-linear Function Activation

In the designed soma, only the synapses that are fired contribute to the output. In this way, the proposed soma avoids to access or to route synapses with a zero contribution. To obtain the non-linearity in the output, the statistical saturation present in the counter is used. This element and the necessary logic to update and to access its content are shown in figure 2.

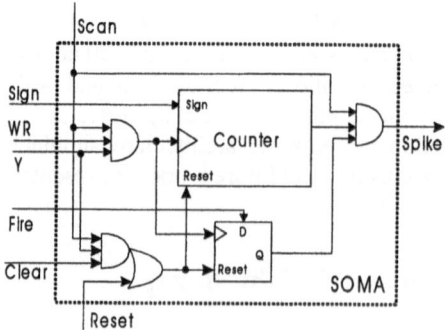

Fig. 2. Soma design. The counter adds the contributions from the previous layer.

The proposed soma counts the signed pulses from an input pulse train. The soma fires when the number of positive pulses exceeds the number of negative pulses and a internal flip-flop contains a one. This flip-flop is set to one when the last synapse associated to the soma fires.

4   Routing of the Activity among Neurons

In the proposed design, the topology can be changed online or offline if the application requires it. Since, there are not direct interconnections among neurons, the topology is mapped to an external memory. This scheme is shown in figure 3.

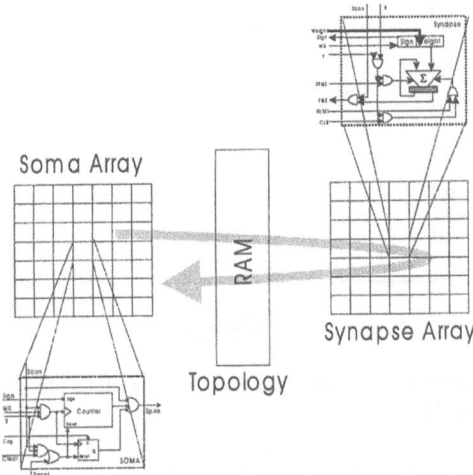

Fig. 3. The proposed neural network design, an external memory is used to map the topology.

In figure 3, two large arrays are defined: the soma and the synapse arrays. Both arrays read the RAM memory to access the corresponding elements in the other array. The RAM memory contains all the module locations, i.e., it contains the *address* from each element inside the arrays.

5   Implementation and Results

The architecture has been modeled in VHDL and synthesized using XILINX ISE 5.1i for an XCV2000BGA560-6. The whole architecture has been tested using FPGA-based board from AlphaData. The statistics of the design are presented in table 1.

Table 1. FPGA Statistics from the proposed design using a XCV2000BGA560-6

	Slices	Flip-Flops	LUTs	FPGA percentage	Frequency
1000 Synapses	13340	21340	25346	70%	-
1120 Somas	5329	6600	9028	27%	-
Total	18669			97%	35.6Mhz

The transfer function from the architecture is shown in figure 4. Its transfer function is approximated from the stochastic response as the equation:

$$f(weight) = \frac{1}{\left(1 + e^{-1.3weight+1}\right)} \cdot \tag{1}$$

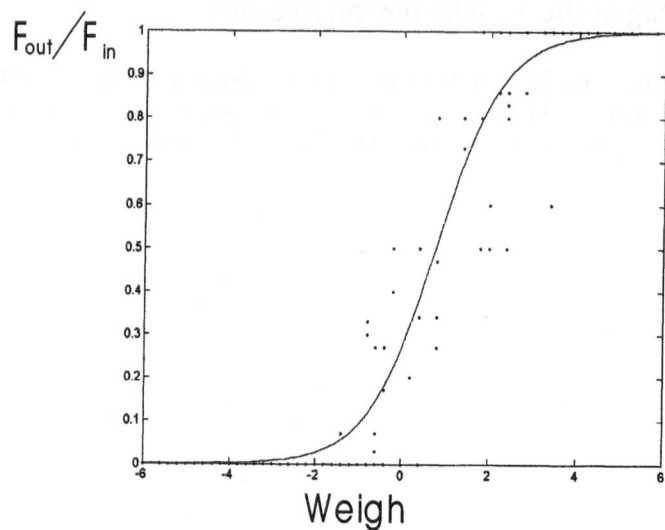

Fig. 4. Transference function. A single soma and a 5x5 synapse array were used. Six synapses are mapped in the external RAM.

Using the PFM technique, the performance in small and large arrays is the same. The bottleneck in the design is the memory, which is used to map the topology. The total number of somas and synapses is a function of the available resources in the FPGA. In small arrays, the memory can be implemented directly in the FPGA.

6   Conclusions and Future Work

An FPGA based high-density neural array was implemented and its functionality has been described. The neuronal arrays are related with an external topology memory. The proposed architecture can implement a massive number of neurons in a single FPGA. Additionally, the proposed design is scalable and flexible. Future work will focus on implementing a learning module and a hierarchy model to avoid the memory bottleneck.

References

1. T. Schoenauer, S. Atasoy, N. Mehrtash and H. Klar, Simulation of a Digital Neuro-Chip for Spiking Neural Networks, International Joint Conference on Neural Networks (IJCNN), Como, Italy, July 2000.
2. M. Schaefer, T. Schoenauer, C. Wolff, G. Hartmann, H. Klar and U. Rueckert, Simulation of Spiking Neural Networks - Architectures and Implementations, Neurocomputing, Elsevier, 2001.

FPGA-Based Computation of Free-Form Deformations

Jun Jiang, Wayne Luk, and Daniel Rueckert

Department of Computing, Imperial College London
180 Queen's Gate, London SW7 2BZ, England

Abstract. This paper describes techniques for producing FPGA-based designs that support free-form deformation in medical image processing. The free-form deformation method is based on a B-spline algorithm for modelling three-dimensional deformable objects. We transform the nested loop in this algorithm to eliminate conditional statements, enabling the development of a fully pipelined design. Further optimisations include precalculation of the B-spline model using lookup tables, and deployment of multiple pipelines so that each covers a different image. Our design description, captured in the Handel-C language, is parameterisable at compile time to support a range of image resolutions and output precisions. An implementation on a Xilinx XC2V6000 device at 67MHz has a throughput which is 12.8 times faster than an Athlon based PC at 1400 MHz.

1 Introduction

This paper describes techniques for producing FPGA-based designs that support free-form deformation (FFD) in medical image processing. Free-form deformations, based on B-splines, are a powerful tool for modelling three-dimensional deformable objects.

The image registration algorithm has been applied in several areas such as remote sensing and three-dimensional computer vision. In medical image processing such as contrast-enhanced breast Magnetic Resonance Imaging (MRI), the FFD method is adopted as an important part of non-rigid registration, a method for analyzing deformable objects. However, there is one disadvantage of this image registration implementation which adopts FFD as the local motion model: the processing time of a three-dimensional image with a resolution of 256 by 256 by 64 voxels takes between 15-30 minutes of processor time on a Sun Ultra 10 workstation.

Previously, we have presented a method for eliminating conditional statements in a nested loop for FFD computation by narrowing the range of the input [2]. This paper presents another method to achieve the same effect by transforming input data. This method has the advantage that all data can be processed by the hardware compared to our previous method.

2 B-Spline Based FFD

In medical image processing, B-spline based free-form deformations (FFD) are frequently used in non-rigid registration, such as 3D contrast-enhanced MRI, to model local deformations. For instance, the motion of the breast is non-rigid so that rigid or

P.Y.K. Cheung et al. (Eds.): FPL 2003, LNCS 2778, pp. 1057–1061, 2003.

affine transformations alone are not sufficient for the motion correction of breast MRI. Therefore a combined transformation T, which consists of a global transformation and a local transformation, is defined as follows [4]:

$$T(x, y, z) = T_{global}(x, y, z) + T_{local}(x, y, z)$$

To define a B-spline based FFD, the domain of the image volume is defined as $\Omega = \{(x, y, z)|0 \leq x < X, 0 \leq y < Y, 0 \leq z < Z\}$. Let Φ denote a $n_x \times n_y \times n_z$ mesh of control points $\phi_{i,j,k}$ with uniform spacing. The FFD can be written as the 3D tensor product of the familiar 1D cubic B-splines:

$$T_{local}(x, y, z) = \sum_{i=0}^{3}\sum_{j=0}^{3}\sum_{k=0}^{3} B_i(u)B_j(v)B_k(w)\phi_{i+l,j+m,k+n} \tag{1}$$

where

$$l = \lfloor \frac{x}{n_x} \rfloor - 1, \quad m = \lfloor \frac{y}{n_y} \rfloor - 1, \quad n = \lfloor \frac{z}{n_z} \rfloor - 1,$$

$$u = \frac{x}{n_x} - \lfloor \frac{x}{n_x} \rfloor, \quad v = \frac{y}{n_y} - \lfloor \frac{y}{n_y} \rfloor, \quad w = \frac{z}{n_z} - \lfloor \frac{z}{n_z} \rfloor$$

and B_i represents the i-th basis function of the B-spline, and $u \in [0, 1)$.

$$\begin{aligned} B_0(u) &= (1 - u)^3/6 \\ B_1(u) &= (3u^3 - 6u^2 + 4)/6 \\ B_2(u) &= (-3u^3 + 3u^2 + 3u + 1)/6 \\ B_3(u) &= u^3/6 \end{aligned} \tag{2}$$

3   Conditional Loop Transformation

In the three nested for-loop of the B-spline based FFD local deformation, there are three conditions which determine whether the loop body is executed or not. All these conditions depend not only on the for-loop variables, but also on the input values of the deformation.

Figure 1 shows the pseudo code of the nested for-loop for the FFD computation. The inner loop body would be executed only when conditions for K, J and I are all satisfied. The variables n, m and l are the fraction part of the input fixed-point numbers. Although one can implement a sequential hardware implementation using the Handel-C language [1], the performance is predictably low.

We use a transformed loop structure (Figure 2) to eliminate the conditional statements by assigning the first and the last elements of the transformed data array to zero. This corresponds to ignoring the effects of those transformed control lattice points outside the grey box shown in Figure 3, which do not have impact on the calculation result. With this method, a pipelined implementation of B-spline based FFD algorithm for processing a 2D image has been successfully compiled for an FPGA.

```
For k=0 to k=3
Begin
  K = k + n - 1
  if (0 <= K < CP_Z)
    For j=0 to j=3
    Begin
      J = j + m - 1
      if (0 <= J < CP_Y)
        For i=0 to i=3
        Begin
          I = i + l - 1
          if (0 <= I < CP_X)
          x = x + Bi(u) * Bj(v) * Bk(w) * Phi_X[K][J][I]
          y = y + Bi(u) * Bj(v) * Bk(w) * Phi_Y[K][J][I]
          z = z + Bi(u) * Bj(v) * Bk(w) * Phi_Z[K][J][I]
        End
    End
End
```

Fig. 1. Pseudo code of B-spline based free-form deformation, where $l = \lfloor x/n_x \rfloor, m = \lfloor y/n_y \rfloor, n = \lfloor z/n_z \rfloor$.

```
For k=0 to k=3 Begin
  K = k + n
  For j=0 to j=3 Begin
    J = j + m
    For i=0 to i=3 Begin
      I = i + l
      x = x + Bi(u) * Bj(v) * Bk(w) * Phi_X[K][J][I]
      y = y + Bi(u) * Bj(v) * Bk(w) * Phi_Y[K][J][I]
      z = z + Bi(u) * Bj(v) * Bk(w) * Phi_Z[K][J][I]
    End
  End
End
```

Fig. 2. Pseudo code of transformed loop, where $l = \lfloor x/n_x \rfloor, m = \lfloor y/n_y \rfloor, n = \lfloor z/n_z \rfloor$. When either (a) $K = 0, J = 0$ or $I = 0$, or (b) $K = CP\_Z, J = CP\_Y$ or $I = CP\_X$, we assign $Phi\_X[K][J][I]$, $Phi\_Y[X][J][I]$ and $Phi\_Z[X][Y][I]$ to zero.

4   Pipelined Design for FFD

Our pipelined design for FFD involves three steps. The first step is to precalculate the four basis functions of a third-order B-spline, as shown in Equation 2, and store the values in four lookup tables. The second step is to design a fully pipelined FFD core. The third step is to deploy multiple pipelines.

Currently our pipelined design uses one input channel. The total number of execution cycles is around $N^d \times (i+1)^d \times M$, where N denotes the resolution, i denotes the order of B-spline, d represents the dimension of an image, and M represents the number of pipeline stages of the floating-point adder (Figure 4).

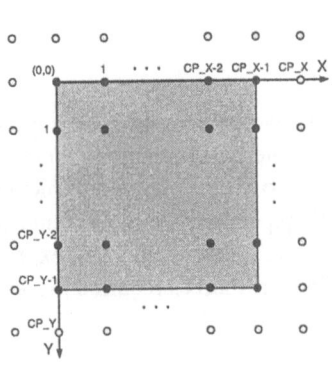

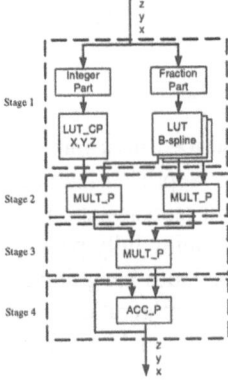

Fig. 3. The arrangement of control lattice on a 2D image.

Fig. 4. Pipelined hardware for free-form deformation computation. MULT_P denotes a pipelined floating-point multiplier. ACC_P denotes a pipelined floating-point adder.

The input data have to be interleaved so that the whole pipeline could be fully used. In the pseudo code (Figure 2), we can see that the values of $Bl(u)$, $Bm(v)$ and $Bn(w)$ need to be accessed simultaneously in order to make a more efficient pipeline. Therefore, we replicate the lookup tables twice to meet the requirement of accessing three independent lookup tables concurrently (Equation 1).

We have also implemented two pipelines in XC2V6000 using our customisable representation with 12-bit mantissa and 8-bit exponent. It costs around 6 percent of Slice resources and 50 percent of Block RAM resources on the XC2V6000.

5   Performance

For a 2D image of resolution 256 by 256, the clock speed after place and route of our current two-pipeline implementation on a Xilinx Virtex II XC2V6000 device (Table 1) with 12-bit mantissa and 8-bit exponent is 67 MHz. Hence the estimated execution time for the XC2V6000 device is $(256 \times 256 \times 16 \times 3)/(67 \times 10^6 \times 2) \simeq 0.023$ second.

Table 1. Performance comparison of FFD processor for images of 256 by 256.

Processor	Area (slices)	Block RAMs (kbit)	Clock Speed (MHz)	Exec.Time (sec)	Throughput (data/second)
XC2V6000 (two pipelines)	2127	504	67	0.023	8375000
XC2V1000 (one pipeline)	1149	252	89	0.035	5562500
XC2V6000 (one pipeline)	1156	252	76	0.041	4750000
AMD Athlon	-	-	1400	0.10	655360
Pentium 4	-	-	1800	0.11	595782

Compared with the software version which runs on an AMD Athlon 1.4 GHz PC, the execution time of our customised system is around 4.3 times faster. Moreover, the throughput of this system is 12.8 times faster than the PC.

6   Summary

We have described hardware techniques for medical image processing. The key elements of our approach include precalculating the B-spline basis function, adopting custom number representation, transforming a nested loop to avoid conditional calculation, and developing fully-pipelined circuits and multiple pipeline designs. Current and future work includes integrating our design with a hardware image warper [3], exploring run-time reconfiguration to reduce the amount of FPGA resources required [5], and automating conditional loop transformations [6].

References

1. http://www.celoxica.com, Celoxica Limited.
2. J. Jiang, W. Luk and D. Rueckert, "FPGA-based computation of free-form deformations", *IEEE International Conference on Field-Programmable Technology*, pp. 407–410, 2002.

3. J. Jiang, S. Schmidt, W. Luk and D. Rueckert, "Parameterizing designs for image warping", *Reconfigurable Technology: FPGAs and Reconfigurable Processors for Computing and Communications, Proc. SPIE*, vol. 4867, 2002.
4. D. Rueckert, L. I. Sonoda, C. Hayes, D. L. G. Hill, M. O. Leach and D. J. Hawkes, "Non-rigid registration using free-form deformations: Application to breast MR images", *IEEE Transactions on Medical Imaging*, vol.18, no.8, pp. 712–721, 1999.
5. N. Shirazi, W. Luk and P.Y.K. Cheung, "Framework and tools for run-time reconfigurable designs", *IEE Proc. Comput. Digit. Tech.*, pp. 147–152, May 2000.
6. M. Weinhardt and W. Luk, "Pipeline vectorization", *IEEE Trans. on Computer-Aided Design*, pp. 234–248, February 2001.

FPGA Implementations of Neural Networks - A Survey of a Decade of Progress

Jihan Zhu and Peter Sutton

School of Information Technology and Electrical Engineering, The University of Queensland,
Brisbane, Queensland 4072, Australia
{jihan, p.sutton}@itee.uq.edu.au

Abstract. The ferst successful FPGA implementation [1] of artificial neural networks (ANNs) was published a little over a decade ago. It is timely to review the progress that has been made in this research area. This brief survey provides a taxonomy for classifying FPGA implementations of ANNs. Different implementation techniques and design issues are discussed. Future research trends are also presented.

1 Introduction

An artificial neural network (ANN) is a parallel and distributed network of simple nonlinear processing units interconnected in a layered arrangement. *Parallelism, modularity* and *dynamic adaptation* are three computational characteristics typically associated with ANNs. FPGA-based reconfigurable computing architectures are well suited to implement ANNs as one can exploit concurrency and rapidly reconfigure to adapt the weights and topologies of an ANN.

FPGA realisation of ANNs with a large number of neurons is still a challenging task because ANN algorithms are "multiplication-rech" and it is relatively expensive to implement multipliers an fine-grained FPGAs. By utilizing FPGA reconfigurability, there are strategies to implement ANNs an FPGAs cheaply and efficiently. It is the goal of this paper to: 1) provide a brief survey of existing ANN implementations in FPGA hardware; 2) highlight and discuss issues that are important for such implementations; and 3) provide analysis an how to best exploit FPGA reconfigurability for implementing ANNs. Due to space constraints, this paper can not be comprehensive; only a selected set of papers are referenced.

2 A Taxonomy of FPGA Implementations of ANNs

2.1 Purpose of Reconfiguration

All FPGA implementations of ANNs attempt to exploit the reconfigurability of FPGA hardware in one way or another. Identifying the purpose of reconfiguration sheds light an the motivation behind different implementation approaches.

P.Y.K. Cheung et al. (Eds.): FPL 2003, LNCS 2778, pp. 1062–1066, 2003.

Prototyping and Simulation exploits the fact that FPGA-based hardware can be rapidly reconfigured an unlimited number of times. This apparent hardware flexibility allows rapid prototyping of different ANN implementation strategies and learning al gorithms for initial simulation or proof of concept. The GANGLION project [1] is a good example of rapid prototyping.

Density enhancement refers to methods which increase the amount of effective functionality per unit circuit area through FPGA reconfiguration. This is achieved by exploiting FPGA run-time / partial reconfigurability in one of two ways. Firstly, it is possible to time-multiplex an FPGA chip for each of the sequential steps in an ANN algorithm. For example, in work by Eldredge et al. [2], a back-propagation learning algorithm is divided into a sequence of feedforward presentation and back-propagation stages, and the implementation of each stage is executed an the same FPGA resource. Secondly, it is possible to time-multiplex an FPGA chip for each of the ANN circuits that is specialized with a set of constant operands at different stages during execution. This technique is also known as *dynamic constantfolding*. James-Roxby [3] implemented a 4-8-8-4 MLP an a Xilinx Virtex chip by using constant coefficient multipliers with the weight of each synapse as the constant. All constant weights can be changed through dynamic reconfiguration in under 69μs. This implementation is useful for exploiting training-level parallelism with batch-updating of weights at the end of each training epoch. The same idea was also previously explored in work by Zhu et al. [4].

As both methods for density enhancement incur reconfiguration overhead, good performance can only be achieved if the reconfiguration time is small compared to the computation time. There exists a break-even point q beyond which density enhance ment is no longer profitable: $q = r/ (s - 1)$ where r is the time (in cycles) taken to re-configure the FPGA and s is the total computation time after each reconfiguration [5].

Topology Adaptation refers to the fact that dynamically configurable FPGA devices permit the implementation of ANNs with modifiable topologies. Hence, iterative construction of ANNs [6] can be realized through topology adaptation. During training, the topology and the required computational precision for an ANN can be adjusted according to some learning criteria. de Garis et al. [7] used genetic algorithms to dynamically grow and evolve cellular automata based ANNs. Zhu et al. [8] implemented the Kak algorithm which supports on-line pruning and construction of network models.

2.2 Data Representation

A body of research exists to Show that it is possible to train ANNs with ***integer*** weights. The interest in using integer weights stems from the fact that integer multipliers can be implemented more efficiently than floating-point ones. There are also special learning algorithms [9] which use powers-of-two integers as weights. The advantage of powersof-two integer weight learning algorithms is that the required multiplications in an ANN can be reduced to a series of shift operations. A few attempts have been made to implement ANNs in FPGA hardware with ***floating-point*** weights. However, no successful implementation has been reported to date. Recent work by Nichols et al. [10] showed that despite continuing advances in FPGA technology, it is still impractical to implement ANNs an FPGAs with floating-point precision weights.

Bit-Stream arithmetic is a method which uses a stream of randomly generated bits to represent a real number, that is, the probability of the number of bits that are "on" is the value of the real number. The advantage with this approach is that the required syn optic multiplications can be reduced to simple logic operations. A comprehensive survey of this method can be found in Reyneri [11] while most recent work can found in Hikawa [12]. The disadvantage of bit-stream arithmetic is the lack of precision. This can severely limit an ANN's ability to learn and solve a problem. In addition, the multiplication between two bit-streams is only correct if the bit-streams are uncorrelated. Producing independent random sources for bit-streams requires large resources.

3   Implementation Issues

3.1   Weight Precision

Selecting weight precision is one of the important choices when implementing ANNs an FPGAs. Weight precision is used to trade-off the capabilities of the realized ANNs against the implementation cost. A higher weight precision means fewer quantization errors in the final implementations, while a lower precision leads to simpler designs, greater speed and reductions in area requirements and power consumption. One way of resolving the trade-off is to determine the "minimum precision" required to solve a given problem. Traditionally, the minimum precision is found through "trial and error" by simulating the solution in software before implementation. Holt and Baker [13] studied the minimum precision required for a class of benchmark classification problems and found that 16-bit fixed-point is the minimum allowable precision without diminishing an ANN's capability to learn these benchmark problems.

Recently, more tangible progress has been made from a theoretical approach to weight precision selection. Draghici [14] relates the "difficulty" of a given classification problem (i.e. how difficult it is to solve) to the required number of weights and the necessary precision of the weights to solve the problem. He proved that, in the worst case, the required weight range $[-p, p]$ is estimated through the

minimum distance between patterns of difference classes $d = (\sqrt{n})/(2p)$ to guarantee a solution, where p is an integer and n is the dimension of the input. This services as an important guide for choosing data precision.

3.2   Transfer Function Implementation

Direct implementation for non-linear sigmoid transfer functions is very expensive. There are two practical approaches to approximate sigmoid functions with simple FPGA designs. **Piece-wise linear approximation** describes a combination of lines in the form of $y = ax + b$ which is used to approximate the sigmoid function. Note that if the coefficients for the lines are chosen to be powers of two, the sigmoid functions can be realized by a series of shift and add operations. Many implementations of neuron transfer functions use such piece-wise linear approximations [15]. The second method is **lookup tables**, in which uniform samples taken from the centre of a sigmoid function can be stored in a table for look up. The regions outside the centre of the sigmoid function are still approximated in a piece-wise linear fashion.

4   Conclusion and Future Research Directions

When implementing ANNs an FPGAs one must be clear an the purpose reconfiguration plays and develop strategies to exploit it effectively. The weight and input precision should not be set arbitrarily as the precision required is problem dependent. Future research areas will include: **benchmarks**, which should be created to compare and ana lyse the performance of different implementations; **Software tools**, which are needed to facilitate the exchange of IP blocks and libraries of FPGA ANN implementations; **FPGA friendly learning algorithms**, which will continue to be developed as faithful realizations of ANN leaming algorithms are still too complex and expensive; and, **topology adaptation** approaches, which take advantage of the features of the latest FPGAs such as specialized multipliers and MAC units.

Referenees

1. Cox, C.E. and E. Blanz, *GangLion - u fast field Programmable gate array implementation of a connectionist classifier.* IEEE Journal of Solid-State Circuits, 1992. **28**(3): p. 288-299.
2. Eldredge, J.G. and B.L. Hutchings. *Density enhancement of a neural network using FPGAs and run-time reconfiguration. in Proceedings of IEEE Workshop an Field-Programmable Custom Computing Machines,* 1994. pp 180-188.
3. James-Roxby, P. and B.A. Blodget. *Adapting constant multipliers in a neural network implementation. in Proceedings of IEEE Symposium an Field-Programmable Custom Computing Machines,* 2000. pp 335-336.
4. Zhu, J.M., G.J.; Gunther, B.K., *Towards an FPGA based reconfigurable computing environment for neural network implementations,* in *Proceedings of Ninth International Conference an Artificial Neural Networks,* 1999. pp. 661-666, vol.2.
5. Guccione, S.A. and M. Gonzalez. *Classification and Performance ofreconfigurable architectures. in Proceedings of the 5th International Workshop an Field-Programmable Logic and Applications.* 1995, pp 439-448, Springer-Verlag, Berlin.
6. Perez-Uribe, A. and E. Sanchez. *FPGA Implementation of an Adaptable-Size Neural Network. in Proceedings of the Sixth International Conference an Artificial Neural Networks.* 1996. pp 382-388, Springer-Verlag.
7. de Garis, H., et al. *Initial evolvability experiments an the CAM-brain machines (CBMs).* in *Proceedings of the 2001 Congress an Evolutionary Computation,* 2001. pp 635-641, vol. 1.
8. Zhu, J. and G. Milne. *Implementing Kak Neural Networks an a Reconfigurable Computing Pla form. in Proceedings of the 10th International Workshop an Field-*

Programmable Logic and Applications - Roadmap to Reconfigurable Computing. 2000. pp 260- 269. Springer Verlag.

9. Marchesi, M., et al., *Fast neural Ntworks without multipliers*. IEEE Transactions an Neural Networks, 1993. **4**(1): p. 53-62.

10. Nichols, K., M. Moussa, and S. Areibi. *Feasibility of Floating-Point Arithmetic in FPGA based Artificial Neural Networks. in Proceedings of the 15th International Conference an Computer Applications in Industry and Engineering*. 2002. pp San Diego, Califomia.

11. Reyneri, L.M. *Theoretical and implementation aspects ofpulse streams: an overview. in Proceedings of the Seventh International Conference an Microelectronics for Neural, Fuzzy and Bio-Inspired Systems*. 1999. pp 78-89.

12. Hikawa, H., *A new digital pulse-mode neuron with adjustable activation function*. IEEE Transactions an Neural Networks, 2003. **14**(1045-9227): p. 236-242.

13. Holt, J.L., T.E. Baker. *Back propagation simulations using limited precision calculations*, in *Proceedings ofInternational Joint Conference an Neural Networks*. 1991. pp 121-126 vol. 2.

14. Draghici, S., *On the capabilities of neural networks using limited precision weights*. Neural Networks, 2002. **15**: p. 395-414.

15. Wolf, D.F., Romero, R. A. F., Marques, E. *Using Embedded Processors in Hardware Models of Artificial Neural Networks. in In proceedings of SBAI - Simpósio Brasileiro de Automao Inteligente*. 2001. pp 78-83.

FPGA-Based Hardware/Software CoDesign of an Expert System Shell

Aurel Neţin, Dumitru Roman, Octavian Creţ, Kalman Pusztai, and Lucia Văcariu

Technical University of Cluj-Napoca, Computer Science Department, ROMANIA
{netin, cret, pusztai, lucia}@bavaria.utcluj.ro,
titiroman@email.ro

Abstract. This paper presents a new method for implementing in hardware expert systems based on belief revision concepts. The expert system's knowledge base is first automatically translated to an equivalent network representation where nodes are facts and links stand for relationships. Then, changes are propagated throughout the network. The conclusions are extracted after no more changes occur in the state of the nodes. The automatic generation of the hardware network structure is described. Finally, the results obtained in this FPGA-based implementation are compared to those yielded by a Java-based implementation, the system's efficiency being thus demonstrated.

1 Introduction

The past years have witnessed a noticeable research effort towards a theory of reasoning under uncertainty. Probability theory was introduced in this area with the emergence of Bayesian belief network [3]. An alternative approach to the probability theory as a tool for modeling uncertainty is the use of belief functions. The research effort has been directed towards the specification of a knowledge representation framework that combines the merits of classical logic and Bayesian belief networks.

We represent uncertainty as a set $E = \{E_i | i \in \{1,2,\dots,9\}\}$ of nine ordered linguistic variables. The natural order induced among the variables holds true: E_1 stands for *impossible*, E_9 - for *certain*. Uncertainty is represented by a set of two parameters varying on E: *support* – the positive evidence for the assertion, and *plausibility* – the difference between the absolute certainty and the support of the negation of the assertion. Support and plausibility are independently updated (they are defined as different kind of information associated to a proposition, separately acquired and conceptually unrelated). The *belief states* partition in six areas the 9×9 table combining all the possible values (see Table 1). Two belief intervals are different only if they belong to different belief states.

Table 1. The belief states intervals (values for support and plausibility)

S \ PL	$E_1 \dots E_4$	$E_5 \dots E_8$	E_9
E_1	Disbelieved (*D*)	Unknown (*U*)	
$E_2 \dots E_5$	Rather Disbelieved (*RD*)		
$E_6 \dots E_9$	Contradictory (*C*)	Rather Believed (*RB*)	Believed (*B*)

P.Y.K. Cheung et al. (Eds.): FPL 2003, LNCS 2778, pp. 1067–1070, 2003.

2 Operators Defined on Belief Intervals

To compute the belief interval associated to a compound expression, *operators* have been defined. The notation |<exp>| is a shorthand for [s(exp), pl(exp)].

- *Negation*: $N(E_i) = E_9 - E_i = E_{9-(i-1)}$ (the negation of a variable)
  $NOT(|a|)=[N(p(a)), N(s(a))]=[s(\bullet a), p(\bullet a)]$ (the negation of a belief interval)
- *Conjunction*
  $[E_1, E_9]$, **if** $|a| \in U, |b| \in C$; $AND(|a|, |b|) = [\min(s(a),s(b)), \min(p(a),p(b))]$, **otherwise**
- *Disjunction*
  $[E_1, E_9]$, **if** $|a| \in U, |b| \in C$; $OR(|a|, |b|) = [\max(s(a),s(b)), \max(p(a),p(b))]$, **otherwise**
- *Aggregation*: Denoted by $\odot$; it aggregates the evidence coming from different sources to a single assertion (a_i stands for the evidence pertaining to a and coming from source i). $AGGR(|a_1|, ..., |a_n|) = [\max(s(a_1), ... ,s(a_n)), \min(p(a_1), ... ,p(a_n))]$
- *Detachment*: Denoted by $\rightarrow$, it propagates the evidence pertaining of an inference rule to its conclusions. The definition of the DET operator, with the belief interval pertaining to the rule's *premises* |h| (hypothesis) and the rule's *strength* |h→t|, expressing the deduction, is given below:
  $AND(|h|, |h{\rightarrow}t|)$, **if** $h \in B, |h| \in RB$; $DET(|h|, |h{\rightarrow}t|) = [E_1, E_9]$, **otherwise**.

3 Translating Rules to Network Representation

For a set of rules, if at least one proposition is inferentially related to itself we face a *circular dependency problem*. In such cases, the network representation contains loops and the belief interval propagation could suffer from termination problems.

Loops can be safely represented by splitting each proposition node belonging to a loop in two sub-nodes. Thus we keep the evidence propagating through loops including p_k, distinct from the evidence propagating through open paths. Evidence coming from loops enters via the T_k links; this node's contribution goes toward cyclic paths via the x_k link. Evidence independent from p_k enters via the S_k links, and propagates to non-cyclic paths via the y_k link.

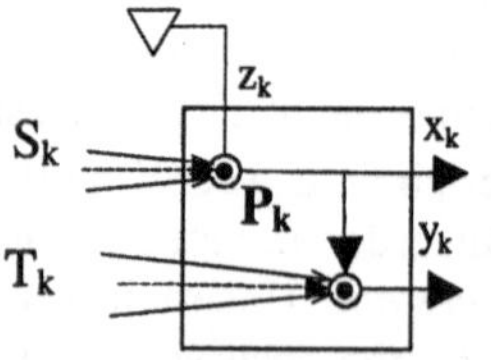

The z_k link is used for initializing the rule. Whenever the global measure of uncertainty is necessary, it can be found at the y_k link. Only the safe part of it (given at the x_k link) is used in loops.

Fig. 1. The node splitting algorithm: the p_k node split example

The best solution is to make the system automatically modify the network's topology without altering the connections' semantic. We represent knowledge by rules in a dependency network, having three types of nodes: *proposition*, *rule* and *operator*. Each *rule* node receives as input the belief intervals of its premises and produces the belief interval of its conclusion. Each *operator* node receives as inputs the belief intervals of its operands and produces the belief interval of its result.

Propositions are modeled as predicate-value pairs. The rule's format is: *name*, used in the explanatory process, *premise*, a compound expression, *conclusion*, and *strength*, a measure of the uncertainty used as a belief interval. By parsing a rule we obtain a node with a *detachment* ($\rightarrow$) operator having the *strength* as parameter.

4 Hardware Implementation Experimental Results

Due to its specific nature, the knowledge base (KB) can be efficiently implemented in hardware (HW): for *propositions*, 8-bits data registers are sufficient to store belief intervals, while the *rules* (DET *operator*) and the other *operators* are implemented by combinational logic. The communication between the system and the external environment is done by means of a Data, an Address and a Control Bus.

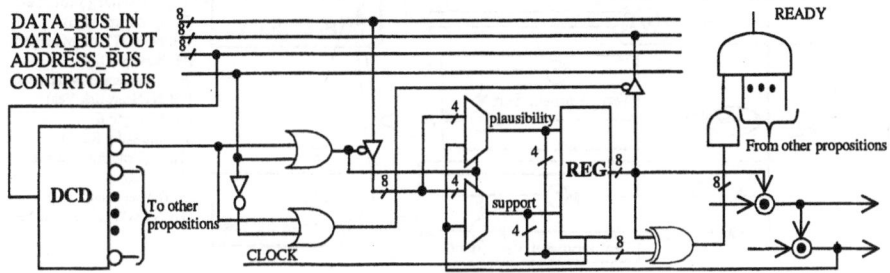

Fig. 2. The hardware structure of a proposition

The KB is given as a text file. This file is parsed and the network representation is automatically generated in a VHDL source code file, by eliminating loops. The operators and the propositions are already described in *predefined* VHDL code (this part is independent of the provided KB), but the actual network's structure is *dynamically generated* at this stage by the software (SW) part of the application.

In the execution phase, information is supplied to the system with evidences as external assumptions for propositions. The observations can be expressed in natural language, then translated in belief intervals and encoded for the network propagation.

The system was tested in two diagnosis fields: the medical and the HW technical support and debugging. A SW (Java-based) and a HW (VHDL and Xilinx FPGA-based) implementations were realized. The HW solution's advantages appeared to be obvious: apart of the intrinsic higher speed of the HW implementation, the *parallel processing* and *parallel propagation* of the belief intervals in the network yielded an increased performance.

The system's intrinsic pipeline-like structure significantly increases its performance; this depends mainly on the number of operators in the most complex rule of the KB, which gives the number of levels of combinational logic between the data buffers. For example, for a rule with 4 levels of logic operators in the KB, the maximal working frequency reported was 109.7 MHz, while for a simple KB (with only one level of operators), it was 120 MHz. For a KB containing 30 rules, with 13 external assumptions, the SW results were obtained after 550 ms, while the HW results were obtained in 142.6 ns (17 clock cycles * 8.39 ns). For a KB containing 60 rules, with 24 external assumptions, the SW results were obtained after 2140 ms, while the HW results were obtained in 234.92 ns (28 clock cycles * 8.39 ns).

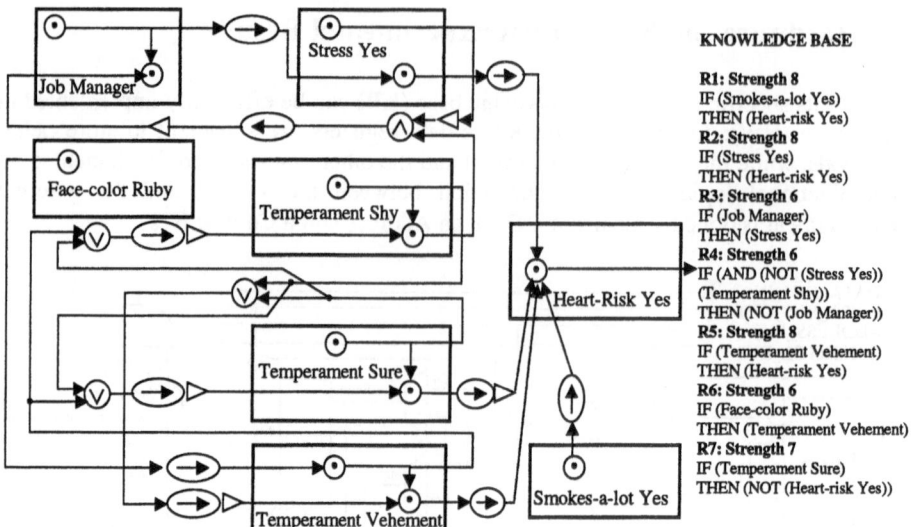

Fig. 3. The medical knowledge base and its network representation

5 Conclusions

A process of belief revision based on uncertainty propagation was proposed. Uncertainty was represented using a theory based on belief intervals defined in terms of subjective linguistic estimates. The KB is translated into an equivalent network representation. The HW architecture is automatically generated by SW, resulting in a VHDL description of the network that corresponds to the KB. After the VHDL synthesis process, the design is downloaded in a Xilinx Virtex FPGA for execution.

The advantages of the HW over the SW implementation, were underlined. They consist mainly of a greater speed, obtained by exploiting the intrinsic parallel features: *parallel processing and propagation* of belief intervals values through the network.

References

[1] A. Bonarini, et. al., "Belief Revision and Uncertainty: a proposal accepting cyclic dependencies", Dipartimento de Elettronica, Politecnico di Milano, Report n.90-067, 1990.
[2] C. Cenan, "An expert system shell based on belief revision concepts". ACAM scientific journal, p.35-45, Cluj-Napoca, 1996.
[3] G. Kleiter, "Bayesian diagnosis in expert systems", Artificial Intelligence, 1992, No. 54, p.1
[4] O. Creţ, et.al., "A HW Implementation of an Expert System Shell Based on Belief Revision Concepts", *4th International Conference on Technical Informatics CONTI' 2000*, Timisoara, Romania,October, 2000.

Cluster-Driven Hardware/Software Partitioning and Scheduling Approach for a Reconfigurable Computer System

Theerayod Wiangtong[1], Peter Y.K Cheung[1], and Wayne Luk[2]

[1] Department of Electrical & Electronic Engineering,
Imperial College, London, UK
{tw1,p.cheung}@ic.ac.uk
[2] Department of Computing,Imperial College, London, UK
wl@doc.ic.ac.uk

Abstract. To achieve a good performance when implementing applications in codesign systems, partitioning and scheduling are important steps. In this paper, a two-phase clustering algorithm is introduced as a preprocessing step to an existing hardware/software partitioning and scheduling system. This preprocessing step increases the granularity in the partition design, resulting in a higher degree of parallelism and a better mapping to the reconfigurable resource. This cluster-driven approach shows improvements in both the makespan of the implementation, and the CPU runtime.

1 Introduction

Coarse grain partitioning can improve the performance of an implementation by increasing parallelism as reported in [1]. It is therefore not surprising that clustering methods, which tend to increase the granularity of tasks, can be applied to the partitioning problem with good effects. Furthermore, after clustering, the size of the task graph (or problem size) is reduced, and this benefits the runtime of the synthesis process.

In this paper, we introduce an algorithm to solve a multiple-objectives clustering problem, called the *two-phase algorithm*. Our clustering algorithm is further applied as a pre-processing step to the partitioning and scheduling algorithm, which maps and schedules tasks to the target system. System constraints including FPGA resources, shared resources conflicts (such as bus or memory contention), as well as reconfiguration time, communication overhead, processing overhead are all taken into account during the partitioning and scheduling process. We employed the tabu search algorithm for partitioning and list scheduling in the scheduler as previously report in [3]. The results of clustering, mapping, and scheduling are then implemented on the UltraSONIC reconfigurable computing platform [4]. In summary, the contributions of this paper are: 1) two-phase clustering algorithm designed for multi-objectives optimization and 2) integration and evaluation of the two-phase clustering algorithm with partitioning and scheduling in codesign systems.

P.Y.K. Cheung et al. (Eds.): FPL 2003, LNCS 2778, pp. 1071–1074, 2003.

2   Two-Phase Clustering Algorithm

In our clustering algorithm, the objectives are set to 1) minimize total communication time and execution time of all the tasks in the DAG[1], and 2) minimize critical path of the DAG. These are subjected to constraints including a maximum cluster size, a maximum number of edges on the new cluster, and the resultant graph must be a DAG.

For such a multiple-objective problem, we need to find a method that can achieve both objectives. Based on preliminary experiments, which reveal that 1) in order to minimize delay on the critical path, tasks on the critical path itself have the most impact, and 2) once a minimum critical path is found, further clustering to reduce system cost will increase the critical path delay from its minimum value, the two-phase clustering algorithm is introduced. During the *first phase*, the delay on the critical path is minimized as much as possible, while the *second phase* is responsible for reducing the system cost without increasing the delay on the critical path.

This algorithm clusters tasks in a hierarchical manner. The first phase combines pairs of task nodes for the best fitness value while shortening the critical path delay on each refinement step. In order to prevent being trapped in a local minimum, the delay on the critical path is allowed to hill-climb. This allowance is limited by a threshold value which is decreasing in each iteration step to continuously force critical path to reduce. Two questions arise at this stage, 1) what is the decrement scheme for the threshold value, and 2) how fast is the decrement rate. Too fast a decrement leads to being trapped in a local optimum; too slow a decrement results in long search time and possible failure to find the minimum point. This is essential the idea of employing a *cooling schedule* found in simulated annealing optimization.

In the second phase, we attempt to reduce the system cost (overall communication time and computation time) without increasing the critical path value obtained in the first phase. This is achieved by only considering nodes and clusters outside the critical path as candidates for merging. The second phase is terminated when no further merging satisfies all the constraints or a maximum number of successful clustering attempts have been met.

3   Experimental Results

3.1   Comparing with Greedy Algorithm

We compare the effectiveness of the two-phase clustering algorithm described above with a straight forward greedy algorithm by applying them to randomly generated task graphs with 100, 200 and 400 tasks nodes with different granularity of tasks varying from 0.1 to 1.0. (Task granularity can be defined in many different ways. In this paper, we use the ratio of the average computation time and communication time as defined in [2].)

---

[1] This is referred to as the *system cost* in the rest of the paper.

Table 1. Comparisons between two-phase approach and greedy approach

Reduction of	Two-Phase algorithm	Greedy algorithm
Critical path delay	30.2%	8.6%
The system cost	14.4%	20.6%
Number of nodes	21.3%	28.1%
Number of edges	16.2%	21.3%
CPU time (PIII 866MHz)	166 sec	362 sec

Table 1 summaries the overall improvement as a result of applying the two clustering methods. From the table, the two-phase method reduces the critical path delay by an average of 30.2%, which is significantly larger than the 8.6% obtained using the greedy algorithm. Since critical path delay has a direct impact on the makespan produced after partitioning and scheduling, this is a significant and useful improvement.

3.2 Combining with the Existing Partitioning and Scheduling Program

The cluster-partition-schedule approach (called CPS) shown in Fig. 1(a) is compared with the one without clustering step (called xPS) shown in Fig. 1(b). As can be seen in Fig. 2, the CPS strategy yields better makespan (between 13% and 17% improvement) and faster runtime (by around 16%) than that without clustering.

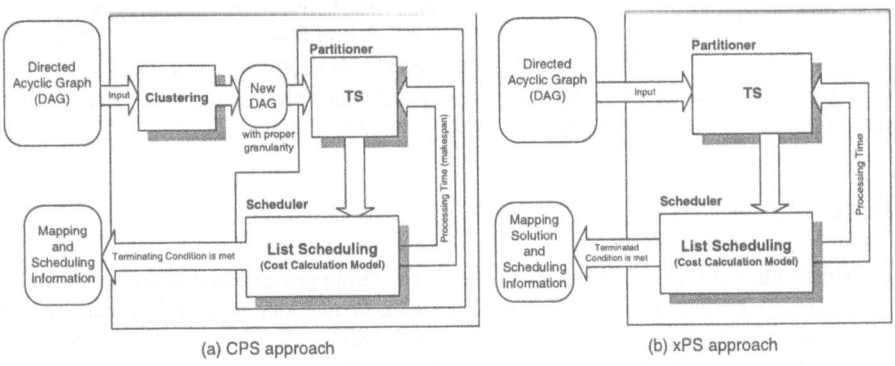

(a) CPS approach   (b) xPS approach

Fig. 1. Algorithm structuring of two different approaches

For a real application, the FFT algorithm is selected as a case study. For each task, the software execution time is obtained by profiling tasks on the PC, while the hardware running time and area are obtained using Xilinx development tools. Parameters such as reconfiguration time, bus speed, FPGA size, are all based on the UltraSONIC platform [4] with two reconfigurable processing elements.

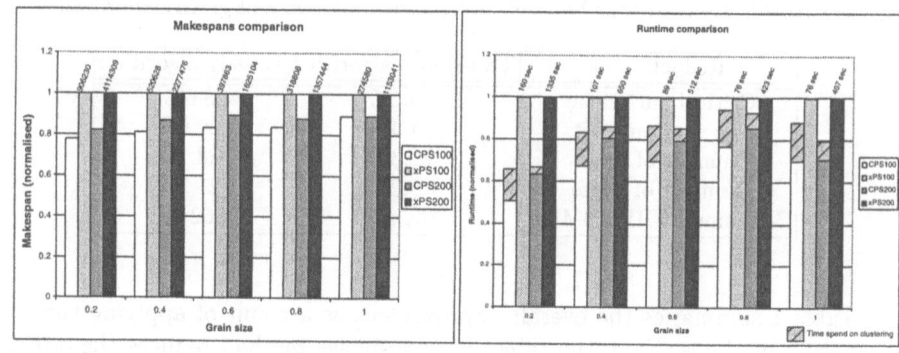

Fig. 2. The comparisons demonstrated in column graphs

Results show that our clustering algorithm can help partitioning and scheduling to reduce the overall makespan of the FFT implementation by around 14%~15% in 8-point and 16-point FFT implementations respectively. This reduction is comparable to average values of improvement getting from random graphs as described earlier.

4   Conclusions

The two-phase clustering algorithm modifies the granularity of the tasks in the DAG in order to improve the critical path delay and the overall communication time and node computation time. The clustering algorithm presented in this paper is successfully combined with our partitioning and scheduling algorithm for mapping abstract task graphs onto a realistic reconfigurable computing system. Using our algorithm to implement the FFT algorithm onto the UltraSONIC reconfigurable computer shows promising results.

References

1. Srinivasan, V.; Govindarajan, S.; Vemuri, R., "Fine-grained and coarse-grained behavioral partitioning with effective utilization of memory and design space exploration for multi-FPGA architectures", *IEEE Transactions on Very Large Scale Integration (VLSI) Systems*, vol. 9, pp. 140 -158, 2001.
2. Palis, M.A.; Liou, J.-C.; Wei, D.S.L., "A greedy task clustering heuristic that is provably good", *Parallel Architectures, Algorithms and Networks*, 1994.
3. Wiangtong, T.; Cheung, P.Y.K.; Luk, W., "Comparing Three Heuristic Search Methods for Functional Partitioning in HW-SW Codesign", *International Journal on Design Automation for Embedded Systems*, vol. 6, pp. 425-449, July 2002.
4. Haynes, S.D.; others, a., "UltraSONIC: A Reconfigurable Architecture for Video Image Processing", *Field-Programmable Logic and Applications (FPL)*, 2002.

Hardware-Software Codesign in Embedded Asymmetric Cryptography Application – A Case Study

Martin Šimka[1], Viktor Fischer[2], and Miloš Drutarovský[1]

[1] Department of Electronics and Multimedia Communications,
Technical University of Košice,
Park Komenského 13, 04120 Košice, Slovakia
{Martin.Simka,Milos.Drutarovsky}@tuke.sk
[2] Laboratoire Traitement du Signal et Instrumentation,
Unité Mixte de Recherche CNRS 5516, Université Jean Monnet,
Saint-Etienne, France
fischer@univ-st-etienne.fr

Abstract. This paper presents a case study of a hardware-software codesign of the RSA cipher embedded in reconfigurable hardware. The soft cores of Altera's Nios RISC processor are used as the basic building block of the proposed complete embedded solutions. The effect of moving computationally intensive parts of RSA into an optimized parameterized scalable Montgomery coprocessor(s) is analyzed and compared with a pure software solution. The impact of the tasks distribution between the hardware and the software on the occupation of logic resources as well as the speed of the algorithm is demonstrated and generalized.

1 Introduction

The protocols in public key cryptography (e. g. RSA [1]) are an excellent example for studying hardware-software codesign concept: the protocol and the key generation have a strong sequential feature, while the algorithm itself can be better realized in parallel and pipelined structures. System on a chip (SOC) offers the best solution: it can consist of an embedded processor and one or more coprocessors. However, the reconfigurable SOC has an extra aspect to be taken into account: both hardware and software part of the system are embedded in the (same) chip. So even entirely software solution occupies hardware resources inside the chip – logic elements for processor implementation and, above all, embedded memory for data and program storing.

2 RSA Algorithm

RSA was proposed by Rivest, Shamir, and Adleman in 1978 [1]. Basic mathematical operation used to encrypt a message X is modular exponentiation [1]:

$$Y = X^E \bmod M \tag{1}$$

P.Y.K. Cheung et al. (Eds.): FPL 2003, LNCS 2778, pp. 1075–1078, 2003.

that a binary or general m-nary methods can break into a series of modular multiplications. All of these computations have to be performed with large k-bit integers (typical $k \in \{1024, 2048, \ldots\}$).

To speed-up modular multiplication required in (1) the well-known Montgomery Multiplication (MM) algorithm [1] is used. It computes the MM product for k-bit integers X, Y: $MM(X, Y) = XYR^{-1} \bmod M$. While the algorithm is simple and can be controlled by software, the MM is an expensive operation suitable for implementation in an algebraic coprocessor.

3   Scalable Montgomery Multiplication Coprocessor and Its Interfacing with the Embedded Processor

We have tested two different approaches to implement scalable processing element (PE) (see Table 1): the first one (called MWR2MM_CSA) is based on a redundant form with Carry-save adders [2], the second one (called MWR2MM_CPA) has a FPLD-optimized architecture based on Carry-propagated structure present practically in all kinds of modern FPLDs. The core of both approaches is a modified Multiple Word Radix-2 Montgomery Multiplication algorithm [2], which imposes no constraints to the precision of operands.

Table 1. Comparison of the PE size and speed for APEX Altera FPLDs

Family	Carry Propagate Adders			Carry Save Adders		
	Length w (bits)	Size (LEs)	Speed (MHz)	Length w (bits)	Size (LEs)	Speed (MHz)
APEX	8	59	161	8	81	232
	16	115	129	16	161	202
	32	229	99	32	321	170

The data path is organized as a cascade chain of PEs (stages) connected to the data memory (implemented in EMBs - Embedded Memory Blocks) (see Figure 1. The maximum degree of parallelism for this organization is found as: $n_{max} = \lfloor \frac{e}{2} \rfloor$. The coprocessor has 3 main parameters (word length w, number of words e, and number of stages n) that can be changed according to the required area of the implemented coprocessor and the required timings for MM computations (n, w) or the security level (e). This approach gives an unusual flexibility to the processor-coprocessor codesign.

As the embedded processor is used a Nios soft-core processor from Altera [4], that includes a CPU optimized for SOC integration. This configurable, general-purpose RISC processor can be combined with user-defined logic and programmed into Altera FPLDs. Nios supports both 16- and 32-bit variants with 16-bit instruction set. Features of an parameterized Avalon bus included in the Nios are used for a flexible connection of the processor and the MM coprocessor(s).

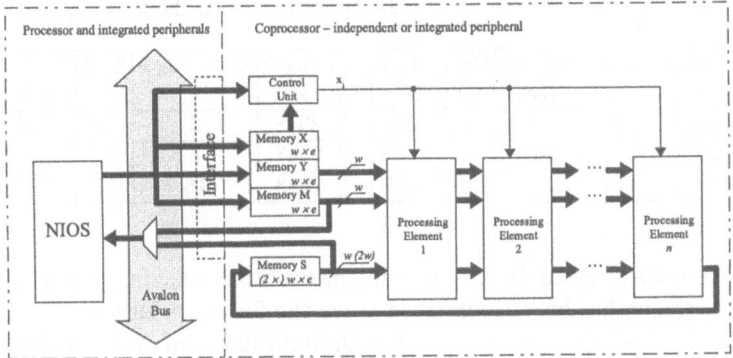

Fig. 1. Block diagram of the Nios processor and the MM coprocessor interconnection

4 Analysis and Discussion of Selected Solutions

To evaluate software/hardware proportion in the solution and its impact on the size and the speed of the system, we have assumed four different representative architectures implemented on Altera Nios development board with APEX EP20K200EFC484-2X [4].

Fully Software Solution. Time-critical part of the software implementation (MM operation) has been programmed in the Nios assembly language using all known optimization techniques. The 32-bit Nios processor has taken 2583 logic elements (LEs) and 45 EMBs including a hardware integer multiplier (used by MUL instruction) occupying 446 LEs. The execution times of the RSA operation ($k = 1024$) in Nios clocked by 50 MHz are: 46 ms for encryption with $E = F_4$, and 845 ms for decryption with CRT algorithm.

Processor with One Pipelined MM Coprocessor. In this version a 16-bit Nios version (occupying only 1275 LEs and 27 EMBs) has been used. Table 2 presents the area occupations and the RSA timings based on the use of the 16-bit MM coprocessor (clocked by 100 MHz) with implemented MWR2MM_CPA algorithm. Note that times includes also pre-computation performed by the Nios processor. Therefore the overall speed is not decreasing linearly with the number of stages. The MM coprocessor requires extra memory resources to store sub-results S, but on the other hand the program code and the program memory is smaller.

Processor with Two Pipelined MM Coprocessors. For typical decryption exponents there are about $k/2$ non-zero bits. Parallel execution on two separate coprocessors can decrease the average execution time to about 66% of the execution time with one coprocessor of the same size. Similarly, during the decryption process based on the CRT algorithm, the computations can be executed in parallel, and thus decrease the execution time to about 50%. However, two coprocessors require two times more hardware resources (LEs and EMBs). When these resources are available, better solution is to add two times more stages to the one coprocessor solution.

Table 2. Execution times of RSA operations with the MM coprocessor

Length $e \times w$	# of stages n	Encr (ms)	Decr (ms)	Size (LEs)	Length $e \times w$	# of stages n	Encr (ms)	Decr (ms)
1024	2	14	155	429	2048	2	55	1146
1024	8	9	56	1462	2048	8	35	354
1024	16	8	39	2837	2048	16	31	222

Fully Hardware Solution. The implementations realizing the whole system as a parallel hardware architecture are the fastest solutions. The disadvantage of this kind of solutions is that all input data are expected to be already stored in a memory before the computation. And in that case even small changes in the implemented protocol may require the remake of the whole design. The software control of the process can allow the user to obtain very flexible and reusable solution. Therefore we do not see the fully hardware solution as a suitable way to implement asymmetric encryption algorithm in FPLDs.

5   Conclusions

Parameterized processors embedded in reconfigurable hardware are becoming a standard building block in complex SOC designs. It was demonstrated that execution of carefully selected parts of the algorithm in properly optimized coprocessors increases considerably the speed of the complete RSA algorithm. Even more, it was shown that hardware resources used in this combined hardware-software design are not more significant than in a pure software solution, because the combined design can use simpler embedded processor.

Acknowledgements

This work has been done as a part of the project CryptArchi (project number CR/02 2 0041) of the French national program ACI Cryptologie.

References

1. J. A. Menezes, P. C. Oorschot, and S. A. Vanstone. *Applied Cryptography.* CRC Press, New York, 1997.
2. A. F. Tenca and C. K. Koc. A scalable architecture for Montgomery multiplication. In C.K. Koc and C. Paar, *Cryptographic Hardware and Embedded Systems,* number 1717 in Computer Science, pages 94–108, Berlin, Germany, 1999. Springer Verlag.
3. M. Šimka and V. Fischer. Montgomery Multiplication Coprocessor for Altera Nios Embedded Processor. *Proceedings of the 5th International Scientific Conference on Electronic Computers and Informatics 2002,* pages 206–211, Kosice, Slovakia, October 2002.
4. Nios Soft Core Embedded processor, www.altera.com/nios

On-chip and Off-chip Real-Time Debugging for Remotely-Accessed Embedded Programmable Systems

Jim Harkin, Michael Callaghan, Chris Peters,
Thomas M. McGinnity, and Liam Maguire

Intelligent Systems Engineering Laboratory,
Faculty of Informatics, University of Ulster, Magee Campus
Northland Rd, Co. Derry, N. Ireland, BT48 7JL, UK
jg.harkin@ulster.ac.uk

Abstract. Embedded programmable systems are becoming common in system designs, resulting in the need for educational institutions to teach advanced embedded systems design and develop debugging competence in students. Remote laboratory experimentation provided as part of a web-based distance learning allows flexible access to on-campus resources free of time or geographical constraints. However, adapting and redeveloping existing software and hardware resources to this purpose is both time consuming and expensive. This paper introduces a remote-access laboratory architecture, which extends current e-learning strategies to provide real-time debugging for embedded programmable systems via the web. An example experiment illustrates the on-chip and off-chip real-time debugging capabilities of the laboratory.

1 Introduction

Electronic engineers are experiencing the need to specialise and re-train in advanced fields, particularly in the area of embedded systems. To provide appropriate education in this field requires practical experience in laboratory experimentation. Distance learning via the web offers professional engineers and students the flexibility to access educational material to suit their lifestyles. Remote experimentation offered as part of a web-based distance learning approach allows remotely located users to develop skills which deal with real systems and instrumentation.

Several remote-access labs are currently available via the web. For example, the University of Zagreb [1] and the Open University [2] provide on-line access experiments in embedded system design. However, the main disadvantage of current laboratories is the inability to provide users with access to debugging instrumentation (logic analysers) and modern embedded systems components (FPGAs).

The Engineering and Physical Sciences Research Council (EPSRC) funded project at the University of Ulster, aims to develop a remote-access laboratory for an Embedded Systems module on a distance learning Masters Degree course [3]. The project aims is to provide experience in embedded systems experiments to supplement the lecture material presented within the distance learning module. This paper presents a remote-access architecture and programming model, which addresses the deficiencies of current remote-access labs and illustrates the integration of modern debugging instrumentation and embedded systems technology. In particular, the

P.Y.K. Cheung et al. (Eds.): FPL 2003, LNCS 2778, pp. 1079–1082, 2003.

ability to perform on-chip an off-chip real-time debugging is demonstrated. Section 2 presents the architecture and programming model, section 3 demonstrates the real-time debugging capabilities of the laboratory and section 4 provides a conclusion.

2   Remote-Access Embedded Systems Laboratory

This section presents the architecture and programming model for the remote-access laboratory. The remote-access lab consists of workstations coupled to various instrumentation and embedded systems. Access to the workstations is via a web-client application [4]. The switching architecture provides a method of inter-connecting the embedded systems boards and the various instrumentation to the workstation. Fig. 1 illustrates the architecture for connecting analysis points on multiple experimental embedded systems boards to the suite of embedded debugging instrumentation. The architecture illustrates 5 main data/control buses: 1) board control, 2) configuration data, 3) diagnostic data, 4) instrument data and 5) instrument SCADA (supervisory control & data acquisition). The board control, configuration data and instrument SCADA buses provide communication with the experimental workstation via parallel, serial and GPIB channels, respectively. Board control provides control/setup information for the embedded system boards' multiplexer devices.

The instrumentation SCADA bus uses the GPIB protocol to send and retrieve information between the workstation and the matrix and instruments. The configuration data bus provides download monitors for program code and FPGA configuration information to target boards. All three buses are connected to the workstation. The diagnostic bus contains 120 analysis points from the target experiment board. Twenty-four of these are fed directly to the logic analyser; the other ninety-six are fed to the switching matrix. The main switching component of the architecture is the 96x16 line switching matrix [4]. The switching connections to be

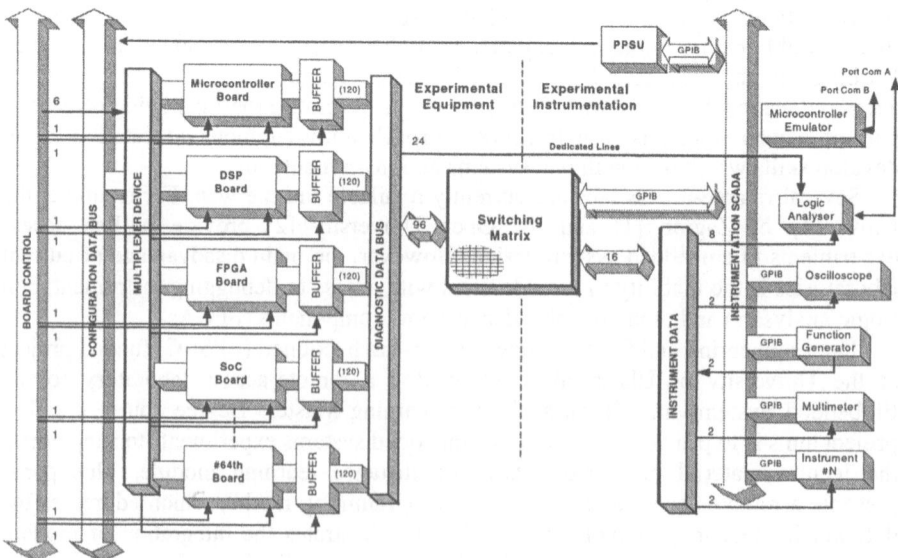

Fig. 1. Remote-access laboratory: Switching architecture

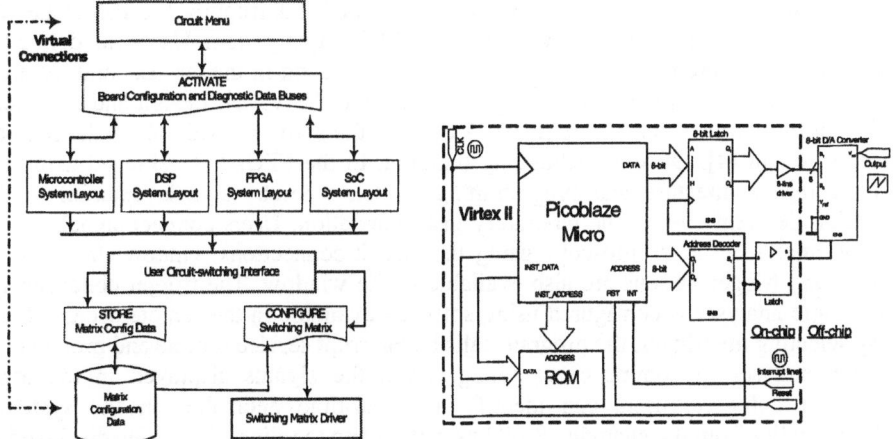

Fig. 3. Programming model   **Fig. 3.** Embedded programmable system

closed, making a link between one of the 96 lines on the left of the matrix with one of the 16 lines on the right, are configured using the instrument SCADA bus (GPIB). Users configure the matrix to connect each of the instrument channels to one of the 96 available analysis points on the embedded systems. Configuration of the architecture is achieved using the programming model illustrated in Fig. 2. Programming can be divided into two tasks: 1) activating the embedded systems board access to the configuration/diagnostic buses and 2) configuring the switching matrix component.

Connecting instruments to the analysis points on the embedded systems is achieved using a circuit-connections window (Flash Macromedia) in the remote-access client application. This provides a visual display of the instrumentation and the points available for analysing. Selecting an experiment from a list of categories produces a layout of the desired Micro/DSP/FPGA embedded system in the circuit window. From the layout, the highlighted analysis points can be connected/wired to the instruments by dragging instrument channel icons to the particular points on the system and using the *Configure* icon. The user's instrument displays are provided in separate individual windows. In the model, the ACTIVATE program is called when an embedded systems experiment is selected. This activates and selects the associated board's multiplexer on the buses. Users highlight analysis points to be connected to the instrumentation by using the drag and drop facility. Clicking on *Configure* calls the CONFIGURE program which reads the matrix configuration data file and configures the switching matrix through the GPIB bus. The STORE program copies the configuration data of the switch matrix to a file. All low-level programming is achieved using C.

3   Real-Time Debugging

Fig 3 illustrates an example embedded programmable system experiment for debugging. The system includes the Virtex II FPGA with the Picoblaze 8-bit microcontroller IP core [5] and an 8-bit digital-analog converter (DAC) external to the FPGA. Users are required to create a program that will execute on the Picoblaze and produce a waveform at the output of the DAC. Interrupting the Picoblaze will cause the program to temporally change the DAC output pattern. Users can debug the hardware-software experiment using the instrumentation in the lab. Laboratory users are able to program the Picoblaze, place and route the VHDL design and download the configuration to the Virtex II FPGA through the software available in the remote-access client [4]. Off-chip debugging (external to the FPGA) is achieved using the switching architecture and programming model, whereby signals external to the FPGA are connected to the laboratory instrumentation. Users connect signals to the logic analyser and oscilloscope using the circuit-connections window. In addition, reset and trigger buttons are also available in the window. The trigger-condition for the logic analyser is configured using software available in the remote-access client. By selecting the trigger, the program calls its interrupt service routine and prompts the logic analyser to commence sampling. From the signals displayed in the logic analyser software, users can identify any real-time anomalies in the system's functionality from the changing 8-bit bus pattern and toggling trigger line displayed in the logic analyser.

On-chip debugging is provided through Xilinx's ChipScope Pro [6]. This allows users to sample signals that are on-chip (internal to the FPGA) in real-time. In this debugging mode, a logic analyser (LA) core is placed inside the FPGA. The switching architecture's configuration data bus enables the transfer of the on-chip data samples from the core to the remote-access client for analysis.

This example experiment illustrates how users can remotely debug systems in real-time. The combination of the architecture and programming model details how users can sample signals in real-time, visualise the information and route it back to their remote desktop. In particular, the remote-access lab demonstrates how commercially available FPGA systems can be programmed, analysed and verified over the Internet.

4   Conclusion

A remote-access architecture and programming model have been presented which address the integration of modern embedded systems experimental equipment and debugging instrumentation for remote access over the web. The laboratory architecture illustrated the method to designing, re-configuring and debugging embedded programmable systems remotely.

References

[1] Muzak, G, Cavrak, I: The Virtual Laboratory Project, Information Technology Interfaces Conference (2000) 241-246
[2] Open University. http://www.open.ac.uk/
[3] Intelligent Systems Engineering Laboratory. http://isel.infm.ulst.ac.uk/
[4] Callaghan, M.J, Harkin, J, McGinnity, T.M, Maguire, L.P: An Internet-based Methodology for Remotely Accessed Embedded Systems, IEEE Conference on Systems, Man and Cybernetics (2002) 157 -162
[5] Xilinx Picoblaze 8-Bit Microcontroller for Virtex-II Devices V1.0 (2002)
[6] Xilinx ChipScope Pro 5.1i User Manual (2002)

Fast Region Labeling
on the Reconfigurable Platform ACE-V

Christian Schmidt and Andreas Koch

Tech. Univ. Braunschweig (E.I.S.), Mühlenpfordtstr. 23, D-38106 Braunschweig, Germany
schmidt,koch@eis.cs.tu-bs.de

1 Introduction

This work will revisit the computer vision application of labeling connected regions in images, and compare results achievable on current configurable architectures with previous work both by our group [1] [2] as well as one of the first attempts targeting the pioneering Splash-2 custom computing machine [3].

2 Algorithm Basics

The algorithm used in our new implementation is fundamentally similar to that of the previous version [1] [2]. For brevity, we will just highlight the changes here.

The algorithm expects as input a black and white image organized as a two-dimensional bit array. The output is a two-dimensional array of 16b words having the same dimensions as the input array, but now labeling all pixels within the same individual object with a unique integer identifier.

In addition to the image itself, we employ two data structures: The *adjacency map* represents an *is-adjacent-to* relation between two object labels. In some cases, data already entered into adjacency map may become invalid when additional adjacency relations are discovered later in the image. To reduce pressure on the memory holding the map, such updates are deferred until the core algorithm itself does not require access to the adjacency map. In the meantime, the adjacency corrections are stored in a separate data structure, called the *adjacency list*, as pairs of 16b integer labels. When the map data structure does become available again, entries in the list are removed and used to correct the corresponding map entries.

As in [1], the labeling procedure itself consists of three distinct phases: 1.) Pixel labeling and adjacency handling, 2.) transitive flattening, and 3.) adjacency-based object merging.

3 Platform Architectures

The algorithm presented in [1] was intended for execution on the Xilinx XC4010-based SPARXIL co-processor [4]. For a valid comparison with current device technology, we ported it to the ACE-V platform.

The configurable computer ACE-V [5] combines a reconfigurable compute unit (RCU, realized by a Xilinx Virtex 1000 FPGA) with a conventional RISC processor

P.Y.K. Cheung et al. (Eds.): FPL 2003, LNCS 2778, pp. 1083–1086, 2003.

(SUN microSPARC-IIep). The RCU has access to four independent 256Kx36b ZBT SSRAM memory banks and via a bus interface unit (BIU) to 64 MB of DRAM shared with the CPU.

The well-known Splash-2 architecture is described in detail in [6]. For our purposes, it consists of 17 XC4010 FPGAs that are connected both in a fixed manner as a systolic array and additionally in a variable manner using a configurable crossbar interconnect.

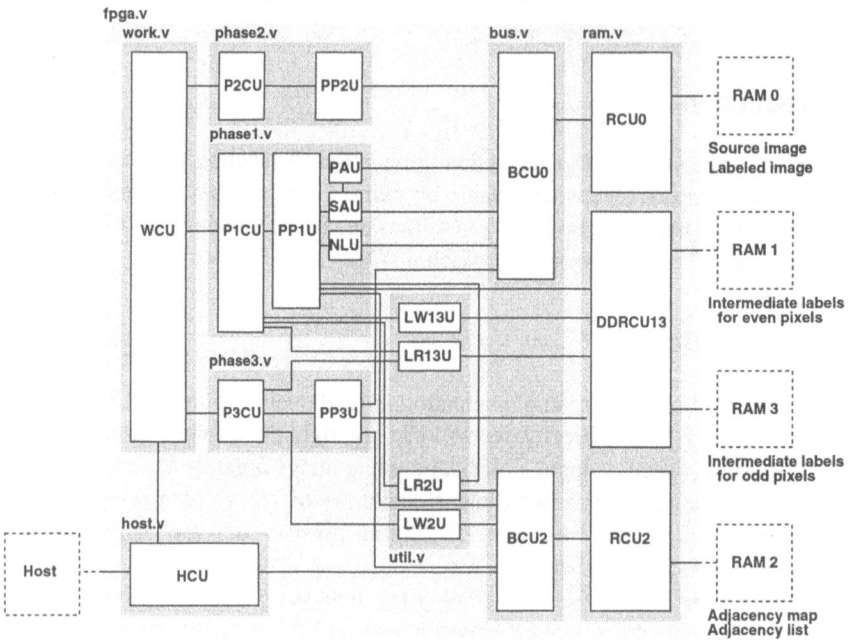

Fig. 1. Hardware architecture

4   Hardware Implementation

Our revised hardware is implemented according to the architecture shown in Figure 1. The Work Control Unit WCU controls the global execution of the algorithm. It accepts a start command from the host CPU and then hands control to the local controllers P1CU, P2CU and P3CU for each processing phase. The WCU also supervises the Host Control Unit HCU, which in turn accepts slave-mode data transfers initiated by the CPU from and to local memories (this is suspended during RCU execution). The HCU also accepts parameters from the host such as the image dimensions and whether to renumber labels consecutively. Furthermore, it also makes per-phase profiling data available to the host for benchmarking (see Section 5).

Each of the three processing phases relies on a highly optimized pipeline unit (PP1U, PP2U and PP3U) to actually perform the computations. In Phase 1, the background task for managing and evaluating the adjacency list is implemented in parallel to the pipeline: The NLU provides the next label to use and enters it in the adjacency map, the SAU stores a adjacency correction in the adjacency list, and the PAU processes list entries for correcting the adjacency map.

Phase 1 and 3 (which actually operate on the image) rely on the dedicated address generators LW13U, LR13U, LR2U and LW2U for storing and retrieving image data. These units also handle special cases such as clipping on the edges of the image. The memory accesses themselves are processed in two bus control units BCU0 and BCU2 that multiplex address and data lines. The associated RAM control units RCU0 (source image, labeled image) and RCU2 (adjacency map and list) handle bidirectional to uni-directional bus conversion, and delay write data according to the ZBT SSRAM access protocol. DDRCU13, the RAM control unit responsible for accesses to the intermediate label array in RAM banks 1 and 3, is a special case: It uses both banks to present an external view of a single bank capable of *simultaneously* reading and writing to alternating even/odd addresses. This access pattern is always used when scanning over the image in Phase 1 and 3.

5   Evaluation

Table 1 compares the various realizations described for processing a 512x512 image. Note that the current implementation can handle any image size of less than 2^{18} total pixels.

Table 1. Area and performance comparison

Algorithm	Platform	RCU	RAMs	Area[LUTs]	Clock[MHz]	Time[ms]
[3]	Splash-2	9x XC4010	9	< 7200	10.0	33+
[1]	Emulated SPARXIL	1x XCV1000	2	2000	33.5	31
New	ACE-V	1x XCV1000	4	2334	36.0	15

The Splash-2 version [3] is not fully comparable with the other solutions, since it is optimized for real-time processing of a stream of video frames at a fixed 30 fps. To this end, it uses *duplicated* hardware for Phase 2 and 3 of the algorithm to process two video frames in parallel. Furthermore, this frame rate is not guaranteed as the hardware can begin to drop frames when they become too complicated to process in the allotted time slot.

Figure 2 shows the execution time in relation to the image size processed. This compares current RISC (SUN UltraSPARC+ 900MHz) and CISC CPUs (AMD Athlon XP 1533MHz) to the ACE-V reconfigurable computer running at 36 MHz. The ACE-V easily beats the RISC even when a highly optimizing compiler is being used and is only slower by 1.16 compared to the AMD CISC (which has 42.6x the clock speed). On average, the improved algorithm requires only 2.5 clocks per pixel. This is a considerable improvement over the design in [1], which also had a higher degree of data-dependence and required up to 6 clocks per pixel.

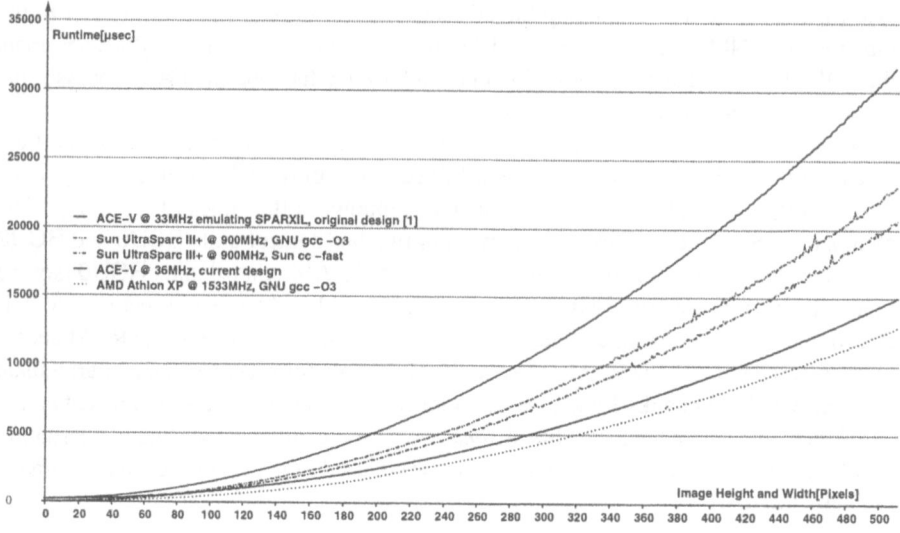

Fig. 2. Performance comparison with RISC and CISC CPUs

6   Potential Improvements

The performance of the new ACE-V based implementation could be improved even further: The factor limiting the clock speed is currently the speed of the HCU and its connection to the *external* BIU. After the redesign presented here, the *internal* computation pipelines PP1U, PP2U and PP3U have much shorter critical paths than in [1], allowing a potential double-clocking of these parts in the 50-60 MHz range even on the slow speed grade -4 Virtex device currently used.

References

1. Koch, A., Golze, U., "Practical Experiences with the SPARXIL Co-Processor", *Proc. 31st Asilomar Conference on Signals, Systems, and Computers,* Pacific Grove (CA), 1997
2. Meyer, K., "Entwurf eines FPGA-basierten Co-Prozessors zur Objekt-Etikettierung in der Bilderkennung", *diploma thesis,* Tech. Univ. Braunschweig (E.I.S.), Germany, 1997
3. Rachakonda, R.V., Athanas, P.M., Abbott, A.L, "High-Speed Region Detection and Labeling using an FPGA-based Custom Computing Platform", *Proc. Field Programmable Logic and Applications (FPL),* Springer, 1995
4. Koch, A., "A Universal Co-Processor for Workstations", in *More FPGAs, eds. Moore, W., Luk, W.,* Oxford 1994
5. Koch, A., Golze, U., "A Comprehensive Prototyping Platform for Hardware-Software Codesign", *Proc. Workshop on Rapid Systems Prototyping,* Paris, 2000
6. Arnold, J.M, Buell, D.A., Davis, E.G., "Splash 2," *Proc. Fourth Annual ACM Symposium on Parallel Algorithms and Architectures,* 1992

Modified Fuzzy C-Means Clustering Algorithm for Real-Time Applications

Jesús Lázaro, Jagoba Arias, José L. Martín, and Carlos Cuadrado

Escuela Superior de Ingenieros,
University of the Basque Country
Alameda Urquijo s/n
48013 Bilbao, Spain
{jtplaarj,jtparpej,jtpmagoj,jtpcuvic}@bi.ehu.es
http://www.ingenierosbilbao.com

Abstract. The fuzzy approach in image processing is taking each day greater importance. It is greatly due to the fact that every new application of artificial vision is closer to human vision. This means that tightly knot algorithms are not always a good solution and a more "imprecise" and fuzzy approach is desirable. This paper describes a modified Fuzzy C-Means algorithm intended to be implemented in hardware. The original algorithm was modified to match the desired level of parallelism, speed and to simplify the hardware implementations.

1 Introduction

The Fuzzy C-Mean (FCM) algorithm is used in a great variety of image processing designs. It is used from satellite image analysis to OCR systems. Since the fuzzy C-Mean algorithm is very time consuming, special attention to performance must be taken in most applications. This is the reason why, since its introduction, several particular implementations have been developed to boost its efficiency [1][2][3][4].

In this paper, a real time C-Mean algorithm is described. The boost of performance of this circuit is far beyond other implementations but, to achieve it, several particularities have been added. One of them is the definition of real-time. The circuit that is proposed in this paper is capable of clustering grey scale video stream with a resolution up to 256x256 pixels per image and 50 images per second (both fields).

2 Fuzzy C-Mean

A clustering approach that involves minimization of some objective function, or error criterion, belongs to a family of objective function clustering algorithms [5]. A common goal of these algorithms is to find an "optimal" partitioning of feature space given a collection of data samples. The algorithms that, in addition to minimizing an error function, estimate the prototypes of resulting

P.Y.K. Cheung et al. (Eds.): FPL 2003, LNCS 2778, pp. 1087–1090, 2003.

classes within a partition, are often referred to as C-Means clustering algorithms, where the integer c stands for the number of classes. If the classes, for which the prototypes are estimated, are allowed to be fuzzy, the Fuzzy C-Means (FCM) clustering algorithm may be used [6].

The fuzzy C-Means algorithm minimizes the least-squares functional that is given by a generalized within-groups sum of square errors function:

$$J_m(U, z) = \sum_{k=1}^{n} \sum_{i=1}^{c} u_{i,k}^m \cdot d_{i,k}^2 ,$$

$$(1)$$

where $U \in M_{fcn}$ is a fuzzy c-partition of X; $z = (z_1, z_2, ..., z_c) \in R^{cp}$, with $z_i \in R_p$ as the cluster center or prototype of the ith class; $d_{ik}^2 = ||x_k - z_i||$, with $|| \cdot ||$ being any inner product induced norm on R^p; and weighting or fuzzy exponent $m \in (1, \infty)$. Clearly, $J_m : M_{fcn} \times R^{cp} \to R^+$. The optimum is reached when the fuzzy partition matrix U^* and a collection of prototypes z^* are found such that J_m is minimized. That is, when the weighted within-groups sum of distances between the samples and the prototypes is the smallest possible.

The solutions of minimization are least-squared error stationary points of J_m. The necessary conditions for minimization of J_m are derived in [5]. The necessary conditions for minimization of J_m can be written as:

$$u_{i,k} = \left(\sum_{j=1}^{c} (d_{i,k}/d_{j,k})^{\frac{2}{m-1}} \right) , \ \forall i, k ,$$

$$(2)$$

$$Z_i = \frac{\sum_{k=1}^{n} u_{i,k}^m \cdot x_k}{\sum_{k=1}^{n} u_{i,k}^m}$$

$$(3)$$

The convergence theory of the FCM algorithm was initially studied in [5][7] and later improved in [8][9].

3   Modified Fuzzy C-Means

The FCM has several problems to be implemented n hardware. One of the first problems is the number of clusters. This number cannot be left as a variable since the amount of memory and circuits depend heavily on it. In fact, the division into two clusters needs the smallest amount of memory.

Moreover, the fuzziness factor m appears as exponent in several points of the algorithm: in equation (2) as the denominator of an exponent and in equation (3) as the exponent itself. The implementations of fractional exponents is such a difficult task that from equation (2) we obtain an "optimum" m of 2. This election makes equation (2) easier to calculate since obtaining the square of a number is a feasible matter.

A second problem is the initialization of the $U \in M_{fcn}$ matrix. This is normally done by using a random number generator (this is the method used by the Matlab algorithm). Such a circuit would increase the size of the resulting circuit and would only we used once every running time. This problem was solved using the input image as the initialized U matrix.

The third problem lays in the loop section. A close analysis shows that it is necessary to iterate through all the input data and the U matrix, in order to obtain the fuzzy centers. In addition, the new U matrix is calculated from the old U matrix, the data and the fuzzy centers. This means that, for each picture, two iterations through the input data are needed or, in other words, the input data should be stored and read twice in the period of time between images. This is practically impossible in a real time application. To solve this problem, the old centers can be used to obtain the new U matrix. This means that the new U matrix and centers can be obtained in the same iteration without needing to store the input image. Another implication of this particularity is that only the value of a single pixel is needed to obtain the corresponding element in the U matrix. Thus, the clustering can be performed as the image arrives, not being necessary to store the whole image to start the processing. This allows us to use a different frame in each iteration instead of storing in memory each image and iterate through it.

As it can be seen, the algorithm performs the same operations over to different data at the same time, one over $U_{i,1}$ and one over $U_{i,2}$. To efficiently use the silicon area, both operations have been performed with the same hardware. To do so, those pipelined parts are clocked at double rate and the special registers (such as the ones in accumulators) have been doubled as well. The ends of these pipes are two registers, one for each different input data.

4   Results

In this section, the results of the proposed algorithm can be seen. The algorithm has been implemented in Matlab, both using floats and integers (any rounding done towards zero). These two algorithms have been tested against the Matlab FCM algorithm with the root mean square as evaluation function.

The *Video 1* is a low motion picture of moving robots. *Video 2* is a video from a crowded corridor with people entering and leaving the scene. The *Video 3* is a medium motion picture with fish swimming in a fish tank.

In table 1 can easily be seen that the lowest motion the better, as it would be expected from an algorithm that instead of iterating over the same data, it uses the new data. Another interesting effect is the initial transitory.

5   Conclusions

This paper presents a modified Fuzzy C-Means algorithm primary intented for real time video applications. This algorithm reduces the needs of memory space and of information movements leading to a highly paralelizable code. This code modifies several important points of the original FCM such as the way of obtaining the U matrix. The algorithm does not iterate over the same data set

Table 1. Root mean square error

Frame	FCM ÷ Float			FCM ÷ Int		
	Video 1	Video 2	Video 3	Video 1	Video 2	Video 3
1	78.3397	208.4966	20.6442	76.1842	209.3727	20.1699
2	36.4761	11.5990	7.0792	31.6272	14.1593	5.7707
3	9.2124	2.9654	1.5493	6.1617	6.3049	1.4090
4	1.6521	0.9975	1.4782	1.2300	4.6446	2.3872
5	0.3508	1.1299	1.7246	1.2843	4.5752	2.3628
6	0.2063	2.301	2.2093	1.2872	4.0168	3.3583
7	0.0235	9.1179	2.1011	1.2875	6.0489	2.9013

but uses a new image for each iteration. This code has been implemented in hardware by means of a programable device.

References

1. Richard J. Hathaway and James C. Bezdek. "Optimization of Clustering Criteria by Reformulation". IEEE Transactions onf Fuzzy Systems, 3(2): 241-245, 1995.
2. M.S. Kamel and S.Z. Selim. "New algortihm for solving the fuzzy clustering problem", Pattern Recognition, 27(3): 421-428, 1994.
3. T.W. Cheng, D-B- Goldgof and L.O. Hall. "Fast clustering with application to fuzzy rule generation" Proc. IEEE Int. Conf. Fuzzy Syst. 2289-2295, 1995.
4. J.F. Kolen and T. Hutcheson. "Reducint the Time Complexity of the Fuzzy C-Means Algorithm". IEEE Transactions onf Fuzzy Systems, 10(2):263-267, 2002.
5. J.C. Bezdek. "Pattern Recognition with Fuzzy Objective Function Algorithms", Plenum Press, New York, 1981.
6. J.C. Bezdek, Robert Ehrlich and William Full, "FCM: The Fuzzy C-Means Clustering Algorithm", Computers and Geosciences, 10:191,203, 1984.
7. J.C. Bezdek. "A Convergence Theorem for the Fuzzy ISODATA Clustering Algorithms", IEEE Transactions on Pattern Analysis and Machine Intelligence, 2(1):1-8,1980
8. J.C. Bezdek, R.J. Hathaway, M.J. Sabin and W.T. Tucker. "Convergence Theory for Fuzzy c-Means: Counterexamples and Repairs". IEEE Transactions on Systems, Man, and Cybernetics, 17(5):873-877, 1987.
9. M.J. Sabin. "Convergence and Consistency of Fuzzy c-means/ISODATA Algorithms". IEEE Transactions on Pattern Analysis and Machine Intelligence, 9(5):661-668, 1987.

Reconfigurable Hybrid Architecture
for Web Applications

David Rodríguez Lozano, Juan M. Sánchez Pérez, and Juan A. Gómez Pulido

Department of Computer Science, University of Extremadura
Campus Universitario s/n. 10071 Cáceres (Spain)
{drlozano, sanperez, jangomez}@unex.es

Abstract. This paper describes a Reconfigurable Hybrid Architecture for the developing, distribution and execution of web applications with high computational requirements. The Architecture is a layered model based on a hybrid device (standard microprocessor and FPGA), for which has been designed and implemented a component as a web browser plug-in. Web applications are divided into two parts: an standard part and a reconfigurable part. The plug-in links the software and hardware applications, implementing an API for the management and access to the FPGA. A real implementation of the proposed architecture has been developed using Handel-C, the RC1000-PP platform, a compatible Intel CPU, and a Visual C++ ActiveX control plug-in.

1 Introduction

Internet has become a great platform for the developing, distributing and exploiting of software applications. Many companies are porting its corporative applications using web architectures and technologies that can be executed on network computers. Many works have demonstrated the viability of developing standalone [2] terminals based on reconfigurable hardware such as Field Programmable Gate Arrays (FGGAs) for the design of reconfigurable Internet platforms [4] with fully networking and multimedia capabilities [3].

Most of desktop and corporate software applications can be distributed and run using web technologies, so the Web Browser may become as the Universal Client of personal and network computers. However, because of limitation in performance of scripting languages and Java virtual machines, applications that require complex computations are actually outside of the web applications scope.

According to these considerations, it seems reasonable the use of a Reconfigurable Hybrid Architecture [1] that will allow a client web browser to request on demand the FPGA as an specific purpose co-processor. In this work we identify the different elements of the proposed reconfigurable hybrid architecture, develop a software component that implements an application programming interface (API) and prototype the hybrid architecture using a standard PC and a FPGA platform.

P.Y.K. Cheung et al. (Eds.): FPL 2003, LNCS 2778, pp. 1091–1094, 2003.

2 Reconfigurable Hybrid Architecture for Web Applications

The proposed architecture is a layered approach Fig. 1. Hybrid web applications are on the top layer. Next layer corresponds to the Reconfigurable Hardware Manager (RHM), which provides to client applications an API for easy access to hardware acceleration. The operating system (OS) is the next layer, it hosts the device drivers for the communication between the RHM and the FPGA. Finally, a standard CPU and a FPGA compose the hybrid processor layer.

Web applications designed for this architecture are divided into two parts: a *soft* part with the program code and a *hard* part with the co-processors descriptions. The program code are the web pages that support the application and user interface, HTML, JavaScript, VBScript or Java may be used. The hardware resources are the FPGA configuration files that describe specific co-processors required by the hybrid web application, Hardware Description Language (HDL) can be used.

In the described architecture the key component is the RHM plug-in, which provides an API that allows the easy development of accelerated web applications. The main functions of the RHM are: to control the correct download of the hardware resources files required by the hybrid web application and to manage the FPGA.

The RHM component can be downloaded as a software plug-in from a web server, different technologies may be used for this plug-in implementation, at this time Microsoft ActiveX, Netscape LiveConect and Sun Java Applets are the most widely used. ActiveX and LiveConect are platform and browser dependent, while Java Applets are limited by the efficiency of the Java Virtual Machine (JVM).

3 Architecture Implementation

An Intelx86 compatible processor is used for the fixed part, and the Celoxica RC1000PP with a Xilinx Virtex for the reconfigurable part. Windows 2000 is the hosting OS. Finally, we use a Microsoft ActiveX control to implement the RHM plug-in functions, and Microsoft Internet Explorer to host the hybrid application.

The Celoxica RC1000 board is a complete platform for the development of reconfigurable computing systems with full support in Handel-C [8]. The RC1000 platform has been used with success in previous works for developing reconfigurable data processing applications [6].

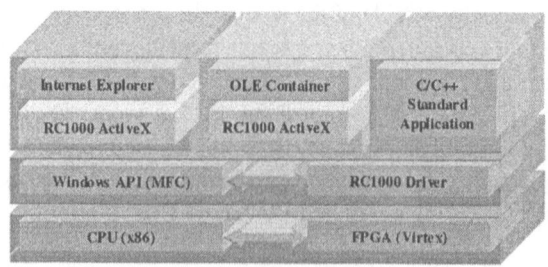

Fig. 1. Components and technologies of the Reconfigurable Hybrid Architecture prototype

Based on the IE components architecture and ActiveX technology [7], the RC1000AX control has been designed and programmed. The RC1000AX control implements all the functions required by the RHM plug-in. The control has been built using Microsoft Visual C++ 6.0, and the Microsoft Foundation Classes (MFC).

The RC1000AX control can be embedded as an OBJECT tag into the HTML pages that implements the hybrid web applications. Any other application that acts as a control container can hosts the RC1000AX control. For example, Microsoft Access or Excel may use the RC1000AX control for accelerating computational operations related with large databases or complex data sheets. The RC1000AX control can be used into an ASP (Active Server Pages) page running on a Windows Internet Information Serve, this would allow hardware accelerating of complex server processes.

4   Application Design

In this section we describe the three phases involved in the development, distribution and execution of hybrid web applications.

Development: The design and implementation of FPGA configuration files is performed using Handel-C and DK1. The RC1000AX is a MFC ActiveX control.

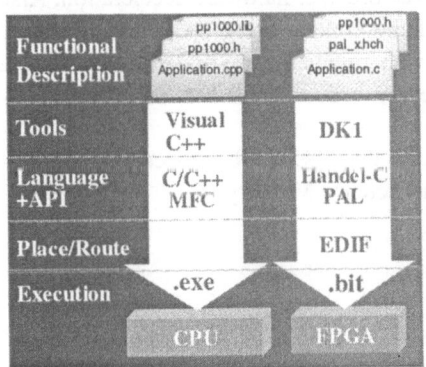
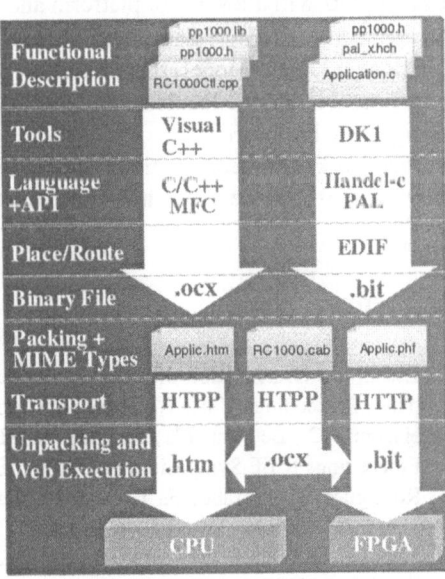

Fig. 5. Tools, languages, protocols and files required in a standard reconfigurable co-processing applications (left) and in a hybrid web application (right)

File Packing: The RC1000AX control and the FPGA description file are prepared for transport over the Internet/Intranet. The control is distributed inside cabinet file, containing all the information necessary to install, and register the control. The binary

FPGA configuration file and additional information is distributed inside a compressed proprietary format with *.phf* (Programmable Hardware File) extension.

Distribution and Execution: The client web browser is the target application for the web pages that support the skeleton and user interface of the hybrid web application, Internet Explorer also acts as container for the RC1000AX control, which is the target application of the new proprietary Programmable Hardware File format. A new MIME Media Type has been defined for indicating to web servers and client web browser how to deal with the files with .phf extension. The new MIME type and subtype are: **application/x-rhmplugin,** and the associated file extension is **phf**.

5   Conclusions and Acknowledgements

In this work we have described the design and implementation a Hybrid Reconfigurable Architecture that allows transparent acceleration of web applications. We have used standard Internet and Web technologies for the programming and distribution of the hybrid architecture components. Web application developers can easily make use of hardware acceleration including the RHM plug-in into HTML pages.The future work includes the design and programming of the Java version of the RHM plug-in, a Java Applets will allow cross-platform and cross-browser hardware acceleration. Also we are working in the distributed version of the control using DCOM (Distributed Component Object Model) that will allow to web clients the use under demand of the reconfigurable device located in a DCOM server, reducing the cost of dedicated reconfigurable hardware.

This work has been partially supported by the TRACER project (TIC- 2002-04498-C05-01) of the Spanish Technology and Science Ministry.

Reference

1. Estrin, G.: Organization of Computer Systems: The Fixed-plus Variable Structure Computer, Proceedings of the Western Joint Computer Conference (1960) 33-40
2. Haenni, J.O., Beuchat, J.L., Sanchez, E.: RENCO: A Reconfigurable Network Computer. Proceedings IEEE Symposium FCCM'98, IEEE Computer Society Press, (1998) 288-289
3. Rodríguez, D., Zarallo, F., Conejo, I.: "Labograph": Graphic Information System over Labomat3 Platform. Proceedings JCRA'01 (2001) 102-110
4. Fallside, H., Smith, M.J.: Internet Connected FPGAs. Proceedings IEEE Symposium FCCM'00, IEEE Computer Society Press (2000) 289-290
5. Chappell, S., Sullivan, C.: Handel-C for co-processing & co-design of Field Programmable System on Chip. Proceedings JCRA'02 (2002) 65-70
6. Styles, H., Luk, W.: Customising Graphics Applications: Techniques and Programming Interface. Proceedings IEEE Symposium FCCM'00 77-88
7. Sankar, K.: Internet Explorer Plug-In and ActiveX Companion. MacMillan (1997)
8. Celoxica Limited: Handel-C Language Reference Manual (v3.1) http://www.celoxica.com

FPGA Implementation
of the Adaptive Lattice Filter

Antonín Heřmánek, Zdeněk Pohl, and Jiří Kadlec*

Department of Signal Processing
Institute of Information Theory and Automation, CAS
Pod vodárenskou věží 4, 182 08 Prague 8, CZE
Tel. +420 266 052 432, xpohl@utia.cas.cz

Abstract. The paper presents the FPGA implementation of a noise canceler with an adaptive RLS-Lattice filter in the Xilinx devices. Since this algorithm requires floating-point computations, Logarithmic Numbering System (LNS) has been used. The pipelined lattice filter macro and input/output conversion routines has been designed. The implementation results are compared with an implementation on 32-bit IEEE floating point signal processor.

1 Introduction

The adaptive filtering has been in focus of DSP research since many years. With the growing computation capacity of modern digital devices, also the requirements on the properties of adaptive algorithms increase, namely on their convergence speed. Algorithms with suitable convergence typically use floating point arithmetic and often do not exist in fixed point version. One of such as algorithm was used in our implementation - an adaptive lattice filter.

The architecture of contemporary digital signal processors is highly optimized for vector operations. However not all DSP algorithms can make efficient use of this operation. In that case the performance of digital signal procesors degrades and the flexibility of FPGAs can provide a promising alternative.

Our team has recently investigated the use of an innovative approach to floating-point arithmetic, so called logarithmic number system (LNS) arithmetic [5, 6]. The proposed solution can provide a very fast floating-point-like multiplications, divisions and square-roots. Additions and subtractions are rather more complicated, but it was shown [3, 5] that it could be very efficient for complicated algorithms with a long critical path.

We have implemented an adaptive noise canceller. At present, a standard LMS filter or one of its variants is often employd at present. In our implementation the lattice filter [1, 2] was selected for the following reasons:

* This work has been partially suported by the Ministry of Education of the Czech Republic under Project LN00B096 and from EU Project RECONF2 (IST-2001-34016).

- very good numerical stability,
- faster convergence than LMS algorithm,
- better theoretical analysis in comparison with some modifications of standard LMS.

On the other hand, the adaptive lattice algorithm is an example of a modern DSP algorithm with high computational requirements and long critical path where the vector processing can not be used effectively. Moreover it requires all basic arithmetic operations – including two divisions on the critical path which degrade the implementation performance (in floating point as well as in fixed point versions). Fortunatelly, the LNS arithmetic is very suitable for such tasks.

The echo canceller implementation will be presented in this paper . The following section presents on overview of LNS arithmetic and its implementation for FPGA. Next, the hardware implementation of a real-time adaptive noise canceller is represented and implementation results are compared with our implementation on digital signal procesor TI TMS320C6711. Finally, the last section will conclude the work.

2   LNS Arithmetic

The design employs the LNS arithmetic as an alternative approach to floating-point. A real number is represented in LNS as the fixed-point value of base two logarithm of its absolute value with a special arrangement to indicate zero and NaN. An additional bit indicates the sign.

Multiplication, division and square-root are implemented as simple as fixed-point addition, subtraction and right shift, but, unfortunatelly, addition and subtraction require more complicated evaluation. The algorithm of add/sub interpolation has been described in [3, 6, 5].

The resulting 19-bit LNS arithmetic library targetted in Xilinx Virtex devices is presented in Table 1. The mul, div and sqrt macros are very small and compact. They finish in two clock cycles, but they are not pipelined due to usage comfort. The add/sub macro is implemented as a dual-issue pipelined 10 stage macro where two add/sub units share the look-up tables.

3   Lattice Adaptive Noise Canceller

A lattice structure of the algorithm is shown on a signal-flow graph on Figure 1. The higher filter order, the more lattice bars (order stages) required. Full description of the update mechanism is beyond the scope of this paper and for a

Table 1. 19 bit LNS arithmetic implementation for Xilinx Virtex-1000

LNS Operation	Latency [Clock cycles]	Number of units in one macro	Pipelined	Macro size [Slices]	BRAMs
add/sub	10	2	Yes	720	6
mul,div,sqrt	2	1	No	≈ 160	0

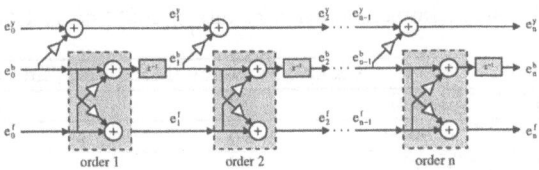

Fig. 1. Lattice computation data flow

reader can found its description on any DSP textbook. The signal-flow graph is shown in Figure 1. The update of the amplifier gain , which is computationally expensive, lies on the critical path of the algorithm. The complete critical path includes one division, four multiplications and two additions. As presented above, the add/sub unit is the most resource consuming part of the LNS arithmetic, therefore it is required to optimize its utilization. We have used pipelining as is shown on the Figure 2. There are four pipeline stages running in parallel. Each stage updates n^{th} order lattice filter sequentially (see Figure 1). Stages 1 and 3 and stages 2 adn 4 share one add/sub unit. To avoid conflicts, the stage 3(4) is started 4 clock cycles later than stage 1(2). The mul and div units are relatively small macros they are not shared between the stages.

The algorithm arrangement provides the lattice filter of order 4n with a latency equal to four input data periods . In other words, four data samples have to be acquired before the first output value will appear.

The noise canceller implementation is composed of the pipelined lattice filter in arrangement shown in the Figure 2 and of the data conversion modules. The conversion modules share the dual add/sub unit with the lattice filter to achieve its higher utilization.

Performance comparison of FPGA echo canceller implementations and our implementation on TI TMS320C6711 (IEEE 32-bit flaoting point device) of 100th order are shown in Table 2. The noise canceller algorithm is optimalized for each platform in a different way, therefore the maximal input data throughput was used as a comparison measure.

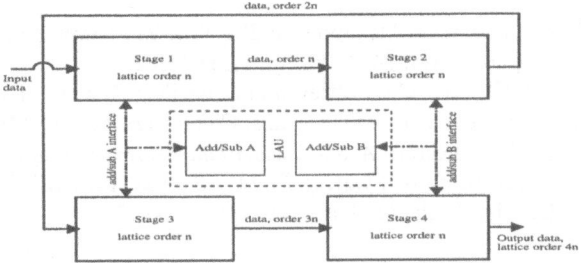

Fig. 2. Structure of the lattice filter implementation

Table 2. Implementation results of 100th order lattice

Arithmetic	Device	Clock Frequency [MHz]	Sampling clock [kHz]
LNS 19 bit	XC2V6000-6	83	48
LNS 19 bit	XCV1000-6	46	27
LNS 19 bit	XCV800-4	36	21
FP 20 bit	XC2V6000-6	140	32
FP 32 bit	TI C6711	225	1

It can be seen that the LNS arithmetic use lower clock frequecy for the same sampling frequenci of the input signal than floating point FPGA implementaiton and thus it comsumes less power. Finally, the architecture of the TI TMS320C6711 device is not able to handle complicated data dependencies and parallel data access and therefore it can not be compared with FPGAs.

4   Conclusions

The lattice noise cancellation application has shown that FPGA can be a suitable platform for complicated DSP algorithms. The presented noise canceller is running at the XSV800 prototyping board in real-time with 16 kHz data frequency for the filter order 160. Thanks to the small overall lattency of LNS arithmetic it propose better solution for the algorithms with difficult data dependencies than floating point solution. In future, we would like to optimize the design for the Virtex-II device family and compare the results with modern FP arithmetic for FPGA.

References

1. B. Friedlander: *Lattice Filters for Adaptive Processing*, Proceedings IEEE, vol. 70, no. 8, August 1982, pp. 829-867.
2. J. Kadlec: *Lattice feedback regularised identification*. In Proc. of 10th IFAC Symposium on System Identification. Preprints. IFAC, Copenhagen 1994, pp. 277-282.
3. R. Matouek, M. Tichý, Z. Pohl, J. Kadlec, C. Softley: *Logarithmic number system and floating-point arithmetics on FPGA*. In Proc. FPL2002, Springer, Berlin 2002, pp. 627-636.
4. E.E. Swartzlander, and A.G. Alexopoulos: *The Sign/Logarithm Number System*. In: IEEE Trans. Computers, vol. 24, 1975, pp. 1,238-1,242.
5. J.N. Coleman, E.I. Chester, C.I. Softley, and J. Kadlec: *Arithmetic on the European Logarithmic Microprocessor*. In: IEEE Trans. Computers, vol. 49, 2000, pp. 702-715.
6. J.N.Coleman and E.I.Chester: *A 32b Logarithmic Number System Processor and Its Performance Compared to Floating Point*. In: Proc. 14th IEEE Symposium on Computer Arithmetic, Adelaide, April 1999, pp. 142-152.

Specifying Control Logic for DSP Applications in FPGAs

J. Ballagh, J. Hwang, H. Ma, B. Milne, N. Shirazi, V. Singh, and J. Stroomer

Xilinx Inc. 2100 Logic Drive, San Jose, CA 95124 (USA)
{Jonathan.Ballagh,Jim.Hwang,Haibing.Ma,Brent.Milne,
Nabeel.Shirazi,Vinay.Singh,Jeff.Stroomer}@xilinx.com

Abstract. New non-HDL programming models for signal processing in FPGAs have focused primarily on building high-performance data paths. Along with the ability to construct sophisticated custom signal processors comes increased requirements for creating complex control circuitry. Recent enhancements to System Generator for DSP begin to address this need by providing mechanisms that include co-simulation interfaces to extend Simulink with HDL semantics, automatic compilation from Matlab m-code into Simulink and VHDL, and embedded microcontrollers. In this paper, we describe how such mechanisms can be used in a QAM receiver designed for a CCSDS standard.

1   Introduction

Field-programmable gate arrays (FPGAs) are widely used in modern digital communication systems in part because of their ability to implement highly parallel custom signal processors. FPGAs are commonly employed for up/down conversion, forward error correction (FEC), adaptive equalization and synchronization, spectral analysis and digital filtering [4].

Recent design tool efforts have focused on providing new programming models and design methodologies for DSP in FPGAs, moving away from the traditional FPGA design flows that begin with hardware description languages (HDLs) and mirror approaches to ASIC design. System Generator for DSP [5] provides a programming model based on Simulink® [6] that allows the signal processing engineer to efficiently target an FPGA without requiring HDL expertise. For example, Dick and Harris have used System Generator to construct a 50 Megabit/s QAM demodulator that includes digital mixing, adaptive equalization, and carrier recovery [1]. We have extended this system to include concatenated FEC and packet framing according to the CCSDS standard for telemetry channels [2]. In addition, we augmented the system with logic that performs carrier quadrant correction and symbol demapping. These extensions require real-time control (e.g. for flow control between the Viterbi and Reed-Solomon decoders) and asynchronous control suitable for a microcontroller (quadrant correction in the QAM demapper).

In this paper, we describe one aspect of the receiver's control circuity, the symbol demapper used for quadrant correction, implemented using the System Generator *m-code* block. This block compiles Matlab code into Simulink for simulation and VHDL during code generation. This bridging of an imperative model of computation (i.e. Matlab semantics) with discrete-time simulation (i.e. Simulink), and discrete-event RTL (which models hardware behavior), is representative of System Generator (which in fact preserves bit and cycle-accuracy) and other heterogeneous computing frameworks [3].

P.Y.K. Cheung et al. (Eds.): FPL 2003, LNCS 2778, pp. 1099–1102, 2003.

2 Symbol Demapping and Quadrant Adjustments

The QAM demodulator computes the phase error between the received quadrature carrier signals and the locally generated carriers, and adjusts the phase of the local carrier to drive the average error to zero. Although this approach suffices when the phase error is between $\pm 90°$, it fails when the error falls outside this range. To correct for this, the receiver provides additional logic to rotate the QAM hard-decisions by 90°, 180° and 270° under the control of a quadrant select signal generated by an embedded microcontroller, also modelled in System Generator and implemented in the FPGA. Figure 1 shows how the demapped symbols are rotated between quadrants in the receiver.

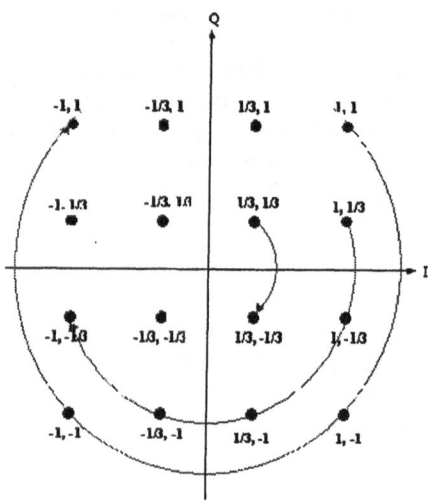

Fig. 1. 16-QAM constellation plot showing 90°, 180° and 270° mapping adjustments

The symbol demapping and rotation logic is shown in Figure 2. A multiplexer selects between phase-shifted hard-decision symbols for the I and Q channels. The multiplexer outputs are converted into 3-bit words that drive the address lines of two ROM blocks that translate hard-decisions into 2-bit output words.

An alternative approach creates a simpler and easier to understand design. The System Generator *m-code block* supports a hardware-centric subset of the Matlab language, including conditional expressions and branching statements, variable assignment, and fixed-point arithmetic. The user provides a Matlab function that the block interprets during Simulink simulation. During code generation, System Generator provides a faithful VHDL translation of the Matlab m-code. Figure 3 shows the Matlab function for symbol demapping and rotation logic.

The m-code block allows us to handle the problem addressed by the circuit shown in Figure 2 using simple, easy to understand Matlab assignments, `switch`, and `if` statements. The m-code block automatically sets its port labels to match the names of the function's input and output parameters (Figure 3).

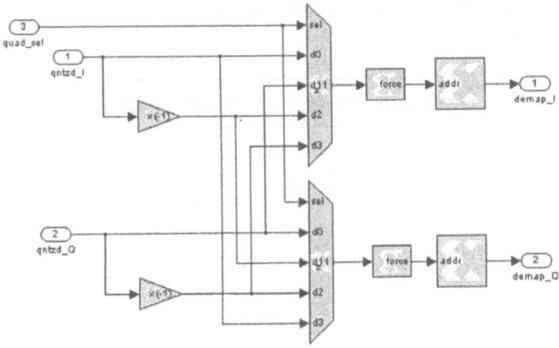

Fig. 2. System Generator block implementation of demapping circutiry

```
function [demapI, demapQ] = demap(quadSelect, qntzdI, qntzdQ)

    plusHalf  = xfix({xlSigned, 3, 1}, .5);
    minusHalf = xfix({xlSigned, 3, 1}, -.5);

    switch quadSelect
        case 0,    ISel = qntzdI;   QSel = qntzdQ;
        case 1,    ISel = qntzdQ;   QSel = -qntzdI;
        case 2,    ISel = -qntzdI;  QSel = -qntzdQ;
        case 3,    ISel = -qntzdQ;  QSel = qntzdI;
        otherwise, ISel = 0;        QSel = 0;
    end

    if (ISel == plusHalf),      demapI = 0;
    elseif (ISel == 1),         demapI = 1;
    elseif (ISel == minusHalf), demapI = 2;
    else,                       demapI = 3;
    end

    if (QSel == plusHalf),      demapQ = 0;
    elseif (QSel == 1),         demapQ = 1;
    elseif (QSel == minusHalf), demapQ = 2;
    else,                       demapQ = 3;
    end
```

Fig. 3. Matlab m-code demapping function (left) with corresponding m-code block (right)

Note that the function inputs qntzdI and qntzdQ are quantized representations of the hard-decisions computed by the demodulator's slicer. By translating hard-decision symbols from the values $1, 1/3, -1$, and $-1/3$ to $1, 1/2, -1$ and $-1/2$, respectively the circuit requires fewer bits for the comparisons.

Although the two simulation models of the demapping circuit are functionally equivalent, their hardware implementations differ. System Generator was used to translate the two circuits into hardware. The hardware was synthesized using XST and run through the Xilinx ISE 5.2i tools to produce the resource costs shown in Table 1. The table shows that the m-code solution is slightly less expensive than the traditional block approach. The difference in size is a consequence of synthesis tool trimming of unused logic from the m-code HDL. The block-based approach uses EDIF-based ROM cores whose logic cannot be further optimized.

Table 1. Virtex-II™ resource costs for the block-based and m-code solutions to the symbol demapping and rotation problem.

Design Approach	Slices	LUTs
Traditional Blocks	22	12
M-Code Block	17	12

3   Conclusions

As modern FPGA-based communication systems become increasingly complex, the control logic required to manage these systems is becoming equally sophisticated. These requirements have impacted FPGA DSP design tools by encouraging design environments in which control can be specified with as much flexibility and abstraction as the data paths. While providing this convenience, the design tool must also ensure an efficient implementation in hardware. We have demonstrated an example where a portion of the control logic in a System Generator QAM receiver design was specified using the Matlab m-code programming language.

References

1. C. H. Dick and F. Harris, "FPGA QAM Demodulator Design," *Field Programmable Logic Conference (FPL)*, Sept. 2-4 Montpellier, France, 2002.
2. Consultative Committee for Space Data Systems, "Telemetry Channel Coding : CCSDS 101.0-B-5 Blue Book,"Recommendation for Space Data Systems Standards, June, 2001.
3. J. Buck, S. Ha, E.A. Lee, D.G. Messerschmitt, "Ptolemy: a Framework for Simulating and Prototyping Heterogeneous Systems," *International Journal on Computer Simulation*, Jan 1994.
4. C. H. Dick and F. Harris, "Configurable Logic for Digital Communications: Some Signal Processing Perspectives," *IEEE Comm. Magazine*, vol. 2, pp 107-111, Aug, 1999.
5. J. Hwang, B. Milne, N. Shirazi, and J. Stroomer, "System Level Tools for DSP in FPGAs," *FPL 2001*, Lecture Notes in Computer Science, pp 534-543, 2001.
6. The Mathworks Inc., "Simulink, Dynamic System Simulation for Matlab, Using Simulink," Natick, Massachusetts, 1999.

FPGA Processor
for Real-Time Optical Flow Computation

Selene Maya-Rueda and Miguel Arias-Estrada

Computer Science Department, INAOE, Apdo. Postal 51 & 216, Mexico
selene@ccc.inaoep.mx, ariasm@inaoep.mx

Abstract. In this work an FPGA-based architecture for optical flow computation in real-time is presented. The architecture is based on an algorithm providing a dense and accurate optical flow at an affordable computational cost. The architecture is composed of an array of processors interconnected under a systolic approach. The array of processors is mainly focused in performing matrix operations to speed up the computations of optical flow. The architecture is being prototyped on an FPGA device. Results are presented and discussed.

1 Introduction

Motion is one of the most important characteristics of an image sequence since it contains the dynamics of a scene through the relationship between spatial characteristics of an image and the temporal changes in a scene [1]. In spite of the research on motion, there are two major limitations to applying motion computation to vision systems in real tasks: robustness in the real world, and the computational resources required for real-time operation [2]. Many current motion detection algorithms and systems require highly specialized hardware or up to several minutes of computing time to process medium resolution images. As a consequence, there exists an inability to obtain reliable motion estimation under real-world conditions, and applications for optical flow algorithms remain scarce [3].

However, the motion estimation problem can be transformed into regular computations to apply parallel processing in a hardware architecture. A systolic FPGA-hardware architecture is proposed for optical flow computation in real-time. The use of FPGA technology allows the design and implementation of parallel and pipelined structures, since FPGA devices are fast enough and provide high internal storage and density for digital logic design.

The structure of the paper is as follows. In section two we describe the problem of motion estimation in general terms and the algorithm used to compute optical flow. In section three the hardware architecture for optical flow computation is presented and discussed in some detail. Implementation results and evaluation of the architecture are presented in section four. Finally, in section five the conclusions and future work are presented.

P.Y.K. Cheung et al. (Eds.): FPL 2003, LNCS 2778, pp. 1103–1106, 2003.

2   Optical Flow Algorithm

Several computational methods have been developed for optical flow computation. A comprehensive survey can be found in [4][5]. The algorithm used in this research was recently developed by Srinivasan [6], and it was selected since it provides a good accuracy-complexity tradeoff at an affordable computational cost. In Srinivasan's algorithm, the optical flow is modeled as a weighted sum of overlapped basis functions. In this approach the motion field is force-fitted to the model and derives the smoothness properties from those of the model basis functions. Equation 1(a) expresses the *gradient constraint*, which forms the basis of gradient-based methods, and equation 1(b) shows the restrictions imposed to 1(a) in the algorithm.

$$\frac{\partial \psi}{\partial t} + u\frac{\partial \psi}{\partial x} + v\frac{\partial \psi}{\partial y} = 0 \qquad \text{(a)}$$

$$u = \sum_{k=0}^{K-1} u_k \phi_k, \quad v = \sum_{k=0}^{K-1} v_k \phi_k \qquad \text{(b)}$$

(1)

where u, v are the velocity vectors, ψ is the image intensity, and ϕ is a family of basis functions.

The basis functions ϕ_k are typically cosine functions. Combining the expressions in equation 1, the problem can be reformulated as a set of linear equations:

$$\int \phi_l \frac{\partial \psi}{\partial x}\frac{\partial \psi}{\partial t} + \sum_k u_k \int \phi_k \phi_l \left(\frac{\partial \psi}{\partial x}\right)^2 + \sum_k v_k \int \phi_k \phi_l \frac{\partial \psi}{\partial x}\frac{\partial \psi}{\partial y} = 0$$

$$\int \phi_l \frac{\partial \psi}{\partial y}\frac{\partial \psi}{\partial t} + \sum_k u_k \int \phi_k \phi_l \frac{\partial \psi}{\partial x}\frac{\partial \psi}{\partial y} + \sum_k v_k \int \phi_k \phi_l \left(\frac{\partial \psi}{\partial y}\right)^2 = 0$$

(2)

Each pair of equations characterizes the solution around the image area covered by the basis functions. The unknowns are the weights of the basis functions, u_l, v_l. To obtain the optical flow vectors it is required to solve equation 2. The iterative method of Preconditioned Biconjugate Gradients (PBG) is employed [7] to solve linear systems of the form $Ax = b$. The matrix size to be inverted is dependent on the image size and the spacing between the basis functions so, there is an accuracy-performance tradeoff. The method involves the computation of matrix operations, such as matrix addition and matrix multiplication in the matrix inversion process.

3   Architecture

A block diagram with a general overview of the proposed architecture for optical flow computation, based on a systolic approach [8], is shown in the figure 1. The main modules of the architecture and a brief description are given in the following.

The *address generator* module provides the addressing sequence for storing pixels in the input memory and reading them when required by the data dispatch module. The external memory organization is based on a Harvard-like scheme. The *data dispatch* module distributes data to the systolic processors for parallel processing, exploiting the data-level parallelism found in image and matrix operators. The *management unit* provides the control signals to synchronize the module operation and data exchange among modules inside the architecture. The *sequence generator* controls the iteration number in the matrix inversion process, to reduce the loop overhead. The *data collector* extracts the computed optical flow vectors from each processor and sends them to the output memory. The *derivatives computation* module computes the temporal and spatial derivatives of the images. The module is composed of a set of registers and adders/substractors. The derivatives are required to setup the equation system for the matrix inversion.

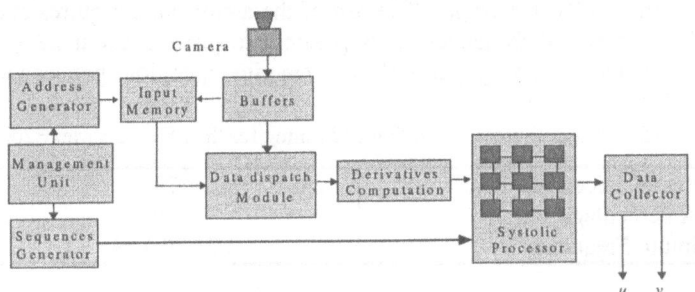

Fig. 1. A block diagram of the hardware architecture for optical flow computation

The *systolic processor* computes the matrix inversion. The systolic processors are arranged in a 2D mesh with local interconnections. Under this approach, matrix multiplication complexity is reduced to a linear order. A closer view of the module is shown in figure 2(a). A set of 8×8 processing elements is employed. The array receives data from previous computed derivatives and the basis functions. A control word from the sequence generator controls the number of iterations. Once the iterative process has converged to a solution, the array transmits the results to the data collector to send them to the output memory bank.

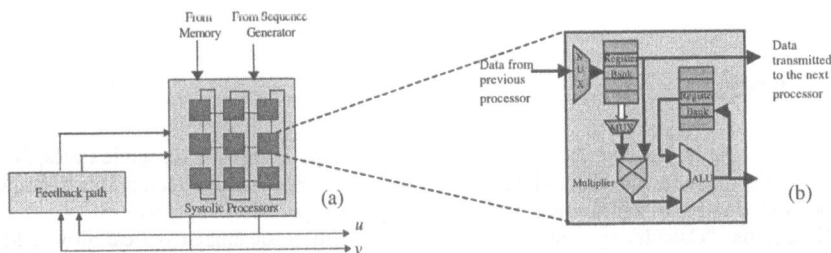

Fig. 2. (a) Systolic processing module and (b) Block diagram of the processing element

The Processing Element (PE) shown in figure 2(b) is based on a specialized Arithmetic Logic Unit (ALU). The PE contains two register banks with four 32-bit registers for temporal storage and pipeline stages. A multiplexer selects data to transmit partial results to neighbor processors. The numerical representation of data is single precision floating point. The PEs perform operations to setup the equation system for optical flow computation, combining information of spatial and temporal derivatives and the basis functions according to equation 2. Then, the PEs establish the equation system (equation 2), perform the matrix inversion and finally calculate the optical flow vectors (equation 1(b)).

4   Implementation and Results

A Handel-C model has been developed and the functional validation of the system has been done. The PE model has been synthesized for a VirtexE2000-6 device. A summary of hardware resources and timing reported for a single PE is shown in table 1. The implementation of an array of 8×8 PEs requires over 50,000 slices, which exceeds the current FPGA capacity. The rest of the architecture requires about 2,000 slices. With the reported frequency, it is possible to process about 22 images per second. The architecture throughput is close to ten giga-operations per second.

Table 1. Hardware resources utilization and timing for the PE on a VirtexE2000-6

Number of Slices	795
FPGA percentage	4%
Maximum Frequency	25.79 MHz

5   Conclusions and Further Work

The proposed architecture is able to compute near real-time the optical flow but it is expected to improve its performance. The design takes advantage of parallel structures and arithmetic logic in the FPGA. The architecture is based on a systolic array of processors interconnected by a 2D mesh and it is modular and scalable. The architecture constitutes a platform for developing motion applications in computer vision, especially for compact and mobile robotic applications. Further goals of this research will be focused on the use of the architecture in applications such as image stabilization, mosaicking, and 3D information recovering.

References

1.   B. Jänhe and H. Haubecker, "Computer Vision and Applications", Academic Press, 2000
2.   M. Arias-Estrada, C. Torres-Huitzil. "A Real-time FPGA Architecture for Computer Vision", Journal of Electronic Imaging, January 2001, pp. 289-296, 2001.
3.   T. Camus. "Calculating time-to-contact using real-time quantized optical flow". Max-Planck-Institut für biologische kybernetik. Tech. Report No. 14. Feb. 1995.
4.   J. L. Barron et al. "Performance of optical flow techniques", Revised tech. Report RPLTR-9107, Queen's University, Jul. 1993.
5.   B. McCane, K. Novins, D. Crannich and B. Galvin. "On Benchmarking Optical Flow" Computer Vision and Image Understanding, Vol. 84, pp. 126-143, 2001.
6.   S. Srinivasan. "Image sequence analysis: estimation of optical flow and focus of expansion, with applications". PhD Thesis, University of Maryland, 1999.
7.   W. H. Press, S. A. Teukolsky, et al. "Numerical Recipes in C", 2nd Edition. Cambridge University Press, Cambridge, UK, 1992.
8.   J. H. Moreno and T. Lang. "Matrix computation on systolic-type meshes", High-performance VLSI Signal Processing Innovative Architectures and Algorithms, Vol. 1. Edited by H. J. Ray Liu and Hung Yao, IEEE Press, 1998.

A Data Acquisition Reconfigurable Coprocessor for Virtual Instrumentation Applications

M. Dolores Valdés[1], María J. Moure[1], Camilo Quintáns[2], and Enrique Mandado[1]

[1] Instituto de Electrónica Aplicada "Pedro Barrié de la Maza",
Universidad de Vigo, España
{mvaldes, mjmoure, emandado} uvigo.es
http://www.dte.uvigo.es
[2] Departamento de Tecnología Electrónica,
Universidad de Vigo, España
{quintans}@uvigo.es

Abstract. Virtual instruments intended for electronic circuits verification arose from the combination of computers supporting advanced graphical interfaces with data acquisition systems providing input/output capabilities. In order to increase the versatility and the operation rate of virtual instruments, we have designed several data acquisition/generation modules based on reconfigurable hardware. By this way, not only the software modules but also the hardware functions are dynamically changed according to the requirements of each specific instrument. The main basis of the software and hardware levels of reconfigurable virtual instruments are described in this paper. This methodology summarize our experience in the design of virtual instrumentation platforms oriented to different measurement applications. Finally, a new data acquisition/generation coprocessor based on FPGAs and optimized for the implementation of portable instruments is described.

1 Introduction

Most of the current virtual instruments are based on a general-purpose data acquisition board, so the data processing and analysis are software implemented. In this way, a wide range of instruments can be programmed using the same hardware resources and the function of the instrument is not completely defined by the manufacturer [1]. Nevertheless, these virtual instruments cannot operate at the same frequency as traditional ones whose hardware is specially oriented to their specific functionality.

Several years ago, we have proposed a technique for electronic instruments design based on a reconfigurable data acquisition hardware [2], [3]. By this way, not only the virtual instrument software modules but also the hardware functions are dynamically changed according to the requirements of each application. The reconfigurable hardware increases versatility of the virtual instruments as well as their bandwith.

P.Y.K. Cheung et al. (Eds.): FPL 2003, LNCS 2778, pp. 1107–1110, 2003.

2 Hardware Level Design

2.1 FPGA Selection

An FPGA intended for virtual instrumentation applications must support the following features:
Standard configuration port. The FPGA must be reconfigured in real time from the computer.
Embedded memory blocks. Memory units in the hardware level make the data acquisition/generation frequency independent from the delays associated with the software levels and the computer interface.
Bidirectional input/output blocks.
Low-skew paths. Fast-capture latches and low skew paths available in some FPGAs contribute to reduce the different delays associated with the capture and propagation of related signals.
Integrated PLLs /DLLs ("Phase Locked Loops"/ "Digital Locked Loops").

2.2 Computer Interface

Standard interfaces commonly used in instrumentation systems were analyzed:
Parallel Port. This protocol is very simple and can be implemented directly in any FPGA using a minimum number of logic blocks.
GPIB. This interface is commoly used in measurement instruments. Its implementation and low speed requirements are similar to the parallel port and both are used for the implementation of low speed measurement instruments.
USB serial bus. This is a more recent standard which allows hot connection, plug&play, and high transfer rates using version 2.0. This is the best approach to portable instruments with medium performance. Nevertheless, the implementation of the USB bus in the FPGA has a great cost in logic resources.
PCI bus. The PCI interface has been chosen as the best way to achieve a high-speed link between the acquition/generation hardware and the computer. Nevertheless its implementation have also a great cost in logic blocks and requires a high speed programmable device.

2.3 Analog Inputs/Outputs

An advantage of using FPGAs to implement the logic control of an analog data input/output system is the availability of a great number of registered I/O pins. In contrast with microcontrollers or DSP based data acquisition systems, characterized by a reduced number of I/O ports, FPGAs allows the simultaneous sample of many analog inputs channels avoiding the use of additional devices like sample&hold circuits, analog or digital multiplexors and the logic control associated with them.

2.4 Hardware/Software Codesign

The following decisions must be taken previously to any implementation:

Hardware/Software functions. At least, the reconfigurable hardware must provide data acquisition/generation services to the software. By the same way, the software application must implement the human interface or control panel of each instrument. The analysis of the basic architecture and functions of the measurement instruments permits to define which parameters must be programmable and which configurable. For example, the reconfiguration must be used when a new instrument is created by the application software or when the main parameters of the current instruments must be modified. By the contrary, the trigger detection logic and timing functions must be programmable because they are prone to be changed during the operation of any instrument.

Hardware/Software synchronization. The tasks previously assigned to the hardware and software levels must carry out in parallel at the maximum possible frequency. Specially, the delays generated in the software levels must not be propagated to the acquisition/generation hardware.

3 Software Level Design

A main process is created when the instrumentation application program starts. This process is called Virtual Instrumentation Manager (VI Manager) and waits for the events generated by the user. These events define the active instruments of the system and their parameters.

When an instrument becomes active, the VI Manager generates a new process (instrument 1). Each instrument process is split into two threads, one manages the user interface and the other controls the data acquisition/generation process. By this way, the delays associated with the user interface (for example due to the graphical representation of data) never stop or delay the acquisition/generation process. When a new instrument becomes active, a new process is created (instrument #). This process/thread based approach guaranties the minimum latency in the active acquisition/generation tasks.

Services provided by the VI Manager are used by the threads in order to communicate with the reconfigurable hardware. The VI Manager encapsulate the particularities of the hardware so the physical system can be reconfigured without software changes at user level. When an application thread from one instrument requests a service, it blocks itself until a callback from hardware is generated. By this way, an optimal resources sharing among active instruments is achieved.

A message from a thread controlling the acquisition/generation process to the thread managing the user interface is generated when new data must be displayed or when some event must be notified to the user.

When a data acquisition/generation is finished, the associated thread or process is killed.

4 A Reconfigurable Data Acquisition/Generation Coprocessor

This section is focused on a reconfigurable data acquisition/generation coprocessor especially oriented to the implementation of portable instruments for electronic measurement. We have chosen the USB 2.0 as the standard interface with the computer and we use the Cypress EZ-USB FX2 controller that includes memory blocks (4KB) and allows four end-points simultaneously [4]. This memory capacity avoids the need of specific memory blocks in the data acquisition hardware. The data transfers can be selected to 8/16 bits and the transfer rate can achieve 48 MHZ using master, slave, synchronous or asynchronous modes. Finally, this chip includes a 8051 core that can be used for some specific tasks; for example we use this microcontroller for the implementation of the JTAG interface intended to the FPGA reconfiguration. Figure 1 shows one of the implementations (patent pending) using an Altera FPGA EP20K100EQC240-2X. Also we have developed additional cards can be combined with this one using the expansion ports in order to implement for example analog/digital and digital/analog conversion.

Fig. 1. Hardware implementation of a reconfigurable virtual coprocessor base on a USB serial interface

References

1. Virtual Instruments in http://zone.ni.com/devzone/
2. Moure, M.J., Valdés, M.D., Mandado, E.: Virtual Instruments based on Reconfigurable Logic. In: Hartenstein, R.W., Keevallik, A. (eds.): Field-Programmable Logic and Applications. Lecture Notes in Computer Science, Vol. 1482. Springer-Verlag, Berlin Heidelberg New York (1998) 505–509.
3. Moure, M.J., Valdés, M.D., Mandado, E.: Virtual Instruments based on Reconfigurable Logic: Design Methodology and Applications. In: Proceedings of the International Workshop on Virtual and Intelligent Instrumentation. IEEE, Annapolis (2000) 44–51
4. In http://www.cypress.com/

Evaluation and Run-Time Optimization of On-chip Communication Structures in Reconfigurable Architectures

T. Murgan, M. Petrov, A. García Ortiz, R. Ludewig, P. Zipf,
T. Hollstein, M. Glesner[1], B. Oelkrug, and J. Brakensiek[2]

[1] Institute of Microelectronic Systems,
Darmstadt University of Technology, Germany
murgan@mes.tu-darmstadt.de
[2] Nokia Research Center, Bochum, Germany

Abstract. With technology improvements, the main bottleneck in terms of performance, power consumption, and design reuse in single chip systems is proving to be generated by the on-chip communication architecture. Benefiting from the non-uniformity of the workload in various signal processing applications, several dynamic power management policies can be envisaged. Nevertheless, the integration of on-line power, performance and information-flow management strategies based on traffic monitoring in (dynamically) reconfigurable templates has yet to be explicitly tackled. The main objective of this work[1] is to define the concept of *run-time functional optimization* of application specific standard products, and show the importance of integrating such techniques in reconfigurable platforms and especially their communication architectures.

1 Introduction

Forthcoming wireless communication standards and ubiquitous computing technologies, together with strict time-to-market, low costs, high performance, and design efficiency requirements pose tremendous pressure on the design process. Driven by those demands reconfigurable architectures are getting an increasing interest in recent times [5]. Thus, a remarkable variety of different platforms have been proposed to trade energy-efficiency, cost and performance [4]. However, as technology improves and the integrable die size enlarges, one of the most important limiting factors for system performance, die area, and power consumption will be generated by the on-chip interconnect networks [2, 7].

This paper addresses the necessity of integrating run-time functional optimization policies into reconfigurable architectures. Additionally, several possible dynamical optimization strategies for on-chip communication architectures are discussed and underlined.

---

[1] The presented work was carried out within the German funded BMBF project IP2, i.e. *Intellectual Property Prinzipien für konfigurierbare Basisband SoCs in Internet-Protokoll basierten Mobilfunknetzen*, No. 01M3059B.

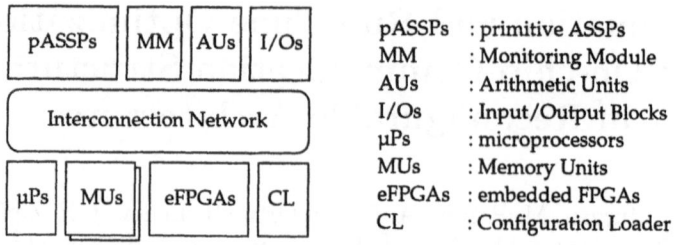

Fig. 1. ASSP Architecture Template

2   Dynamic Optimization of Reconfigurable Architectures

A reconfigurable architecture template or platform as in figure 1 is build around a reconfigurable interconnection network and may consist of processing units, memory blocks, specialized arithmetic units, reconfigurable and dedicated hardware, I/O blocks, monitoring modules, and configuration loaders. The so-called *Application Specific Standard Products (ASSPs)* are obtained through different instantiations of such modules and their application class is influenced by the design of what we call *primitive ASSPs (pASSPs)*.

The majority of battery powered electronic devices, like portable computers and mobile phones, exhibit a non-uniform workload. Peak or very high performance levels are generally needed only during particular operation states. In an attempt to extend battery life and reduce the cost and noise of cooling systems, *Dynamic Power Management (DPM)* techniques reduce the power consumption of electronic systems by dynamically adjusting performance levels to the workload through shut-down or at least slow-down of active components [1].

The decisive drawbacks of reconfigurable architectures in comparison to ASICs are the required die area, performance, and power consumption. Power consumption and dissipation become critical issues and the extension of the inter-charging operations is of increasing interest. Consequently, we believe that dynamic power management strategies emerge as an inextricable part of dynamically reconfigurable architectures, especially with the advent of VDSM (very deep sub-micron) technologies.

We call *run-time functional optimization policies* the set of those dynamic power and performance management strategies that can be efficiently applied in primitive ASSPs and especially in interconnection structures of reconfigurable architectures during operation time in order to dynamically reduce the power consumption under performance constraints, or vice-versa.

3   On-chip Communication Architectures Requirements

In order to interconnect the processing and memory blocks in a complex System-on-Chip, various interconnection networks can be employed. In both fine- and coarse-grained reconfigurable templates, the programmable interconnect architecture has a vital influence in the total area, performance and power consumption of the system. To achieve high computational performance and flexibility, the different modules have to be richly interconnected.

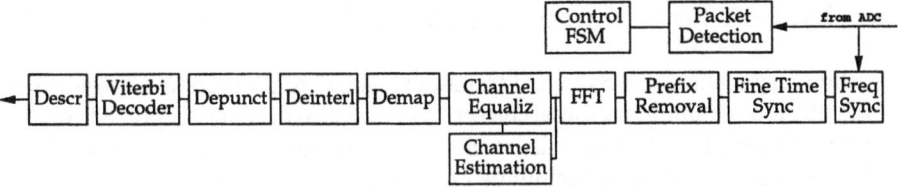

Fig. 2. Basic block scheme of a baseband OFDM receiver

Several interconnection strategies for coarse-grained reconfigurable architectures have been reported. Taking into account the locality of the data in several applications, local architectures are generally composed by point-to-point like structures devoted to provide high performance communication of closed units. They use to be restricted to the first level of neighbors (e.g *DReAM, KressAray*) or they can include a second level as in *MATRIX*. Other employed structures can be divided into three categories: multi-bus (e.g *RaPID*), crossbar (e.g. *PADDI-2*), and mesh; either regular (e.g. *DReAM, KressArray, Matrix, Garp*) or irregular (*Pleiades*).

We investigate dynamically reconfigurable architectures tailored for communication systems envisaging mainly OFDM-based wireless applications. The aforementioned project focuses on run-time functional optimization for OFDM-based wireless systems. Since the IEEE 802.11a wireless LAN standard allows for multiple data rates - from 6MB/s to 54MB/s - the data flow between modules will be variable, depending on the selected data rate. Several studied modules can be found in figure 2 representing a basic OFDM receiver.

However, the effective sample rate $f_s = 20MHz$ at the output of the sender is specified by the standard as a function of the symbol duration and the number of samples, thus not depending on the selected data rate. Consequently, the sample rates after and before the cyclic prefix extension are $20\,Ms/s$ and $16\,Ms/s$, respectively. The IFFT also increases the data rate, because only 48 of the 64 samples are data samples, the others being either pilots or zero signals. Knowing that the data rate at the output of the modulator is constant, i.e. $12\,Ms/s$, the possible data rates at the output of the sender as functions of the type of modulation and puncturing can be computed.

For the receiver, the data rate requirements can be determined in a similar manner. The FFT block operates at $16\,Ms/s$, exactly like the IFFT on the transmitter side. Assuming that 8 bits are used for sample quantization and knowing that each sample is a complex value, the effective bit rate will then be $256\,Mbps$. After the FFT, the data rate slightly decreases with the ratio 52/64, resulting in an effective data rate of $13\,Ms/s$. Thus, for 8-bit samples, the channel correction blocks operates at a data rate of $208\,Mbps$. The same data rate can also be found at the input of the phase tracker, which uses the 4 pilot signals to perform further adjustments of the constellation. Thereafter, the pilot signals are removed, leading thus to a decrease of the data rate to $192Mbps$. Furthermore, the traffic that has to be supported in a reconfigurable template increases due to the activity of the monitoring module and the configuration loader.

4   System Level Communication Evaluation

Preliminary results of a SystemC framework allowing for high level performance evaluation of communication architectures have already been reported in [6]. In order to ameliorate the overall latency of communication architectures with a high number of routers, a technique called *wormhole routing* is used [3].

In order to develop an application independent evaluation platform, we use a parameterizable hierarchical Markov model for both producer and consumer processes. Basically, the producer and the consumer may switch in three basic states: an idle one, and two active ones. While in the idle one no data can be send or received, during the other two active states data can be produced or consumed with a different probability. At their turn, both active states are hierarchical, and the probabilities can be slightly modified. Hence, both abrupt and slight data traffic modifications can be modeled stochastically.

5   Optimization Opportunities and Future Work

Several optimization strategies can be envisaged. In a hardware module that monitors the transition activity of signals in a digital system for estimating the power consumption has been developed. Such a task may be included together with a traffic controller in a DPM observer and used at run-time for monitoring and functional optimization. For example, the queuing strategy can be dynamically adapted. The buffers can be either clock gated or completely shut down whenever registering sparse communications. Thus, a dynamic trade-off between performance and power is realized. Additionally, the routing algorithm can also be dynamically changed or adjusted under the same circumstances.

References

1. L. Benini, A. Bogliolo, and G. De Micheli.   A Survey of Design Techniques for System-Level Dynamic Power Management. *IEEE Trans. on VLSI Systems*, 8(3):299–316, June 2000.
2. L. Benini and G. De Micheli. Networks on Chips: A New SoC Paradigm. *IEEE Computer*, pages 70–78, January 2002.
3. J. Duato, S. Yalamanchili, and L. Ni. *Interconnection Networks. An Engineering Approach.* Morgan Kaufmann Publishers, 2003.
4. R. Hartenstein. A Decade of Reconfigurable Computing: A Visionary Retrospective. In *Proc. of DATE*, pages 642–649, 2001.
5. K. Keutzer, A. R. Newton, J. M. Rabaey, and A. Sangiovanni Vincentelli. System-Level Design: Orthogonalization of Concerns and Platform-Based Design. *IEEE Trans. on CAD of Int. Circuits and Systems*, 19(12):1523–1543, December 2000.
6. T. Murgan, A. García Ortiz, M. Petrov, and M. Glesner. A Stochastic Framework for Communication Architecture Evaluation in Networks-on-Chip. In *IEEE Intl. Symposium on Signals, Circuits and Systems*, July 2003.
7. C. Seitz. Let's Route Packets Instead of Wires. In *Advanced Research in VLSI: Proc. of the 6th MIT Conf.*, pages 133–138, 1990.

A Controlled Data-Path Allocation Model for Dynamic Run-Time Reconfiguration of FPGA Devices

Dylan Carline and Paul Coulton

Department of Communication Systems, Lancaster University, Lancaster,
Lancashire, U.K., LA1 4YR
d.carline@lancaster.ac.uk, p.coulton@lancaster.ac.uk

Abstract. Although methods for dynamic run-time FPGA reconfiguration have been proposed, few address the problems associated with increasing data-path delays due to a full or partial reconfigurations. In this paper, a method is proposed that enables specific timing requirements to be maintained within a reconfigurable architecture, by using logic-module partitioning and *known-delay* interconnection modules. This system allows data-paths of varying widths to be routed effectively between device modules along paths that are fixed in both length and position. Further, the technique may be regarded as extending the Xilinx Modular Design tools methodology to support run-time scenarios.

1 Introduction

Reconfigurable computing is currently enjoying immense popularity in both communications and computing fields, through systems capable of near-software flexibility with a near-hardware performance. This coupled with the *dynamic* handling of device functionality, makes reconfigurable systems extremely attractive. Although new device architectures are being developed for this role [1,2], the Field Programmable Gate Array (FPGA) remains the most applicable target currently available. Using the FPGAs for this research also allows the concepts described to be applied *generically* to other emergent architectures. The research is based on the natural partitioning of internal FPGA logic and the sending of data though paths of set position and length in order to maintain synchronicity in a *run-time* reconfigurable environment. Serious problems can be encountered when elements are reconfigured in terms of their specific timing requirements and using static routing elements can maintain the module's interconnectivity. In this paper, Section 2 describes the benefits gained from device partitioning, whilst in Section 3 the module interconnections are described. Section 4 details the Xilinx Partial Reconfiguration model and an implementation of this idea is given in section 5, before our conclusions are finally drawn.

2 Device Partitioning

As FPGA devices have increased in functional size, the algorithms used in their design and operational control have become increasingly complex. In an environment

P.Y.K. Cheung et al. (Eds.): FPL 2003, LNCS 2778, pp. 1115–1118, 2003.

where dynamic reconfiguration may be responsible for the movement of individual logical elements, it is imperative that such movement does not prevent signals maintaining synchronicity across the device. Traditionally, device partitioning has been used in the placement of logic in FPGA devices [3,4]. Rather than using the partitioning to define the initial placement however, it is useful to first consider the *logical content* of the design. As System on Chip (SoC) implementations begin to proliferate and their target devices become commonplace, a natural separation occurs between specific logical groups or 'objects' within a design. Furthermore, as the functional size of devices increase, individual modules may take the form of complete systems (e.g. modulators or coders). The use of modular techniques is comparable to the way in which software-programming benefits from the use of object oriented techniques. These separate objects form the basis of the natural partitioning utilised in our study.

Not only are the specific architectures of FPGAs evolving in terms of reconfiguration support, new software tools are also emerging. Among these are the now well-known JBits Application Program Interface (API) [5] developed by Xilinx, which offers advantages to reconfigurable computing by providing direct manipulation of the Xilinx Virtex FPGA bitstream [6] thus reducing the normally considerable re-design times and enabling a finer granularity of device control [7].

3   Module Interconnections

In a practical environment, where a reconfigurable unit is adapting both its functionality *and* connectivity at high-speed, serious problems can occur when logical units are placed far from its connecting logic, ultimately preventing clock synchronicity. One method of overcoming this problem is to maintain a *fixed* route with *fixed* port locations between the modules with fixed elements remaining largely static over time. This coupled with a pre-defined boundary for logic-element placement within the modules, reduces complexity for any placement algorithms used thereby helping decrease the required time between possible reconfiguration events.

To create this system, register-modules are defined that allow data to pass between the functional elements of the design with registers having *fixed* port positions *within* each partition of the design. The delay time associated with these interconnections can be used to ensure that synchronicity between logical objects is maintained. The initial register modules were designed and implemented using the Xilinx Modular Design tools [8].

4   Partial Reconfiguration

The Partial Reconfiguration (PR) model by Xilinx [9] provides support for altering the functionality of parts of a design whilst the remainder maintains operation. Two modes are provided: '*Multi-Column PR, Independent Designs*' and '*Multi-Column PR, Communication Between Designs*', both allowing individual *column-based* reconfiguration through the device's SelectMAP [10] interface. Currently, the macro will only support *one* bit communication between modules, per Tri-State Buffer

(TBUF) longline, of which four can be supported per CLB row. Although this system provides a means of inter-module communication, the routing resources themselves are largely static. In terms of a reconfiguration event, this is a serious limitation as it is likely that the routing will need to be altered when the structure of the modules change or another module is added to the design. The system previously proposed avoids this problem by designing the routing resources as individual modules *themselves*, giving them freedom to be placed as per particular requirement. Furthermore, if CLB resources are remaining, routing resources can be added without overburdening the more scarce resource of tri-state buffers.

5   Implementation

The system proposed takes in a serial input to the module '*framer*' which separates the input stream into 8-bit words. These 8-bit words are subsequently passed to the '*detector*' module, which searches for a particular 8-bit synchronization sequence. Upon receiving this word, the module sends a copy of the synchronization word to both the '*inverter*' and '*concatenate*' modules. The '*inverter*' module sends an inverted copy of the synchronisation word to the '*concatenate*' module, which finally concatenates the two inputs before outputting the 16-bit word. The placed and routed version of the design can be seen in figure 4.

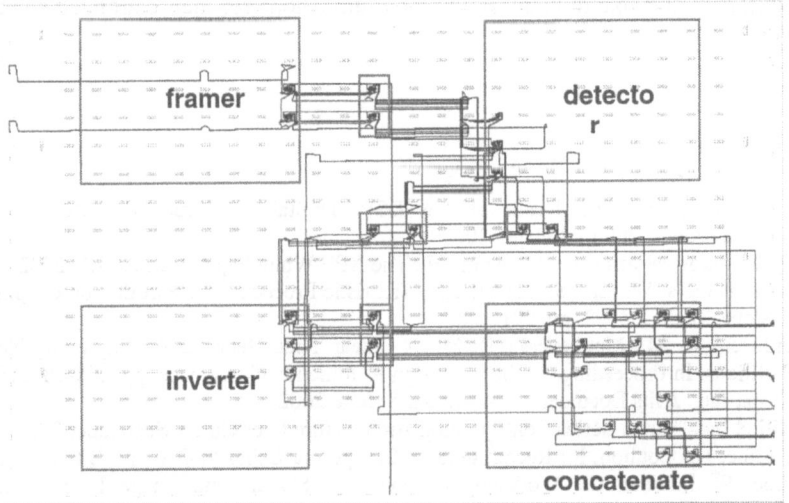

Fig. 1. The placed-and-routed layout of the four-module implementation

6   Conclusion

As the operating environments in modern communications and computing become increasingly demanding in terms of both protocol diversity and algorithm complexity, reconfigurable computing using devices capable of *run-time* reconfiguration is becoming increasingly desirable.

As previously stated, a new breed of device family is in development, which will offer enhanced reconfigurability with low power and memory requirements. Currently, these devices seem far from commercial availability and in the interim period other technologies must be considered. FPGA's cannot be overlooked in this regard, as the recent evolution of these devices has given further possibilities for their long-term integration. This is demonstrated by their increased support for partial reconfiguration and modular design techniques and ever-increasing functional sizes and operating bandwidths. One hardware aspect that must be improved upon however is the FPGA's large power requirements, which negate it from use in many portable implementations. If run-time FPGA reconfiguration is to be achievable, provision must be made for increased support of placement and routing processes. A method has been proposed that provides a structure enabling major logical groups to be considered independently thus reducing the time taken by optimisation processes and directly aiding the placement of logic. Further, by providing a mechanism of data transfer along paths of known delay the management of routing within a design is simplified. We believe the techniques suggested in this paper contribute to the evolution of the field of run-time reconfigurable computing and its eventual implementation.

Acknowledgement

The authors would like to thank H W Communications for their continued support of this work.

References

1. Tredennick, N.: The Death of the DSP. Digital Infrastructures: Megaflops and Microwonders. Dublin. Ireland. 2001
2. PACT Informationstechnologie GmbH.: The XPP White Paper. Release 2.1. 2002
3. Kernighan, B. W., Lin, S.: An Efficient Heuristic Procedure for Partitioning Graphs. Bell Systems Tech. J. 49. 2. pp. 291-308
4. Fiduccia, C. M., Mattheyeses, R. M.: A Linear Time Heuristic for Improving Network Partitions. In Proceedings of the 9th Design Automation Conference. pp. 175-181. 1982
5. Xilinx Inc.: JBits Documentation. JBits 2.8. June 2002
6. Xilinx Inc.: Virtex Series Configuration Architecture User Guide. September 2000
7. Carline, D., Coulton, P.: Reconfigurable Computing Using FPGAs: is JBits the Answer. Proc. 9th International Workshop on Systems, Signals and Image Processing. UK. November 2002
8. Xilinx Inc.: Development System Reference Guide – ISE 5. February 2003
9. Xilinx Inc.: Two Flows for Partial Reconfiguration: Module Based or Small Bit Manipulations. May 2002
10. Xilinx Inc.: Using a Microprocessor to Configure FPGAs Slave Serial or SelectMAP Modes. November 2002

Architecture Template and Design Flow to Support Application Parallelism on Reconfigurable Platforms

Sergei Sawitzki[1] and Rainer G. Spallek[2]

[1] Philips Research Laboratories
Prof. Holstlaan 4
5656AA Eindhoven, The Netherlands
Sergei.Sawitzki@philips.com
[2] Institute of Computer Engineering
Department of Computer Science
Dresden University of Technology
01062 Dresden, Germany
rgs@ite.inf.tu-dresden.de

Abstract. This paper introduces the ReSArT (Reconfigurable Scalable Architecture Template). Based on a suitable design space model, ReSArT is parametrizable, scalable, and able to support all levels of parallelism. To derive architecture instances from the template, a design environment called DEfInE (Design Environment for ReSArT Instance Generation) is used, which integrates some existing academic and industrial tools with ReSArT-specific components, developed as a part of this work. Different architecture instances were tested with a set of 10 benchmark applications as a proof of concept, achieving a maximum degree of parallelism of 30 and an average degree of parallelism of nearly 20 16-bit operations per cycle.

1 Introduction

Parallel processing and reconfigurable computing are two very powerful techniques of computation, which can be advantageously combined, as has been proven by dozens of successful projects [1, 2]. Some of the proposed architectures are introduced in the form of templates, which can be parameterized to suit the needs of the particular application. The question, however, remains, if it is possible to combine a design space model, a universal architecture template, and a tooling environment to support both instruction and data parallelism at different grain sizes.

This paper introduces the ReSArT (Reconfigurable Scalable Architecture Template). Based on a suitable design space model, ReSArT is parameterizable, scalable and able to support all levels of parllelism. To derive architecture instances from the template, a design environment called DEfInE (Design Environment for ReSArT Instance Generation) is used, which integrates some existing academic and industrial tools with ReSArT-specific components, developed as a part of this work.

P.Y.K. Cheung et al. (Eds.): FPL 2003, LNCS 2778, pp. 1119–1122, 2003.

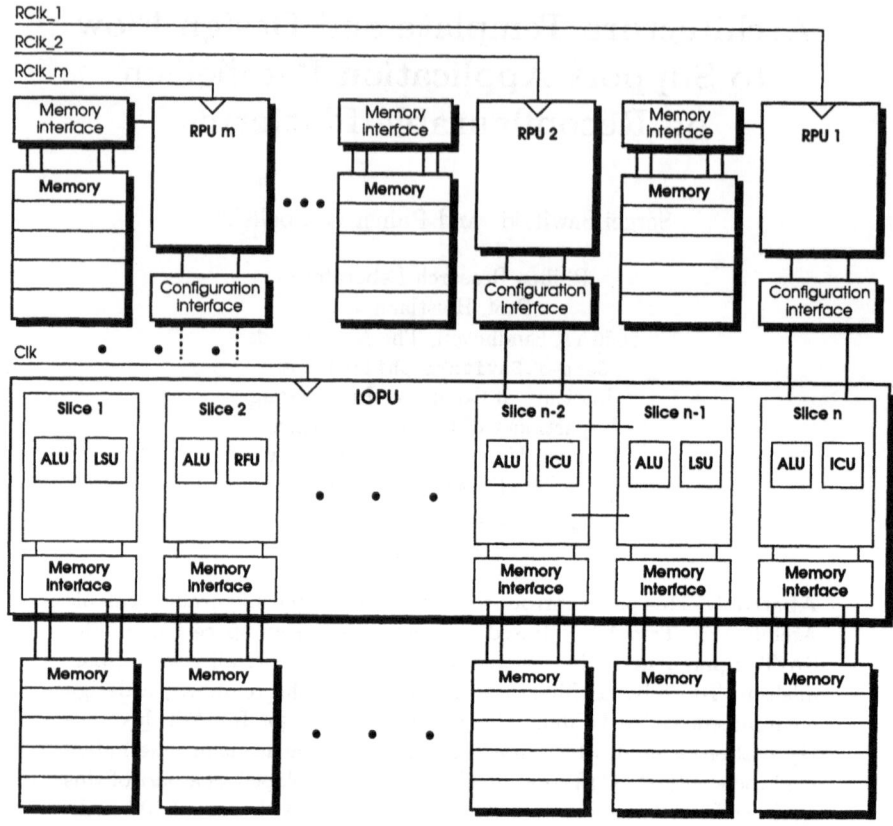

Fig. 1. ReSArT Top Level View

2   ReSArT and DEfInE

The detailed description of the design space can be found in [3]. According to this model, ReSArT leaves most dimensions open at the template level. In general, ReSArT can be split up in three views: a template, describing a family of reconfigurable architectures, a set of microarchitectures of reconfigurable resources (RPU), and a configuration context (as well as program code) to derive instances from these general descriptions and program them to complete particular tasks.

ReSArT consists of an instruction oriented processing unit (IOPU) and N_{RPU} reconfigurable processing units ($N_{\mathrm{RPU}} \geq 0$), see Fig. 1. The IOPU executes RISC-like instructions. It is constructed out of slices, each one being a simple double-issue processor core. Every slice includes one instruction fetch unit (IFU), one or two register files, one decode unit (DU) and two execution units. One execution unit is always an ALU, whereby the second one could be also an ALU or load-store unit (LSU), reconfigurable functional unit (RFU) or interface control unit (ICU). RFU is a small reconfigurable unit with much less logic capacity than RPU. Still, it can be used to implement simple single-cycle instructions which are not covered by the other hardwired units. The ICU is used to implement the communication between IOPU and RPU via memory-mapped I/O.

The RPU is used to implement computationally intensive data-flow-oriented parts with less synchronization. Each RPU has its own memory interface to further reduce the communication overhead. For each application, the RPU can be generated in one of three different flavors: pipelined, static with shared I/O ports, and dynamically reconfigurable. The hardware structure of the RPU fits into the model for SRAM-based, clustered, island-style FPGAs [4]. For every application, a different LUT and cluster size can be used. The optimal values are determined by the DEflnE (Design Environment for ReSArT Instance Generation).

The input language of DEflnE is a special class of directed acyclic graphs, so called $\mathcal{F}$-DAG. Such graphs can describe both fine and coarse-granular kernels and can be easily produced based on other formalisms. Each node of the $\mathcal{F}$-DAG represents an operation whereby edges represent the data dependencies.

In the first step, the flow_opt module reads the $\mathcal{F}$-DAG description in and generates an initial schedule according to a "as soon as possible" (ASAP) principle. In the second step, this initial schedule is optimized according to one of the four different strategies (minimize area function of the stages, minimize delay difference of the nodes in the same stage, minimize input or output data stream of the stages). After the optimization, the critical path of the schedule is computed and (based on this value) the stage numbers are mapped to the pipeline stages (or subcircuits, in case of the dynamically reconfigurable PU). Finally, the statistics of the flow_opt run are created and stored.

The best $\mathcal{F}$-DAG schedule is converted to the BLIF format [5] and transferred to the rec_map module, which consists of the SIS, TVpack, VPR [4] and par_extract steps. SIS and TVpack are responsible for technology mapping for a set of 36 different RPU microarchitectures (featuring different LUT and cluster sizes). VPR produces the place and route statistics, which are used by the par_extract tool to find the best microarchitecture for the given application. The par_check tool compares the circuit parameters with the requirements and (in case they are conform) transfers the processing to the flow2vhdl converter, which produces the appropriate VHDL code for the RPU. This code is forwarded to the arch_gen module, which adds the IOPU description and produces one design library for the selected ReSArT instance. This library is synthesizable and can be further processed by traditional back-end tools.

3   Experiments and Results

For the proof of concept some experiments were carried out with ReSArT and DEflnE using the Xilinx Virtex II architecture as a prototyping platform. The scalability of the IOPU was proven with respect to bitwidth and the number of slices (as well as some other parameters). The area scales linearly with all parameters. A prototype with 8 IOPU slices on the XC2V10000 device is able to execute $6 \cdot 10^8$ 16 bit operations per second, consuming only about 35% of the logic resources of the chip (thus, leaving a lot of space for RPU implementation).

For the RPU tests, a set of 10 benchmark applications was translated into $\mathcal{F}$-DAG format. This set includes a simple FFT module (8 points transform as a building block for bigger transforms), a soft-in soft-out decision subblock of a turbo decoder, a linear predictive codec for speech compression, as well as a couple of other applications. The DEflnE performance was tested for all optimization strategies. The improvement of 15.38–100% in the schedule quality in 6 of 10 cases results in the clock frequency improvement of 3.2–17.6% after mapping to the Virtex II chip (the difference is due to the fact that Virtex II was not selected by DEflnE as an optimal architecture for all benchmarks).

Some results can be compared with Xilinx benchmarks documented in [6]. The 1024pt 16 bit FFT requires about 1 μs on Xilinx Virtex II. Using ReSArT and DEflnE, the same transform runs in about 1.2 μs, given the programmability of the architecture and possibility to run additional tasks on the same device. A TMS340C64 family DSP needs slightly less than 8 μs for the same benchmark.

4   Conclusions

This paper introduced the ReSArT architecture template developed to support instruction and data parallelism at various levels and the DEflnE design environment to create ReSArT architecture instances. Tests with a set of 10 benchmark applications prove that ReSArT is scalable with respect to the bitwidth and the number of slices. Different RPUs running the benchmarks achieved the maximum degree of parallelism of 30 with an average of nearly 20 operations.

ReSArT and DEflnE build a powerful platform for the exploration of reconfigurable parallel processing systems. Considering tighter time-to-market margins in the modern SoC world resulting in the requirements for shorter design cycles, template-based architectures with embedded reconfigurable resources and integrated software environment will become increasingly important in the future.

References

1. Hartenstein, R.W., Hirschbiel, A., Weber, M.: A Novel Paradigm of Parallel Computation and its Use to Implement Simple High Performance Hardware. Int. Conf. memorizing the 30th Ann. of Computer Society Japan, Tokyo (1990)
2. Gokhale, M.B., Holmes, W. et.al.: Building and Using a Highly Parallel Programmable Logic Array. IEEE Computer **24**(1) (1991) 81–89
3. Sawitzki, S., Spallek, R.G.: A Concept for an Evaluation Framework for Reconfigurable Systems. In: Lysaght, P., Irvine, J., Hartenstein, R.W. (eds.): Field-Programmable Logic and Applications: 9th Intl. Workshop. LNCS, Vol. 1673. Springer-Verlag, Berlin Heidelberg New York (1999) 475–480
4. Betz, V., Rose, J., Marquardt, A.: Architecture and CAD for Deep-Submicron FPGAs. Kluwer Academic Publishers, Boston (1999)
5. Sentovich, E., Singh, K.J. et.al.: SIS: A System for Sequential Circuit Synthesis. Electronics Research Laboratory, Department of Electrical Engineering and Computer Science, University of California, Memorandum UCB/ERL M92/41 (1992)
6. R.T. Olay III: Xilinx XtremeDSP Initiative Meets the Demand for Extreme Performance and Flexibility, Xcell 40 (2001) 30–33

Efficient Implementation of the Singular Value Decomposition on a Reconfigurable System

Christophe Bobda, Klaus Danne, and André Linarth

Heinz Nixdorf Institute / Paderborn University
Fuerstenallee 11,D-33102 Paderborn, Germany
{bobda,danne,linarth}@upb.de

Abstract. We present a new implementation of the singular value decomposition (SVD) on a reconfigurable system made upon a Pentium processor and a FPGA-board plugged on a PCI slot of the PC. A maximum performance of the SVD is obtained by an efficient distribution of the data and the computation across the FPGA resource. Using the reconfiguration capability of the FPGA help us implement many operators on the same device.

1 Introduction

The Singular Value Decomposition of a matrix has many important scientific and engineering applications [4, 1, 5]. The SVD implementation on parallel processors is usually done using the Brent and Luk[4] parallelization of the Hestenes method explained in section 2. Using hardware to compute the SVD of a matrix have been done by Cavallaro et al [10, 11]. A CORDIC processor have been used here to compute the SVD of a 2x2 matrix. No indication have been given on the application of the method to solve larger matrices. In this paper we present an efficient implementation of the SVD on a reconfigurable system made upon a general purpose processor and a FPGA board.

Section 2 introduces the SVD problem and explain how a solution due to Hestenes is computed. It also presents the Brent and Luk parallel implementation of the Hestenes method. Section 3 explains the SVD implementation on our taget architecture while the performance of the system is given in section 4. Finally, we give an overview of our work in section 5.

2 The Singular Value Decomposition

The Singular Value Decomposition of an $m \times n$ real matrix A is its factorisation into a product $A = U \times \Sigma \times V^T$, where U is an $(m \times n)$ matrix with orthogonal columns, $\Sigma = diag(\sigma_1, ..., \sigma_n)$ and V is an $(n \times n)$ orthogonal matrix. The values σ_i of the diagonal matrix Σ are the singular values of the matrix A, the columns of the matrices U and V are the left and right singular vectors of A. The SVD can be computed using the Hestenes method based on the classical one side Jacobi iteration for digitalization of real symmetric matrices. The idea is to generate an

P.Y.K. Cheung et al. (Eds.): FPL 2003, LNCS 2778, pp. 1123–1126, 2003.

orthogonal matrix V such that the transformed matrix $AV = W$ has orthogonal columns. Having the matrix W, the Euclidean length of each non-zero column $W_{(:,i)}$ will be normalised to unity. The singular values and vectors are then computed as follows: $\sigma_i = \|W_{(:,i)}\|$, $U_{(:,i)} = \frac{W_{(:,i)}}{\sigma_i}$ and $W = U\Sigma$. The SVD of the matrix A is then given by: $AV = W \rightarrow AV = U\Sigma \rightarrow A = U\Sigma V^T$. Plane rotations represented by a matrix Q are incrementally applied to the matrix A to compute the matrix W. At the k-th step with the rotation matrix $Q^{(k)}$, we have: $A^{(k+1)} = Q^{(k)}A^{(k)}, 0 \leq k \leq k_r$. With a sweep defined to be a series of $\frac{n(n-1)}{2}$ pairwise column-orthogonalizations of the matrix $A^{(k)}$, the convergence of the matrix $A^{(k)}$ to W is guaranteed for $k_r = S\frac{n(n-1)}{2}$, where S is the number of sweeps, $A^0 = A$ and $W = A^{k_r}$. The multiplication of $A^{(k)}$ by $Q^{(k)}$ affects only the column-pair $(A_{(:,i)}^{(k)}, A_{(:,j)}^{(k)})$. The computation of $A^{(k+1)} = Q^{(k)}A^{(k)}$ is reduced to:

$$\begin{pmatrix} A_{(:,i)}^{(k+1)} \\ A_{(:,j)}^{(k+1)} \end{pmatrix} = \begin{pmatrix} \cos\Theta_{ij}^{(k)} & \sin\Theta_{ij}^{(k)} \\ -\sin\Theta_{ij}^{(k)} & \cos\Theta_{ij}^{(k)} \end{pmatrix} \begin{pmatrix} A_{(:,i)}^{(k)} \\ A_{(:,j)}^{(k)} \end{pmatrix} \tag{1}$$

The rotation angle $\Theta_{ij}^{(k)}$ is chosen such a way that the new column pairs are orthogonal. Using the formulas of Rustishauser[12], we set $\Theta_{ij}^{(k)} = 0$ if $\gamma_{ij}^{(k)} = 0$; Otherwise we compute $\xi^{(k)} = \frac{\alpha_j^{(k)} - \alpha_i^{(k)}}{2\gamma_{ij}^{(k)}}$ and $\tan\Theta_{ij}^{(k)} = \frac{sign(\xi^{(k)})}{|\xi^{(k)}| + \sqrt{1 + \xi^{(k)2}}}$ with $\alpha_i^{(k)} = A_{(:,i)}^{(k)} \bullet A_{(:,i)}^{(k)}$, $\alpha_j^{(k)} = A_{(:,j)}^{(k)} \bullet A_{(:,j)}^{(k)}$ and $\gamma_{ij}^{(k)} = A_{(:,i)}^{(k)} \bullet A_{(:,j)}^{(k)}$. This rotation angle always satisfies: $|\Theta_{ij}^{(k)}| \leq \frac{\pi}{4}$.

The main operations of this method are the generation of the rotation angle and the updating of the columns elements which rely on multiply accumulate (MAC) operations. Their dataflow nature as well as the large sizes of the matrices considered makes this computation a nice candidate for hardware implementation.

3   Implementation on a Reconfigurable System

Since the orthogonalization of column-pairs $(A_{(:,i)}^{(k)}, A_{(:,j)}^{(k)})$ are independent, they can be done in parallel. Brent and Luk [4] suggested the use of a set of n/2 processors connected together in a ring topology to orthogonalize a matrix of dimension n. Each processor has an exclusive access to a memory segment which stores the pairs of columns to be orthogonalize by that processor. After the orthogonalization of their column-pairs, the processors exchange data with their left and right neighbours. This process is repeated until all column pairs are orthogonalize, thus completing a sweep. The platform used is made upon a personal computer equipped with a RC-1000 FPGA board of the company Celoxica. The board contains up to four memory banks independently connected around the FPGA. Therefore we can implemented up to four processing element (PE) in the FPGA. Because the size of the FPGA is limited, it will not be possible to implement 8 PEs (4 for the dot product and 4 for the column-pairs update). We use the reconfiguration to implement the 8 PEs. The first reconfiguration loads the FPGA with the 4 PEs PE_11, PE_12, PE_13 and PE_14 (figure 1 a) which are used for the dot-products computation. The second configuration loads the 4 PEs PE_21, PE_22, PE_23 and PE_24 (figure 1 b)) which are used for the updating of the columns. The computation of the rotation angle is too complex for a hardware implementation. Since this computation is not often executed, it is left to the processor which collects the dot-product values from the FPGA and returns the rotation angle.

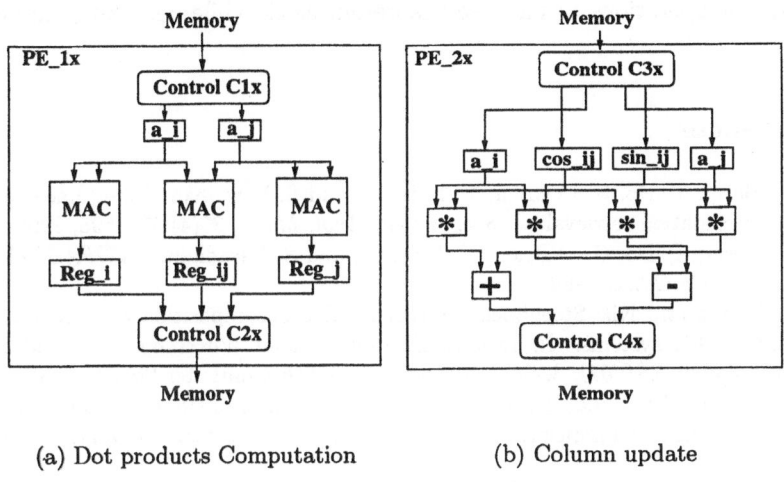

(a) Dot products Computation   (b) Column update

Fig. 1. Structure of the processing elements

4   Performance

The floating-point implementation of the PEs can be run with a clock frequency of maximal 20MHZ. Therefore the total time needed by the 4 PEs PE_11, PE_12, PE_13 to compute all the dot products required in to sweep for 4 block matrices is given by: $(number\_of\_rows \times number\_of\_columns \times (number\_of\_columns - 1) \times (cycle\_period))/2$. For 4 block matrices with 100 columns and 50.000 rows each, using a 20MHz clock, the time need to complete the computation of the dot products required by a sweep is: $(50.000 \times 100 \times (10 - 1) \times (1/20.000.000))/2 = 12.375s$. The rotation process is much faster, since it does not need to sweep all the columns 2 by 2. So the computation time can be done as follow: $number\_of\_columns \times number\_of\_rows \times cycle\_period$. The time needed to update a matrix of 50.000 x 400 with the same clock is 250ms. With the reconfiguration overhead of one second, our implementation will compute the complete SVD of a 400 by 50.000 matrices in 13.625s which is much higher than the time need on a Pentium processor with up to 450 MHZ.

5   Conclusion

In this paper we have dealt with an efficient implementation of the singular value decomposition of big matrices on a reconfigurable system made upon a PC and a FPGA-Board. With the structure of the RC 100-PP, it is possible to implement 4 PEs in parallel for each computation step. Because the complete function to compute the SVD could not fit on the FPGA, we made use of the hardware reconfiguration to implement the complete function needed for the SVD. The main bottleneck of the system is the floating point computation which occupy large space and slow down the clock. We are working on a solution based on the fixed-point operations to increase the design clock while decreasing the design area.

References

1. M. Berry, T. Do, G. O'Brien, V. Krishna, and S. Varadhan. Using linear algebra for information retrieval. *J. Soc. Indust. Appl. Math.*, 37(4):573–595, 1995.
2. M. Berry, T. Do, G. O'Brien, V. Krishna, and S. Varadhan. *SVDPACK(Version 1.0) User's Guide*, 1996.
3. C. Bobda and Nils Steenbock. Singular value decomposition on distributed reconfigurable systems. In *12th IEEE International Workshop On Rapid System Prototyping(RSP'01), Monterey California*. IEEE Computer Society, 2001.
4. Richard P. Brent and Franklin T. Luk. The solution of singular-value and eigenvalue problems on multiprocessor arrays. *SIAM J. Sci. Stat. Comput.*, 6(1):69–84, 1985.
5. S. Deerwester, S. Dumai, G. Furnas, T. Landauer, and R. Harshmann. Indexing by latent semantic analysis. *Journal of American Society for Information Science*, 41(6):391–407, 1990.
6. G. E. Forsythe and P. Henrici. The cyclic jacobi method for computing the principal values of a complex matrix. *Trans. Amer. Math. Soc.*, 94:1–23, 1960.
7. Gene H. Golub and Charles F. Van Loan. *Matrix Computations*. North Oxford Academic Publisching, 1983.
8. Eldon R. Hansen. On cyclic jacobi methods. *J. Soc. Indust. Appl. Math.*, 11(2):448–459, 1963.
9. M. R. Hestenes. Inversion of matrices by biorthogonalization and related results. *J. Soc. Indust. Appl. Math.*, 6(1):51–90, 1958.
10. F. Luk J. Cavallaro. CORDIC arithmetic for an svd processor. *Journal of Parallel and Distributed Computing*, 5(3):271–290, 1998.
11. J. Cavallaro N. Hemkumar. Efficient complex matrix transformations with CORDIC. In *IEEE Symposium on Computer Arithmetic*, pages 122–129. IEEE, 1993.
12. H. Rustishauser. The jacobi method for real symetric matrices. *Handbook for Automatic Computation*, Vol 2 (linear Algebra):202–211, 1971.
13. G. Salton. *The SMART Retrieval System*. Prentice Hall,Inc, 1971.
14. J. H. Wilkinson. *The algebraic eigenvalue problem*. Oxford University Press, 1965.

A New Reconfigurable-Oriented Method for Canonical Basis Multiplication over a Class of Finite Fields GF(2^m)

José Luis Imaña[1] and Juan Manuel Sánchez[2]

[1] Dpto. Arquitectura de Computadores y Automática, Universidad Complutense, 28040 Madrid, Spain
jluimana@dacya.ucm.es
[2] Dpto. Informática, Escuela Politécnica, Universidad de Extremadura, 10071 Cáceres, Spain
sanperez@unex.es

Abstract. A new method for multiplication in the canonical basis over GF(2^m) generated by an all-one polynomial (AOP) is introduced. The theoretical complexities of the bit-parallel canonical multiplier constructed using our approach are equal to the smallest ones found in the literature for similar methods, but the multiplier implementation over reconfigurable hardware using our method reduces the area requirements.

1   Introduction

Galois or finite fields have several applications in communication systems, such as satellite links, computer networks, or compact disks [1]. They use arithmetic operations in the Galois field for cryptography, error correction or algebraic codes. Finite fields with q elements are represented as GF(q), and the fields with fundamental interest for technical applications are the *extension fields* of GF(2), denoted as GF(2^m). The representation of the field elements has crucial role in the efficiency of the architectures for arithmetic operations. Considering a basis representation of the elements, the *addition* is relatively inexpensive, whereas the *multiplication* is the most important operation. There are different basis representations for elements of GF(2^m), and the most popular are the *canonical* [2], *normal* and *dual* bases. The complexity of the multiplier depends on the basis and the defining irreducible polynomial selected for the field.

The field GF(2^m) can be considered as a vector space of dimension m over GF(2), so it can be represented using any basis of m linearly independent elements. Therefore, elements of GF(2^m) are represented by m-bit vectors. *Addition* is realized by a bit-wise XOR operation, whereas the *multiplication* is determined by the basis. *Canonical* basis Ω is the set $\Omega = \{1, \omega, \dots, \omega^{m-1}\}$, where ω is a root in GF(2^m) of an irreducible polynomial of degree m over GF(2). Using this basis, the elements of GF(2^m) are polynomials of degree at most $m-1$ over GF(2), and arithmetic is carried out modulo the irreducible polynomial. We denote as $\underline{\alpha}_\Omega = (a_{\Omega_0}, \dots, a_{\Omega_{m-1}})^t$ the coordinate vector of α with respect to Ω.

P.Y.K. Cheung et al. (Eds.): FPL 2003, LNCS 2778, pp. 1127–1130, 2003.

In this contribution, a new method for multiplication in the canonical basis for the field $GF(2^m)$ generated by an irreducible *all-one-polynomial* (AOP) is introduced. Our approach, named the *transpositional* method, uses the permutations given by the value m, and the aim of the method is to *group* the coordinates of the operands in order to the reconfigurable synthesis tool could find a good mapping on the reconfigurable device selected for the implementation of the multiplier. The theoretical space and time complexities of the bit-parallel multipliers constructed using our approach are equal to the smallest ones found in the literature for similar approaches based on generating AOPs [3][4][2][5][6], but the practical implementation over reconfigurable hardware (FPGAs and CPLDs) using our method reduces the area requirements of the multipliers.

2 The Transpositional Method

The *transpositional* method [7] for multiplication in the canonical basis over $GF(2^m)$ is based on the computation of 1-cycles and 2-cycles which determine *product* and *sum-of-product* terms, respectively, of the operands coordinates.

Let $\alpha, \delta, \chi \in GF(2^m)$ and $\underline{\alpha}_\Omega, \underline{\delta}_\Omega, \underline{\chi}_\Omega$ be their coordinate vectors, respectively, with respect to Ω. The multiplication $\delta = \alpha \cdot \chi$ in Ω involves the presence of *inner products* [7], such as $\underline{\alpha}_\Omega^t \cdot \underline{\chi}_\Omega^r = a_{\Omega_0} c_{\Omega_{m-1}} + a_{\Omega_1} c_{\Omega_{m-2}} + \cdots + a_{\Omega_{m-1}} c_{\Omega_0}$, where the c_{Ω_i}s are the coordinates of χ with respect to Ω and where $\underline{\chi}_\Omega^r = (c_{\Omega_{m-1}}, \ldots, c_{\Omega_0})$. These *sum-of-product* expressions defined over $GF(2)$ can be represented using the notation given in group theory for the *permutations*, in which the upper row contains the subscripts of the coordinates of α which are multiplied by the coordinates of χ with subscripts given in the lower row [7]. It can be proved that the *permutations* so defined involve the presence of 1-*cycles* (k) and 2-*cycles* (i,j), where (k) represents a *product* term $x_k = (a_k c_k)$ and where (i,j) represents a *sum of products* $x_{ij} = (a_i c_j + a_j c_i)$. The 2-cycles (i,j) are called in group theory as *transpositions*. We can define functions $\mathbf{EC_i^m}$, $\mathbf{OC_i^m}$ and $\mathbf{MC^m}$ which give the cycles for the permutation corresponding to the values i and m. Defining the functions $\mathbf{E_i}$, $\mathbf{O_i}$ and $\mathbf{M}$ as the addition of the terms x_{ij}'s and x_k's represented by the cycles (i,j)'s and (k)'s given by $\mathbf{EC_i^m}$, $\mathbf{OC_i^m}$ and $\mathbf{MC^m}$, respectively, the following expressions [7] for the coordinates of the product δ can be given

$$d_{\Omega_i} = \mathbf{E_0} + \begin{cases} \mathbf{O_{i+1}} & i \ even \\ \mathbf{E_{i+1}} & i \ odd, \ i \neq m-1 \\ \mathbf{M} & i = m-1 \end{cases} \qquad (1)$$

where d_{Ω_i}s, with $i = 0, 1, \ldots, m-1$, are the coordinates of δ with respect to Ω. We have named the multiplication method given by equation 1 as *transpositional* because it is based on the computation of 1-cycles and 2-cycles (*transpositions*).

In Table 1 a comparison for the theoretical complexities obtained with our approach and with other methods is given, for canonical multipliers with generating AOPs. It can be observed that the *theoretical* complexities obtained using our method are equal to the lowest ones obtained using other similar approaches.

Table 1. Theoretical complexities of bit-parallel canonical basis multipliers

	#XOR	#AND	Delay
Itoh-Tsu.[2]	$m^2 + 2m$	$m^2 + 2m + 1$	$T_{AND} + \lceil log_2 m + log_2(m+2)\rceil T_{XOR}$
Hasan [4]	$m^2 + m - 2$	m^2	$T_{AND} + (m + \lceil log_2(m-1)\rceil)T_{XOR}$
Koç-Sun.[5]	$m^2 - 1$	m^2	$T_{AND} + (2 + \lceil log_2(m-1)\rceil)T_{XOR}$
Halbut.[3]	$m^2 - 1$	m^2	$T_{AND} + (1 + \lceil log_2(m-1)\rceil)T_{XOR}$
Zhang[6]	$m^2 - 1$	m^2	$T_{AND} + (1 + \lceil log_2(m-1)\rceil)T_{XOR}$
Transposit.	$m^2 - 1$	m^2	$T_{AND} + (1 + \lceil log_2(m-1)\rceil)T_{XOR}$

3   Implementations over FPGAs and CPLDs

The theoretical complexity given in Section 2 is not an exact predictor of the area consumption if reconfigurable hardware is used [8]. We have used *Xilinx Foundation F2.1i* for the implementation of canonical multipliers over FPGAs and CPLDs using our *transpositional* method and using the method given in [3].

For FPGAs, 4013XLPQ160 devices from XC4000XL family have been used. In Table 2, the experimental results obtained for the multipliers implemented using the approach given in [3] and using our method are showed. The total CLB count using transpositional method is 6.6% lower than using the other approach, whereas the total maximum combinational path delay of our method is 3.6% slower, but this is because we have performed optimization for area. The reduction of the CLB count is due to the devices used are LUT-based, and gates with smaller number of inputs can be easily included in a LUT, which increases the possibility of obtaining a better mapping solution [9]. *Transpositional* method groups the coordinates of the operands by means of the 2-cycles and the 1-cycles, therefore helping to the mapping tool to reduce the CLB count.

Table 2. Experimental results for FPGA implementations

	Halbutogu.& Koç		Transpositional	
	CLBs	Max.Path(ns)	CLBs	Max.Path(ns)
$GF(2^4)$	6	14.7	5	14.8
$GF(2^{10})$	40	25.0	37	25.8
$GF(2^{12})$	54	27.3	53	28.6
$GF(2^{18})$	116	33.0	106	36.8
$GF(2^{28})$	281	45.3	259	43.0
$GF(2^{36})$	454	51.7	428	55.1
Total	951	197.0	888	204.1

For CPLDs, XC95288XV devices from XC9500XV family have been used. In Table 3, the experimental results obtained for the multipliers implemented using the method given in [3] and using the *transpositional* method are showed. The total MC count using our method is 12.8% lower than using the other approach, and for the total delay, the transpositional method is 29.7% faster.

Table 3. Experimental results for CPLD implementations

	Halbutogu.& Koç		Transpositional	
	MCs	T_{PD}(ns)	MCs	T_{PD}(ns)
$GF(2^4)$	10	9.8	9	9.1
$GF(2^{10})$	55	17.3	57	15.9
$GF(2^{12})$	89	24.1	79	20.7
$GF(2^{18})$	213	38.4	175	17.3
Total	367	89.6	320	63.0

Experimental results given in Tables 2 and 3 seem demonstrate, therefore, that the *grouping* technique provided by the *transpositional* approach is a good method when *any* reconfigurable platform is used for the implementation.

4   Conclusions

A new *transpositional* method for canonical basis multiplication over finite fields $GF(2^m)$ generated by irreducible AOPs has been presented. The theoretical complexities of the bit-parallel multipliers constructed using our method are equal to the lowest ones found in the literature, but the FPGA and CPLD implementations of multipliers using our transpositional approach lead to a lower count of CLBs and MCs, respectively, than using other similar approaches.

References

1. Menezes, A.J. (ed.): Applications of Finite Fields. Kluwer Academic (1993)
2. Itoh, T., Tsujii, S.: Structure of Parallel Multipliers for a Class of Finite Fields $GF(2^m)$. Information and Computation, Vol.83 (1989) 21–40
3. Halbutogullari, A., Koç, Ç.K.: Mastrovito Multiplier for General Irreducible Polynomials. IEEE Trans. Computers, Vol.49, No.5 (2000) 503–518
4. Hasan, M.A., Wang, M.Z., Bhargava, V.K.: Modular Construction of Low Complexity Parallel Multipliers for a Class of Finite Fields $GF(2^m)$. IEEE Trans. Computers, Vol.41, No.8 (1992) 962–971
5. Koç, Ç.K., Sunar, B.: Low-Complexity Bit-Parallel Canonical and Normal Basis Multipliers for a class of Finite Fields. IEEE Trans. Computers, Vol.47, No.3 (1998) 353–356
6. Zhang, T., Parhi, K.K.: Systematic Design of Original and Modified Mastrovito Multipliers for General Irreducible polynomials. IEEE Trans. Computers, Vol.50, No.7 (2001) 734–749
7. Imaña, J.L., Sánchez, J.M., Fernández, M.: Método de multiplicación canónica sobre campos $GF(2^m)$ generados por AOPs orientado a hardware reconfigurable. II Jornadas sobre Computación Reconfigurable y Aplicaciones, JCRA (2002) 215–220
8. Imaña, J.L.: Bit-Parallel Arithmetic Implementations over Finite Fields $GF(2^m)$ with Reconfigurable Hardware. Acta Applicandae Mathematicae, Vol.73, No.3. Kluwer Academic Publishers (2002) 337–356
9. Kao, C.C., Lai, Y.T.: A Routability Driven Technology Mapping Algorithm for LUT Based FPGA Designs. IEICE Trans. Fundamentals, Vol.E84-A, No.11 (2001) 2690–2696

A Study on the Design of Floating-Point Functions in FPGAs

Fernando E. Ortiz[1], John R. Humphrey[1], James P. Durbano[2], and Dennis W. Prather[1]

[1] University of Delaware, 140 Evans Hall, Newark, DE 19716
{ortiz, humphrey, dprather}@ee.udel.edu
[2] EM Photonics, Inc. 102 East Main St., Newark, DE 19711
durbano@emphotonics.com

Abstract. Floating-Point Operations represent a common task in a variety of applications, but such operations often result in a bottleneck, due to the large number of machine cycles required to compute them. Even though the FPGA community has developed advanced algorithms to improve the speed of FLOPs, floating-point transcendental functions are still underdeveloped. In this paper, we discuss some of the tradeoffs faced when implementing floating-point functions in FPGAs. These techniques, including lookup tables, and CORDIC algorithms, have been used in the past for the implementation of fixed-point analytic functions. This paper seeks to apply those methods to floating-point functions. The implementation results from different versions of a floating-point sine function are summarized in terms of speed, area, and accuracy to understand the effect of different architectural alternatives.

1 Introduction

Nearly every modern microprocessor-based system is capable of processing non-integer values using floating-point methods to represent real numbers. In addition, floating-point operations (FLOPs) represent a common task in modern applications, ranging from 3D graphics, to simulations. To increase performance, floating-point operations have been highly researched and optimized for use in FPGAs. In comparison, more advanced operations, such as trigonometric functions, have not received as much attention. Although trigonometric functions have been realized in FPGAs, these implementations typically have been based on fixed-point, rather than floating-point, formats. As we will show, the advantages of the techniques used in fixed-point do not apply directly to floating-point designs, and therefore novel schemes are required. In this paper, we examine the various tradeoffs associated with developing floating-point trigonometric functions in FPGAs, using one of two general schemes: adapt the well known fixed-point methods to accept and produce floating-point numbers, or use techniques that use floating-point arithmetic to approximate arbitrary functions.

P.Y.K. Cheung et al. (Eds.): FPL 2003, LNCS 2778, pp. 1131–1134, 2003.

2   Design Alternatives

In this section, we discuss several alternatives involved in implementing trigonometric functions in reconfigurable hardware. Even though there exists a considerable amount of published work in this area [1-5], it is all based on fixed-point formats. First, we will show that the advantages of the techniques used in fixed-point arithmetic do not apply directly to floating-point designs (and then propose modifications to these methods). Then, techniques for implementing other well known functional approximation techniques in hardware will be presented, using only floating-point arithmetic throughout the computations.

2.1   Adapting the Fixed-Point Approximation Methods to Floating-Point Arithmetic

This section will study the feasibility of modifying the currently available (fixed-point) function approximation techniques that will enable them to accept and produce floating-point numbers. Every fixed-point implementation of a function can be trivially turned into floating-point one by the use of conversion modules between fixed-point and floating point-numbers at the input and output ports, as shown in figure 2. However, this alternative should be avoided because of the high extra latency that results from the addition of these modules, typically on the order of five clock cycles (each) for fully pipelined, high-speed versions [6].

LUT. The LUT method can be extended for use in floating-point systems with two steps. First, the addition of an FP2Int function that maps the floating-point input into an integer that can be used for the table lookup. The second modification applies to the values stored in the table; these should be stored in floating-point notation, eliminating the need for an output conversion module, but increasing the memory usage, due to redundant exponents.

Bipartite LUT. This method is unsuitable for floating-point conversion because it is based on a second order Taylor interpolation around a point derived from the decomposition of the fixed-point input argument. This technique cannot be applied for floating-point numbers, since the exponent bits cannot be interpreted in the same way as mantissa bits and the sign bit.

CORDIC. The CORDIC algorithm can be executed using floating-point numbers, but doing so would result in large inefficiencies. The simplicity of these operations remains no longer true for floating-point systems, particularly in the case of floating-point addition, where multiple variable-length shifters are required. The only comparative advantage of using CORDIC methods to implement floating-point functions in hardware is that the additional latency introduced by the floating-point conversions is small compared with the intrinsic latency of high-precision CORDIC units.

2.2 Floating-Point Approximation Methods

As was shown, the use of the fixed-point solutions for floating-point problems results in high latencies and large resource usage. In this section, we propose the use of interpolation techniques with small LUTs embedded in the hardware and computations being carried out in floating-point arithmetic.

Linear Interpolation. This method uses a small set of known values of the function and extrapolates to the complete interval of interest using a straight line approximation

In order to implement linear interpolation in hardware, three steps are required. First, the input is converted to an integer (n) that will serve as an index for the lookup. Then, the actual table lookup takes place and the values of $f(x)$, $f'(x)$ and $n\Delta x$ are output. Finally, the interpolation takes place using two floating-point adders and two floating-point multipliers (see Figure 1). Note that the $f'(x_0)$ and $n\Delta x$ can be calculated instead of stored in the table, and doing so reduces the memory usage significantly, but also adds drastically to the latency.

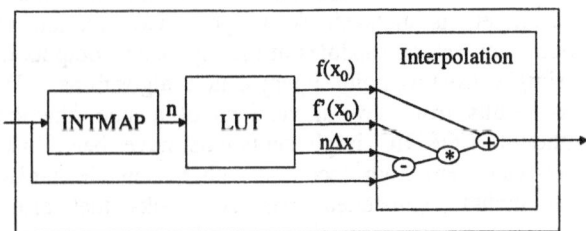

Fig. 1. Hardware Implementation of a Floating-Point Linear Interpolator

3 Example: Floating Point Sine

The previous sections presented several alternatives available for implementing floating-point functions in hardware. In this section, we will present the design of a floating-point sine function, used in a computationally intensive electromagnetic simulation algorithm, where the high speed and low latency are of primary importance. Low error is considered in the cost function, with a smaller weight factor.

Periodic functions require additional argument reduction, to force the input to be in a smaller interval where the function is defined. This interval is $[0, 2\pi]$ in the case of the sine function. The alternatives to consider were a full LUT, the CORDIC algorithm, and linear interpolation. Because of the symmetry of the function, only one quarter of a wavelength needs to be stored in the LUT. The remainder of the values can be determined with the appropriate sign and phase changes. This technique reduces the size of the table by 75% without any loss in precision. To better understand the impact of these factors in hardware, several of these choices were implemented in a Xilinx Virtex-II 6000-4 FPGA. The results are provided in Table 1.

1134   Fernando E. Ortiz et al.

Table 1. Implementation Results of Floating-Point Sine. This table compares the results of several implementation alternatives of a sine function. Sine Output = 32-bit, IEEE 754 floating-point number

Comp. Method	Area(Slices+BRAM)	Maximum Error(%)	Latency	Speed (MHz)
CORDIC-8bit	428	2.45	20	137.6
CORDIC-16 bit	889	1.00E-02	28	121
LUT-8bits-no Interp	222	2.45	7	226.2
LUT-16bits-no Interp	183+32	1.00E-02	7	177.8
LUT-8 bits-Interp	1431	7.53E-05	18	104.5

4   Conclusion

Several design alternatives exist when implementing floating-point functions in FPGA-based systems. In this paper, we have discussed these approaches and presented the relative tradeoffs involved with each. We showed that the traditional solutions, such as CORDIC and bipartite tables, do not offer the best solution when floating-point arithmetic is intended. We proposed two solutions to this problem: (1) Use floating-point conversion modules at the inputs and outputs of these algorithms and (2) use simple floating-point interpolation algorithms. We then presented implementation results from several versions of a sine function to quantify our analysis. Although CORDIC implementations have been historically preferred because they provide very high accuracy with reasonable hardware requirements, current FPGAs include embedded memory blocks that afford accurate LUT implementations with significantly higher speeds and reduced latencies. CORDIC algorithms should be used when the memory blocks are unavailable, if the required precision makes the size of the LUT impractical, or reduced-area requirements outweigh the long latency. Ultimately, the specific requirements of the end application dictate the architecture that provides the best complement in terms of area, speed, and precision.

References

[1] J. Volder, "The CORDIC Trigonometric Computing Engine," *IRE Transactions on Electronic Computers*, vol. EC-8, pp. 330-334, 1959.
[2] J. Duprat and J. M. Muller, "The CORDIC algorithm: new results for fast VLSI implementation," *IEEE Transactions on Computers*, vol. 42, pp. 168-178, 1993.
[3] D. S. Phatak, "Double Step Branching CORDIC : A New Algorithm for Fast Sine and Cosine Generation," *IEEE Transactions on Computers*, vol. 47, pp. 587-602, 1998.
[4] R. Andraka, "A survey of CORDIC algorithms for FPGA based computers," presented at ACM International Symposium on Field-Programmable Gate Arrays, Monterey, California, 1998.
[5] T. Vladimirova and H. Tiggeler, "FPGA Implementation of Sine and Cosine Generators Using the CORDIC Algorithm," presented at Military and Aerospace Applications of Programmable Devices and Technologies, Laurel, Maryland, 1999.
[6] P. Belanovic, "Library of Parameterized Hardware Modules for Floating-Point Arithmetic with An Example Application," in *Electrical and Computer Engineering*. Boston, Massachusetts: Northeastern University, 2002, pp. 83.

Design and Implementation
of RNS-Based Adaptive Filters

Javier Ramírez[1], Uwe Meyer-Bäse[2], Antonio García[1], and Antonio Lloris[1]

[1] Department of Electronics and Computer Technology
University of Granada
javierrp@ugr.es {agarcia,lloris}@ditec.ugr.es

[2] Department of Electrical and Computer Engineering
FAMU-FSU College of Engineering
umb@eng.fsu.edu

Abstract. This paper presents the residue number system (RNS) implementation of reduced complexity and high performance adaptive FIR filters on Altera APEX20K field-programmable logic (FPL) devices. Index arithmetic over Galois fields along with a selection of a small wordwidth modulus set are keys for attaining low-complexity and high-throughput. The replacement of a classical modulo adder tree by a binary adder with extended precision followed by a single modulo reduction stage improved area requirements by 10% for a 32-tap FIR filter. A block LMS (BLMS) implementation was preferred for the update of the adaptive FIR filter coefficients. RNS-FPL merged filters demonstrated its superiority when compared to 2C (two's complement) filters, being about 65% faster and requiring fewer logic elements for most study cases.

1 Introduction

Residue number system (RNS) based [1, 2] digital filter designs can be effective for realizing high speed sum-of-products kernels. One of the main properties of the RNS is the inherent modularity which induces efficient implementations and excellent levels of performance. These characteristics hold for a variety of technologies, from cell-based integrated circuits (CBIC) [3] to field-programmable logic (FPL) [4].

FPGAs have intrinsically weak arithmetic capabilities when compared to ASICs. In addition, FPL deficiencies increase geometrically with precision, as a result of architectural limitations. FPL device families, such as Altera FLEX10K or Xilinx Virtex, are organized in channels (typically 8-bits wide) with local short delay propagation paths; also dedicated memory blocks to synthesize small RAM and ROM functions are included. Performance rapidly suffers when carry bits and/or data have to propagate across a channel boundary. We call this the *channel barrier* problem [4]. New trends in FPL device design tend to add dedicated hardware for multiplication to the new FPL device families, as has happened with Altera APEX II and Virtex II.

P.Y.K. Cheung et al. (Eds.): FPL 2003, LNCS 2778, pp. 1135–1138, 2003.
© Springer-Verlag Berlin Heidelberg 2003

An alternative design paradigm is advocated in this paper with the development of efficient structures for RNS-based adaptive FIR filters. The RNS advantage is gained by reducing arithmetic to a set of concurrent operations that reside within small wordlength non-communicating channels. This attribute makes the RNS potentially attractive for implementing these systems with FPL technology.

2 RNS Background

In the RNS, numbers are represented in terms of a relatively prime basis set (moduli set) $P=\{m_1, \ldots m_L\}$. Any number $X \in Z_M=\{0, \ldots, M-1\}$, where $M=\Pi m_i$, has a unique RNS representation $X \leftrightarrow \{X_1, \ldots, X_L\}$, where $X_i=X \bmod(m_i)$. RNS arithmetic is defined modulo M by pair-wise modular operations:

$$Z = X \pm Y \leftrightarrow \left\{ \langle X_1 \pm Y_1 \rangle_1, \ldots, \langle X_L \pm Y_L \rangle_L \right\} \tag{1}$$

$$Z = X \times Y \leftrightarrow \left\{ \langle X_1 \times Y_1 \rangle_1, \ldots, \langle X_L \times Y_L \rangle_L \right\}$$

where $<Q>_j$ denotes $Q \bmod(m_j)$. It is the ability of the RNS to do arithmetic within independent small wordlength channels that makes it particularly attractive for FPL insertion, as has been referred in the literature [5-7].

On the other hand, index arithmetic [2] constitutes an efficient means for enhancing and reducing the complexity of RNS-based DSP applications. All the non-zero elements in a Galois field can be generated exponentiating a primitive element, denoted g_l. Thus, multiplication in GF(m_l) can be implemented as:

$$\left| q_1 q_2 \right|_{m_l} = g_l^{\left| i_1 + i_2 \right|_{m_l - 1}} \qquad l = 1, 2, \ldots, L \tag{2}$$

where $q_1, q_2 \in \{1, \ldots m_l\}$ and $i_1, i_2 \in \{0, \ldots m_l-1\}$ are the indexes of q_1 and q_2, respectively. This multiplication scheme just requires LUTs for index computation, a modulo m_l-1 adder for index addition and a LUT for the inverse index transformation.

3 Design of RNS-Based Adaptive FIR Filter

An adaptive filter processes a digital input $x(n)$, obtaining an output sequence $y(n)$ through adjustable parameters whose values affect how $y(n)$ is computed. The output is compared to a second signal $d(n)$, called the *desired response* signal, thus getting the *error signal* $e(n)= d(n)-y(n)$ This is used to adapt the parameters of the filter, so the output matches the desired response signal, i.e., the magnitude of $e(n)$ must decrease with time. The LMS algorithm is simple to implement [8, 9] and powerful enough to evaluate the practical benefits of adaptation. It only requires the error signal, the input signal vector and a step size μ for adjusting the coefficients:

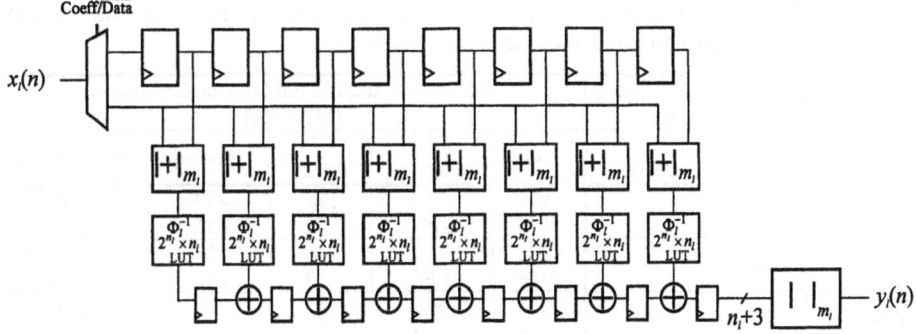

Fig. 1. Index-based Adaptive FIR filter design.

$$W(n+1) = W(n) + \mu e(n)x(n) \tag{3}$$

LMS computation requires multiplications and additions to be implemented, nearly in the same number as that of the FIR filter structure with fixed coefficient values, which is one of the reasons for its popularity. Also block (BLMS) implementations are possible, thus reducing further its computational costs.

An RNS-based BLMS adaptive FIR filter has been developed. The internal modular processing engine is intended to work externally in a 2C format in order to communicate with the LMS block that performs the FIR coefficient update. Efficient block decomposition 2C-to-index [3, 4] and ε-CRT-based [10] or CRT-based RNS-to-2C conversions are used within the modular processing engine, while Fig. 1 shows the structure of one of the L internal index-based channels for an 8-tap FIR filter example. The channel accepts index inputs, so coefficient multiplication is implemented with modulo m_i-1 adders, while the filter product summation is efficiently implemented with an enhanced modulo m_i adder chain consisting of conventional adders (with precision extension) and a modulo m_i reduction stage.

Design examples were implemented using Altera APEX20K for both 2C arithmetic and RNS. Table 1 compares 2C FIR filters ranging from 8 to 32 taps with the filters proposed in this paper. Input signal and coefficients are 16-bit wide and, for the 2C design, the 16×16-bit multipliers are designed with five pipeline stages. The table shows the number of taps (N), the system dynamic range (W), the number of LEs, the maximum frequency and the modulus set used for the RNS study cases.

5   Conclusions

This paper has shown the benefits provided by the RNS for the implementation of adaptive FIR filters using FPL devices, with a BLMS structure for the update of the adaptive FIR filter coefficients that reduces the computational load while retaining good convergence properties. The proposed RNS filters are about 65% faster than 2C designs and require fewer logic elements in most cases. Concretely, complexity is reduced up to 13% when ε-CRT converters are used.

Table 1. Resource reduction and speed-up achieved by an adaptive FIR filter built in RNS-FPL technology when compared to the equivalent 2C system.

N	W	2C Adaptive FIR Filter Implementation		RNS-based Adaptive FIR filter implementation CRT/ε-CRT				
		LEs	F (MHz)	Les	Resource reduction	F (MHz)	Speed-up	Modulus set
8	34	4255	91	5525/ 4276	-33% -1%	130/ 149	43% 64%	11,13,17,19, 23,29,31,32
16	35	8121	84	9070/ 7348	-12% 10%	128/ 137	52% 63%	11,13,17,19, 23,29,31,64
24	36	12102	80	13258/ 11009	-9% 10%	122/ 133	53% 66%	7,11,13,17, 19,23,29,31,32
32	36	16104	79	16637/ 14140	-3% 13%	119/ 131	51% 66%	7,11,13,17, 19,23,29,31,32

Acknowledgements

Part of this work was supported by Ministerio de Ciencia y Tecnología (Spain) under project TIC2002-02227. CAD tools were provided by Altera Corp., San Jose CA.

References

1. Szabo, N. S., Tanaka, R. I.: Residue Arithmetic and its Applications to Computer Technology. McGraw-Hill, New York, 1967.
2. Soderstrand, M., Jenkins, W., Jullien, G. A., Taylor, F. J.: Residue Number System Arithmetic: Modern Applications in Digital Signal Processing. IEEE Press, 1986.
3. Ramírez, J., Meyer-Bäse, U., Taylor, F., García, A., Lloris, A.: "Design and Implementation of High-Performance RNS Wavelet Processors Using Custom IC Technologies", Journal of VLSI Signal Processing, vol. 34, pp. 227-237, 2003.
4. Ramírez, J, García, A., Meyer-Bäse, U., Lloris, A.: "Fast RNS-based FPL Communications Receiver Design and Implementation", Lecture Notes in Computer Science, vol. 2438, pp. 472-481, 2002.
5. Hamann, V., Sprachmann, M.: "Fast Residual Arithmetic with FPGAs," Workshop on Design Methodologies for Microelectronics, 1995.
6. Safiri, H., Ahamadi, H., Jullien, G., Dimitrov, V.: "Design of FPGA Implementation of Systolic FIR Filters Using Fermat Number ALU," Asilomar Conference on Signals, Systems and Computers, Pacific Grove, 1997.
7. Meyer-Bäse, U., García, A., Taylor, F.: "Implementation of a Communications Channelizer Using FPGAs and RNS Arithmetic," Journal of VLSI Signal Processing, vol. 28, no. 1/2, pp. 115-128, 2001.
8. Jenkins, W. K., Schnaufer, B. A.: "Fault tolerant adaptive filters based on the block LMS algorithm", 1993 IEEE International Symposium on Circuits and Systems, vol.1, pp. 862-865, 1993.
9. Liu, C. M., Jen, C. W.: "A parallel adaptive algorithm for moving target detection and its VLSI array realization", IEEE Transactions on Signal Processing, vol. 40, no. 11 , pp. 2841-2848, 1992.
10. Griffin, M., Sousa, M., Taylor, F.: "Efficient Scaling in the Residue Number System", 1989 International Conference on Acoustics, Speech and Signal Processing, pp. 1075-1078, 1989.

Domain-Specific Reconfigurable Array
for Distributed Arithmetic

Sami Khawam[1], Tughrul Arslan[1,2], and Fred Westall[3]

[1] School of Electronic and Engineering, The University of Edinburgh, KB,
Mayfield Road, Edinburgh EH9 3JL, UK,
S.Khawam@ee.ed.ac.uk
[2] Institute for System Level Integration, Livingston, EH54 7EG, UK
[3] EPSON Scotland Design Centre, Livingston, EH54 7EG, UK

Abstract. Distributed Arithmetic techniques are widely used to implement Sum-of-Products computations such as calculations found in multimedia applications like FIR filtering and Discrete Cosine Transform. This paper presents a flexible, low-power and high throughput array for implementing distributed arithmetic computations. Flexibility is achieved by using an array of elements arranged in an interconnect mesh similar to those employed in conventional FPGA architectures. We provide results which demonstrate a significant reduction in power consumption in addition to improvements in timing and area over standard FPGA architectures.

Keywords: Embedded reconfigurable array, programmable, distributed arithmetic, FPGA, domain specific, low-power.

1 Introduction

The arrival of portable devices processing audio and video data endorses the need for solutions providing high-speed and low-power consumption for implementing the compute-intensive multimedia calculations. Hardwired implementations of such algorithms are not suitable, as a margin of flexibility is required due to the constantly changing algorithms and DSPs provide low-throughput and high power-consumption. In the past years reconfigurable hardware has emerged as a low-cost and flexible solution for high-throughput custom hardware at the cost of increased power consumption and area.

As reported earlier in [1], a reconfigurable array specific to one type of calculation provides a good compromise between flexibility, power-consumption, area and performance when compared to DSPs, FPGAs and hardwired solutions. This paper presents a reconfigurable architecture specific to computations that can be implemented in Distributed Algorithms [2]; this includes computations such as DCT and FIR filtering used in video and audio systems.

Previous domain-specific and coarse-grain reconfigurable architectures are more processor based; e.g. [3] provides simple programmable processors interconnected together for the execution of complex algorithms. In [4] the datapath of a processor can be reconfigured during run-time to adapt to calculations. The architecture proposed in this paper is based on a heterogeneous array with a mesh of interconnects

P.Y.K. Cheung et al. (Eds.): FPL 2003, LNCS 2778, pp. 1139–1144, 2003.

that is able to provide more parallel computations and a higher throughput at a lower frequency. Previous programmable and configurable architectures for DCT such as the one presented in [5] provide limited flexibility in the wide range of possible implementations that could be suitable. By using FPGA-style interconnects and elements we can provide greater flexibility at a lower-level.

The paper is organized as follows: In section 0 the algorithms to be supported are overviewed. Section 0 describes the reconfigurable system and the proposed array, and in section 0 the performance of the proposed array is assessed.

2   Target Algorithms

Distributed Arithmetic (DA) [2] is a technique used to compute the inner product of two vectors, where one of the vectors is a constant. Multiplications by fixed coefficients are replaced by ROM tables and shift-accumulate operations. The basic DA scheme has shift-registers to convert the N parallel input coefficients into bit-serial data. The bits at the output of the N shift-registers are combined to form the N bits wide address for the ROM table. N different ROM tables are provided. The output coefficients are found by shift-accumulation of the output of the ROM tables (see Fig. 3 below).

Discrete Cosine Transform (DCT) [7] is an algorithm used in many compression standards like MPEG and is suitable to be implemented using DA. A number of DA implementations of DCT exists, each having different features and compromises [6]. The array presented in this paper is flexible enough to support a number of DCT implementations with features such as: CORDIC based DCT [8], digit-serial implementations [9], memory reduction using the odd-even decomposition [9], DCT size and precision change.

3   Reconfigurable System

The authors recently introduced a System-on-Chip (SoC) architecture compromising domain-specific reconfigurable arrays in [1]. A number of configurable arrays can be embedded in the system, each specific to a computation, such as Motion Estimation or DCT. The arrays are configured dynamically by the processor or the DSP. The input data to the arrays is fed either by the DSP or the processor and the output is read back from the array. The elements of the array are described below.

3.1   Clusters for Distributed Arithmetic

A general DA implementation of a Sum-of-Product requires the following elements:
- Shift registers to convert bit-parallel input coefficient to bit-serial or digit-serial data.
- Memory elements to store the content of the ROM replacing the multiplication.
- Shift-accumulators for calculating the result.

Additionally, adders and subtracters are needed to implement techniques such as the odd-even DCT decomposition. It was thus chosen to use two types of elements in the array: A memory element and an element for add/sub/shift and accumulation, as detailed below. A cluster is formed by combining four such elements together using configurable switches. The clusters are arranged in an array as shown in Fig. 1. More add-shift clusters are used than memory clusters, due to the needs of the application. The columns are arranged uniformly to simplify manual routing and placement.

3.1.1 Memory Element

In this initial architecture the memory element used is a dual-port 1-Kbit RAM. Four such elements are packed into a *memory cluster*. The cluster contains logic, similar to the one found in [10], allowing to configure the memory in a number of geometries as shown in Table 1. The size of the memory was chosen according to the most used memory sizes in DCT calculations.

Table 1. Possible geometries of a memory cluster

Word Size	Bits per word			
	4-bit	8-bit	12-bit	16-bit
256	x	x	x	x
512	x	x		
768	x			
1024	x			

3.1.2 Add and Shift Element

To support most DA calculations, the add-and-shift element can be configured to act as the following:

- Adder/Subtractor
- Loadable shift-register useful for parallel-to-serial conversion. Right and left shifts supported.
- Accumulator. The adder in the accumulator can be dynamically configured to subtract.
- Shift-accumulator to be used at the ROM table output in DA calculations.

The elements have a programmable register at the output that can be enabled to support pipelined implementations. Every element is 4-bits wide; four elements are grouped into a cluster with interconnects provided to support cascading in order to allow wider bit ranges (up to 16-bit).

3.2 Mesh Interconnects

As described earlier in [1], symmetrical mesh interconnects are used to provide the connections between the clusters. Two types of tracks are provided: Six 8-bit wide tracks for data and six 1-bit tracks for control lines. Interconnects are composed of connection-boxes (C-Boxes) that connect the pins of a cluster to the tracks and switch-boxes (S-Boxes) that connect together the intersections of tracks. As in [1], the C-Boxes have a flexibility of Fc=6 and the S-Boxes of Fs=3, as defined in [11].

The configurable switches are implemented using tri-state buffers, which greatly increases the area, timing and power consumption of the array [12], when compared to using pass-transistors. Using tri-state buffers makes the architectures designed synthesizable, portable to any process and compliant with the design-flow used for the rest of the SoC

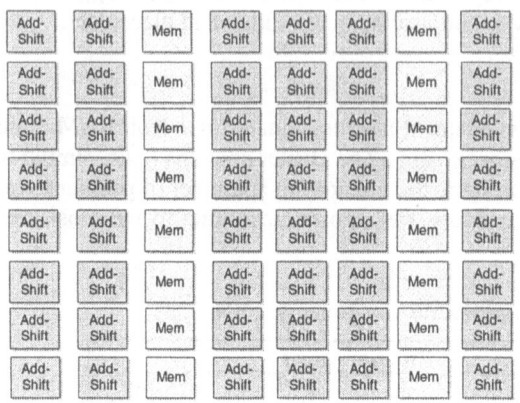

Fig. 1. Clusters arrangement in the array. More add-shift clusters are used due to their need.

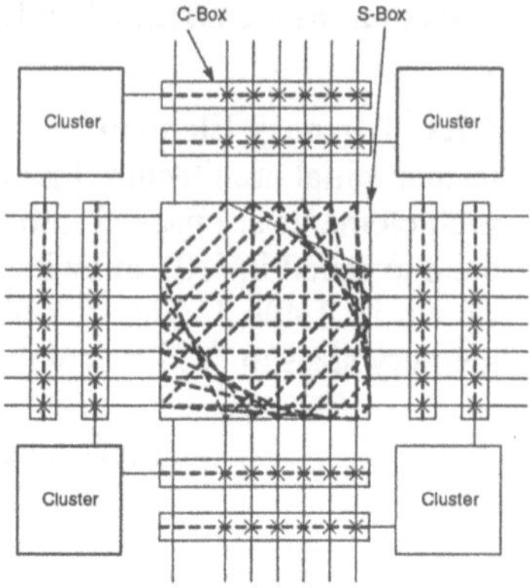

Fig. 2. Connection Boxes and Switch Boxes connect the clusters together [11].

4   Array Performance

The benchmark circuit used is a simple 8-point 1-D DCT using bit-serial DA without memory compression, as shown in Fig. 3. This circuit is manually mapped to the array as follows:

- Each 12-bits shift register is mapped to three add-and-shift elements.
- Each 2-Kbit memory is mapped to two 1-Kbit memories preprogrammed and used as ROMs.
- Each 16-bit shift accumulator is mapped to four add-shift modules from one cluster.

By implementing this benchmark circuit, *our array* was compared to a standard hardwired ASIC specially optimized for DCT and a commercial Xilinx Virtex-E FPGA. The measured power consumption, area and maximum frequency is shown in Table 2. All of these systems use .18µm CMOS technology and run at 1.8V and 10MHz.

The area of the Virtex-E and our implementations does not include the area used by the configuration memory, but includes the area used by interconnects and reconfigurable switches. The Xilinx area estimation is based on the assumption that a *slice* and its belonging C- and S-boxes have an area of 3303 µm^2 ; 71 slices are needed to implement one row. It can be seen that our implementation is 14% smaller than the Virtex-E area, however, this stays significantly larger than the hardwired implementation. This is partly caused by the fact that in the ASIC implementation no

RAMs are used for storing the coefficient but hardwired logic is used to implement the coefficient ROM tables, which greatly decreases the area and limits the flexibility.

The power consumption values measured for our array and for the hardwired implementation are obtained using post-routing simulation with typical switching activity. In the case of the Virtex-E FPGA, the power consumption is obtained with typical estimations provided by Xilinx. In both array cases, the power values include the power consumed by the configuration circuit. Our array consumes 38% less power than the Xilinx implementation since it has less interconnects and operates using larger clusters. Our reconfigurable array consumes 277% more power than ASIC due to the added switches, and also partly to the use of RAM over hardwired-LUTs.

With respect to timing, our array has a maximum frequency around 54% higher than that of Virtex-E., This is still 63% less than the maximum frequency achievable with hardwired ASIC due to the delays added in reconfigurable switches and to the higher-loads and longer routing.

Table 2. Performance comparison between hardwired ASIC, our array and a commercial Xilinx FPGA on one row of the array.

	.18μm ASIC	*Our array*	Xilinx's Virtex-E
Area (μm²)	17 483	202 366	234 510
Power cons. (mW)	0.52	1.965	3.2
Max Freq. (MHz)	210	77	50

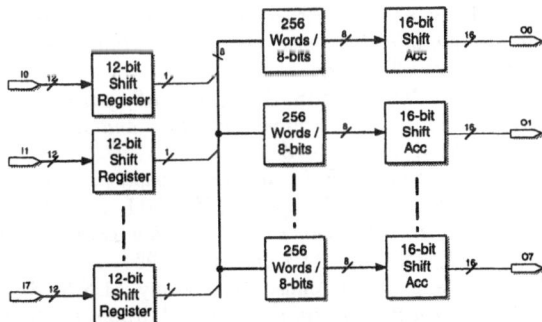

Fig. 3. 1-D DCT implemented using DA. For 8-points DCT 8 rows or elements are required.

When comparing the DCT implementation on our array to that on Virtex-E, it should be noted that the routing was done manually in the case of our array, while it was done with automatic software on Virtex-E, meaning that the performance of our array could have been more optimized if software was used for generating the routing and placement.

5 Conclusion

In this paper, we have introduced an embedded reconfigurable array targeting Distributed Arithmetic (DA) operations. The architecture is based on programmable clusters of add-shift and memory elements arranged in an array. Different levels of reconfigurable interconnects are provided to allow flexible mapping of diverse DA algorithms, such as a wide range of DCT computations, to the array.

The performance measured shows that DA implementations on the proposed architecture provide considerable improvements over standard low-level FPGAs ones: Power consumption is reduced by 38%, occupied area is decreased by 14% and the maximum operating frequency is increased by 54%.

When comparing the array with standard ASIC implementations, it becomes clear that this architecture provides a compromise between hardwired ASIC and generic FPGA solutions in terms of flexibility, area, timing and power consumption when used in portable multimedia devices.

References

1. S. Khawam, T. Arslan, F. Westall, *Embedded reconfigurable array targeting motion estimation applications*, 2003 IEEE International Symposium on Circuits and Systems (ISCAS 2003)
2. White, S.A, *Applications of distributed arithmetic to digital signal processing: a tutorial review*, ASSP Magazine, IEEE, Volume: 6 Issue: 3 , Jul 1989, Page(s): 4 -19
3. A. Abnous, J. M. Rabaey. *Ultra-low-power domain-specific multimedia processors.* IEEE VLSI Signal Processing. 1996
4. D. C. Cronquist, P. Franklin C. Fisher M. Figueroa and C. Ebeling. *Architecture Design of Reconfigurable Pipelined Datapaths*, 20th Anniversary Confe. on Advanced Research in VLSI, 1999
5. Burleson, W.; Jain, P.; Venkatraman, S., *Dynamically parameterized architectures for power-aware video coding: motion estimation and DCT*, 2nd International Workshop on DVC 2001, Pages: 4- 12
6. Sungwook Yu; Swartzlander, E.E., Jr., *DCT implementation with distributed arithmetic*, IEEE Transactions on Computers , Vol. 50 Issue 9 , Sept. 2001
7. N. Ahmed, T. Natarajan, K.R. Rao, *Discrete Cosine Transform*, IEEE Trans. On Computers, Vol. C-23, No. 1, pp.90-93, December 1984
8. Yi Yang; Chunyan Wang; Omair Ahmad, M.; Swamy, M.N.S., *An on-line CORDIC based 2-D IDCT implementation using distributed arithmetic*, Sixth ISSPA 2001 , Volume: 1
9. Kyeounsoo Kim; Jong-Seog Koh, *An area efficient DCT architecture for MPEG-2 video encoder* ,Consumer Electronics, IEEE Transactions on , vol. 45 Issue: 1 , Feb. 1999
10. Wilton, S.J.E.; *Embedded memory in FPGAs: recent research results*, Communications, Computers and Signal Processing, 1999 IEEE Pacific Rim Conference on , 1999 , Page(s): 292 -296
11. Rose J., Brown S., *Flexibility of interconnection structures for field-programmable gate arrays*, Solid-State Circuits, IEEE Journal of , Vol.26, Iss.3, 1990, Pages: 277-282
12. V. George, H. Zhang J. Rabaye. *The design of low energy FPGA*, Proceedings. 1999 International Symposium on Low Power Electronics and Design, pp. 188-193. 1999

Design and Implementation of Priority Queuing Mechanism on FPGA Using Concurrent Periodic EFSMs and Parametric Model Checking

Tomoya Kitani[1], Yoshifumi Takamoto[1], Isao Naka[2], Keiichi Yasumoto[3], Akio Nakata[1], and Teruo Higashino[1]

[1] Graduate School of Information Science and Technology, Osaka University
{t-kitani,takamoto,nakata,higashino}@ist.osaka-u.ac.jp
[2] Dept. of Tourism, Osaka Seikei University naka@osaka-seikei.ac.jp
[3] Graduate School of Information Science, Nara Institute of Science and Technology
yasumoto@is.aist-nara.ac.jp

Abstract. In this paper, we propose a design and implementation method for priority queuing mechanisms on FPGAs. First, we describe behavior of WFQ (weighted fair queuing) with several parameters in a model called *concurrent periodic EFSMs*. Then, we derive a parameter condition for the concurrent EFSMs to execute their transitions without deadlocks in the specified time period repeatedly under the specified temporal constraints, using parametric model checking technique. From the derived parameter condition, we can decide adequate parameter values satisfying the condition, considering total costs of components. Based on the proposed method, high-reliable and high-performance WFQ circuits for gigabit networks can be synthesized on FPGAs.

1 Introduction

Due to recent progress of IP telephony and video/audio streaming systems, it has been very important to provide QoS (Quality of Service) in wide area networks.

In typical situations, it is required for high-end routers for backbones to process up to 10 Gbps traffic, while SOHO routers at most 100 Mbps traffic at cheaper cost. In order to reduce development costs in hardware implementation, it is desirable to use the same architecture for both high-end and SOHO routers, and to synthesize circuits with the specified performance only by adjusting parameters such as CPU/memory speed, circuit size, etc.

In this paper, we propose a flexible and reliable hardware design and implementation method using concurrent periodic EFSMs [3] and parametric model checking [4].

2 High-Reliable Design and Implementation Method

In the proposed method, we design and implement hardware circuits as follows: (1) describe behavior of a target system with several parameters in concurrent periodic EFSMs; (2) derive parameter conditions for deadlock freeness in the system using a parametric model checking technique; (3) derive a scheduler that allows all EFSMs to

P.Y.K. Cheung et al. (Eds.): FPL 2003, LNCS 2778, pp. 1145–1148, 2003.

execute only schedulable paths (satisfying time constraints, synchronization conditions, and so on) by assigning appropriate values to parameters; (4) derive VHDL-description which correspond the each EFSM and the scheduler module; and (5) implement the VHDL-description on FPGA with a commercial tool.

Concurrent Periodic EFSMs: EFSM is an extended FSM which has registers to deal with variables. Each transition rule is defined as $s_{cur} \xrightarrow{a[guard]} s_{next}$. Here, $guard$ is a transition condition. If the value of the transition condition $guard$ is true at state s_{cur} and event a is executed, then the EFSM moves to state s_{next}. By using time variables in transition conditions, we can give a constraint for execution time of each event.

We assume that every path (event sequence) from the initial state has the special dummy transition ψ as the last event of the path where the transition condition of ψ is specified so that the path can be executed in the specified time interval T. In our model, the multi-way synchronization mechanism [2] can be specified among EFSMs so that any subset of EFSMs can synchronize with each other by exchanging data when some conditions hold among the subset. By using multi-way synchronization, we can easily describe the real-time hardware system consisting of multiple parallel modules which frequently interact with each other by exchanging messages.

Parametric Model Checking: In [4], we have proposed a parametric model checking method for a periodic EFSM. In our method, temporal properties are written in *RPCTL* (Real-time and Parametric extension of Computation Tree Logic). Since the model is restricted to be periodic, our method can derive parameter conditions efficiently by analyzing at most three periods' behavior of a given periodic EFSM (see [4] for details).

Here, we use the parametric model checking method in [4] for obtaining the parameter condition. However, other model checkers such as Ref. [1] can be also used when the specifications are restricted in a class which the model checker can treat.

Hardware Synthesis of Concurrent Periodic EFSMs: In [3], we have proposed a tool to generate RT-level VHDL descriptions from given system specifications in concurrent periodic EFSMs. In the derived VHDL description, EFSMs are implemented as sequential circuits working with the same clock, and the multi-way synchronization among EFSMs is implemented as AND gates with priority encoders. Moreover, the scheduler which controls EFSMs to execute only schedulable paths (i.e., path satisfying time constraints) is implemented (see [3] for details).

3 Application and Evaluation

Specifying Priority Queuing Mechanism

In WFQ mechanism, each packet has its own priority called *class*, and different queues are used for the classes. We can assign priorities among classes so that total output amounts from queues are proportional to the fixed rates given to the corresponding classes.

As shown in Fig. 1, we compose the WFQ mechanism of four parts that are Pi, $Q(i)$, Po and Sch. Here, each $Q(i)$, $1 \leq i \leq CMAX$ is responsible for storing and extracting

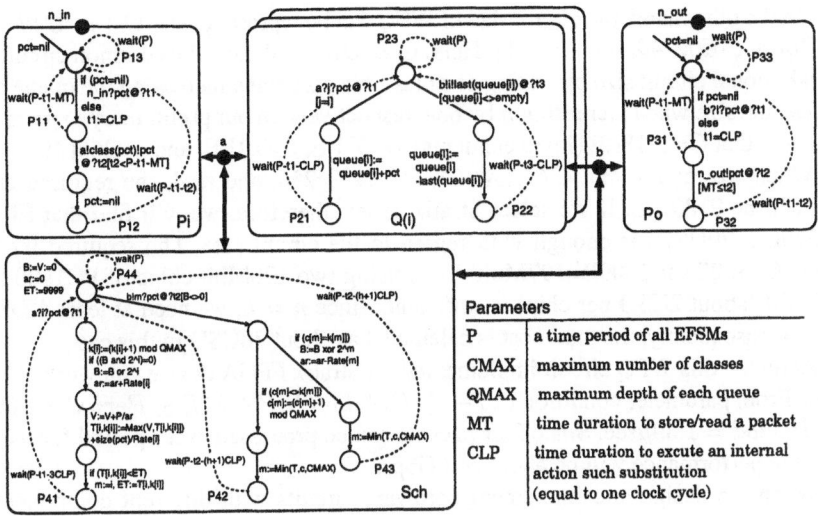

Fig. 1. Modules for WFQ in model concurrent periodic EFSM

packets with class i to/from the i-th queue. *Sch* is the WFQ algorithm described based on [5]. As a total, CMAX+3 periodic EFSMs are executed in parallel. In Fig. 1, n_in and n_out are gates which correspond to an input link from the network to the router and an output link from the router to the network, respectively. And, a and b are internal gates which are used as interaction points between Pi and one of $Q(1), ..., Q(CMAX)$, between one of $Q(1), ..., Q(CMAX)$ and Po, respectively.

3.1 Experimental Results and Evaluation

Using our tools proposed in [3], we have synthesized the RT-level VHDL description from the specification of WFQ in Fig. 1. Before synthesizing circuits, we must assign appropriate values to constant parameters explained in the previous section. The parameter condition derived by the technique in Sect. 2 was

$$2n(h+1) \cdot CLP \leq Period \quad and \quad (2n-1)(h+1) \cdot CLP + MT \leq Period$$
$$and \quad 6n \cdot CLP \leq Period \quad and \quad (2n-1) \cdot 3CLP + MT \leq Period$$

Here, $h = \lfloor log_2 CMAX \rfloor$. We also assume that n is the number of parallel execution of P_i, P_o and $2n \leq CMAX$ (When $n = 2$, two EFSMs are executed in parallel for P_i and P_o, respectively, that is, two packets can be processed in a period).

Parameter Decision for High-end Routers: In this case, it may be considered that performance is more important than component costs. So, we choose components to minimize *Period*, that is, make values of MT and CLP as small as possible. When we select Stratix series of Altera whose maximum clock frequency is 420MHz. Thus, the lower bound of CLP is 1/420MHz=2.38nsec. Stratix series support PC1600 DDR SDRAM memory. If we suppose that packet size is 1500B (bytes) and data bus size is 64 bit, we obtain that $MT = 1500 \times 8/64/200MHz=938nsec$.

On the other hand, the circuit size of WFQ can be represented by $C + \alpha \cdot CMAX + \beta \cdot CMAX \cdot QMAX + \gamma \cdot (n-1)$. Here, C, α, β and γ denote the common circuit size, the additional circuit size by adding one class, the size when increasing the queue depth by one, the size when increasing n by one, respectively. In our preliminary experiments, we know that C=2718LE (logic elements), α=57LE, β=0.1LE, and γ=266LE.

When we require that $CMAX$=64, $QMAX$=256, and n=4, the required circuit size will be 9180 LE. If we select Stratix series, therefore, we will find that EP1S10 (10570LE, 70USD) is enough with regard to the circuit size. The required memory size is $64 \times 256 \times 1500$B=197Mbit. Then using two 128Mbit chips of PC1600 DDR SDRAM (about 2USD per chip) is sufficient. Since $n = 4$, we need at least 4 DRAM chips. Consequently, the total cost is calculated as about 78USD in this case.

Actually, this WFQ circuit implemented on Stratix FPGA device can work at 108.9 MHz. From parameter condition $(2n - 1) \cdot (h+1) \cdot CLP + MT \leq Period$, we obtain that $Period = 1.39\mu$sec. Since four packets can be processed every period for the best case, the performance will be about 34.6Gbps.

When we design and implement hardware circuits with different components to achieve various performance, we have to describe the corresponding specification for each circuit. In our method, based on the same specification, we can derive parameter conditions for satisfying a specified property such as deadlock freeness and synthesize the hardware circuits for satisfying different requirements by deciding parameter values.

4   Conclusion

In this paper, we have proposed a design and implementation method for a WFQ algorithm using concurrent periodic EFSMs and a parametric model checking. In our method, we can derive parameter conditions for deadlock free property with several parameters such as maximum input/output link speed of a router, the number of classes, memory speed, and so on. From experimental results, we believe that our method can be used for design and implementation of QoS routers for various environments by only adjusting parameter values considering cost and performance of components.

References

1. T. A. Henzinger, P-H. Ho and H. Wong-Toi : "HYTECH: A Model Checker for Hybrid Systems", International Journal on Software Tools for Technology Transfer, vol. 1, no. 1-1, pp.110–122 (1997).
2. ISO : "Information Processing System, Open Systems Interconnection LOTOS", ISO 8807 (1989).
3. H. Katagiri, M. Kirimura, K. Yasumoto, T. Higashino and K. Taniguchi, K. : Hardware Implementation of Concurrent Periodic EFSMs, Proc. of Joint Intl. Conf. on 13th Formal Description Techniques and 20th Protocol Specification, Testing, and Verification (FORTE/PSTV2000), pp. 285 - 300 (2000).
4. A. Nakata, and T. Higashino: "Deriving Parameter Conditions for Periodic Timed Automata Satisfying Real-Time Temporal Logic Formulas," in Proc. of 21st IFIP Int'l Conf. on Formal Techniques for Networked and Distributed Systems (FORTE2001), pp. 151-166, 2001.
5. A. Parekh and B. Gallager : "A Generalized Processor Sharing Approach to Flow Control in Integrated Services Networks — The Single Node Case —", IEEE/ACM Trans. on Networking, Vol. 1, No. 3, pp. 344–357 (1993).

Custom Tag Computation Circuit
for a 10Gbps SCFQ Scheduler

Brendan McAllister, Sakir Sezer, and Ciaran Toal

School of Electrical and Electronic Engineering
Queen's University Belfast
Ashby Building, Stranmillis Road
Belfast, BT9 5AH, N. Ireland, U.K.

Abstract. This paper details the architecture and implementation of a tag computation circuit for a Self-Clocked Fair Queuing (SCFQ) Scheduler. The core objectives of the presented project is the implementation of a custom accelerator circuit that is optimized to process tag values for terabit router nodes operating at 10 Gbps per link. The system is implemented using FPGA technology and provides extended programmability to adapt the tag computation to a range of custom scheduling schemes.

1 Introduction

One of the major shortcomings of the current Internet is the best effort services. On the other hand, the traffic diversity is increasing with new emerging services producing variable and burst traffic patterns requiring extremely low network delay. It has become a necessity to differentiate Internet and service based traffic on their type and priority. The QoS issue with regards to the Internet has been one of the main research topics of the industrial and academic research community over the last six years. Numerous protocols, traffic handling schemes and techniques have been proposed and implemented

The research presented in this paper is based on IP QoS research and investigates hardware architectures for accelerator circuits and network processing elements for terabit core routers. It examines parallel processing architectures for programmable scheduling and presents an architecture and implementation of a SCFQ scheduler using a FPGA.

2 Weighted Fair Queuing

WFQ is probably the most well known fair queuing algorithm for fair scheduling of variable size packets. It allows an arbitrary number of end-to-end connections having a fair access to a link. WFQ is computationally complex and causes a significant implementation problem for high throughput rates. The computation complexity of

P.Y.K. Cheung et al. (Eds.): FPL 2003, LNCS 2778, pp. 1149–1152, 2003.

virtual time occurs, as whenever there is a change from busy to idle or reverse in a class, the algorithm requires further computation.

Using the same principles of WFQ, Self-Clocked Fair Queuing (SCFQ) is an approximation of the same calculation. The virtual time used in SCFQ is a measure of the progress of the system itself. Whenever the system changes state from busy to idle, virtual time resets to zero. In fact, SCFQ computation of virtual time is much simpler than that of WFQ and is therefore much more practical solution. It does not always achieve the delay and fairness properties of WFQ but these disadvantages are easily outweighed by the ease of implementation.

3 Tag Computation Architecture

Tag computation plays a vital role in the scheduling procedure of many fair queuing scheduling schemes. The method for computing finishing tags determines not only the fairness of the scheduling process but also the throughput rate i.e. how many finishing tags can be computed per second. Our research investigates architectures that are optimized for a specific technology, in this case FPGAs. The presented finishing tag computation circuit is highly parallel and pipelined. It is composed of a range of distributed on chip memory blocks.

Figure 1 shows the block level description of the finishing tag computation block. It is composed of four main blocks; R_k Evaluation, IP Acquire, Tag Value Calculation and Virtual Time. Two lookup tables, one for R_k lookup and another for tag data lookup, are implemented using Stratix embedded block RAMs

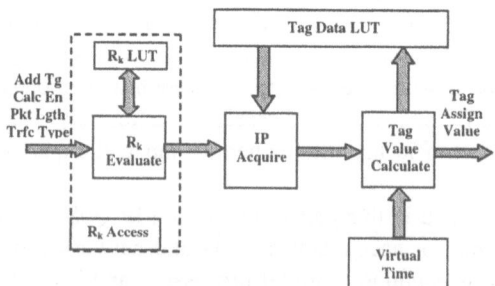

Fig. 1. Block Level Description

3.1 R_k Lookup and IP Acquire

The R_k lookup table is composed of 8 Stratix M4K RAM blocks of 4kbits per block accommodating up to 1024 lookup entries. Figure 2 shows the data flow of the R_k lookup. The lookup table can be addressed by the *TrafficType*, or by the *AddTg* which are determined by the packet classifier at the switch/router input port. This feature allows the tag computation block to be programmable to support per class or per flow queuing or a mixture of both. The lookup table translates the traffic type into an equivalent R_t which will be used to calculate the individual finishing tag for each IP packet. The lookup table size determines the number of flows simultaneously supported by the SCFQ scheduler. The presented implementation supports up to 1024 flows. The QoS adjustment of each individual traffic type or flow is accomplished with the assigned R_t value.

The IP Acquire block is similar to the R_t lookup block and is composed of 8 Stratix M4K RAM blocks. It looks up the *PrvTgValue* (previous finishing tag value) and the *PktNo* (previous packet number) for finishing tag computation. Figure 3 shows the block diagram of the IP acquire block.

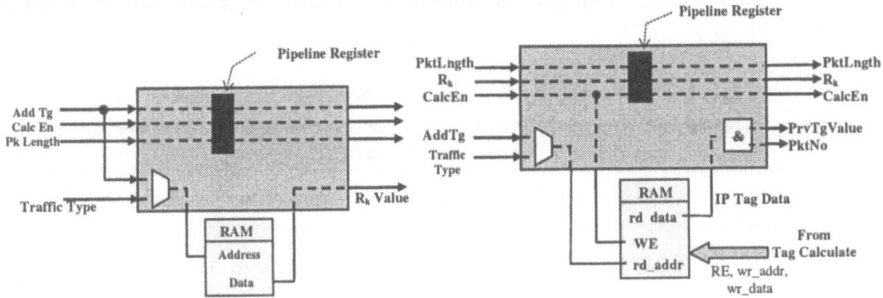

Fig. 2. R_t Lookup   **Fig. 3.** IP Acquire

3.2 Virtual Time and Tag Calculator

The Virtual Time block is a counter circuit producing an integer count value to emulate the motion of time for the SCFQ scheduler. The finishing tag may run for infinite time whereas the CVT will only be a finite value. Tag computation is carried out by the tag calculator block. Figure 4 illustrates the data path of the tag calculator block.

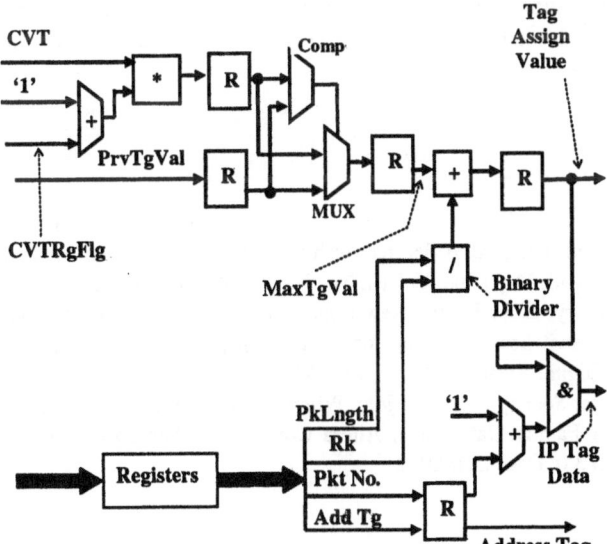

Fig.4. Tag Calculator Data Path

The tag computation circuit is a pipelined parallel map of the tag computation algorithm [1] and is able to compute one tag value per clock cycle.

4 Circuit Study and Conclusions

The finishing tag computation circuit is synthesized and targeted to the smallest Altera Stratix device [4] using Synplify Pro and Altera Quartus II tools. The speed and area performances are examined. The post-layout synthesis results are included in Table 1.

Table 1. Post-Layout Synthesis Results for Altera EPIS10F484C5

Tag Assign Circuit Post-layout Synthesis				
Device	Clock Speed	Logic Cells	Register Bits	RAM Blocks
EPIS10F484C5	17.65 MHz	1,662 (16%)	358 (4%)	16 M4K RAM (27%)

Although the latest high performance FPGA technology with embedded high speed memory has been used, the maximum speed of the circuit is significantly lower than expected. A circuit speed of 17.65 Mhz means 17.65 million tag computation per second can be achieved. The circuit is optimized to perform one finishing tag computation every clock cycle. Assuming a minimum IP packet length of 80 bytes, the tag computation circuit is able to service a link with a throughput rate of 10 Gbps, fulfilling the minimum requirement.

The presented architecture demonstrates that new generation FPGA architectures with a range of embedded peripherals and memories are an ideal platform for high throughput programmable network processing allowing the implementation of Terabit network nodes.

References

1. Mark W. Garrett, Bellcore. A Service Architecture for ATM: From Application to Scheduling, IEEE Network Magazine, Vol.10, No. 3, pp 6-14, May/June 1996.
2. A. K. Choudhury, E. L. Hahne. New implementation of multi-priority pushout for shared memory ATM switches, Computer Communications, vol. 19, pp. 245-256, March 1996.
3. K.C.Chang, Digital Systems Design with VHDL and Synthesis. An Integrated Approach, IEEE Computer Society Press and John Wiley & Sons, 1999, ISBN 0-7695-0023-4.
4. Stratix FPGA Family Data Sheet, Altera Corporation, San Jose, CA 95134, USA, Data Sheet Version 3.0, December 2002.

Exploiting Stateful Inspection of Network Security in Reconfigurable Hardware

Shaomeng Li, Jim Tørresen, and Oddvar Søråsen

Department of Informatics, University of Oslo, N-0316 Oslo, Norway
{shaomenl,jimtoer,oddvar}@ifi.uio.no

Abstract. One of the most important areas of a network intrusion detection system (NIDS), stateful inspection, is described in this paper. We present a novel reconfigurable hardware architecture implementing TCP stateful inspection used in NIDS. This is to achieve a more efficient and faster network intrusion detection system as todays' NIDSs show inefficiency and even fail to perform while encountering the faster Internet. The performance of the NIDS described is expected to obtain a throughput of 3.0 Gbps.

1 Introduction

"Stateful inspection" is applied in Network Intrusion Detection Systems (NIDS) and is a more advanced network security tool than firewalls. It is used for checking the handshakes in a communication session by using detailed knowledge of the rules of the communication protocol. This is to make sure that it is completed in an expected and timely fashion. By checking a connection (packet by packet) – not just one single packet, and knowing what has just happened and what should happen next, stateful inspection detects incorrect or suspicious activity and alerts flags to the system administrator [1].

1.1 TCP Connection Stateful Inspection

TCP (Transmission Control Protocol) [2] is an important Internet protocol. It provides a full duplex reliable stream connection between two end points in the TCP/IP network. The approach of using stateful inspection will be one of the best ways (maybe the only way) to monitor a TCP connection.

In NIDS Snort, a software based STREAM4 preprocessor with 3000 lines software code is designed to conduct the TCP stateful inspection performing two functions: Stateful inspection sessions (monitoring handshakes) and TCP stream reassembly (collecting together packets belonging to one TCP connection). Testing Snort on various networks has shown that the STREAM4 preprocessor leads to a bottleneck in Snort for some network traffic environments (details can be found in [3]). Thus, to improve the performance of Snort, we would like to explore how reconfigurable hardware might be used to replace the STREAM4 TCP stateful inspection.

P.Y.K. Cheung et al. (Eds.): FPL 2003, LNCS 2778, pp. 1153–1157, 2003.

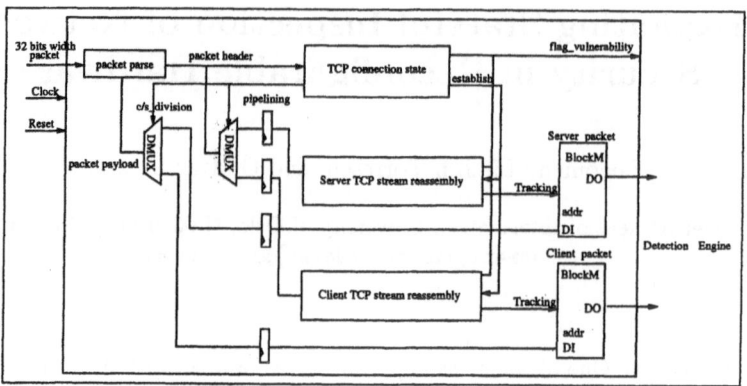

Fig. 1. Block diagram of reconfigurable hardware on TCP connection.

2   Exploiting New Implementation Methods for TCP Stateful Inspection

The new approach would be to process the stateful inspection in hardware rather than software as usual. Implementation in Field Programmable Gate Arrays (FP-GAs) is appropriate to make such explorations. The new hardware architecture is proposed in Fig. 1. This unit will be an add-on unit for the computer running the Snort software.

Incoming packet data (32 bit width) is input to the reconfigurable hardware unit which processes the TCP three way handshake and the Server and Client TCP stream reassembly. The information of the packet header will be stored in some registers based on the libpcap library which is used in Snort to get a packet off from the wires.[1] The basic packet header information most frequently referenced are the sequence number, acknowledge number, window size and TCP flags such as the SYN and ACK bit.

The TCP connection state unit is implemented as a state machine to check the three way handshake of the TCP connection. After establishing the proper connection (by TCP three way handshake), the data over a TCP connection can be exchanged between the Client and the Server. The processing of data flowing to the Server side and the Client side can be performed separately and in parallel, even if the Server and the Client TCP stream reassembly units conduct the same function. This means that packets sent to the Server and the Client side are reconstructed individually in independent hardware units. By doing this, the processing of TCP stream reassembly units in a NIDS is accelerated, of course, at a cost of extra FPGA resources. Two 32 bit DMUXs (one for header and one for payload) are added to separate incoming packets into the Server and the Client packets. The reason for doing this is to feed incoming packets into the Server TCP stream reassembly unit and the Client TCP stream reassembly

---

[1] Registers for the packet header are not shown in Fig.1.

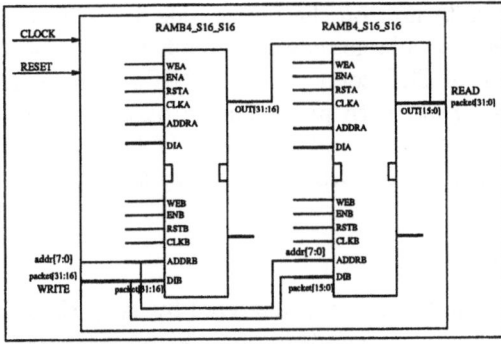

Fig. 2. The Server (or the Client) packet memory.

unit, respectively. The TCP stream reassembly units are running in parallel and determine which packets need to be stored in the "Client packet" or the "Server packet" memory. This avoids the need for large TCP stream reassembly buffers.

Two 32 bit comparators and one 32 bit adder are needed to implement one TCP stream reassembly unit. If the sequence number of an incoming packet is outside of the band size (band size is decided by the initial sequence number (ISN) and window number), the packet will be dropped. The payload of the packet is otherwise stored into the "Server packet" memory or "Client packet" memory respectively, to reconstruct the data for a succeeding detection engine. By pipelining the TCP stream reassembly, the "Server packet" memory and the "Client packet" memory unit, the total performance can be enhanced.

The size of the packet memories are 5x32 bit (16 bit data bus). 5x32 bit is required as the signature pattern can be matched at a maximum of 5x32 bits in the succeeding detection engine [4]. However, a dual port (write/read) memory is required for the Server/Client packet units with minimum size of 5x32 bits. Using a dual port RAM for the packet memory is important to be able to receive new data when matching (reading) is concurrently undertaken.

Virtex XCV1000-6 FPGA to be used in this work contains RAM blocks called SelectRAMs. Each has a capacity of full synchronous dual ported 4096-bit memory and is ideal to implement the Server/Client packet unit. One such block SelectRAM can be configured as a memory with different data widths and depths. However, since dual port RAM is required, the maximum data width is limited to 16 bits. The library primitives, the RAMB4-S16-S16 is dual ported where each port has a width of 16 bits and a depth of 256 bits which is available in the XCV1000-6. By considering the size of the RAMB4-S16-S16 and the packet which has 32 bit data width, two such block SelectRAMs are therefore needed to implement *one* 32 bit data bus packet memory – see Fig.2. Since there are two packet memories (the Client and the Server), a total of four block SelectRAMs are therefore needed to implement the "Server packet" and the "Client packet" memory units.

Processing the data flow on the Server side and Client side in parallel and eliminating the need for a large reassembly buffer are our main contributions to improve the process of TCP connection in a NIDS. This makes it different from the approach in [5]. Data path processing in parallel is the main feature used when implementing the Server and the Client TCP stream reassembly in FPGA. Thereby the performance of NIDS facilitated by the method could be enhanced.

3   Experiments

Our implementation of this study is analyzed by using the ISE FPGA tool from Xilinx [6]. Designs are to be mapped onto a Virtex XCV1000-6 FPGA.

All individual modules such as TCP connection state, TCP stream reassembly unit and DMUX are implemented in VHDL. The simulation of those functions were conducted by the Modelsim XE II v6.5a simulator [6].

Except for the packet parsing, the whole system has been placed and routed into a XCV1000-6 FPGA. The minimum clock period for data from input to output is 10.467 ns which corresponds to a throughput of 3.06 Gbps.

However in IP/TCP networks, the Server often needs to be able to handle multiple connections simultaneously. Hence, multiple TCP connections have to be considered in this study. The process which consumes most SLICEs in the FPGA is the module which does doing the TCP three way handshake. Although there are 12288 SLICEs in one XCV1000-6 FPGA, the possibility of having multiple TCP connections is limited to the capability of implementing units of the "Server packet" memory and the "Client packet" memory in one such FPGA. The reason for this is that the height of the CLB array in one FPGA decides the number of block SelectRAMs, consequently determining the size of the packet units. One XCV1000-6 FPGA with the amount of 32 block SelectRAMs can therefore implement only 8 TCP connections. Although the size of the SLICES of such an FPGA should be checked to see if it is enough to implement remaining modules of 8 TCP connections simultaneously.

By using a Virtex XCV812E FPGA which has 280 block SelectRAMs as used in [5], 70 multiple TCP connections can be expected to be implemented in one FPGA.

4   Conclusions

Stateful inspection over a TCP connection is studied and implemented in FPGA based hardware to remove the bottleneck of TCP connection in a network traffic environment. A novel approach using reconfigurable hardware is introduced. Experiments show that the performance could be improved by this implementation to a throughput of 3.0 Gbps.

References

1. Michael Clarkin. "*Comparison of CyberwallPLUS Intrusion Prevention and Current IDS technology*". NETWORK-1, Security Solutions, Inc., White Paper.

2. J. Postel. *"Request For comment 793, Transmission control Protocol"*. 1998.
3. Sergei et al. *"SNORTRAN: An Optimizing Compiler for Snort Rules"*. Fidelis Security Systems, Inc., 2002.
4. Shaomeng Li et al. *"Exploiting Reconfigurable Hardware for Network Security"*. in Proc. of 11th Annual IEEE Symposium on Fiels-Programmable Custom Computing Machines (FCCM'03), 2003.
5. Marc Necker et al. *"TCP-Stream Reassembly and State Tracking in Hardware"*. in Proc. of 10th Annual IEEE Symposium on Fiels-Programmable Custom Computing Machines (FCCM'02), School od Electrical and computer Engineering, Georgia Institute of Technology, Atlanta, GA, 2002.
6. *http://www.xilinx.com*

Propose of a Hardware Implementation for Fingerprint Systems

Vanderlei Bonato[1], Rolf Fredi Molz[1], João Carlos Furtado[1],
Marcos Flôres Ferrão[1], and Fernando G. Moraes[2]

[1] UNISC – Departamento de Informática
Av. Independência, 2293 Bairro Universitário. CEP: 96815-900
Santa Cruz do Sul/RS – Brazil
rolf@unisc.br, vbonato74@hotmail.com
[2] PUCRS – Faculdade de Informática
Av. Ipiranga, 6681 - Prédio 30.
CEP: 90619-900 - Porto Alegre - Brazil
moraes@inf.pucrs.br

Abstract. Fingerprint is graphical flow-like ridges presents on human fingers. Each fingerprint is unique, offering a clear and unambiguous method to identify an individual. The uniqueness of each fingerprint is determined by fine details embedded in its overall structure, named *minutiae*. Fingerprint classification system is CPU time intensive, usually implemented in software. This paper presents an alternative way to identify the *minutiae* from fingerprints, aiming real-time processing. In the first part is implemented the alternative algorithm in software (Delphi) and after this is presented an architecture to be implemented using a configurable devices (FPGA). The performance of this algorithm, in hardware and software, are analyzed, presenting the spent time within each system block.

1 Introduction

Fingerprint is graphical flow-like ridges presents on human fingers. They have been widely used in personal identification for several centuries. The uniqueness of each print is determined by fine details embedded in its overall structure that is known as minutiae. In figure 1 is showed some fingerprint details, where are presented minutiae, core and axis.

2 Proposed Algorithm

The algorithm proposed in this paper is considered the state-of-the-art and well different of the existing solutions found in automated fingerprint systems. The main goal is to implement a system in a hardware environment to obtain high performance. To obtain this performance the algorithm used is based on simple tasks.

P.Y.K. Cheung et al. (Eds.): FPL 2003, LNCS 2778, pp. 1158–1161, 2003.
© Springer-Verlag Berlin Heidelberg 2003

Fig. 1. Some fingerprint details

The algorithm is divided in the following processing steps: (i) input filter represented by a Gaussian filter; (ii) gradient and direction computation; (iii) ridge detection; (iv) minutiae detection. These steps are illustrated in Figure 2. In the Figure 3 we can see the results of the processing.

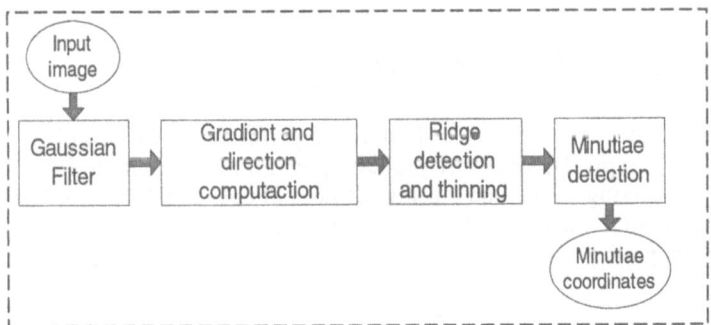

Fig 2. Blocks of the algorithm proposed

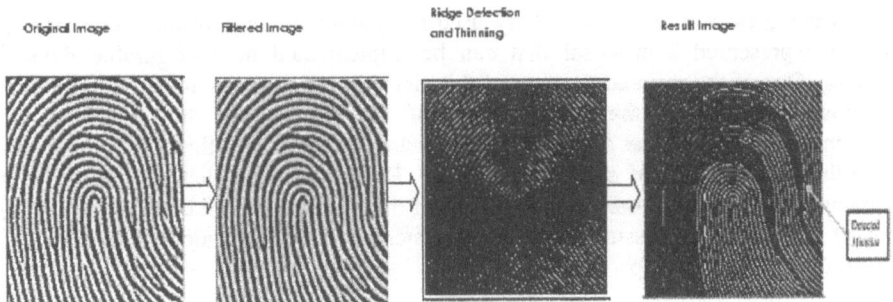

Fig. 3. Results after steps processed.

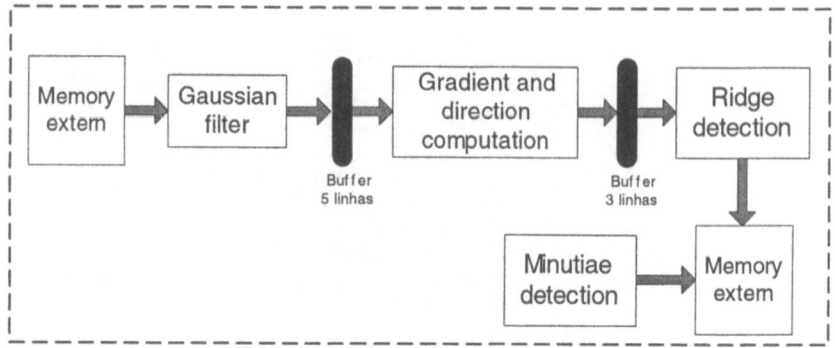

Fig. 4. Hardware partition.

3 Proposed Architecture

The block diagram presented in Figure 4 corresponds to the hardware partition of the system. The system is organized as a pipeline, to increase the final throughput. The tasks executed in hardware, by the FPGA device are the Gaussian filter, the gradient and direction computation, the ridge detection and thinning, and the minutiae detection.

4 Hardware x Software Results

In Table 1 are presents software and hardware performance for complete fingerprint image (256 x 256 pixels). Software processing was made by PentiumII 233 MHz and hardware processing was made using device FPGA EPF10K50EQC208-1 from FLEX10KE family. The performance of this device was 27.65Mhz, that means 35.9ns for each clock period.

Conclusion and Future Works

This paper present a techniques study used in systems of fingerprints processing. It was also presented a proposal that can be implemented in configurable devices (FPGA). One of the main advantages of this algorithm of minutiae location in relation to those presented in the section State of the Art, is that the functions are accomplished without use complex calculations, that is to say, that allows that this algorithm is implemented easily in hardware. For future work that can be made, is implementation of the classification method of the fingerprints. Like this being, at the end of the whole processing a complete system will be implemented in a device FPGA (System-on-to-chip - SOC).

Table 1. Performance between Software x Hardware

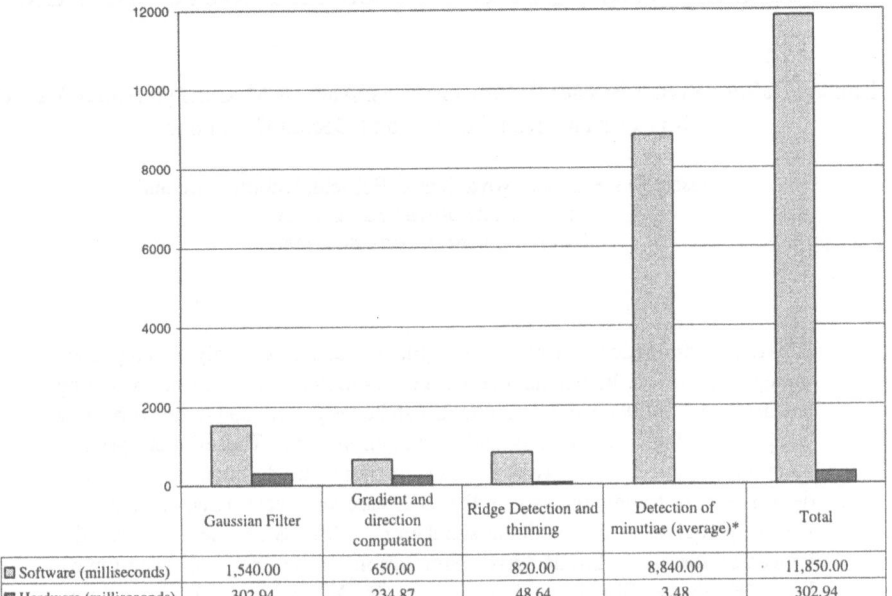

	Gaussian Filter	Gradient and direction computation	Ridge Detection and thinning	Detection of minutiae (average)*	Total
Software (milliseconds)	1,540.00	650.00	820.00	8,840.00	11,850.00
Hardware (milliseconds)	302.94	234.87	48.64	3.48	302.94

* Due scale range, is not possible to see the hardware time in the chart

Acknowledgement

The author Rolf Fredi Molz would like to gratefully acknowledge the support of FAPERGS project number 01/0168.5-Brazil

APPLES:
A Full Gate-Timing FPGA-Based Hardware Simulator

Damian Dalton, Vivian Bessler, Jeffery Griffiths, Andrew McCarthy, Abhay Vadher,
Rory O'Kane, Rob Quigley, and Declan O'Connor

Neosera Systems Ltd, Nova Centre, Belfield, Dublin 4, Ireland
applesinfo@neosera.com
http://www.neosera.com

Abstract. Verification of large VLSI digital circuits is primarily accomplished through simulation. In general, there is a trade-off between speed of processing and accuracy. Software simulation tools can be very accurate but are very slow compared to logic accelerators and emulation systems. These latter systems, many FPGA based, while two to three orders of magnitude faster than software, deliver inferior timing analysis, in the latter case and cycle-based simulation it is merely equivalent to functional simulation. **APPLES (Associative Parallel Processor for Logic Event-driven Simulation)** is the first Full Gate-timing Logic Hardware Simulator, implemented in Xilinx Virtex-II technology. APPLES is a true simulator, delivering timing analysis with the accuracy of a software simulator, but has the distinction that processing is executed entirely in hardware devoid of any machine code. This has the potential to permit APPLES to be one to two orders of magnitude faster than equivalent software systems.

1 Introduction

In the testing and verification of digital circuits Logic Simulation plays a pivotal position, occupying an area where speed of computation is as an important consideration as the accuracy of the results. Parallel processing has been investigated extensively as a means to accelerate computational speed, but has had limited success. Compiled code and Event-driven simulation [1],[2],[3],[4],[5] are the two strategies from the sequential environment that have been employed in parallel logic simulation. Particularly important are synchronous event-driven MIMD. To obtain optimal performance, consideration must be given to Global synchronisation between processors. Unfortunately some of these tasks contribute significantly to the Communication overhead. Soule and Blank [6] and Mueller-Thuns et al [7] have studied these systems and under optimal conditions the best speedup figures were between 3 and 5 on an 8-processor iPSC-Hypercube. Fundamental communication and synchronisation issues in parallel logic simulation can be found in [8],[9],[10],[11],[12] and [13]. State of the art accelerators can be found at various commercial websites [14].

P.Y.K. Cheung et al. (Eds.): FPL 2003, LNCS 2778, pp. 1162–1165, 2003.
© Springer-Verlag Berlin Heidelberg 2003

2 The APPLES Architecture

In APPLES (Associative Parallel Processor for Logic Event-driven Simulation), a succession of signal values that have appeared on a particular wire over a period of time are stored in a specific word in an Associative memory in a time ordered sequence.

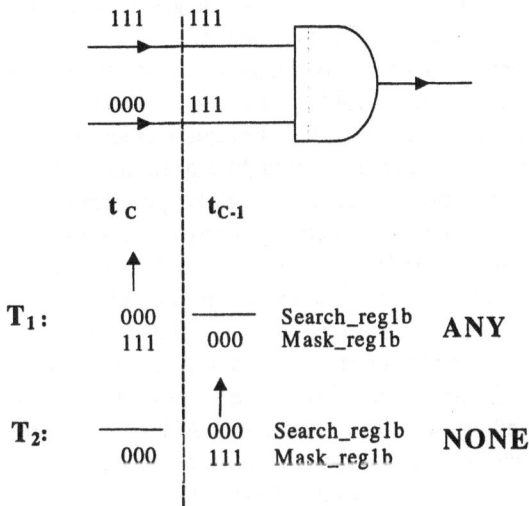

Fig. 1. The APPLES Algorithm for a Unit Delay AND Gate Transitioning to Logic 0

This associative memory structure permits gate evaluation to be performed through a number of parallel test patterns. Therefore, all gates of a particular type can be evaluated in the time it takes to apply the test patterns. For example, as shown in Fig. 1., if a unit-delay AND gate output is going to transition from logic 1 to logic 0, it is necessary to know the signal values at the current time t_c and the previous time interval t_{c-1}. Assume we are using 8-state logic, where 3 bits are required for each signal value. Two simple bit pattern tests will suffice. If ANY current input value is logic 0 (Test T1) and NONE of the previous input values are logic 0 (Test T2), then the output will change to logic 0. Different delay models are constituted by alternative test patterns.

Active gates are identified as gates that have passed all necessary tests. When all tests for the current time period are concluded the fan-out gate inputs affected by these active gates are updated. All signal values are time incremented by shifting the old and new input values into the least time position in the words of the associative array. A more detailed description of the APPLES processor can be found in [15],[16].

2 Benchmark Results

To evaluate the performance of the APPLES prototype, a cycle accurate Verilog version was designed and several ISCAS-85 benchmarks simulated. The gate count of these benchmarks ranged from 622 gates (C880) to 4392gates (C7552). Furthermore, it was assumed that all circuit data was accommodatable by the APPLES' arrays and no memory caching system was necessary. The number of cycles required to evaluate each active gate including all overhead tasks ranged from 3.6 to 5.2 cycles, where a cycle is defined as a register transfer to memory.

Initial implementation of the APPLES processor to accommodate the ISCAS benchmarks and designs up to 150K gates was realised on a Xilinx V1000BG560 (0.22 micron technology) FPGA with a 16Mbyte memory system. This design had a simple paging system, which simply brought sequentially from memory all pages to be processed at every time interval. This APPLES processor implementation running at 10MHz, simulated a circuit containing approximately 140K gates running at the same processing rate as Modelsim running on a 1.2 GHz Pentium processor. The number of clock cycles required for the Modelsim software simulator to compute each active gate was found to be in the range 1,500 to 2,100.

The V1000 FPGA implementation imposes a limited on the APPLES processor clock frequency at 21 Mhz. Transferring the same design to a Virtex-II XC2V1000 raises the clock ceiling to 56Mhz. Additional APPLES modifications such as the introduction of a cache memory system and transferring the design to the Virtex-II XC2V8000 will move this limit towards 100MHz.

3 Integrating the APPLES Processor into Existing CAD Tools

To be acceptable in existing design flows the APPLES system has been designed so that it integrates transparently. Physically, APPLES is on a board which connects to the PCI-bus of a PC. APPLES extracts its netlist information through the standard Verilog PLI interface.

During simulation runs, APPLES transmits information via the PLI interface to the host Verilog simulator. This structure enables any existing netlist Verilog files to be excuted on APPLES and the output to be displayed by the GUI of the Verilog simulator. Only one line of modification is required in a standard netlist file to execute it on APPLES. Essentially, the host Verilog simulator acts as a front-end.

Analysis of APPLES system performance with circuits having gate counts of 1 million gates or more indicates that there are many issues and factors that contribute to the overall processing rate. Important contributors to simulation speed are similar to conventional software simulators and emulators, composition of testbench, idiosyncracies of the circuit being simulated and memory access rates. Nevertheless, initial benchmarking of large, million gate circuits, have mantained APPLES speed advantage.Furthermore, through the PLI interface, access is made to the Verilog testbench stimulating the circuit and other behavioural/RTL modules in the circuit.

3 Conclusion

The APPLES design indicates that the speed and capacity of current FPGA technology can be considered as a target structure not merely for prototypes but also for the eventual implementation technology of the processor it-self. Decomposing standard algorithms partially or fully into hardware may be sensible economically and in terms of performance, but greater benefits may be derived when radically different approaches are considered as exemplified by the APPLES processor.

References

1. MacMillan et al: An Industrial View of Electronic Design Automation: IEEE Trans CAD of ICs and Systems, Vol 19, No 12, Dec (2000) 1428-1449
2. Darringer et al: EDA in IBM: Past, Present and Future: IEEE Trans CAD of ICs and Systems, Vol 19, No 12, Dec (2000) 1476-1498
3. Breuer et al: Fundamental CAD Algorithms: IEEE Trans CAD of ICs and Systems, Vol 19, No 12, Dec (2000) 1449-1476
4. Dunn: IBM's Engineering Design System Support for VLSI Design and Verification, IEEE Design and Test of Computers, Feb (1984) 30-40
5. Agrawal et al: Logic Simulation and Parallel Processing, IEEE Proc Intl Conf on CAD (ICCAD) 1990.
6. Soule et al: Parallel Logic Simulation on General Purpose Machines, IEEE Proc Design Automation Conf, June (1988) 166-171
7. Mueller-Thuns et al: Benchmarking Parallel Processing Platforms: An Application Perspective, IEEE Trans on Parallel and Distributive Systems, Vol 4, No 8, Aug(1998)
8. Chandy, Misra: Asynchronous Distributed Simulation via Sequence of Parallel Computations. Comm ACM 24(ii), April (1981)
9. Bryant: Simulation of Packet Communications Architecture Computer Systems. Tech Rept MIT-LCS-TR-188. MIT Cambridge, USA (!977)
10. Briner: Parallel Mixed Level Simulation of Digital Circuits Virtual Time. PH.D Thesis, Dept of Elec Eng, Duke University, (1990)
11. Jefferson: Virtual Time. ACM Trans Programming Languages Systems, July (1985), 404-425
12. Soule, Gupta: Characterisation of Parallelism and Deadlocks in Distributed Digital Logic Simulation, Proc 26th Design Automation Conf, June (1989) 81-86
13. Ghosh, Lu: An Asynchronous Distributed Approach for the Simulation of Behavior-level Models on Parallel Processors, IEEE Trans on Parallel and Distributed Systems, Vol 6, No 6, June(1995)
14. www.cadence:com,   www.memtor_graphics.com,   www.aldec.com,   www.aptix.com, www.axis.com , www.tharas.com, www.eve.com
15. Dalton: The Speedup Performance of an Associative Memory Based Logic Simulator, Proc 5th Intl Conf on Parallel Computation Technologies (PaCT-99), St Petersburg, Russia, LNCS Springer-Verlag, Sept(1999)
16. Dalton: Avoiding Conventional Overheads in Parallel Logic Simulation: A New Architecture, ACM/IEEE Proc Intl Conf on High Performance Computing, Calcutta, India, Dec(1999)

Designing, Scheduling, and Allocating Flexible Arithmetic Components

Vinu Vijay Kumar and John Lach

Department of Electrical and Computer Engineering
University of Virginia
351 McCormick Road, P.O. Box 400743
Charlottesville, VA 22904 USA
{vv6v,jlach}@virginia.edu

Abstract. This paper introduces new scheduling and allocation algorithms for designing with hybrid arithmetic component libraries composed of both operation-specific components and flexible components capable of executing multiple operations. The flexible components are implemented primarily in fixedlogic with only small amounts of application-specific reconfigurability, which provides the flexibility needed without the negative area and performance penalties commonly associated with general-purpose reconfigurable arrays. Results obtained with hybrid library scheduling and allocation on a variety of digital signal processing (DSP) filters reveal that significant area savings are achieved.

1 Introduction

The optimal schedule and allocation of components during high-level synthesis are hardware dependent, requiring algorithms to be altered based on the target component library. This paper introduces algorithms for scheduling operations and allocating components based on a novel hybrid arithmetic library composed of both fixed-logic components and flexible components capable of performing multiple operations. Applications implemented with such a hybrid library can reap significant area benefits.

However, hardware flexibility must not be gained at the expense of performance and area, as is the case with general-purpose reconfigurable fabrics such as field programmable gate-arrays (FPGAs). This paper introduces a new technique for designing area- and delay-efficient flexible components. *Small-scale reconfigurability* minimizes area and delay penalties by inserting into fixed-logic only the amount of reconfigurable logic and interconnect required to achieve the desired component flexibility. Therefore, arithmetic components designed with this technique have the flexibility to perform multiple operations but are ASIC-like in their efficiency. This enables the area gains provided by the hybrid component library scheduling and allocation algorithms to be maintained.

P.Y.K. Cheung et al. (Eds.): FPL 2003, LNCS 2778, pp. 1166–1169, 2003.

2 Small-Scale Reconfigurability

A flexible component could simply be the multiplexed set of fully implemented individual components. To receive any area benefit, however, the flexible component must be smaller than the total size of all individually implemented components. Partitioning operations across time instead of space, each operation can be implemented in the same physical space, invoking the proper component configuration at the necessary time. Chiricescu describes an implementation of such a 'morphable' unit utilizing common sub functions to implement addition and multiplication [1].

This approach can be extended using reconfigurable logic to replace fixed-logic gates in the circuit. This would even enable circuits without common sub functions to be shared spatially. Taken to an extreme, the entire circuit would be implemented much like an embedded FPGA core, with the associated area and performance penalties of general-purpose reconfigurable fabric. The key to improving efficiency is to limit the added reconfigurable logic and interconnect. The highly optimized nature of arithmetic components requires custom design of flexible components, but logic synthesis techniques such as Boolean matching [2] can be used to find the minimum distance between the set of functions to be implemented. In addition, design decisions must be made as to how to add flexibility, such as multiplexing a set of fixed-logic gates or using a single lookup table (LUT) capable of implementing each gate.

We have designed a bit-sliced flexible component capable of executing 4-bit fixedpoint addition, multiplication, and comparison. The design of the adder/multiplier part is similar to that in [1]. The base arithmetic structures used are a carry lookahead adder and a parallel array multiplier, with sections of the multiplier partial product summation network utilized for addition when in adder mode. Using small-scale reconfigurability, flexibility is added to the logic cones of the output bit-lines to implement the comparator operation.

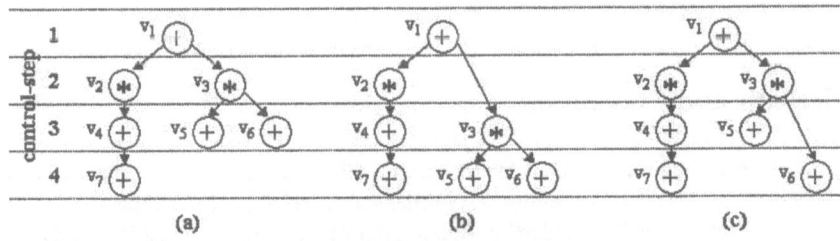

Fig. 1. Scheduling example: (a) Original DFG, (b) Conventional force directed list schedule, (c) Hybrid force directed list schedule

3 Hybrid Library Scheduling and Allocation

Given that optimality depends on the target component library, high-level scheduling and allocation algorithms have been modified with the introduction of flexible arithmetic components. The well-known force directed list scheduling algorithm is used, with a modified force calculation and node selection criteria. For a detailed description of the conventional algorithm, refer to [3, 4]. To demonstrate the modifications to the algorithm for hybrid scheduling, consider the example data flow graph (DFG) in Fig. 1a. The conventional algorithm produces the schedule shown in Fig. 1b, with the intent to reduce operator concurrency. Hence, nodes v2 and v3 will not be scheduled in the same time step, as they are both multiply operations with high area costs. This schedule requires 3 adders and 1 multiplier. The hybrid scheduling algorithm produces the schedule shown in Fig. 1c. The force calculations are modified such that total number of operations per cycle is minimized, rather than individual operator concurrency. This schedule requires 2 adders and 2 multipliers if only fixed-logic components are used, leading to higher area cost than the earlier conventional sched- ule. However, if we allocate on this schedule using a hybrid library, only 2 flexible components are needed, which has a lower area cost than the previous solutions.

As with the conventional algorithm, the number and type of components is first initialized based on the nodes along the critical path. Operations are scheduled from the ready list of nodes, using modified force as the metric for scheduling efficiency. An additional parameter, slack, which is a measure of urgency of a node, is used to guide the selection of candidate nodes. If no resources are available to schedule a node with zero slack, a component is added and the scheduling process is iterated. When choosing between equal slack nodes, the selection is made such that individual operation concurrency is equalized with the previous step. This is to reduce the probability that a flexible component would need reconfiguration when used in that particular step. Scheduling is continued until either all of the nodes are scheduled or the time budget is exceeded, in which case the process iterates with the addition of a component.

Allocation is trivial for a case with only two operation types. The two types can be considered individual sets, with their intersection representing the nodes to be implemented by the flexible component. Hence, the formula from set theory for calculating the cardinality of the intersection of two sets given their individual cardinalities and that of their union is directly applicable here. For cases with more than two disparate operations, more complicated techniques (e.g. linear programming) are needed to derive the optimum allocation.

4 Results and Conclusion

The area efficiency of small-scale reconfigurability is revealed by the component implementations described in Section 2. Normalized to the area of a comparator X: Adder - 1.44X, Multiplier - 4.5X, Adder/Multiplier Limited Flexible Unit (LFU) - 4.81X, and Comparator/Adder/Multiplier Full Flexible Unit (FFU) - 5.31X. In comparison, an equivalent multiplier on an FPGA has an area of approximately 126X.

Normalized to the delay of the multiplier (the slowest fixed component) Y, the delays of the LFU and FFU are 1.27Y and 1.36Y, respectively, and the length of each control-step must be increased accordingly. These increases (which are significantly less than that of FPGAs) must be traded off against the area savings obtained. The reconfiguration time is not considered as only a small number of configuration bits need to be set and reconfiguration will not happen every control-step. In addition, the scheduling algorithm minimizes the flexible component reconfiguration frequency.

Using a library of these components, DSP examples from the high-level synthesis literature [5,6,7] have been scheduled and allocated with our modified algorithms. The examples were scheduled with the smallest number of control-steps, and the area results (normalized to a single comparator) are revealed in Table 1. FDLS AL is the component allocation via traditional force directed list scheduling and allocation using fixed components. FDLS HAL is the allocation of hybrid library components on the exact same schedule. HFDLS HAL uses our hybrid scheduling and allocation algorithms. The last column shows the area savings obtained, which are maximized by the combination of the hybrid scheduling and allocation algorithms and the efficient flexible component implementations provided by small-scale reconfigurability.

Table 1. Area savings of hybrid scheduling and allocation with flexible components

CIRCUIT	FDLS AL		FDLS HAL			HFDLS HAL			FDLS AL - HFDLS HAL
	#fixed	area	#fixed	#flex	area	#fixed	#flex	area	FDLS AL
ELLIP	5	13.3	3	1 LFU	12.2	1	2 LFU	10.6	20.27%
EDGE	3	10.4	1	1 LFU	9.31	1	1 LFU	9.31	10.82%
ARFILT	6	20.9	6	0	20.9	1	3 LFU	18.9	9.34%
FIRFILT	4	11.9	4	0	11.9	2	1 LFU	10.8	9.51%
DIFFEQ	5	12.9	2	1 FFU	11.3	2	1 FFU	11.3	12.66%

Acknowledgements

This work is supported in part by the National Science Foundation under grant No.CCR-0105626.

References

1. Chiricescu, S., et al.: Morphable Multipliers. In: Glesner, M., et al. (eds.): FPL 2002, Lecture Notes in Computer Science, LNCS 2438. Springer-Verlag, Berlin Heidelberg (2002) 647–656
2. Savoj, H., et al.: Boolean Matching in Logic Synthesis. Proceedings of the European Design Automation Conference (1992) 168-174
3. Paulin, P.G, and Knight, J.P: Force-Directed Scheduling for the Behavioral Synthesis of ASIC's. IEEE Transactions on Computer Aided Design of Integrated Circuits and Systems, vol. 8, no. 6 (1989) 661-679
4. Verhaegh, W.F.J., et al.: Improved Force-Directed Scheduling in High-Throughput Digital Signal Processing. IEEE Transactions on Computer-Aided Design of Integrated Circuits and Systems, vol. 14, no. 8 (1995) 945–960
5. Haralick, R.M., and Shapiro, L.G.: Computer and Vision. Addison-Wesley, Reading, MA (1992) 6. Högstedt, K., and Orailoglu, A.: Integrating Binding Constraints in the Synthesis of Area-Efficient Self-Recovering Microarchitectures. Proceedings of the International Conference on Computer Design (1994) 331-334
7. Karri, R., and Orailoglu, A.: High-Level Synthesis of Fault-Secure Microarchitectures. Proceedings of the Design Automation Conference (1993) 429-433

UNSHADES-1: An Advanced Tool
for In-System Run-Time Hardware Debugging

M.A. Aguirre, J.N. Tombs, A. Torralba, and L.G. Franquelo

Electronic Engineering Dpt.
Escuela Superior de Ingenieros. University of Sevilla
c/Camino de los Descubrimientos s/n 41092 Sevilla (SPAIN)
{aguirre,jon,torralba,leopoldo}@gte.esi.us.es

Abstract. The aim of a Rapid Prototyping System for electronic circuit design is to obtain a physical model as similar as possible to the final system as the hosting technology can allow. Large digital integrated circuits are substituted by complex and advanced Field Programmable Gate Arrays (FPGA's) which emulate the whole circuit functionality. These devices can provide more information than the pure circuit emulation itself, they provide a special scheme to access the device configuration and execution time information of the design state registers. This paper describes the UNSHADES-1 system and is focused on the set of software tools that provide easy management and access to this execution time information.

1 Hardware Debugging

Software debugging normally provides a set of tools that can help the programmer to look into the running code and inspect the contents of the variables during the execution (on a step-by-step basis). All this information is linked to the high level source code (such as C, C++, BASIC, ...). Software debuggers provide a means of selecting breakpoints where the execution will be stopped and the code can be inspected. When a breakpoint is reached, then the software can be run step by step, run to the next breakpoint or run until the conclusion of the execution.

In our hardware debugging concept, different than that found in [1], but not opposed to, we try to reproduce this software debugging model, providing a closer link between the designer and the running code. Other approaches are Altera Signal Tap, tor Xilinx Internal Logic Analyzer, that are intrusive methods for registering preselected internal signals. In our approach, the main objective consists in obtaining a hardware scenario that can interchange the information between the emulator system and the man-machine interface in a comprehensive way, but restricted to one snapshot. Once the hardware problem is solved, the development of a set of tools to providing control and manage hardware data is our task.

This paper presents the platform UNSHADES-1 that stands for University of Sevilla HArdware DEbugging System. UNSHADES consists of a hardware platform based on a Xilinx Virtex device and a set of software tools running on a personal computer. Other Tools for inner inspection that compete with UNSHADES-1 are Jbits, LabView, JHDL, Xilinx ILA, but none is able of forcing a value to a single register.

P.Y.K. Cheung et al. (Eds.): FPL 2003, LNCS 2778, pp. 1170–1173 2003.

2 UNSHADES-1 Hardware. Highlights of the Emulation System

The UNSHADES-1 is described in [2] hardware consists of a board with two FPGAs: The emulation system (the VIRTEX FPGA) called S-FPGA and a smaller FPGA with a fixed configuration, called the C-FPGA. The C-FPGA performs the transfer tasks (protocol adaptation and others) with the host PC and certain control functions. One of the tasks of the C-FPGA is related to the control of some general purpose IO lines that can be used for to provide, if needed, extra control over the emulation system. These IO lines are useful for some particular tasks related to the debugging system.

3 UNSHADES-1 Software.
The Run-Time Capture and Scheme System

From software point of view UNSHADES is fully integrated into the Xilinx standard design flow. Two files are needed for integration of UNSHADES into any Xilinx design flow: the bitstream file and the bit allocation file. The first is necessary because the initial configuration process is controlled by the UNSHADES software and the second provides the map that about the location of every register placed across the S-FPGA core and its design level name. Together these two file provide a link between the physical information and their names given during the high level design stages, in other words, it provides a method for closing the loop back. The execution time information can be displayed using comprehensive names. The primary task of the software consists of reading the low-level bitstream information and associating it to busses and registers. The UNSHADES software provides a graphic user interface (GUI) that presents a scheme of the registers and bits, associated with their last captured value.

3.1 External Snapshot

The simplest debugging tool is to take snapshots using an external line (figure 1, b). The PC sends a signal that requests that the Virtex launch the capture mechanism. An external IO line is used to initiate the capture and the S-FPGA continues to operate at all times at design speed. After capture, the UNSHADES software uploads the information to the PC and presents it in the GUI. A natural extension of this tool is to launch the capture macro every certain amount that is programmed. The designer will have information about the evolution of selected signals. Also the information can be recorded into a file to provide display in a waveform viewer. The aim of this tool is to

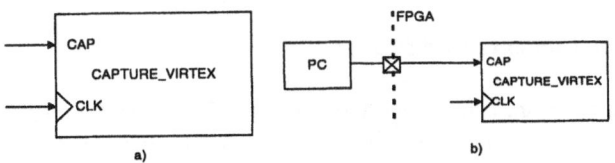

Fig. 1. Capture macro and external snapshot

observe the evolution of the system that has 'slow variables' like state variables in power system controllers or to capture the system state once a design under emulation has begun to operate incorrectly.

3.2 Single Clock Cycle Evolution

The system evolution can be frozen when a run-time event is satisfied. Run-time conditions have to be foreseen at design time. They can refer to register contents that have enough interest that it should be studied when it occurs. Usually they're comparisons with bus values or bits. More complex conditions can be introduced that are combined with time conditions.

All Virtex flip-flops that belong to the design under debug can be controlled by means of their *'clock enable'* input. Using this pin the system evolution can be totally or partially frozen without any risk of malfunctioning due to glitches or clock skew problems. If the *'clock enable'* input is only asserted during a single clock cycle, the system will perform a 'single step' evolution. In the UNSHADES system an external IO line, called *'debug clock'* is used to allow the software perform this single stepping of the S-FPGA. For this option, a small circuit that detects changes on this control line must be included in the S-FPGA design. We use 160 system gates for this circuit. Using a second line, called the *'resume'* line, normal execution can be re-launched until the next run-time condition.

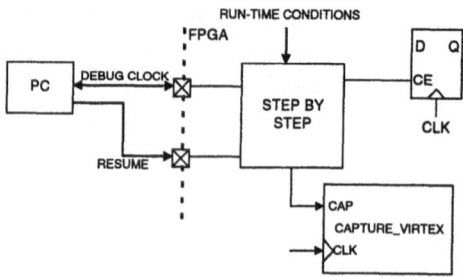

Fig. 2. Single step scheme

After each rising edge of the *'debug clock'* line the flip-flop *'clock enable'* input is asserted during a single clock cycle, and a *'capture'* is launched. After this, the software uploads the information to be represented in the GUI. UNSHADES can lunch a sequence of a configurable number of steps and record the information to be displayed in a classical waveform viewer.

3.3 Flip-Flop Contents Modification

UNSHADES software can change the contents of the S-FPGA registers during run-time. This task has never been reported before in the literature for the Virtex technologies. Previous work and vendor information affirm that changes cannot be performed on selected bits. To make the change correctly the process can only be performed when the system is in a frozen state, this is because a sequence of steps is required and no single access is sufficient. The main difficulties for inducing a new value in a flip-flop is determined by the Configurable Logic Block (CLB) architecture which doesn't permit direct modification of the design Flip-Flops. An original read-modify-write-unmodify scheme has been developed for this purpose and is able to change the flip-flop state within a modified CLB and then return the CLB back to the original design configuration.

4 UNSHADES-1 Graphic User Interface

A good interface human machine is necessary for a comprehensive interaction between the emulation hardware and the information displayed. During synthesis stage all combinational parts are collapsed into look-up tables and cannot be rebuilt into their original schematics. Only registers can display their information.

5 Conclusions

A new technique for hardware debugging has been presented exploiting features of a commercial field programmable FPGA. This new debugging environment brings software debugger techniques into the field of hardware debugging. The FPGA devices chosen to emulate the complete digital design can provide observability, controllability and insertion of desired states into internal registers. UNSHADES tools have been presented as a set of software tools in spite of the fact that a custom hardware system has been designed to match with the software. The software has been developed in a hardware independent way and can be adapted to other commercial hardware platforms with few requirements.

References

[1] Hutchings B., Nelson B. and Wirthlin M.J.. "Designing and Debugging Custom Computing Applications". IEEE Design & Test of Computers. Jan-March 2000. pp 20-28.
[2] M.A. Aguirre, J. Tombs, A. Torralba and L.G. Franquelo. "Improving the Design Process of VLSI Circuits by Means of a Hardware Debugging System: UNSHADES-1 Framework". To be `published in Proceedings of the 28th IEEE Industrial Electronics Conference. IECON'02. Sevilla November 2002.
[3] Xilinx Data book. 2002.
[4] Xilinx XC6200 series datasheet.
[5] J. Faura, C. Horton, Bernd Krah, J. Cabestany, M.A. Aguirre and J.M. Insenser. "A New Field Programmable System On-a-Chip for Mixed Signal Integration" European Design & Test Conference 1997.
[6] Xilinx application notes number xapp138 and xapp151.

4. DYNADES-1 Graphic User Interface

References

[1] ...
[2] ...
[3] ...
[4] ...
[5] ...

Author Index

Lecture Notes in Computer Science

For information about Vols. 1–2710
please contact your bookseller or Springer-Verlag